Recycling of Plastics, Metals, and Their Composites

Emerging Materials and Technologies
Series Editor: Boris I. Kharissov

Bioengineering and Biomaterials in Ventricular Assist Devices
Eduardo Guy Perpétuo Bock

Semiconducting Black Phosphorus: From 2D Nanomaterial to Emerging 3D Architecture
Han Zhang, Nasir Mahmood Abbasi, and Bing Wang

Biomass for Bioenergy and Biomaterials
Nidhi Adlakha, Rakesh Bhatnagar, and Syed Shams Yazdani

Energy Storage and Conversion Devices: Supercapacitors, Batteries, and Hydroelectric Cell
Anurag Gaur, A.L. Sharma, and Anil Arya

Nanomaterials for Water Treatment and Remediation
Srabanti Ghosh, Aziz Habibi-Yangjeh, Swati Sharma, and Ashok Kumar Nadda

2D Materials for Surface Plasmon Resonance-Based Sensors
Sanjeev Kumar Raghuwanshi, Santosh Kumar, and Yadvendra Singh

Functional Nanomaterials for Regenerative Tissue Medicines
Mariappan Rajan

Uncertainty Quantification of Stochastic Defects in Materials
Liu Chu

Recycling of Plastics, Metals, and Their Composites
R.A. Ilyas, S.M. Sapuan, and Emin Bayraktar

Viral and Antiviral Nanomaterials
Synthesis, Properties, Characterization, and Application
Devarajan Thangadurai, Saher Islam, and Charles Oluwaseun Adetunji

Drug Delivery Using Nanomaterials
Yasser Shahzad, Syed A.A. Rizvi, Abid Mehmood Yousaf, and Talib Hussain

For more information about this series, please visit: https://www.routledge.com/Emerging-Materials-and-Technologies/book-series/CRCEMT

Recycling of Plastics, Metals, and Their Composites

Edited by
R.A. Ilyas, S.M. Sapuan, and Emin Bayraktar

CRC Press
Taylor & Francis Group
Boca Raton London New York

CRC Press is an imprint of the
Taylor & Francis Group, an **informa** business

First edition published 2022
by CRC Press
6000 Broken Sound Parkway NW, Suite 300, Boca Raton, FL 33487-2742

and by CRC Press
4 Park Square, Milton Park, Abingdon, Oxon, OX14 4RN

© 2022 Taylor & Francis Group, LLC

CRC Press is an imprint of Taylor & Francis Group, LLC

Reasonable efforts have been made to publish reliable data and information, but the author and publisher cannot assume responsibility for the validity of all materials or the consequences of their use. The authors and publishers have attempted to trace the copyright holders of all material reproduced cin this publication and apologize to copyright holders if permission to publish in this form has not been obtained. If any copyright material has not been acknowledged please write and let us know so we may rectify in any future reprint.

Except as permitted under U.S. Copyright Law, no part of this book may be reprinted, reproduced, transmitted, or utilized in any form by any electronic, mechanical, or other means, now known or hereafter invented, including photocopying, microfilming, and recording, or in any information storage or retrieval system, without written permission from the publishers.

For permission to photocopy or use material electronically from this work, access www.copyright.com or contact the Copyright Clearance Center, Inc. (CCC), 222 Rosewood Drive, Danvers, MA 01923, 978-750-8400. For works that are not available on CCC please contact mpkbookspermissions@tandf.co.uk

Trademark notice: Product or corporate names may be trademarks or registered trademarks and are used only for identification and explanation without intent to infringe.

Library of Congress Cataloging-in-Publication Data
Names: Ilyas, Rushdan Ahmad, editor. | Sapuan, S. M., editor. | Bayraktar, Emin, editor.
Title: Recycling of plastics, metals, and their composites / edited by R.A. Ilyas, S.M. Sapuan, and Emin Bayraktar.
Description: First edition. | Boca Raton, FL : CRC Press, 2022. | Series: Emerging materials and technologies | Includes bibliographical references and index. | Summary: "This authoritative reference work provides a comprehensive review of the recycling of waste polymer and metal composites. It provides readers with the latest advances and covers fundamentals of recycled polymer and metal composites such as preparation, morphology, and physical, mechanical, thermal, and flame-retardancy properties. This work targets technical professionals working in the metal and polymer industries, as well as researchers, scientists, and advanced students. It is also of interest to decision makers at material suppliers, recycled metal and polymer product manufacturers, and governmental agencies working with recycled metal and polymer composites"-- Provided by publisher.
Identifiers: LCCN 2021036751 (print) | LCCN 2021036752 (ebook) | ISBN 9780367708474 (hbk) | ISBN 9780367709747 (pbk) | ISBN 9781003148760 (ebk)
Subjects: LCSH: Recycling (Waste, etc.)
Classification: LCC TD794.5 .R4375 2022 (print) | LCC TD794.5 (ebook) | DDC 363.72/82--dc23/eng/20211008
LC record available at https://lccn.loc.gov/2021036751
LC ebook record available at https://lccn.loc.gov/2021036752

ISBN: 978-0-367-70847-4 (hbk)
ISBN: 978-0-367-70974-7 (pbk)
ISBN: 978-1-003-14876-0 (ebk)

DOI: 10.1201/9781003148760

Typeset in Times
by MPS Limited, Dehradun

Contents

Preface ix
Editors xi
Contributors xiii

Chapter 1 Introduction to Recycling of Polymers and Metal Composites 1

R.A. Ilyas, S.M. Sapuan, Abdul Kadir Jailani,
Amir Hamzah Mohd Yusof, Mohd Nurazzi Norizan,
Mohd Nor Faiz Norrrahim, M.S.N. Atikah, A. Atiqah,
and Emin Bayraktar

Chapter 2 Preparation of Metal Matrix Composites by Solid-State
Recycling from Waste Metal/Alloy Chips 37

Debasis Chaira

Chapter 3 A Comprehensive Study on the Recycled Aluminum
Matrix Composites Reinforced with NiAl Intermetallics
and TiB_2–TiC Ceramic Powders 59

H. Murat Enginsoy, Özgür Aslan, Emin Bayraktar,
Dhurata Katundi, and Fabio Gatamorta

Chapter 4 Recycling for a Sustainable World with Metal Matrix
Composites 75

Uğur Aybarç and M. Özgür Seydibeyoğlu

Chapter 5 Properties of Recycled Metal Matrix Composites 93

A. Atiqah, N. Ismail, K.K. Lim, A. Jalar, M.A. Bakar,
M.A. Maleque, R.A. Ilyas, and A.B.M. Supian

Chapter 6 Morphology of Recycled Metal Composites 109

V. Anandakrishnan and S. Sathish

Chapter 7 Performance of Natural Fiber Reinforced Recycled
Thermoplastic Polymer Composites under
Aging Conditions 127

M. Chandrasekar, T. Senthil Muthu Kumar, K. Senthilkumar,
Sabarish Radoor, R.A. Ilyas, S.M. Sapuan, J. Naveen,
and Suchart Siengchin

Chapter 8 Physical and Mechanical Properties of Recycled Metal
Matrix Composites 141

Pradeepkumar Krishnan and Ramanathan Arunachalam

Chapter 9 Thermal Properties of Recycled Polymer Composites 163

Marwah Rayung, Min Min Aung, and Hiroshi Uyama

Chapter 10 Thermal Properties of Recycled Polymer Composites 185

*Havva Hande Cebeci, Korkut Açıkalın, and
Aysel Kantürk Figen*

Chapter 11 Flame Retardancy of Recycled Polymer Composites 197

*Maryam Jouyandeh, Henri Vahabi, Fouad Laoutid,
Navid Rabiee, and Mohammad Reza Saeb*

Chapter 12 Mechanical and Tribological Properties of Scrap
Rubber/Epoxy-Based Composites 221

*L.M.P. Ferreira, I. Miskioglu, E. Bayraktar,
and D. Katundi*

Chapter 13 Design for Recycling Polymer Composites 235

*Siti Norasmah Surip, Hakimah Osman, and
Engku Zaharah Engku Zawawi*

Chapter 14 Effect of Heat Treatment Modification on the Tensile
Strength and Microstructure of X7475 Al-Alloy
Fabricated from Recycled Beverage Cans (RBCs)
for Bumper Beam Applications 251

*A. Kazeem, N.A. Badarulzaman, W.F.F. Wan Ali,
M.Z. Dagaci, and S.S. Jikan*

Chapter 15 Recycling of Multi-Material Plastics in the Example
of Sausage Casings Wastes 263

Marek Szostak and Pawel Brzek

Chapter 16 Influence of Recycled Steel Scrap in Nodular Casting
Iron Properties 287

*Marcelo Luis Siqueira, Sebastião Bruno Vilas Boas,
Fabio Gatamorta, Claudney de Sales Pereira Mendonça,
and Mirian de Lourdes Noronha Motta Melo*

Contents

Chapter 17 Optimization of Surface Integrity of Recycled Ti-Al Intermetallic-Based Composite on the Machining by Water Jet Cutting via Taguchi and Response Surface Methodology 295

M. Douiri, M. Boujelbene, E. Bayraktar, and S. Ben Salem

Chapter 18 Wear Behavior Analysis of a AlMg1SiCu Matrix Syntactic Foam Reinforced with Boron Carbide Particles and Recycled Fly Ash Balloons 315

J.P. Paschoal, J.J. Thottathil, E. Daniel, R.C. Moraes, F. Gatamorta, E. Bayraktar, and T.V. Christy

Chapter 19 Procedures for Additions of Wastes to Cementitious Composites – A Review 325

M.A. de B. Martins, F.B. Pinto, D. Werdine, L. Ramon, C.V. Santos, P.C. Gonçalves, M.L.M. Melo, and R.M. Barros

Chapter 20 Analysis of the Scientific Production of Cementitious Composites with Recycled Polymeric Materials 357

L.R. Roque-Silva, P.M. Alves, M.H.B. Souza, G.Z. Costal, R.M. Martins, P.C. Gonçalves, P. Capellato, M.G.A. Ranieri, R.G. Torres, M.L.N.M. Melo, and V.C. Santos

Chapter 21 Cementitious Composites for Civil Construction Made with Marble and Granite Waste... 387

M.G.A. Ranieri, P. Capellato, M.A. de B. Martins, V.C. dos Santos, P.C. Gonçalves, L.R.R. da Silva, M.L.M. Melo, and A. da S. Mello

Chapter 22 Influences of the Ceramic Inclusions on the Toughening Effects of Devulcanized Recycled Rubber-Based Composites 409

A.B. Irez and Emin Bayraktar

Chapter 23 Evaluation of Mechanical and Microstructural Properties of Low-Density Concrete with Residual (Scrap) Vegetable Fiber and Blast Furnace Slag 423

K.M.A. Silva, C. Alves, L.M.P. Ferreira, and E. Bayraktar

Chapter 24 Evaluation of Reinforced Concrete (RC) with Different
Scrap Coarse Aggregates 433

*E.S. Fonseca, K.M.A. Silva, S.H.S. Santana,
L. M. Policarpio, and E. Bayraktar*

Chapter 25 Influence of Iron Content on the Microstructure and
Properties of Recycled Al–Si–Cu–Mg Alloys 443

*A.J. Vasconcelos, R.S.M. Silva, P.J. Oliveira,
M.L.N.M. Melo, and O.F.L. Rocha*

Chapter 26 Polymer Recycling in Malaysia: The Supply Chain and
Market Analysis 463

K. Norfaryanti and Z.M.A. Ainun

Chapter 27 Life Cycle Assessment (LCA) of Recycled
Polymer Composites 487

*H.N. Salwa, S.M. Sapuan, M.T. Mastura,
M.Y.M. Zuhri, and R.A. Ilyas*

Index 503

Preface

Recycling of Plastics, Metals, and Their Composites provides a detailed review on advanced recycled polymers and metal composites. It discusses the latest research on the fundamentals of recycled polymers and metal composites such as preparation, physical properties, morphological study and mechanical, thermal as well as flame retardancy properties. This book also comprehensively covers a state-of-the-art review of recycling of polymer composites and recycling of metal composites for sustainability and design for recycling composites. Besides that, this book includes the performance of natural fiber reinforced recycled thermoplastic polymer composites under aging conditions, recycling of multi-material plastics on the example of sausage casings wastes, bumper beams, etc. Moreover, this book covers the analysis of the scientific production of cementitious, concrete and ceramic composites with recycled polymeric and metallic materials. The market and sustainability of recycled polymers and metal composites, as well as the life cycle analysis (LCA) of recycled polymers and metal composites, were also discussed in detail in this book. It includes a commentary from leading industrial and academic experts in the field who present cutting-edge research on advanced recycled polymers and metal composite materials for various industries. Besides that, end users can upgrade their current traditional system to advanced technology. A lot of problems will be solved by integrating the current system and advanced technology system from extensive research. This book also reviews the increasingly important issues of recycling and reuse as a result of the increased use of composites in many industries.

Editors

R.A. Ilyas, PhD, is a senior lecturer in the School of Chemical and Energy Engineering, Faculty of Engineering, Universiti Teknologi Malaysia, Malaysia. He received his diploma in forestry at the Universiti Putra Malaysia, Bintulu Campus (UPMKB), Sarawak, Malaysia from May 2009 to April 2012. In 2012, he was awarded the Public Service Department (JPA) scholarship to pursue his bachelor's degree (BSc) in chemical engineering at the Universiti Putra Malaysia (UPM). Upon completing his BSc program in 2016, he was again awarded the Graduate Research Fellowship (GRF) by the Universiti Putra Malaysia (UPM) to undertake a PhD degree in the field of biocomposite technology and design at Institute of Tropical Forestry and Forest Products (INTROP) UPM. R.A. Ilyas's research interests include polymer engineering, material engineering, natural fibers, biocomposites and nanocomposites. R.A. Ilyas was the recipient of the MVP Doctor of Philosophy Gold Medal Award UPM 2019, for Best PhD Thesis and Top Student Award, INTROP, UPM. In 2018, he was awarded outstanding reviewer by Carbohydrate Polymers, Elsevier United Kingdom, Best Paper Award (11th AUN/SEED-Net Regional Conference on Energy Engineering), Best Paper Award (Seminar Enau Kebangsaan 2019, Persatuan Pembangunan dan Industri Enau Malaysia), and National Book Award 2018. R.A. Ilyas also was listed and awarded Among World's Top 2% Scientist (Subject-Wise) Citation Impact during the Single Calendar Year 2019 and PERINTIS Publication Award 2021 by Persatuan Saintis Muslim Malaysia. His main research interests are (1) polymer engineering (biodegradable polymers, biopolymers, polymer composites, polymer-gels) and (2) material engineering (natural fiber reinforced polymer composites, biocomposites, cellulose materials, nano-composites). To date, he has authored or co-authored more than 242 publications (published/accepted): 84 journals indexed in JCR/Scopus, 14 books, 69 book chapters, 51 conference proceedings/seminars, 2 research bulletins, 10 conference papers (abstract published in book of abstract), 6 guest editor of journal special issues and 6 editor/co-editor of conference/seminar proceedings on green materials related subjects.

S.M. Sapuan, PhD, is a professor of composite materials at the Universiti Putra Malaysia. He earned his BEng degree in mechanical engineering from the University of Newcastle, Australia in 1990, MSc from Loughborough University, UK in 1994 and PhD from De Montfort University, UK in 1998. His research interests include natural fiber composites, materials selection and concurrent engineering. To date, he has authored or co-authored more than 1,521 publications (730 papers published/accepted in national and international journals, 16 authored books, 25 edited books, 153 chapters in books and 597 conference proceedings/seminar papers/presentation (26 of which are plenary and keynote lectures and 66 of which are invited lectures). S.M. Sapuan was the recipient of the Rotary Research Gold Medal Award 2012, The Alumni Medal for Professional Excellence Finalist, 2012 Alumni Awards, University of Newcastle, NSW, Australia, Khwarizmi International Award (KIA). In 2013, he was awarded the 5 Star Role Model Supervisor award by UPM. He has been awarded "Outstanding Reviewer" by Elsevier for his contribution in reviewing journal papers. He received the Best Technical Paper Award in UNIMAS STEM International Engineering Conference in Kuching, Sarawak, Malaysia. S.M. Sapuan was recognized as the first Malaysian to be conferred fellowship by the U.S.-based Society of Automotive Engineers International (FSAE) in 2015. He was the 2015/2016 recipient of SEARCA Regional Professorial Chair. In the 2016 ranking of UPM researchers based on the number of citations and h-index by SCOPUS, he is ranked 6th out of 100 researchers. In 2017, he was awarded the IOP Outstanding Reviewer Award by the Institute of Physics, UK, National Book Award, The Best Journal Paper Award, UPM, Outstanding Technical Paper Award, Society of Automotive Engineers International, Malaysia, and Outstanding Researcher Award, UPM. He also received in 2017 the Citation of Excellence Award from Emerald, UK, SAE Malaysia the Best Journal Paper Award, IEEE/TMU Endeavour Research Promotion Award, Best Paper Award by Chinese Defence Ordnance and Malaysia's Research Star Award (MRSA), from Elsevier. In 2019, he was awarded Top Research Scientist Malaysia (TRSM 2019) and Professor of Eminence Award from AMU, India.

Emin Bayraktar, Habil., Dr (PhD), DSc- Dr es Science, is an academic and research staff-member at mechanical and manufacturing engineering at ISAE-Supmeca-Paris, France. His research areas include manufacturing techniques of new materials (basically composites – hybrid), metal forming of thin sheets (Design + test + FEM), static and dynamic behavior and optimization of materials (Experimental & FEM – Utilization and design of composite-based metallic and non-metallic, powder metallurgy, energetic materials aeronautical applications) based metallics, non-metallic, powder metallurgy and metallurgy of steels, welding, heat treatment as well as processing of new composites, sintering techniques, sinter and forging, thixoforming, etc. He has authored more than 200 publications in international journals and international conference proceedings, and has also authored more than 90 research reports (European = Steel Committee projects, Test + Simulation). He already advised 32 PhDs and 120 MSc Theses already advised and 7 are going on. He is a fellow of WAMME (World Academy of Science in Materials and Manufacturing Engineering), editorial board member of JAMME (International Journal of Achievement in Materials and Manufacturing Engineering) and advisory board member of AMPT-2009 (Advanced Materials Processing technologies), APCMP-2008 and APCMP-2010. He was a visiting professor at Nanyang Technology University, Singapore in 2012, Xi'an Northwestern Technical University, Aeronautical Engineering, in 2016, University of Campinas, UNICAMP-Brazil in 2013 until 2023. He is the recipient of Silesian University Prix pour "FREDERIK STAUB Golden Medal-2009" by Academy of WAMME, World Academy of Science-Poland, material science section, and recipient of William Johnson International Gold Medal-2014, AMPT academic association.

Contributors

Korkut Açıkalin
Department of Energy Systems
  Engineering
Yalova University
Yalova, Turkey

Z.M.A. Ainun
Institute of Tropical Forestry and
  Forest Products (INTROP)
Universiti Putra Malaysia
Selangor, Malaysia

W.F.F. Wan Ali
Faculty of Mechanical Engineering
Universiti Teknologi Malaysia
  (UTM)
Johor, Malaysia

C. Alves
Faculty of Civil Engineering
Federal University of Southern and
  Southeastern Pará (UNIFESSPA)
Marabá, Brazil

P.M. Alves
Institute of Physics and
  Chemistry
Federal University of Itajubá
  (UNIFEI)
Itajubá, Brazil

V. Anandakrishnan
Department of Production
  Engineering
National Institute of Technology
Tiruchirappalli, India

Ramanathan Arunachalam
Mechanical and Industrial
  Engineering Department
Sultan Qaboos University
Muscat, Sultanate of Oman

Özgür Aslan
Department of Mechanical
  Engineering
Atilim University
Ankara, Turkey

M.S.N. Atikah
Department of Chemical and
  Environmental Engineering
Universiti Putra Malaysia
Selangor, Malaysia

A. Atiqah
Institute of Microengineering
  and Nanoelectronics
  (IMEN)
Universiti Kebangsaan Malaysia
Selangor, Malaysia

Min Min Aung
Institute of Tropical Forestry and
  Forest Products
and
Department of Chemistry
Faculty of Science
and
Chemistry Department
Center of Foundation Studies and
  Agricultural Science
Universiti Putra Malaysia
Selangor, Malaysia

Uğur Aybarç
CMS Wheel Company
and
N.A. Badarulzaman
Nanostructure and Surface
  Modification Focus Group
  (NANOSURF)
and
Faculty of Mechanical and
  Manufacturing Engineering
Universiti Tun Hussein Onn
  Malaysia (UTHM)
Johor, Malaysia

M.A. Bakar
Institute of Microengineering and
  Nanoelectronics (IMEN)
Universiti Kebangsaan Malaysia
Selangor, Malaysia

R.M. Barros
Natural Resources Institute
Federal University of Itajubá
Itajubá, Brazil

Emin Bayraktar
School of Mechanical and
  Manufacturing Engineering
Supméca Institute of Mechanics
  of Paris
Saint-Ouen, France

Sebastião Bruno Vilas Boas
Department of Mechanical
  Engineering
Federal University of Itajubá
Itajubá, Brazil

M. Boujelbene
School of Mechanical and
  Manufacturing Engineering
Supméca Institute of Mechanics of Paris
Saint-Ouen, France
and
College of Engineering
University of Ha'il
Ha'il, Saudi Arabia

Pawel Brzek
Institute of Materials Technology
Department of Mechanical Engineering
Poznan University of Technology
Piotrowo, Poland

P. Capellato
Institute of Mechanical Engineering
and
Institute of Physics and Chemistry
Itajubá, Brazil

Havva Hande Cebeci
Department of Chemical Engineering
Yildiz Technical University
Istanbul, Turkey

Debasis Chaira
Department of Metallurgical and
  Materials Engineering
National Institute of Technology
Rourkela, India

M. Chandrasekar
School of Aeronautical Sciences
Hindustan Institute of Technology and
  Science
Chennai, India

Contributors

T.V. Christy
PRIST University
Vallam, India

G.Z. Costal
Institute of Mechanical Engineering
Federal University of Itajubá (UNIFEI)
Itajubá, Brazil

M.Z. Dagaci
Department of Chemistry
Ibrahim Badamasi University
Lapai, Nigeria
and
Faculty of Applied Science and
  Technology
Universiti Tun Hussein Onn Malaysia
Johor, Malaysia

E. Daniel
Karunya Institute of Technology
Coimbatore, India

M. Douiri
Supméca Institute of Mechanics of Paris
Saint-Ouen, France
and
University of Tunis El Manar
Ecole Nationale d'Ingénieurs de Tunis
Tunis, Tunisia

H. Murat Enginsoy
Mechanical and Manufacturing
  Engineering Supmeca
Paris, France
and
Department of Industrial Engineering
Canakkale Onsekiz Mart University
Canakkale, Turkey

L.M.P. Ferreira
Faculty of Civil Engineering
Federal University of Southern and
  Southeastern Pará (UNIFESSPA)
Marabá, Brazil
and
Supméca Institute of Mechanics of Paris
Saint-Ouen, France

Aysel Kantürk Figen
Department of Chemical Engineering
Yildiz Technical University
Istanbul, Turkey

E.S. Fonseca
Faculty of Civil Engineering
Federal University of Southern and
  Southeastern Pará (UNIFESSPA)
Marabá, Brazil

Fabio Gatamorta
School of Mechanical Engineering
State University of Campinas
São Paulo, Brazil

P.C. Gonçalves
Institute of Natural Resources
and
Institute of Production and
  Management Engineering
Federal University of Itajubá (UNIFEI)
Itajubá, Brazil

R.A. Ilyas
School of Chemical and Energy
  Engineering
Faculty of Engineering
and
Centre for Advanced Composite
  Materials (CACM)
Johor, Malaysia

A.B. Irez
Department of Mechanical Engineering
Faculty of Mechanical Engineering
Istanbul Technical University (ITU)
Istanbul, Turkey

N. Ismail
Department of Applied Physics
Faculty of Science and Technology
Universiti Kebangsaan Malaysia
Selangor, Malaysia

Abdul Kadir Jailani
Advanced Engineering Materials and Composites (AEMC)
Department of Mechanical and Manufacturing Engineering
Faculty of Engineering
Universiti Putra Malaysia
Selangor, Malaysia

A. Jalar
Institute of Microengineering and Nanoelectronics (IMEN)
and
Department of Applied Physics
Faculty of Science and Technology
Universiti Kebangsaan Malaysia
Selangor, Malaysia

S.S. Jikan
Faculty of Applied Science and Technology
Universiti Tun Hussein Onn Malaysia
Johor, Malaysia

Maryam Jouyandeh
Université de Lorraine, Centrale Supélec
Metz, France

Dhurata Katundi
Supméca Institute of Mechanics of Paris
School of Mechanical and Manufacturing Engineering
Saint-Ouen, France

A. Kazeem
Nanostructure and Surface Modification Focus Group (NANOSURF)
and
Faculty of Mechanical and Manufacturing Engineering
Universiti Tun Hussein Onn Malaysia (UTHM)
Johor, Malaysia
and
Department of Science Policy and Innovation Studies
National Centre for Technology Management (NACETEM)
North Central Zonal Office
Abuja, Nigeria

Pradeepkumar Krishnan
Mechanical and Industrial Engineering Department
National University of Science and Technology
Muscat, Sultanate of Oman

T. Senthil Muthu Kumar
Department of Mechanical Engineering
Kalasalingam Academy of Research and Education
Krishnankoil, India

Fouad Laoutid
Laboratory of Polymeric and Composite Materials
Materia Nova Research Center
Mons, Belgium

Contributors

K.K. Lim
Pusat Citra Universiti
Universiti Kebangsaan Malaysia
Selangor, Malaysia

M.A. Maleque
Department of Materials and
  Manufacturing
Kulliyah of Engineering
International Islamic University
  Malaysia
Kuala Lumpur, Malaysia

M.A. de B. Martins
Institute of Physics and Chemistry
Federal University of Itajubá (UNIFEI)
Itajubá, Brazil

R.M. Martins
Institute of Physics and Chemistry
Federal University
  of Itajubá (UNIFEI)
Itajubá, Brazil

M.T. Mastura
Faculty of Mechanical and
  Manufacturing Engineering
  Technology
Universiti Teknikal Malaysia Melaka
  (UTeM)
Melaka, Malaysia

A. da S. Mello
Institute of Production and
  Management Engineering
Federal University of Itajubá
  (UNIFEI)
Itajubá, Brazil

M.L.N.M. Melo
Institute of Mechanical
  Engineering
Federal University of Itajubá (UNIFEI)
Itajubá, Brazil

Claudney de Sales Pereira Mendonça
Department of Mechanical Engineering
Federal University of Itajubá
Itajubá, Brazil

I. Miskioglu
Mechanical
  Engineering—Engineering
  Mechanics Department
Michigan Technological University
Houghton, Michigan

R.C. Moraes
Department of Materials
State University of Campinas
São Paulo, Brazil

J. Naveen
School of Mechanical Engineering
Vellore Institute of Technology
Vellore, India

K. Norfaryanti
Institute of Tropical Forestry
  and Forest Products
  (INTROP)
Universiti Putra Malaysia
Selangor, Malaysia

Mohd Nurazzi Norizan
Centre for Defence Foundation
  Studies
Universiti Pertahanan Nasional
  Malaysia (UPNM)
Kuala Lumpur, Malaysia

Mohd Nor Faiz Norrrahim
Research Center for Chemical
  Defence
Universiti Pertahanan Nasional
  Malaysia (UPNM)
Kuala Lumpur, Malaysia

P.J. Oliveira
Institute of Mechanical Engineering
Federal University of Itajubá
   (UNIFEI)
Itajubá, Brazil

Hakimah Osman
Faculty of Chemical Engineering
   Technology
Universiti Malaysia Perlis
Perlis, Malaysia

J.P. Paschoal
Department of Materials
State University of Campinas
São Paulo, Brazil

F.B. Pinto
Mechanical Engineering Institute
Federal University of Itajubá
Itajubá, Brazil

L.M. Policarpio
Faculty of Civil Engineering
Federal University of Southern
   and Southeastern Pará
   (UNIFESSPA)
Marabá, Brazil

Navid Rabiee
Department of Chemistry
Sharif University of Technology
Tehran, Iran

Sabarish Radoor
Department of Materials and
   Production Engineering
The Sirindhorn International Thai-
   German Graduate School of
   Engineering (TGGS)
King Mongkut's University of
   Technology North Bangkok
Bangkok, Thailand

L. Ramon
Physical and Chemical Institute
Federal University of Itajubá
Itajubá, Brazil

M.G.A. Ranieri
Institute of Production and
   Management Engineering
Federal University of Itajubá
   (UNIFEI)
Itajubá, Brazil

Marwah Rayung
Institute of Tropical Forestry and
   Forest Products
Universiti Putra Malaysia
Selangor, Malaysia

O.F.L. Rocha
Federal Institute of Pará
Pará, Brazil

L.R. Roque-Silva
Institute of Mechanical Engineering
Federal University of Itajubá (UNIFEI)
Itajubá, Brazil

Mohammad Reza Saeb
Université de Lorraine,
   Centrale Supélec
Metz, France

S. Ben Salem
University of Tunis El Manar
Ecole Nationale d'Ingénieurs de Tunis
Tunis, Tunisia

H.N. Salwa
Institute of Tropical Forestry and
   Forest Products (INTROP)
Universiti Putra Malaysia
Selangor, Malaysia

S.H.S. Santana
Faculty of Civil Engineering
Federal University of Southern and
  Southeastern Pará (UNIFESSPA)
Marabá, Brazil

C.V. Santos
Natural Resources Institute
Federal University of Itajubá
Itajubá, Brazil

V.C. Santos
Institute of Natural Resources
Federal University of Itajubá
  (UNIFEI)
Itajubá, Brazil

S.M. Sapuan
Laboratory of Biocomposite Technology
Institute of Tropical Forestry and Forest
  Products (INTROP)
and
Advanced Engineering Materials and
  Composites (AEMC)
Department of Mechanical and
  Manufacturing Engineering
Faculty of Engineering
Universiti Putra Malaysia
Selangor, Malaysia

S. Sathish
Department of Production
  Engineering
National Institute of Technology
Tiruchirappalli, India

K. Senthilkumar
Center of Innovation in Design and
  Engineering for Manufacturing
  (CoI-DEM)
King Mongkut's University of
  Technology
Bangkok, Thailand

M. Özgür Seydibeyoğlu
Department of Materials Science and
  Engineering
Izmir Katip Çelebi University
İzmir, Turkey
and
Advanced Structures and
  Composite Center
University of Maine
Orono, Maine, USA

Suchart Siengchin
Department of Materials and Production
  Engineering
The Sirindhorn International Thai-
  German Graduate School of
  Engineering (TGGS)
King Mongkut's University of
  Technology
Bangkok, Thailand

K.M.A. Silva
Faculty of Civil Engineering
Federal University of Southern and
  Southeastern Pará
  (UNIFESSPA)
Marabá, Brazil

R.S.M. Silva
Institute of Mechanical
  Engineering
Federal University of Itajubá (UNIFEI)
Itajubá, Brazil

Marcelo Luis Siqueira
Department of Mechanical
  Engineering
Federal University of Itajubá
Itajubá, Brazil

M.H.B. Souza
Institute of Natural Resources
Federal University of Itajubá
  (UNIFEI)
Itajubá, Brazil

A.B.M. Supian
Advanced Engineering Materials and
　Composites Research Centre
Department of Mechanical and
　Manufacturing Engineering
Universiti Putra Malaysia
Selangor, Malaysia

Siti Norasmah Surip
Eco-Technology Program
Faculty of Applied Sciences
Universiti Teknologi MARA
Selangor, Malaysia

Marek Szostak
Institute of Materials Technology
Department of Mechanical
　Engineering
Poznan University of Technology
Piotrowo, Poland

J.J. Thottathil
Amaljyothi College of Engineering
Kerala, India

R.G. Torres
Institute of Physics and Chemistry
Federal University of Itajubá (UNIFEI)
Itajubá, Brazil

Hiroshi Uyama
Department of Applied Chemistry
Graduate School of Engineering
Osaka University
Osaka, Japan

Henri Vahabi
Université de Lorraine, Centrale Supélec
Metz, France

A.J. Vasconcelos
Institute of Mechanical Engineering
Federal University of Itajubá
　(UNIFEI)
Itajubá, Brazil

D. Werdine
Physical and Chemical Institute
Federal University of Itajubá
Itajubá, Brazil

Amir Hamzah Mohd Yusof
Advanced Engineering Materials and
　Composites (AEMC)
Department of Mechanical and
　Manufacturing Engineering
Faculty of Engineering
Universiti Putra Malaysia
Selangor, Malaysia

Engku Zaharah Engku Zawawi
Polymer Technology Program,
　Faculty of Applied Sciences
Universiti Teknologi MARA
Selangor, Malaysia

M.Y.M. Zuhri
Advanced Engineering Materials and
　Composites Research Centre (AEMC)
Department of Mechanical and
　Manufacturing Engineering
Universiti Putra Malaysia
Selangor, Malaysia

1 Introduction to Recycling of Polymers and Metal Composites

R.A. Ilyas[1,2], S.M. Sapuan[3,4],
Abdul Kadir Jailani[4],
Amir Hamzah Mohd Yusof[4],
Mohd Nurazzi Norizan[5],
Mohd Nor Faiz Norrrahim[6],
M.S.N. Atikah[7], A. Atiqah[8],
and Emin Bayraktar[9]

[1]School of Chemical and Energy Engineering, Faculty of Engineering, Universiti Teknologi Malaysia, Johor, Malaysia
[2]Centre for Advanced Composite Materials (CACM), Universiti Teknologi Malaysia, Johor, Malaysia
[3]Laboratory of Biocomposite Technology, Institute of Tropical Forestry and Forest Products (INTROP), Universiti Putra Malaysia, Selangor, Malaysia
[4]Advanced Engineering Materials and Composites (AEMC), Department of Mechanical and Manufacturing Engineering, Faculty of Engineering, Universiti Putra Malaysia, Selangor, Malaysia
[5]Centre for Defence Foundation Studies, Universiti Pertahanan Nasional Malaysia (UPNM), Kuala Lumpur, Malaysia
[6]Research Center for Chemical Defence, Universiti Pertahanan Nasional Malaysia (UPNM), Malaysia
[7]Department of Chemical and Environmental Engineering, Universiti Putra Malaysia, Selangor, Malaysia
[8]Institute of Microengineering and Nanoelectronics, Universiti Kebangsaan Malaysia, Selangor
[9]Mechanical and Manufacturing Engineering, SUPMECA, Saint-Ouen, France

CONTENTS

1.1 Introduction 2
1.2 Properties of Recycled Metal Composites 4
    1.2.1 Mechanical Properties 5
    1.2.2 Chemical Properties 5
    1.2.3 Physical Properties 6
    1.2.4 Thermal Properties 6
1.3 Recycling Method of Polymer and Metal Composite 7
    1.3.1 Mechanical Recycling 7
    1.3.2 Chemical Recycling 7
    1.3.3 Thermal Recycling 8
1.4 Impact of Recycling Polymer and Metal Composites on Environment, Industry and Economy 8
1.5 Recycling of Polymer Composites 9
1.6 Metal Recycling Composite 16
1.7 Conclusions 29
Acknowledgment 29
References 29

1.1 INTRODUCTION

Composite products and structural components are one of the strongest manifestations of this interrelated mechanism in the creation of materials, structures and technology (Suriani, Radzi et al., 2021; Suriani, Rapi et al., 2021; Suriani, Sapuan et al., 2021; Suriani, Zainudin et al., 2021; Vasiliev & Morozov, 2018). As human technologies advance, the progress of material production follows the same route, and as this advancement is occurring humans should never forget their responsibility towards the environment and the fact that resources can be limited (Ilyas & Sapuan, 2019, 2020). Polymers' further application after use has been constantly coupled to their production; as the development of polymers improve, their recycling technology is lagging in comparison (Ignatyev et al., 2014). Recycling of this composite's material is important in the engineering field as it will promote the sustainability and continuous growth of industrial processes. The reason behind the stagnant progress in recycling technologies of composite (matrix and reinforcement) material is due to its heterogenous nature, making it a poor material to recycle (Yang et al., 2012). Another reason for the slow progress in composite recycling technologies is the added value after recycling is rather low; for example, a large amount of used plastic and synthetic textile is partly sent back to the economic cycle. Furthermore, compared to the recycling of metal, recycled polymer has some downgraded properties (Ignatyev et al., 2014). Thus, as the problem of recycling has become more common, it will not be impossible to discover a way to recycle polymer and metal composite with an efficient method.

Introduction

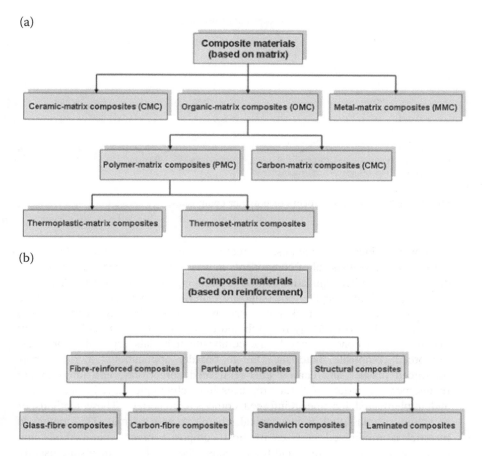

FIGURE 1.1 Classification of composite material based on matrix and reinforcement material. Adapted with copyright permission from Yang et al. (2012).

Figure 1.1 shows the classification of composite material based on matrix (Figure 1.1a) and reinforcement material (Figure 1.1b). As recycling issues are becoming common, the industries and researchers are operating towards a circular economy where they focus on recycling, reusing, and remanufacturing products at the end-of-life (EOL) stage (Zhang et al., 2020). In addition, a tool that recognized the environmental burden of a composite material over its lifetime has been developed; it is called the life cycle assessment (LCA). LCA is beneficial to understand the ecological "pinch-points," savings possibilities and method trade-offs (Tapper et al., 2020). Hence, it is not impossible for other methods and strategies to emerge and develop to counter the recycling issues, specifically recycling of polymer and metal composites, which are not easy to salvage.

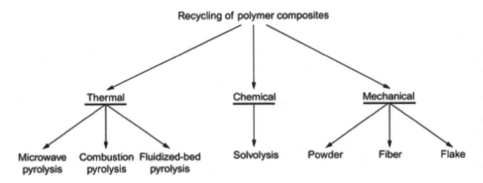

FIGURE 1.2 Recycling process of polymer composites.

However, increased composite structure production has meant that alternatives to simple disposal of components in waste sites have been required, with directives and other legislation aimed at reducing composite waste. A readily visible approach to recycle the composite materials is to replace the composite polymer structure and break the fibers in small lengths (e.g., carbon fiber) (Huntley et al., 2018). Many research and development (R&D) and technology that have been done to recycle the composite materials especially for fiber-reinforced composite materials. Due to broader applications in glass fiber-reinforced plastics and carbon fiber-reinforced composites, and longstanding requirements, a variety of distinct reprocessing technologies have been advanced and implemented for fiber-reinforced composite materials. These methods of recycling are classified as mechanical, chemical, and thermal. The composite material is minimized in sizes in most composite processing methods and comprises of fiber, polymer, and filler blends. The smallest elements are usually powdered, containing a higher amount of polymer and filler percentage while the grainier elements often have fibrous constitution, in which the products acquire a greater aspect ratio (Asmatulu et al., 2014). Figure 1.2 shows the recycling process of polymer composites and Figure 1.3 shows the methods of recycling for carbon fiber reinforced polymer (CFRP), respectively.

The objective of this chapter is to introduce the topic of recycling polymer and metal composite material, whether it is the recycling method, the problem related to recycling of polymer and metal composite material and progress in recycling polymer and metal composite material.

1.2 PROPERTIES OF RECYCLED METAL COMPOSITES

A recent study by Ustundag and Varol (2019) compared the impact of recycled and commercially available PM titanium alloy properties; due to their high-strength properties and low aspect ratio, titanium alloys are used in many unique applications, such as pharmaceutical, military, athletic products and aerospace

Introduction

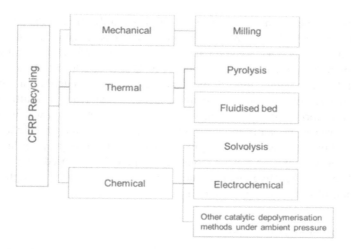

FIGURE 1.3 Methods of recycling CFRP recycling (Zhang et al., 2020).

industries. This paper has reported from their experiments that the powder conduct during the milling process, the powder morphology of the recycled Ti–6Al–4V, is more rounded than the commercially available Ti–6Al–4V powder. Thus, commercially available alloys are more ductile and prevent intra-granular crack prolongation during a fracture, as spherical and rounded, formed powders have a lower impact fracture toughness, and material microstructures and phase distribution are influenced by temperature levels and durations during the sintering process (Ustundag & Varol, 2019).

1.2.1 MECHANICAL PROPERTIES

In recent research, Rady et al. (2020) have studied the physical and mechanical behavior of direct recycled aluminum alloy (AA6061) under heat treatment, where they have experimented on chipped aluminum in a high heat extrusion process, with varying temperature variables. The results have shown that the preheating temperature gives a more significant effect than the preheating time on the mechanical properties of the aluminum alloys (AA6061), where the increased temperature has given a higher tensile strength and lower microhardness. The heat treatment process was carried out with quenching temperatures reaching 530°C in 2 hours and aging processes at 175°C in 4 hours for optimal situations, where the tensile strength and microhardness of extruded samples have been greatly increased by heat treatment (Rady et al., 2020).

1.2.2 CHEMICAL PROPERTIES

These approaches have been influential in the field because of the studies in the chemical properties of recycled metal composites. Recently, Liu et al. (2017)

have experimented on mild chemical recycling and utilization of the decomposed resin of aerospace fiber/epoxy composite waste, where using a $ZnCl_2$/ethanol catalyst method, an effective approach to mild chemical recycling of carbon fiber reinforces polymers with a T_g of around 210°C was developed. As a result, the research has concluded that at moderate temperature degradation (less than 200°C) the recovered fibers were undamaged, while reactive multifunctional groups were formed in the decomposed matrix polymer, which was in the form of an oligomer. In addition, the high modulus and strength might also be maintained by the cross-linked polymers relative to the tidy polymer lacking the inclusion of decomposed matrix polymer, when the decomposed matrix polymer was utilized as a reactive element and 15 weight percentage was applied to the production of new epoxy materials (Liu et al., 2017).

1.2.3 Physical Properties

More recent evidence has shown the physical properties of recycled metal composites. In the last few years, research by Mohd Joharuddin et al. (2020) on the physical constitution and properties of treated amorphous silica, which was applied as the reinforcement for recycled chipped aluminum (AA7075), where it was prepared using the powder metallurgy technique. The experiment from this research has reported the physical characteristic of the recycled aluminum changes as the metal matrix composite (MMC) has an improved porosity and water absorption as the composition of treated rice husk ash increases, while when up to 5 weight percentage of the treated rice husk ash was applied, there is a decline in density. Thus, when processed amorphous silica from rice husk ashes was used as the reinforcement it was shown that there is a development in the physical constitution and characteristic of the recycled chipped aluminum specimen (Mohd Joharudin et al., 2020).

1.2.4 Thermal Properties

Bhouri and Mzali (2020) have analyzed the physical and thermo-elastic characteristics of recycled aluminum composites; investigating the thermal control systems by studying their thermo-elastic behaviors in the use of graphite in an aluminum matrix composite. The result shows the sintered composite sample densities were analyzed using a gravimetric approach and it is observed that the density of composites, determined by SEM analysis, decreases as graphite content increases. The displacement difference of the aluminum composites with 5 weight percentage and 10 weight percentage of graphite is calculated for a single thermal period at temperatures ranging from 42 to 500°C. Therefore, the dimensional shift of the samples indicate a temperature rise, but the inclusion of graphite particles substantially decreases the composite CTE (Bhouri & Mzali, 2020).

1.3 RECYCLING METHOD OF POLYMER AND METAL COMPOSITE

Composite material is being used in a broad number of applications such as aerospace, industry and the renewable energy sector. However, they have not been sufficiently recycled because of their inherent nature, particularly for thermoset-based polymer composites. Current and prospective environmental and waste management legislation require that all technical elements of end of life (EOL) products, including cars, aircraft and wind turbines be properly reused and recycled. Recycling would potentially result in cuts in capital and electricity. Several technologies have been developed that are primarily based on recycling and yet to be marketed: chemical recycling, thermal recycling and mechanical recycling (Yang et al., 2012).

1.3.1 MECHANICAL RECYCLING

Mechanical recycling was the most popular for the method of recycling because mechanical recycling did not change the molecular properties of the composites. Firstly, mechanical recycling starts by breaking off or grinding at low speeds by reducing the composite scrap size (up to 50–100 mm). Furthermore, by hammer mill or other high-speed milling, the size can be decreased to 10 mm to 50 μm for fine molting. The waste products are consequently divided into fiber-rich (coarser) and matrix-rich (finer) fractions by cyclones and strings. Most mechanical processing is relatively easy, but it can be energized and can be produced only in short milled fibers, where the mechanical properties used as reinforcing fillers are weak (Yang et al., 2012). The mechanical recycling will decrease the mechanical properties of the composites.

1.3.2 CHEMICAL RECYCLING

Chemical recycling requires a chemical de-polymerization or matrix elimination procedure by using fiber release chemical dissolution reagents. The chemical recycling process will recycle the clean fibers and fillers in a form of monomers and petrochemical feedstock and the depolymerized matrix. The method of dissolution is also referred to as solvolysis and can be categorized in hydrolysis (water), glycolysis (glycols) and acid digestion depending on the solvent (using acid). High temperatures and high pressures of alcohol or water are usually used to improve dissolution and productivity in sub- and supercritical environments. Acid digestion typically requires ambient conditions, but the reaction time can be very slow. Solvolytic processes like glycolysis will break down epoxy resin to create a potential chemical feedstock in its original monomers. Even potential mediums for recycling fibers and resin are supercritical fluids (SCFs) and particularly supercritical alcohols and supercritical water (SCW) (Yang et al., 2012).

1.3.3 THERMAL RECYCLING

For the thermal recycling, the composites are recycled at high temperatures and usually consist of fluidized bed and pyrolysis recycling process. Pyrolysis is a process of fiber recovery by break down the resin matrix of the composite with inert gas heat into organic small molecules. The pyrolysis will interact with the combustion process during which the polymer resins are oxidized into water vapor and CO_2 to release steam, as a combination of solid carbon dioxide, gas and liquid into a low molecular weight of organic compounds. This produces products that can be used to further refine chemicals as feedstock (Yang et al., 2012).

1.4 IMPACT OF RECYCLING POLYMER AND METAL COMPOSITES ON ENVIRONMENT, INDUSTRY AND ECONOMY

A research on plastics waste to commercial raw materials were made by Gu et al. (2017), where a study on plastic recycling corporation in China has been made, to research on the life cycle environmental effects of mechanically recycled plastic practice. The findings of life cycle analyses suggest that using fillers, additives and the extrusion process has the highest environmental impacts, while mechanical recycling has been found to be a superior solution in most facets of the world relative to the production of new plastics and composites. Thus, the most important contributors to the environmental benefits of recycled composites are impact modifiers which is among the fillers and additives (Gu et al., 2017).

In 2012, an evaluation of the environmental effects of composites containing recycled plastics was made by S. Rajendran et al., where the potential for resource degradation and global warming potential for recycled plastics packed with fiber glass and flax was analyzed and examined with the virgin alternatives, respectively. It is also mentioned in this paper that increases in prices of virgin plastics has forced suppliers or buyers of plastic products search for alternative resources to compete in a tough business environment that has made recycled plastics a potential option (Rajendran et al., 2012).

Vo Dong et al. (2015) conducted a study on environmental effect modeling and economic effects of composite recycling pathways for fiber-reinforced polymers, where the recycling pathway creation for growing amount of reinforced fiber polymer scrap produced has been motivated by both environmental and economic factors. Both the glass and CFRP recycling method (mechanical recovery, pyrolysis and fluidized beds method) were compared to low-value EOL options (co-incineration, landfill and incineration), where pyrolysis appears to be an attractive solution to recycling carbon fiber reinforce polymers that meet both environmental and economic benefits, while glass fiber reinforce polymers appear to be more successful for co-incineration (Vo Dong et al., 2015).

1.5 RECYCLING OF POLYMER COMPOSITES

In a study conducted by Huntley et al. (2018) on recycling composite materials using a water-jet tape deposition method, it was shown that one strategy to recycle composite material is to strip the composite polymer matrix, and then break the carbon fibers into small (discontinuous) pieces. These short fibers will then be loaded on a water suspension and sprinkled between two parallel panels, and placed on a movable belt in order to create a new recycled composite tape for new parts. This paper was simulation to study the performance of the method. In this study, it was reported that a 2D nozzle angle analysis has shown that the small distance between parallel plates is the core component driving fiber alignment. This analysis also shows that the nozzle angle determines the quantity and position in which the fluid reaches the plates. This may have additional consequences for the entire 3D model.

Extensive research has been conducted on the recycling of polymer composites. In a review conducted by Kaiser et al. (2018) on recycling polymer-based multilayer packaging, it was shown that there are many method to recycle polymer-based multilayer packaging. In general, two different methods of recycling multilayered packaging products are possible. The first approach is to isolate and recycle different components of multilayer in various recovery sources. The second approach is to process the components in one consistent step together. The first method shows a good result and is easy to be implemented compared to the second method, with the addition of other molecules that function as compatibilizers. It is the best way to enhance the consistency of partially degraded polymers or the reuse of post-industry packages waste on the market today. It is a strategy that has been widely applied. For example, a method to treat improperly sorted PP or PE fractions from recycled packaging waste compatibility is used. However, this method is only limited to some specific polymers.

In research conducted by Cruz Sanchez et al. (2017) on recycling polymer in an open-source additive manufacturing (AM) context in terms of mechanical issues, it has been shown that the integration of open-sourcing 3D printers and polymer manufacturing will possibly be a new model of distributed recycling. The journal was to research the conditions for the reuse of polylactide acid (PLA), a material commonly used in open-source AM using the technology for the development of fused filaments. In order to determine the mechanical degradation after five recycling cycles, four separate processes of recycling chains (feedstock, reference, 3D printing and 3D printing (reference) were proposed). The process chain recycling allows a regular process and 3D open-source printing to compare the mechanical performance of the material and it helps to determine the material degradation effect of the printing process. The result shows that mechanical properties of the polymer decrease due to 3D printing process. This study shows that polymer that have been use for 3D printing process cannot be recycled as much as in the injection process.

Recently, there has been renewed interest in the recycling of polymer composites. In a research conducted by Hunt et al. (2015) on recycling polymer codes for distributed manufacturing with 3D printers, it was shown that there are some solutions for the uncoded recycled polymer by creating a code model of recycling based on resin recognition codes produced in China that can be expanded by incorporating more advanced 3D printing materials. This journal offers a way to print resin identifying codes onto items, using OpenSCAD scripts based on a recycle model. The step of the method proposed a voluntary recycling code, developed using an OpenSCAD recycling symbol script and code incorporation tests. Test prints were very good in overall quality, with some variations because of the paper, the printer and the printed object. The recycling icon was designed deep enough that its honesty could be readily seen, but not too profoundly. The prosumer also can adjust the thickness of the embedding to display the recycling symbol at observed angles.

In recent years, there has been a dramatic increase in the study of recycling polymer composites. In a study by Bhadra et al. (2017) on recycling polymer composites, technologies and principals involved and their properties and advantages and disadvantages in recycling various forms of polymer composites has been discussed. The study also analyzes several case studies on micro- and nanocomposite recycling and the latest challenges. There are three ways of recycling: chemical recycling, thermal recycling and mechanical recycling. These methods rely on the application and properties of the polymer composites. The polymer composites are ground, reprocessed and compounded for a mechanical approach to construct a new product or raw material to manufacture a new product intended for the same, or new, use. Chemical recycling is the process for the chemical breakdown into its monomers or other important chemical components of polymer wastes. There can be three forms of thermal recycling: pyrolysis, fluidized beds pyrolysis and microwave pyrolysis. The recovery of all fibers or fillers and other inserts at the detriment of the useful matrix is underlined by all three methods. Polymer composite recycling is seen as an effective way of handling scrap plastic scrap waste end of life. Research on plastic waste recovery and recycling has increased with increasing environmental and nonrenewable resources concerns.

Various studies have evaluated the effectiveness of recycling polymer composites. In a study by Datta and Kopczyńska (2016) on a journal title from waste of polymer to potential major industrial products: In the actual state of recovering and recycling, it is shown that polymer waste processing helps to protect our environmental resources and polymer materials that are made from oil and gas. There are some primary types of recycling: reuse, chemical recycling, mechanical recycling and recovery of energy. Chemical recycle turns polymer waste into monomers, chemical feedstock or fuels manufacturing. Useful compounds like oil, plastics or monomers may be obtained by chemical recycling of polymer materials. Recycling of feedstock should be further produced and strengthened to reduce waste volumes in waste sites and to conserve natural resources. Recycling of feedstock should be further produced and strengthened to reduce waste volumes in waste sites and to conserve natural resources. After that, chemical recycling will look for new methods or for new components. For PU decomposition, researchers used beaver oil and fish oil. Researchers are likely to concentrate on decomposers of natural or renewable resources.

To date, several studies have investigated the recycling of polymer composites. In a study conducted by Dertinger et al. (2020) on distributed polymer composite recycling technological paths for distributed manufacturing: windshield wiper blades. It addressed the use of a windshield blade case study to explore distributed recycling and additive production technical pathways. These blades are a soft (flexible) and a hard (less flexible) thermoplastic composite. The dispersed methods in this journal included the mechanical molding of granular fusion, heated syringe print, 3D impressed mold, injection molding output and filaments in a boot to recycle the fusion filament. A good way has been turning used windshield wipers into useful, high-value, high-quality biomedical handheld hand with fingertip grips and reflex hammer products. This recycling model can be used to expand the range of solutions for the distributed recycling additives in a circular economy.

Several studies have used longitudinal data to examine the recycling of polymer composites. In a study conducted by Alberto et al. (2017) on the recycling of polymer and AM in an open-source context on optimization of methods and processes, recycle waste polymer into open-source 3D printer feedstocks was discussed. The recycling of high-density polyethylene (HDPE) from bottles of milk jugs using a RecycleBot open-source filament system has been tested. In this review, the mechanical recyclability of the commonly used material PLA in open-source 3D printing was chosen with the goal of determining the feasibility of this recycled material for use on open-source 3D printers. The deterioration of the mechanical and rheological properties of the material is measured after several cycles in various extrusion and printing processes. The results show that the recycled material has a minor effect on its mechanical properties during the recycling process. The elastic module is constant, but after 5 cycles, the pressure in breaks shows a decrease of 10.63%. The findings of the Melt Flow Index (MFI) indicate a significant improvement of approximately 6.05 times in comparison to the virgin value following recycling. The present findings demonstrate the feasibility for PLA recycling using 3D open-source printing technologies.

Previous studies have explored the recycling of polymer composites. In a study conducted by Coates et al. (2020) on chemical recycling to monomer for an ideal, circular polymer economy, it shows a desirable option to convert polymers back into monomers for re-polymerization. This is a form of chemical recycling that has been called Chemical Recycling Monomer (CRM). We analyze, briefly, CRM efforts with commodities polymers, including polyolefin thermolysis and nylon six-ring-setting de-polymerization, and examine the latest flora of CRM with modern global polymers. In this paper, the chemical and structural characteristics and the properties, advantages and liabilities of these

CRM-compliant ring-opening polymerization monomers are used to be identified. The smallest amount is made of individually numbered resin (PS and PET); while CRMs of the former have selectivity problems, monomers can still be recovered well in excess of 90% yield, and PET remains the highest resin recycled. CRM is chemically difficult (PVC) or highly difficult (polyolefins) for three polymers generated in the greatest amount: PP, PE and PVC. Since this study focuses on one resin, the PCW polymer flow on several axes is complex.

Several studies have used longitudinal data to examine the recycling of polymer composites. In a study conducted by Pinho et al. (2020), 3D printing goes greener: properties analysis of recycled polymers for the manufacture of engineering components after consumption shows the assessment of different characteristics/properties of 3D polymers and acrylonitrile-butadienestyrene (ABS), recycled from food packets and vehicle dashboards. This paper discusses the ability to use materials/components for add-value applications through recycled polymers recovered from solid polymer waste (SPW). The recycled polymer was printed by using 3D printing to test the properties of the recycled polymer. The method of recycling causes a chain scission with greater effects on the properties/characteristics in semi-cristalline polymer (PLA) printed specimens than in amorphous copolymer (ABS). Some features, like cleaner printed surfaces, may be treated as valuable as it prevents another post-processing. This has been done due to the decreased viscosity caused by lower molecular weight than virgin fiber during extrusion of recyclable filaments. In addition, the recycled materials analyzed met the mechanical property criteria if the same use of polymer materials was considered.

In a review conducted by Valerio et al. (2020) on the strategies for polymer recycling from waste and current trends and opportunities for increasing the circular economy of polymers in South America, it shows that because of the outstanding cost to performance and toughness, polymers are one of the most flexible fabrics, so widespread use of polymers in many different applications which sometimes surpass product requirements. This produces sources of polymer waste that maintain large quantities of their original properties and can then be recycled. There are three methods to recycle polymer that have been discussed in this review: mechanical, chemical and thermal recycling. Current technologies allow some items to be recycled. However, recycling polymers to manufacture new products in certain situations is costlier than using virgin polymers. Recycling processes also must be advanced in order to manufacture products that are competitive in cost to virgin polymers. New compatibilizers or compatibility strategies are required for mechanical recycling, which can permit melting of mixed polymers to manufacture high-performance goods. However, for chemical products and enzyme recycling, new mild processes are required with green solvents and enzyme catalysts that can be easily recycled for several processing loops.

In research conducted by Chen et al. (2020) on recycling of thermosetting polymers for digital light processing 3D printing, their study shows that due to the growing use of DLP 3D printing, the recycling of printed thermosetting materials is strongly desired. A two-stage treatment method includes complex reactions with the use of 3D thermosetting hybrid acrilate-epoxy resin (DLP). Excellent mechanical properties of the resulting polymers have been demonstrated by the tensile and dynamic mechanical analysis (DMA) tester results. The thermosetting components in soluble oligomers may be depolymerized by the bond exchange response (BER) process, which can repolymerize with reverse BERs. Thus, the recovered solvent can be reused during the next printing round after excess EG removal and the mixture with fresh acrylate resins. The samples repeated were still high in module and power equal to virgin samples.

Recently, there has been renewed interest in recycling of polymer composites. In research conducted by Mumbach et al. (2020) on a closed-loop process design for recycling waste of expanded polystyrene by dissolution and polymerization, it is shown that the dissolution technique is used to recycle extended polystyrene (EPS) wastes in a closed loop. The method is dissolved in styrene (its monomer) a maximum rate of EPS waste followed by a suspension polymerization of the solution in order to integrate the monomer (the solvent) in the polymer chains to prevent divergence between the polymer and the solvent. In order to classify and compare the thermal, chemical and rheological features of polymers that are recycled with the regular polymer, Fourier transform infrared spectroscopy (FTIR), TGA, DMA and gel permeation chromatography have been performed. For the recycling of EPS waste, the dissolution process combined with polymerization is successful. A substantial reduction of EPS waste and thus transport effects and expenses can be accomplished when dissolved in styrene. The resulting recycled material has appropriate features to produce new EPS. The process does not involve the loss of chemical products, which saves the costs of treating secondary pollution. In a study that has been done by Hatti-kaul et al. (2019) on designing bio-based recyclable polymers for plastics, the transition to a renewable plastics system requires not only a change to fossil-free feedstock and producing electricity but also the rational nature of material-free polymers and their versatility for the purposes of recycling. We are outlining strategies to improve polymers' performance and recyclability, to increase monomer degradability and to improve chemical recyclability through the development of polymers with various chemical functionalities.

In a study conducted by Hart et al. (2018) on recycling of meal ready to eat (MRE) pouches into polymer filament for material extrusion AM, it was shown that soldiers have consumed a lot of MRE, and tons of the residual polymeric packing waste was left behind, recycled or otherwise incinerated in the atmosphere. The method that has been used to recycle the MRE pouch products was by combining manufacturing protocol for combined compounding, fused filament products and filament extrusion. The findings indicate that additive structures were compared to native MRE sack bag material. The result was showing that the mechanical characteristics and barrier properties were comparable. MRE wastes are used to preserve, additively generated materials using good chemical resilience, barrier properties, and hardness of native low-density polyethylene content. The MRE wastes are the most useful applications. A short financial comparison was made with similar case studies of our recycling process, which revealed the numerous economic and environmental advantages of on-site recycling.

Extensive research has studied the recycling of polymer composites. A review conducted by Lundquist et al. (2020) on reactive compression molding post-inverse vulcanization studies a method to recycle, assemble and repurpose sulfur polymers and composites. It was shown that in this analysis, reactive compression molding is seen in polymers prepared by reverse vulcanization. Mechanical compression is used to contact the reactive interfaces of sulphur polymers. This method is fast and technically straightforward, requiring a simple hot press to coalesce the polymer's reactive faces. Temperatures are lower than the initial polymerization. The theoretical analysis of the S–S metathesis reaction in the middle of the process has shown that the weakest of the S–S connections in an organic polysulfide decreases from di- to tetra, but then decreases in the higher ranks of sulphur. This paper proves that reactive compression molding can be combined with a variety of useful and mechanical properties in the synthesis of compounds. It expects such results in many applications to influence renewable materials and their use.

In recent years, there has been increased interest in the recycling of polymer composites. In research conducted by Shanmugam et al. (2020) on recycling of polymers in AM, there is an opportunity for the circular economy. It was shown that the development of various products by AM of different polymers has been on the rise. In order to promote efficiency and development in a circular economy, AM will provide a stronger manufacturing approach. While polymer waste disposal requires different techniques, polymer recycling is an efficient process. For example, incineration creates destructive greenhouse gases and sites and consume space and impact the atmosphere adversely. The evolution of filament deposition modeling (FDM) results in an expanded use in various applications of recycled polymers. The recycled polymers in FDM provide routes for successful polymer waste management. However, the performance of recycled polymer products produced through AM must be understood to meet application requirements. One method of analyzing the recycled ability of waste plastics is to reintegrate them into composites and to test their mechanical properties. As a result, compared to printed recycled polymers, the production of thoroughly recycled composites through FDM will lead to greater strength.

In a study done by Byard et al. (2019) on green fab lab applications of large-area waste polymer-based AM, it is shown that using industrial 3D printers to print directly from waste plastics sources for fused component manufacturing/fused particulate manufactured (FPF/FGF), green fab labs may serve as a default hub for recycling waste goods in their communities. In this research, Gigabot X, a 3D open-source industrial FPF/FGF printer, demonstrated considerable economic value in using recyclable feed stocks in a green laboratory environment, as a distributed recycling/manufacturing system. The findings demonstrate that FPF/FGF 3D print is able, through three case studies, to generate high-value professional sports products. The device is sufficiently energy efficient such that electricity costs for such processed goods were not extremely high because the products were small. These findings illustrate explicitly that the economic advantages of dispersed recycled goods and the demand output of large usable items are assisted by an integrated FPF/FGF method.

Various studies have evaluated the effectiveness of recycling polymer composites. In a study by Zander (2019) on recycled polymer feedstocks for material extrusion AM, it shows that the reuse and recycle of thermoplastics must be regarded with the quickly growing space of polymer AM. Chemical recycling, mechanical recycling and incineration are typically recycled in three distinct ways. In mechanical recycling, the polymer is segregated into granulates from the pollutants and then converted into the desired component by a mechanical means. Chemical recycling allows polymer chains to be broken, either by the complete depolymerization of monomer units or by the oligomers. In relation to the manufacture of new material, recycling plastics often saves resources. Around 100 million kJ of energy can be saved for recycling by tons of plastic. Barriers to recycling are better value for recycled plastics, better storage and shipping processes, modern washing and processing equipment and mixed or polluted feedstock techniques for recycling and innovative methods for recycling a large variety of plastics.

To date, several studies have investigated the recycling of polymer composites. In a study conducted by Reich et al. (2019) on mechanical properties and applications of recycled polycarbonate particle material extrusion-based AM, it shows that the scope for the expanded use of polymers in 3D printing is that of particulate extrusion (fused granular fabrication (FGF)/fused particle fabrication (FPF)). For the printing of the discarded PC particles, a Gigabot X prototype was used. Slic3r sliced 3D templates, and Marlin firmware powered the printer. The tensile strength of recycled PC particles (64.9 MPa) components has been shown to be competitive with that of industrial filaments printed with 3D (62.2 MPa) and large-format (66.3 MPa) desktop printers. Recycled PC 3D particle printing findings demonstrate that heat-resistant products and high strength can be manufactured at low capital. Thus, the use of recycled PC particles on PME/FPF 3D printers has proven cheap and useful for AM. The sustainability of recycled PC particles provides a wide variety of applications that can substitute both prosumers and industry with costly solutions.

Several studies have used longitudinal data to examine the recycling of polymer composites. In a study conducted by Jin et al. (2016) on the improvement of self-healing, tensile properties and recycle of thermoset styrene/2-vinylfuran copolymers via thermal triggered rearrangement of covalent crosslink, it shows that random styrofoam (S-co-2Vf) 2-vinylfuran copolymers were cross-related with 1,1′-bismaleimide (BMI), which allows recyclable and self-healing polymers with enhanced mechanical properties. They have been retroactively interrelated with the reaction of Diels-Alder (DA). A method for the measurement of the tensile effects of nanoscale thin films using the HarmoniX-Atomic Force Microscopy nanomechanical test was suggested and tested in contrast to the volume properties of normal tensile testing. Treatment with BMI increases the modulus elasticity by about one order of magnitude, leaving all mechanical properties impacted. After reshaping, compression molding of 150°C, the S-co-2VF treated with the BMI can then be recycled and diet cutting of 110°C; in reality, the tensile properties of 3-BMI–T–R–T and 5-BMI–R–T were in line with the properties of the respective pristine specim 3-BMI–T and 5-BMI–T except for a significantly improved Young module. The recycled samples display a higher degree of functionalization, which improves the intensity after reworking.

Previous studies have explored the recycling of polymer composites. In a study conducted by (Ayre, 2018) on a review of technology advancing polymers and polymer composites towards sustainability, it shows that much of these wastes are packing materials; however, the retention and reuse of this oil-based resource could be possible through recycling routes and options. Polymer recycling methods can be categorized into four groups: primary, secondary, tertiary and incineration. For primary, mechanical plastic reprocessing to make the same plastic (e.g., closed-loop HDPE bottles recycled). For secondary, lower-performance products mechanical reprocessing (downgrading). For tertiary, recycling chemical for the regeneration of petrochemical plastic components and for incineration, the benefit is reduced because heat-recovered energy constitutes a portion of recycled energy. Sadly, the recycle of polymer did not get much attention to the world. Global amounts of recycled plastic waste could be expanded dramatically only by the implementation of current recycling technology. The current recycling of composites will provide the appropriate actions and one route for composite recycling involves the "solubilization" of the plastic portion.

1.6 METAL RECYCLING COMPOSITE

A recent study by Jacobs (2020) involved innovative methods for recycling composites by pyrolysis, and it was revealed that in a stable setting, as cloth woven from carbon fibers is joined with a thermoset resin, it results in a very solid substance, precisely tested against metals and ceramics on a pound-for-pound basis. Pyrolysis was carefully chosen as a preferred method, then the samples were inspected with SEM. Next, the samples fibers were trimmed down to prepare for extrusion and the tensile strength was tested. The fibers can be reused and further processed in useful ways by isolating the dried fibers by burning off the resin (a procedure called pyrolysis). Overall, there seems to be some evidence to indicate that these studies showcase its environmentally friendly capability by constructing new composite structures to minimize landfill waste with fibers salvaged through pyrolysis (Jacobs & Jacobs, 2020).

In 2019, Mendonça et al. published a paper in which they described the recycling of chipped stainless steels by experimenting with all possible combinations of factor settings. In order to manufacture metal or carbide composites from a high-energy mechanical milling process, this research was carried out to provide an experimental investigation into the novel method for recycling chips of duplex stainless steel, with the addition of vanadium carbide. In this case, the experimental preparation considered in the literature as acceptable approaches for this reason are the complete factorial designs. Rotation speed, milling time, powder to ball weight ratio and carbide percentage were the four variables examined in this report and for each variable or factor, experiments were carried out at two levels in order to statistically approximate the internal behavior between them. As a result, the study reported that with the incorporation of carbide in milling method, it has been examined that there is an average reduction in particle size in contrast to material absent of carbide (Mendonça et al., 2019).

A great deal of previous investigation into metal composites has focused on recycling. A significant analysis and discussion on the efficient dissociation of matrix and reinforcements alloy by supergravity technology from aluminum composites was presented by Sun et al. (2020). The process of separation was studied by evaluating the effects of the coefficient of gravity, temperature and time of separation on the efficiency of separation. In order to extract silicon carbide and matrix of alloy particles from scrap aluminum matrix composites, a separation induced by supergravity process was proposed. The interfacial response between silicon carbide and liquid aluminum formed by aluminum carbide and free silicon was restricted at 710°C. In the waste, a slight quantity of aluminum carbide (Al_4C_3) was preserved. The silicon carbide (SiC) particles, when thoroughly retrieved after scraping aluminum carbide and additional metallic aluminum from the residue by an acid pickling process, could be directly salvaged as new reinforcements. The study has reported that the filtered silicon carbide and aluminum alloy particles originated from waste aluminum silicon carbide phosphate composites were reclaimed effectively by using a separation method induced with supergravity (Sun et al., 2020).

A review by Al-Alimi et al. (2020) described in a review on the advancements of MMCs and its associated processing methods by directly salvaging light metals. As an alternative way to minimize and reuse discarded materials, recycling of manufactured waste such as aluminum chips has become a key concern for producers today. This will minimize the cost of generating finished products and save the world from toxic pollutants (e.g., CO_2, CO, NO_X) and reduce the use of electricity in all industries of production. The key contribution of this work was a thorough analysis of a broad range of sever plastic deformation (SPD) strategies applied in the production of reinforced MMC-based waste materials, as many SPD approaches were chosen in this analysis to be selected for solid-state recycling. It addresses the various uses of MMCs in the industry and associated production method, with more focus on the processing parameters that specifically define the characteristics of the components. This thesis was investigated to analyze the work of previous researchers on particle strengthening that improved the properties of materials (Al-Alimi et al., 2020).

Recently, a study by Cherrington et al. defined the encouragement in salvaging wind turbine blades composites in Europe. An increasing problem in the elimination processed of used material associated with wind turbine blades composites that is out of service in the future is generated by a recent expansion of the wind power generation, even though the average recyclability of components in a modern wind turbine has been measured to be 80% by mass, this has created a waste management issues. Research on the existing and future wind turbine blade removal practices and prospect of carbon fiber could contribute to the blade recycling process. Currently, regulations in place to control the handling of EOL waste in the wind power sector in Europe are lacking. An analysis on the waste management strategy in Europe, however, has demonstrated that landfill restrictions have avoided waste from being dumped into landfills and contribute positively to energy reclamation. The result of the review has evaluated the producer accountability scenarios on fabricators to recycle wind turbine blades and the effect of producer transparency on the renewable energy sector, where both economic and environmental viewpoints were considered in this study (Cherrington et al., 2012).

One study by Karuppannan Gopalraj and Kärki (2020) was on the recycling of carbon fiber or glass fiber-reinforced composites including fiber retrieval, characteristics and analysis on their life-cycles. This thesis provides an analysis of the current methods of processing carbon fiber and glass fiber cumulative composite waste, with a focus on fiber repossession and recognizing their conserved properties. Under critical conditions, mechanical recycling, thermal recycling (including fluidized beds and pyrolysis) and chemical recycling (solvolysis) on carbon and glass fibers are tested by determined subjects, relevant to the respective processing method. There is a critical need to recognize and further refine such an industrial-scale process into a sustainable approach for achieving a higher fiber recovery yield; by briefly analyzing the previous research, this can be done progressively. This study has reported the different recycling strategies focused on economic and environmental principles on carbon fiber-reinforced polymer and glass fiber-reinforced polymer waste, which prioritize sustainably defined recycling approaches (Karuppannan Gopalraj & Kärki, 2020).

Previous research on recycling carbon fiber composites by Shehab et al. (2020) has established the challenges in cost modeling. Carbon fiber composites are commonly used in various industries because of their superior properties in contrast to traditional materials. Increasing demand in accordance with ecological legislation has encouraged numerous advancements in various carbon fiber composites salvaging strategies, such as thermal, chemical and mechanical recycling processes, where the respective process of reprocessing has its criteria and outputs, along with certain financial effects, and cost modeling is used to validate it. Thanks to their lightweight qualities, carbon fiber composites are commonly used in the aviation industry and contribute to reducing emissions from flights, where up to half of the weight of modern aircraft, such as the Airbus A350, consists of carbon fiber composites. The studies presented that potentially the core issues found may be a beneficial approach to understanding cost drivers by offering an analysis of fundamental conditions that could influence costs (Shehab et al., 2020).

Recently, Wu et al. (2020) have researched suitable sources of silicon or silicon oxide composites for high-capability lithium-ion battery anodes by recycling silicon-based industrial waste. The composites display 992.8 milliampere hour per gram at 0.5 ampere per gram after 400 cycles, where the capacity decay is low, due to the inheritance the integral benefit of silicon oxide and silica such as cycling stability and high capacity, respectively. To compensate for the permanent deficiency of power in the first step, a controllable pre-lithiation process is further added to the silicon or silicon oxide composites, causing an increase of initial electrostatic performance of more than 90%. The silicon or silicon oxide composites with an optimized molarity balance of one to one, offer a compromise among cycling durability and reversible capacity when added as an anode material in a lithium-ion battery by taking advantage of each constituent. This study has demonstrated that with a fundamental, expandable and energy-effective pathway, waste from silicon-based industrial can become cost-efficient tools for high-capability lithium-ion battery anodes (Wu et al., 2020).

Based on a recent study by Bulei et al. (2018) about a recycled processing technique of MMCs with salvaged aluminum alloy, as with other composites, the characteristics of aluminum matrix composites can be adjusted, as their behavior and characteristics predicted, and there is a great deal of attention to considerations toward the structural constitution, characteristics and the relationship among the components. The stir casting method, which started from specially recycled aluminum alloy, is widely accepted as a promising route among different production methods due to the low price and small blemish to the reinforcement not constrained by the shape and size. With outstanding mechanical characteristics, such as strong wear resistance, increased hardness, low density, high strength to weight ratio and low coefficient of thermal expansion, these MMCs are valuable engineering resources. This study prepared a synopsis of the comparatively inexpensive stir casting process for application in the manufacture of aluminum alloy or silicon carbide MMCs by using salvaged aluminum (Bulei et al., 2018).

Quite a few academic projects have started to study metal composite recycling. One study by Karayannis et al. (2012) examined the recycled nickel-based composites from lignite highly calcareous fly ash. In order to minimize weight for energy conservation, improved insulation, reduced contraction and increased corrosion impedance, adding hardness and wear characteristics as well as reducing manufacturing costs of the composites, fly ash may be integrated into upgraded metal-ceramic composites. In the application of x-ray diffraction and SEM-energy dispersive x-ray analyses, together with apparent density and Vickers microhardness measurements, composite specimens were successfully prepared and analyzed. Lightweight composites based on nickel, with no noticeable decrease in nickel hardness, are created by introducing fly ash into the nickel matrix. It has been reported that in this study, starting with powder blends of nickel and up to 50 weight percentage recycled lignite, extremely chalky fly ash and applying powder metallurgy techniques, the possibility of developing advanced nickel fly ash bulk composites has been explored (Karayannis et al., 2012).

A study by Lopez-Urionabarrenechea et al. (2020) examined the recovery of added-value gases and carbon fibers in the pyrolysis-based composites reprocessing method, where the development is progressing fast. Pyrolysis requires thermal disintegration of polymer resin to extract the fibers, but the substance worth of these resins has not yet been improved. The tests were done in a laboratory-scaled system comprising a sequence of two vessels, where the pyrolysis of the waste material is executed at 500°C in the first vessel, while the vapors from the first vessel are treated at 900°C in the second vessel. The result of the experiment allows oxidation and pyrolysis of CFRP-type poly-benzoxazine waste to facilitate the retrieval of carbon fibers with largely clean surfaces and a mechanical characteristic equivalent to an industrial fresh carbon fiber. This study has reported that the findings demonstrate that it is possible to recycle material from polymer resin in the CFRP waste salvaging process, which signifies a major development in that field of studies (Lopez-Urionabarrenechea et al., 2020).

A few experiments so far have explored the recycling of polymer and metal composites. A qualitative study by Dennis (2020) described the recycling of carbon fiber composite powertrain and feed control, where it explains excess material from discarded manufacturing process waste contains valuable carbon fiber that can be recovered and recycled. To recover the carbon fiber from the discarded waste, a machine was designed to improve the pyrolysis preparation where it consists of three main system components. The progress of the research has been successful in providing useful data to improve the design of the delamination machine. This research has proposed that a redesign of the machine was determined to contain the force essential to deconstruct the composite material effectively (Dennis, 2020).

A study by Kaewunruen and Liao (2020) has investigated the recyclabilities and sustainabilities of composite materials used in railway turnout systems. The findings of the railway turnout system's recyclability and recoverability rates are subject to a data-related overview retrieved from a specific reference (e.g., UNIFE), where the outcome indicates a value of 92.23% and 93.50%, correspondingly. The EOL management appraisal focused on the method of waste disposal and regeneration, with three steps followed while evaluating the railway turnout recycling process: pre-treatment period, dismantling stage and shredding stage. This study assessed impacts of the environmental life cycle established on the estimation of carbon emissions and energy usage over the life span of 75 years and the sensitivity examination reveals that the mass recovery factor has a great consequence on the values of recyclability and recovery relative to the energy recovery factor. The paper has stated the latest trends in railway turnouts for fiber-reinforced foamed urethane sleeper and it determines the mechanisms of re-cycling and recovery of this infrastructure scheme (Kaewunruen & Liao, 2020).

The proficiency of recycling metal composites has been investigated. Based on recent research by Memon et al. (2020) on epoxy vitrimer containing imine was determined to have a multipurpose recyclability and implementation in completely salvageable carbon fiber reinforced composites. The research stated that it is important to establish recycling solutions with the increasing amount of discarded CFRP composites by environmental economics. It is claimed that the production of reprocessing, mendable and degradable thermosets is a viable solution in solving the recycling dilemma of carbon fiber reinforced composites that are at the end of their service life. Specifically, after interlaminar shear loss, 92% of the strength is recovered for the restored carbon fiber reinforced composites, and when deteriorating matrix resin in amine solvent, non-destructive carbon fibers are extracted from carbon fiber reinforced composites and the degraded materials can be reclaimed to prepare new epoxy resins; thereby, a complete recovering mechanism for carbon fiber reinforced composites is achieved. This research has stated that the results of this study provide a promising solution of recycling carbon fiber reinforced composites (Memon et al., 2020).

A research by Pietroluongo (2020) examined the composite component mechanical recycling EOL automotive. The study suggests that mechanical reprocessing adds to the degradation of the fibers, minimizing its influence of mechanical strength. In contrast to composite made of PA66 reinforced with 35 weight percentage glass fiber, the component was put through three stages of grinding and injection molding. The degradation of the mechanical characteristic can be mostly attributed to the reduction of the fiber lengths when induced to the milling and molding steps involved in the reprocessing method, and improvement in processing the fiber breakage can be seen during the first and second recycling phases, while it becomes less important during further salvaging processes as the breakage of fibers decreases. This research paper has stated that recycling EOL products by means of a mechanical approach and determining the potential uses of the materials is obtained by contrasting the rheological, morphological and mechanical characteristics of the salvaged materials with those of the compared materials (Pietroluongo et al., 2020).

The feasibility of metal matrix or reinforced composites has been tested in different trials. Based on a previous investigation by Bakshi et al. (2020) involving carbon nanotube reinforced MMCs, this composite was being used in structural applications as it has high specific strength and it has exciting electrical and thermal characteristics. In order to attain a uniform disposition of carbon nanotubes in the matrix, the processing methods applied to the amalgamation of composites have been carefully appraised. The changes in mechanical properties obtained by the inclusion of carbon nanotubes in different metal matrix structures are summarized. As the physical and chemical stability of carbon nanotubes in various metal matrices and the relevance of the carbon nanotube and metal interface have been checked, the factors deciding reinforcement obtained by carbon nanotubes reinforcement are elucidated. This review paper has reported that the essential challenges of carbon nanotubes reinforced MMCs are the manufacturing processes, distribution of nanotubes, boundary, mechanism reinforcement and mechanical characteristics (Bakshi et al., 2010).

A variety of experiments to date have studied metal matrix or reinforced materials. A review by Sharma et al. (2020) discussed the study of application opportunities and improvement of aluminium metal matrix composites (AMMCs), where in accordance with recent technical advances, new technologies are constantly produced and material properties are enhanced in order to achieve safety and operating requirements. In a wide range of industries, the way to produce stronger lightweight materials that have high quality and performance was created by having a continuous observation. For some wide range implementation of reinforcement materials, AMMCs show a tremendous prospect in manufacturing composites, with necessary properties and it has also been designed to achieve good, lightweight mechanical and tribological properties based on implemented and specified demand. Overall, manufacturers typically require lightweight, medium strength and lower cost, for which AMMCs has an advantage, so it is used as potential materials for many engineering applications in a new generation (Sharma et al., 2020a).

There has been several studies and research on metal matrix or reinforced composites. Based on a recent study by Dhanesh et al. (2020) about the recent progresses in hybrid AMMCs, due to the superior strength to weight ratio and high tolerance temperature, MMCs have recently gained a rising interest in potential implementation in the automotive and aerospace industries. The study was done by analyzing the experiments and studies on mechanical and production characterization of hybrid AMMCs carried out by different academics, organizations and institutions. The different reinforced AMMCs were tested, where mechanical property improvements of the construction techniques are studied, such as silicon carbide, fly ash, tungsten, boron carbide, etc. as significant reinforcements, although some experiments have concentrated on the production of organic materials through non-conventional and hybrid reinforcements. Thus, this review has reported that MMCs based on aluminum are commonly recognized as some of the best structural engineering components because of their high strength, rigidity, thermal resilience, wear resistance and many advantages that suit their purpose and often differ with the type and quantity of reinforcements used (Dhanesh et al., 2020a).

More than a few scientific works on metal matrix or reinforced composites have begun. One study by Hayat et al. (2019) examined research on titanium MMCs, where it was explained that in contrast to steel- and nickel-based materials, titanium matrix composites have good basic strength and rigidity; in contrast to monolithic superalloys, high-temperature titanium matrix composites will give up to 50% weight reduction while also retaining their strength and stiffness when applied in jet engine propulsion systems. In the past two decades, growth of AM technology has opened up new possibilities for application-oriented titanium matrix composites to expand, such as selective laser sintering or melting, laser engineered net forming, fused deposition modeling, 3D printing and ultrasonic consolidation, which are the most widely used AM technologies for composites. This paper has stated that this analysis has extensive information on common re-enforcements, development procedures and assesses certain common TMCs' static and dynamic properties (Hayat et al., 2019).

In some experiments to date, metal matrix reinforced composites have been researched. Recent research on the progress in hybrid MMCs by Zhou et al. (2020) has established that compared to their traditional counterparts, hybrid MMCs possess better overall functional and mechanical performance, and have a greater prospect for use of applications in functional systems and structural engineering. According to various forms, shapes and proportions, the hybrid reinforcements are graded, and an experimental manufacturing method is proposed to achieve uniform disposition of hybrid reinforcements and constructing special constitutions was also reviewed. Hybrid MMCs display outstanding overall mechanical, tribological and physical characteristics, where more and more interest in this area has been drawn to the extraordinary combinations of these properties. Several forms of hybrid reinforcements have been suggested, and multiple production processes (ex-situ and in-situ) have been produced in recent years. The studies presented thus far have suggested that while variable types of hybrid reinforcements have been applied, different types based on performance ought to be developed and recommended, while in order to produce hybrid MMCs with the optimal delivery of hybrid reinforcements, innovative manufacturing paths, such as modern additive processing processes, should be considered (Zhou et al., 2020).

Kazemi et al. published a paper in which they have reviewed hybrid titanium composite laminates with an emphasis on fabrications, mechanical characteristics and surface treatments. Here they have stated that fiber metal laminates are greater functioning hybrid systems grounded on fiber reinforced polymer composite ply, and a metal alloy layer with alternating layered configurations. Different surface treatment processes, mechanical properties, fabrications and classifications of hybrid titanium composite laminates have been investigated in this study. In order to develop better mechanical properties and reduce the weight of the structures, the mechanical strength ability of fiber metal laminates inspired an inquiry into new metals, adhesives and composite systems. Compared with conventional fiber metal laminates, fiber-reinforced polymer composites and hybrid titanium composite laminates provide more advantages, particularly in aeronautical, maritime, military, and offshore applications, either at room or elevated temperatures, as well as in unforgiving environmental surroundings (Kazemi et al., 2020).

Some experiments have been performed to analyze metal matrix and reinforced composites. One study by Casati and Vedani (2014) reviewed nano-particle reinforced MMCs, where the composites comprise a nano-particle filled a metal matrix possessing mechanical and physical properties that are somewhat distinct from those of the matrix; and in terms of damping properties, mechanical efficiency and wear resistance, the nano-particles could be used to boost the base material. The goal of this work is to examine the important processing methods used to synthesize nanocomposites in the bulk metal matrix. The development of

TABLE 1.1
Report Work on Recycling of Polymer Composites

Polymer	Fiber/Monomer/ Recycle Product	Effect of Reinforcement	References
Low-density polyethylene (LDPE)	Millions of meals ready to eat (MREs)	• Mechanical properties and barrier properties show similar performance from the native MRE bag material.	(Hart et al., 2018)
Composite polymer matrix	Carbon fiber	• The nozzle angle of the water jet needs to adjust in the recycling process.	(Huntley et al., 2018)
High-density polyethylene (HDPE)	Bottles of used milk jugs	• Elastic modulus remains constant. • The strain of the recycle polymer has decreased.	(Alberto et al., 2017)
Hard polymer and soft polymer	Windshield wiper blades	• Recycled polymer was less brittle compared to the windshield wiper blades. • The recycled polymer has a similar behavior for tensile strain less than 100%.	(Dertinger et al., 2020)
Poly(lactic acid) (PLA) polymer	Food packages	• Semi-crystalline polymer revealed losses of 33 percent both in tensile stress and flexural strength. • The recycled ABS did not reveal any improvements in the mechanical properties.	(Pinho et al., 2020)
Polycondensation polymers	Poly(ethylene terephthalate) (PET)	• The recycle polymer was suitable to manufacture.	(Valerio et al., 2020)
Expanded polystyrene (EPS)	Styrene	• Recycled polymer maintains its chemical, thermal and rheological properties.	(Mumbach et al., 2020)
Polylactic acid (PLA)	Lactide	• The elastic modulus of the recycled polymer increased. • Decrease in mechanical properties	(Cruz Sanchez et al., 2017)
Polyurethanes (PU)	Gasoline	• High-yield products • Lighter product	(Datta & Kopczyńska, 2016)
Thermosetting polymers	Digital light processing (DLP) product	• Increased strength • Increased level of modulus	(Chen et al., 2020)

(Continued)

TABLE 1.1 (Continued)
Report Work on Recycling of Polymer Composites

Polymer	Fiber/Monomer/ Recycle Product	Effect of Reinforcement	References
Polyhydroxyalkanoates (PHAs)	Compost bags	• Increased ratio of aromatic units led to polymers with decreasing molecular weights and increasing thermal stability.	(Hatti-kaul et al., 2019)
LDPE	Postconsumer oil	• Low-quality recycled materials	(Kaiser et al., 2018)
Carbon nanotube (CNT) polymer	Plastic	• The quality of the recycled polymer decreased.	(Bhadra et al., 2017)
Polyethylene	Polymer solid wastes	• Increased mechanical strength • Increased shear stress	(Bukhkalo et al., 2018)
Polysulfides	Sulfur	Recycled polymer has tunable mechanical properties.	(Lundquist et al., 2020)
Acrylonitrile butadiene styrene (ABS)	recycled ABS	• Decrease in mechanical properties for each cycle of recycle.	(Byard et al., 2019)
Polycarbonate (PC)	Plastic waste	• Recycled polymer has high strength. • Recycled polymer has high heat resistance.	(Reich et al., 2019)
Polypropylene	Cellulose waste	• Increased elastic modulus • Constant tensile strength	(Zander et al., 2019)
Polyethylene terephthalate (PET)	Plastic bottles	• Increased tensile strength	(Shanmugam et al., 2020)

compounds stabilized by nano-ceramic particles, such as carbides, nitrides, oxides and carbon nanotubes, different kinds of metals, mainly aluminum, magnesium and copper, have been utilized. This paper has reported that the reinforcing mechanisms responsible for enhancing the mechanical characteristics of nano-reinforced composites of metal matrix have been studied and the main potential utilization for this emerging class of materials are envisaged (Casati & Vedani, 2014).

Research by Munir et al. (2020) examined the superior mechanical and corrosion capabilities of graphene nanoplatelet-reinforced magnesium metal matrix nanocomposites in biomedical implementations, where the MMCs of magnesium reinforced with graphene nanoplatelets were formed with powder metallurgy. Via high-energy ball-milling methods, graphene nanoplatelets of varying concentrations of 0.1, 0.2 and 0.3 weight percentage, 5 nm and 9 nm layer thicknesses and

TABLE 1.2
Report Work on Metal Composites

Metal Composites	Fiber/Filler/Process	Effect of Reinforcement	References
Graphene	Metal oxide	Better capacity and rate capability. Superior cycling stability.	(Wu et al., 2012)
CNT	CNT-reinforced MMCs	High basic strength for their fascinating thermal and electrical properties as well as practical materials	(Bakshi et al., 2010)
Boron	Reinforced epoxy Aluminum	Properties such as modulus, strength and resistance to fatigue, especially on a standardized basis of weight	(Lynch & Kershaw, 2018)
Aluminum Magnesium Copper	Carbides Nitrides Oxides Carbon nanotubes	Improved wear resistance. Enhance damping properties. Better mechanical strength.	(Casati & Vedani, 2014)
Gold Silver	Binding a photoactive molecule (e.g., pyrene)	Improves photochemical behavior and allows the organic-inorganic hybrid nano-assemblies ideal for optoelectronic and light-harvesting applications	(Kamat, 2002)
Copper matrix composites	Ceramic fiber or particle reinforced light metals Cemented carbides	Yield a novel substance that has its own appealing engineering characteristics	(Mortensen & Llorca, 2010)
Graphene-metal particle nanocomposites	Metal nanoparticles (Au, Pt and Pd)	Potential application in direct methanol fuel cells	(Xu et al., 2008)
Reduced graphene oxide Metal/metal oxide composites	Manganese dioxide Silver	Greater distribution coefficient (Kd)	(Sreeprasad et al., 2011)
Aluminum (Al) metal matrix composites (AMMCs)	SiC or Al2O3 or B4C or WC	Enhanced hardness, fatigue, stiffness, corrosion, strength, creep and wear properties in materials	(Kumar et al., 2020)

(*Continued*)

Introduction

TABLE 1.2 (Continued)
Report Work on Metal Composites

Metal Composites	Fiber/Filler/Process	Effect of Reinforcement	References
Carbonaceous materials, metal oxides, metals	Conductive polymers (CPs), consist of polyaniline, polypyrrole and poly(3,4-ethylenedioxythiophene), etc.	High sensing performance due to synergistic effect of the components	(Y. Wang et al., 2020)
Colloidal liquid metal alloys of gallium	Liquid metal droplet and graphene co-fillers	Good electrical conductivity. The electrical conductance remains nearly constant, with changes less than 0.5%.	(Saborio et al., 2020)
FGMs joining metals to ceramics and metal-matrix composites	Stainless, nickel, titanium and copper alloys	Eliminating the need for dissimilar-metal welds and joints	(Reichardt et al., 2020)
AluminumMagnesiumTitaniumCopperIronNickel	Hybrid reinforcements	Superior overall functional and mechanical greater prospect to be widely utilized for structural engineering and functional device implementations	(Zhou et al., 2020)
Aluminum metal matrix composites	Wide variety of reinforcing materials	High wear resistance Good damping capacities High specific rigidity Suitable resistance to corrosion	(Sharma et al., 2020b)
Copper matrix	Reinforced polymer and ceramic matrix composites Graphene-like nanosheets	High level interfacial shear stress. Improved load transfer strengthening and crack-bridging toughening simultaneously. Constructs additional three-dimensional hyper channels for electrical and thermal conductivity.	(Zhang et al., 2020)
Titanium dioxide	Metal-organic frameworks	Outstanding thermal and chemical stability. Wide bandgap with suitable band edge. Low priced. Non-toxic. Corrosion resistance.	(Wang et al., 2020)

(*Continued*)

TABLE 1.2 (Continued)
Report Work on Metal Composites

Metal Composites	Fiber/Filler/Process	Effect of Reinforcement	References
Magnesium metal matrix composites	Graphene nanoplatelets	Enhanced mechanical properties via synergetic strengthening modes. Better ductility, amended compressive strength. Corrosion resistance. No significant toxicity detected.	(Munir et al., 2020)
Aluminum metal matrix composites	Many combinations of reinforcement particles	Improved mechanical property. Reduced weight. Increased strength. Reduced production cost.	(Arunkumar et al., 2020)
Hybrid titanium composite laminates (HTCLs)	Fiber-reinforced polymer composite pliesMetal alloy sheets	Outstanding stiffness. Excellent yield stress. Exceptional fatigue properties. Good high-velocity impact properties.	(Kazemi et al., 2020)
Hybrid Aluminum metal matrix composites	Silicon carbide Boron carbideFly ashTungsten	Greater strength. High stiffness. Excellent thermal stability. Good wear resistance.	(Dhanesh et al., 2020b)
Titanium matrix composites	Continuous fibersDiscontinuous particulates	High specific strength. High stiffness.	(Hayat et al., 2019)

µm and 5 µm particle sizes were distributed into the magnesium residue. The findings suggest that graphene nanoplatelets are ideal candidates in the development of biodegradable magnesium-based composite implantations as reinforcements in magnesium matrices, and by synergetic strengthening modes, the graphene nanoplatelets addition enhanced the mechanical properties of magnesium. This research shows that for load-bearing applications, graphene nanoplatelet-reinforced magnesium metal matrix nanocomposites (Mg-xGNPs) with less than 0.3 weight percentage establish novel decomposable implant materials (Tables 1.1 and 1.2). Table 1.1 and Table 1.2 show the report work on recycling of polymer composites and metal composite.

1.7 CONCLUSIONS

Polymers and metal waste are highly resistant to violent environmental factors and raise the risk of environmental and economic harm by their aggregation. This will further worsen the environmental and economic issues created by a higher level of polymer waste. There are many ways that have been used to overcome with this problem and one of them was recycling. Current technologies have been developing to recycle this composite. There are three methods to recycle polymers and metal composites: chemical recycling, thermal recycling and mechanical recycling. Most of the recycled polymer and metal composites show positive results: the recycled polymer has good mechanical and thermal properties, as good as virgin material. In certain cases, recycling polymers for the manufacture of new products is more costly than using virgin polymers. This is because the need for initial recycled waste preparation and treatment nearly doubles the recycled waste costs, in addition to the feedstock. Recycling processes also need to be improved to manufacture products that can compete against virgin polymers at cost. Besides that, a variation of polymer waste also needs a technologically challenging and costly disposal investment. The vast majority of this form of waste is then deposited superficially, exceeding 10–15% of the solid waste composition of sites. Thus, each party needs to be taken seriously and involve the recycling of polymers and metal composites.

ACKNOWLEDGMENT

The authors would like to thank Universiti Teknologi Malaysia for the financial support through the grant CRG 30.3: Retardant coating using graphene/bamboo aerogel mixtures on SAR robotics system (PY/2020/03495).

REFERENCES

Al-Alimi, S., Lajis, M. A., Shamsudin, S., Chan, B. L., Mohammed, Y., Ismail, A. E., & Sultan, N. M. (2020). Development of metal matrix composites and related forming techniques by direct recycling of light metals: A review. *International Journal of Integrated Engineering*, *12*(1), 144–171.

Alberto, F., Sanchez, C., Lanza, S., Boudaoud, H., Hoppe, S., Alberto, F., Sanchez, C., Lanza, S., Boudaoud, H., Hoppe, S., Camargo, M., Cruz, F., Lanza, S., Boudaoud, H., Hoppe, S., & Camargo, M. (2017). Polymer recycling and additive manufacturing in an open source context: Optimization of processes and methods. In *Annual International Solid Freeform Fabrication Symposium*, ISSF 2015.

Arunkumar, S., Subramani Sundaram, M., Suketh Kanna, K. M., & Vigneshwara, S. (2020). A review on aluminium matrix composite with various reinforcement particles and their behaviour. *Materials Today: Proceedings*. doi: 10.1016/j.matpr.2 020.05.053.

Asmatulu, E., Twomey, J., & Overcash, M. (2014). Recycling of fiber-reinforced composites and direct structural composite recycling concept. *Journal of Composite Materials*, *48*(5), 593–608. doi: 10.1177/0021998313476325.

Ayre, D. (2018). ScienceDirect Technology advancing polymers and polymer composites towards sustainability: A review. *Current Opinion in Green and Sustainable Chemistry*, *13*, 108–112. doi: 10.1016/j.cogsc.2018.06.018.

Bakshi, S. R., Lahiri, D., & Agarwal, A. (2010). Carbon nanotube reinforced metal matrix composites – a review. *International Materials Reviews*. doi: 10.1179/095 066009X12572530170543.

Bhadra, J., Al-Thani, N., & Abdulkareem, A. (2017). Recycling of polymer-polymer composites. In S. Thomas, R. Mishra, & N. Kalarikkal (Eds.), *Micro and Nano Fibrillar Composites (MFCs and NFCs) from Polymer Blends* (1st ed., pp. 263–277). Elsevier. doi: 10.1016/B978-0-08-101991-7.00011-X.

Bhouri, M., & Mzali, F. (2020). Analysis of thermo-elastic and physical properties of recycled 2017 aluminium alloy/Gp composites: Thermal management application. *Materials Research Express*, *7*(2). doi: 10.1088/2053-1591/ab5eeb.

Bukhkalo, S. I., Klemeš, J. J., Tovazhnyanskyy, L. L., Olga, P., Kapustenko, P. O., & Perevertaylenko, O. Y. (2018). Eco-friendly synergetic processes of municipal solid waste polymer utilization. *Chemical Engineering Transactions*, *70*, 2047–2052.

Bulei, C., Todor, M. P., & Kiss, I. (2018). Metal matrix composites processing techniques using recycled aluminium alloy. *IOP Conference Series: Materials Science and Engineering*, *393*(1). doi: 10.1088/1757-899X/393/1/012089.

Byard, D. J., Woern, A. L., Oakley, R. B., Fiedler, M. J., Snabes, S. L., & Pearce, J. M. (2019). Green fab lab applications of large-area waste polymer-based additive manufacturing. *Additive Manufacturing*, *27*(March), 515–525. doi: 10.1016/j.addma.2019.03.006.

Casati, R., & Vedani, M. (2014). Metal matrix composites reinforced by nano-particles – A review. *Metals*. doi: 10.3390/met4010065.

Chen, Z., Yang, M., Ji, M., Kuang, X., Qi, H. J., & Wang, T. (2020). Recyclable thermosetting polymers for digital light processing 3D printing. *Materials & Design*, 109189. doi: 10.1016/j.matdes.2020.109189.

Cherrington, R., Goodship, V., Meredith, J., Wood, B. M., Coles, S. R., Vuillaume, A., Feito-Boirac, A., Spee, F., & Kirwan, K. (2012). Producer responsibility: Defining the incentive for recycling composite wind turbine blades in Europe. *Energy Policy*, *47*(2012), 13–21. doi: 10.1016/j.enpol.2012.03.076.

Coates, Geoffrey W & Getzler, Y. D. (2020). Chemical recycling to monomer for. *Nature Reviews Materials*. doi: 10.1038/s41578-020-0190-4.

Cruz Sanchez, F. A., Boudaoud, H., Hoppe, S., & Camargo, M. (2017). Polymer recycling in an open-source additive manufacturing context: Mechanical issues. *Additive Manufacturing*, *17*, 87–105. doi: 10.1016/j.addma.2017.05.013.

Datta, J., & Kopczyńska, P. (2016). From polymer waste to potential main industrial products: Actual state of recycling and recovering. *Critical Reviews in Environmental Science and Technology*, *46*(10), 905–946. doi: 10.1080/10643389.2016.1180227.
Dennis, W. (2020). Carbon Fiber Composite. All Undergraduate Projects, 121, 1–59.
Dertinger, S. C., Gallup, N., Tanikella, N. G., Grasso, M., Vahid, S., Foot, P. J. S., & Pearce, J. M. (2020). Resources, conservation & recycling technical pathways for distributed recycling of polymer composites for distributed manufacturing: Windshield wiper blades. *Resources, Conservation & Recycling*, *157* (December 2019), 104810. doi: 10.1016/j.resconrec.2020.104810.
Dhanesh, S., Kumar, K. S., Fayiz, N. K. M., Yohannan, L., & Sujith, R. (2020a). Recent developments in hybrid aluminium metal matrix composites: A review. *Materials Today: Proceedings*. doi: 10.1016/j.matpr.2020.06.325.
Dhanesh, S., Kumar, K. S., Fayiz, N. K. M., Yohannan, L., & Sujith, R. (2020b). Recent developments in hybrid aluminium metal matrix composites: A review. *Materials Today: Proceedings*. doi: 10.1016/j.matpr.2020.06.325.
Gu, F., Guo, J., Zhang, W., Summers, P. A., & Hall, P. (2017). From waste plastics to industrial raw materials: A life cycle assessment of mechanical plastic recycling practice based on a real-world case study. *Science of the Total Environment*, *601–602*, 1192–1207. doi: 10.1016/j.scitotenv.2017.05.278.
Hart, K. R., Frketic, J. B., & Brown, J. R. (2018). Recycling meal-ready-to-eat (MRE) pouches into polymer fi lament for material extrusion additive manufacturing. *Additive Manufacturing*, *21*(February), 536–543. doi: 10.1016/j.addma.2018.04.011.
Hatti-Kaul, R., Nilsson, L. J., Zhang, B., Rehnberg, N., & Lundmark, S. (2019). Designing biobased recyclable polymers for plastics. *Trends in Biotechnology*, 1–18. doi: 10.1016/j.tibtech.2019.04.011.
Hayat, M. D., Singh, H., He, Z., & Cao, P. (2019). Titanium metal matrix composites: An overview. *Composites Part A: Applied Science and Manufacturing*. doi: 10.1016/j.compositesa.2019.04.005.
Hunt, E. J., Zhang, C., Anzalone, N., & Pearce, J. M. (2015). Polymer recycling codes for distributed manufacturing with 3-D printers. *Resources, Conservation and Recycling*, *97*, 24–30. doi: 10.1016/j.resconrec.2015.02.004.
Huntley, S., Rendall, T., Longana, M., Pozegic, T., Potter, K., & Hamerton, I. (2018). Recycling composite materials using a water-jet tape deposition method. In *13th SPHERIC International Workshop Galway, Ireland*, June 26–28, 2018.
Ignatyev, I. A., Thielemans, W., & Vander Beke, B. (2014). Recycling of Polymers: A Review. *ChemSusChem*, *7*(4), 1579–1593. https://doi.org/10.1002/cssc.201300898
Ilyas, R. A., & Sapuan, S. M. (2019). The preparation methods and processing of natural fibre bio-polymer composites. *Current Organic Synthesis*, *16*(8). doi: 10.2174/1570179416082001205616.
Ilyas, R. A., & Sapuan, S. M. (2020). Biopolymers and biocomposites: Chemistry and technology. *Current Analytical Chemistry*, *16*(5), 500–503. doi: 10.2174/157341101605200603095311.
Jacobs, M. (2020). *Novel Methods For Composites Recycling Via Pyrolysis. Undergraduate Honors Theses. 129.*
Jin, K., Li, L., & Torkelson, J. M. (2016). Recyclable crosslinked polymer networks via one-step controlled radical polymerization. *Adv Mater*, *28*, 6746–6750. doi: 10.1002/adma.201600871.
Kaewunruen, S., & Liao, P. (2020). Sustainability and recyclability of composite materials for railway turnout systems. *Journal of Cleaner Production*, 124890. doi: 10.1016/j.jclepro.2020.124890.

Kaiser, K., Schmid, M., & Schlummer, M. (2018). Recycling of polymer-based multilayer packaging: A review. *Recycling, 3*(1). doi:10.3390/recycling3010001.

Kamat, P. V. (2002). Photophysical, photochemical and photocatalytic aspects of metal nanoparticles. *Journal of Physical Chemistry B.* doi:10.1021/jp0209289.

Karayannis, V. G., Moutsatsou, A. K., & Katsika, E. L. (2012). Recycling of lignite highly-calcareous fly ash into nickel-based composites. *Fresenius Environmental Bulletin, 21*(8B), 2375–2380.

Karuppannan Gopalraj, S., & Kärki, T. (2020). A review on the recycling of waste carbon fibre/glass fibre-reinforced composites: Fibre recovery, properties and life-cycle analysis. *SN Applied Sciences, 2*(3), 433. doi:10.1007/s42452-020-2195-4.

Kazemi, M. E., Shanmugam, L., Yang, L., & Yang, J. (2020). A review on the hybrid titanium composite laminates (HTCLs) with focuses on surface treatments, fabrications, and mechanical properties. *Composites Part A: Applied Science and Manufacturing.* doi:10.1016/j.compositesa.2019.105679.

Kumar, A., Singh, R. C., & Chaudhary, R. (2020). Recent progress in production of metal matrix composites by stir casting process: An overview. *Materials Today: Proceedings.* doi:10.1016/j.matpr.2019.10.079.

Liu, T., Zhang, M., Guo, X., Liu, C., Liu, T., Xin, J., & Zhang, J. (2017). Mild chemical recycling of aerospace fiber/epoxy composite wastes and utilization of the decomposed resin. *Polymer Degradation and Stability, 139*, 20–27. doi:10.1016/j.polymdegradstab.2017.03.017.

Lopez-Urionabarrenechea, A., Gastelu, N., Acha, E., Caballero, B. M., Orue, A., Jiménez-Suárez, A., Prolongo, S. G., & de Marco, I. (2020). Reclamation of carbon fibers and added-value gases in a pyrolysis-based composites recycling process. *Journal of Cleaner Production, 273.* doi:10.1016/j.jclepro.2020.123173.

Lundquist, N. A., Tikoalu, A. D., Worthington, M. J. H., Shapter, R., Tonkin, S. J., Stojcevski, F., Mann, M., Gibson, C. T., Gascooke, J. R., Karton, A., Henderson, L. C., Esdaile, L. J., & Chalker, J. M. (2020). Reactive compression molding post-inverse vulcanization: A method to assemble, recycle, and repurpose sulfur polymers and composites. *Chemistry a European Journal, 10035–10044.* doi:10.1002/chem.202001841.

Lynch, C. T., & Kershaw, J. P. (2018). Metal matrix composites. *Metal Matrix Composites.* doi:10.1201/9781351074445.

Memon, H., Wei, Y., Zhang, L., Jiang, Q., & Liu, W. (2020). An imine-containing epoxy vitrimer with versatile recyclability and its application in fully recyclable carbon fiber reinforced composites. *Composites Science and Technology, 199*(May), 108314. doi:10.1016/j.compscitech.2020.108314.

Mendonça, C., Capellato, P., Bayraktar, E., Gatamorta, F., Gomes, J., Oliveira, A., Sachs, D., Melo, M., & Silva, G. (2019). Recycling chips of stainless steel using a full factorial design. *Metals, 9*(8). doi:10.3390/met9080842.

Mohd Joharudin, N. F., Latif, N. A., Mustapa, M. S., Badarulzaman, N. A., & Mahmod, M. F. (2020). Physical properties and hardness of treated amorphous silica as reinforcement of AA7075 recycled aluminum chip. *IOP Conference Series: Materials Science and Engineering, 824*(1). doi:10.1088/1757-899X/824/1/012015.

Mortensen, A., & Llorca, J. (2010). Metal matrix composites. *Annual Review of Materials Research.* doi:10.1146/annurev-matsci-070909-104511.

Mumbach, G. D., Bolzan, A., Antonio, R., & Machado, F. (2020). A closed-loop process design for recycling expanded polystyrene waste by dissolution and polymerization. *Polymer, 209*(March), 122940. doi:10.1016/j.polymer.2020.122940.

Munir, K., Wen, C., & Li, Y. (2020). Graphene nanoplatelets-reinforced magnesium metal matrix nanocomposites with superior mechanical and corrosion performance for biomedical applications. *Journal of Magnesium and Alloys*. doi:10.1016/j.jma.2019.12.002.

Pietroluongo, M., Padovano, E., Frache, A., & Badini, C. (2020). Mechanical recycling of an end-of-life automotive composite component. *Sustainable Materials and Technologies*, *23*, e00143. doi:10.1016/j.susmat.2019.e00143.

Pinho, A. C., Amaro, A. M., & Piedade, A. P. (2020). 3D printing goes greener: Study of the properties of post-consumer recycled polymers for the manufacturing of engineering components. *Waste Management*, *118*, 426–434. doi:10.1016/j.wasman.2020.09.003.

Rady, M. H., Mustapa, M. S., Wagiman, A., Shamsudin, S., Lajis, M. A., Alimi, S. Al, Mansor, M. N., & Harimon, M. A. (2020). Effect of the heat treatment on mechanical and physical properties of direct recycled aluminium alloy (AA6061). *International Journal of Integrated Engineering*, *12*(3), 82–89. doi:10.30880/ijie.2020.12.03.011.

Rajendran, S., Scelsi, L., Hodzic, A., Soutis, C., & Al-Maadeed, M. A. (2012). Environmental impact assessment of composites containing recycled plastics. *Resources, Conservation and Recycling*, *60*, 131–139. doi:10.1016/j.resconrec.2011.11.006.

Reich, M. J., Woern, A. L., Tanikella, N. G., & Pearce, J. M. (2019). Mechanical properties and applications of recycled polycarbonate particle material extrusion-based additive manufacturing. doi:10.3390/ma12101642.

Reichardt, A., Shapiro, A. A., Otis, R., Dillon, R. P., Borgonia, J. P., McEnerney, B. W., Hosemann, P., & Beese, A. M. (2020). Advances in additive manufacturing of metal-based functionally graded materials. *International Materials Reviews*. doi:10.1080/09506608.2019.1709354.

Saborio, M. G., Cai, S., Tang, J., Ghasemian, M. B., Mayyas, M., Han, J., Christoe, M. J., Peng, S., Koshy, P., Esrafilzadeh, D., Jalili, R., Wang, C. H., & Kalantar-Zadeh, K. (2020). Liquid metal droplet and graphene co-fillers for electrically conductive flexible composites. *Small*. doi:10.1002/smll.201903753.

Shanmugam, V., Das, O., Neisiany, R. E., Babu, K., & Singh, S. (2020). Polymer recycling in additive manufacturing: An opportunity for the circular economy. doi:10.1016/j.jclepro.2020.121602.

Sharma, A. K., Bhandari, R., Aherwar, A., Rimašauskiene, R., & Pinca-Bretotean, C. (2020a). A study of advancement in application opportunities of aluminum metal matrix composites. *Materials Today: Proceedings*, *26*, 2419–2424. doi:10.1016/j.matpr.2020.02.516.

Sharma, A. K., Bhandari, R., Aherwar, A., Rimašauskiene, R., & Pinca-Bretotean, C. (2020b). A study of advancement in application opportunities of aluminum metal matrix composites. doi:10.1063/5.0036141.

Sharma, A. K., Bhandari, R., Aherwar, A., Rimašauskiene, R., & Pinca-Bretotean, C. (2020). A study of advancement in application opportunities of aluminum metal matrix. *Materials Today: Proceedings*. doi:10.1016/j.matpr.2020.02.516.

Shehab, E., Meiirbekov, A., & Sarfraz, S. (2020). Challenges in cost modelling of recycling carbon fiber composites. *Advances in Transdisciplinary Engineering*, *12*(October), 594–601. doi:10.3233/ATDE200120.

Sreeprasad, T. S., Maliyekkal, S. M., Lisha, K. P., & Pradeep, T. (2011). Reduced graphene oxide-metal/metal oxide composites: Facile synthesis and application in water purification. *Journal of Hazardous Materials*. doi:10.1016/j.jhazmat.2010.11.100.

Sun, N., Wang, Z., Guo, L., Wang, L., & Guo, Z. (2020). Efficient separation of reinforcements and matrix alloy from aluminum matrix composites by supergravity technology. *Journal of Alloys and Compounds*, *843*, 155814. doi: 10.1016/j.jallcom.2020.155814.

Suriani, M. J., Radzi, F. S. M., Ilyas, R. A., Petrů, M., Sapuan, S. M., & Ruzaidi, C. M. (2021). Flammability, tensile, and morphological properties of oil palm empty fruit bunches fiber/pet yarn-reinforced epoxy fire retardant hybrid polymer composites. *Polymers*, *13*(8), 1282. doi: 10.3390/polym13081282.

Suriani, M. J., Rapi, H. Z., Ilyas, R. A., Petrů, M., & Sapuan, S. M. (2021). Delamination and manufacturing defects in natural fiber-reinforced hybrid composite: A review. *Polymers*, *13*(8), 1323. doi: 10.3390/polym13081323.

Suriani, M. J., Sapuan, S. M., Ruzaidi, C. M., Nair, D. S., & Ilyas, R. A. (2021). Flammability, morphological and mechanical properties of sugar palm fiber/polyester yarn-reinforced epoxy hybrid biocomposites with magnesium hydroxide flame retardant filler. *Textile Research Journal*, 1–12. 0.1177/00405175211008615

Suriani, M. J., Zainudin, H. A., Ilyas, R. A., Petrů, M., Sapuan, S. M., Ruzaidi, C. M., & Mustapha, R. (2021). Kenaf fiber/pet yarn reinforced epoxy hybrid polymer composites: Morphological, tensile, and flammability properties. *Polymers*, *13*(9), 1532. doi: 10.3390/polym13091532.

Tapper, R. J., Longana, M. L., Norton, A., Potter, K. D., & Hamerton, I. (2020). An evaluation of life cycle assessment and its application to the closed-loop recycling of carbon fibre reinforced polymers. *Composites Part B: Engineering*, *184*(June 2019), 107665. doi: 10.1016/j.compositesb.2019.107665.

Ustundag, M., & Varol, R. (2019). GERİ Dönüştürülmüş VTicari OlarakTeminEdilen TmTitanyum Alaşimlarinin Darbe DayaniminiİncelenmesI. *Mühendislik Bilimleri ve Tasarım Dergisi*, *7*(2), 232–237. doi: 10.21923/jesd.461882.

Valerio, O., Muthuraj, R., & Codou, A. (2020). Strategies for polymer to polymer recycling from waste: Current trends and opportunities for improving the circular economy of polymers in South America. *Current Opinion in Green and Sustainable Chemistry*, 100381. doi: 10.1016/j.cogsc.2020.100381.

Vasiliev, V. V., & Morozov, E. V. (2018). Introduction. *Advanced Mechanics of Composite Materials and Structures*, xvii–xxv. doi: 10.1016/b978-0-08-102209-2.00022-0.

Vo Dong, P. A., Azzaro-Pantel, C., Boix, M., Jacquemin, L., & Domenech, S. (2015). Modelling of environmental impacts and economic benefits of fibre reinforced polymers composite recycling pathways. In *Computer Aided Chemical Engineering* (Vol. 37). Elsevier. doi: 10.1016/B978-0-444-63576-1.50029-7.

Wang, C. C., Wang, X., & Liu, W. (2020). The synthesis strategies and photocatalytic performances of TiO2/MOFs composites: A state-of-the-art review. *Chemical Engineering Journal*. doi: 10.1016/j.cej.2019.123601.

Wang, Y., Liu, A., Han, Y., & Li, T. (2020). Sensors based on conductive polymers and their composites: A review. In *Polymer International*. doi: 10.1002/pi.5907.

Wu, H., Zheng, L., Zhan, J., Du, N., Liu, W., Ma, J., Su, L., & Wang, L. (2020). Recycling silicon-based industrial waste as sustainable sources of Si/SiO2 composites for high-performance Li-ion battery anodes. *Journal of Power Sources*, *449*(August 2019), 227513. doi: 10.1016/j.jpowsour.2019.227513.

Wu, Z. S., Zhou, G., Yin, L. C., Ren, W., Li, F., & Cheng, H. M. (2012). Graphene/metal oxide composite electrode materials for energy storage. *Nano Energy*. doi: 10.1016/j.nanoen.2011.11.001.

Xu, C., Wang, X., & Zhu, J. (2008). Graphene – metal particle nanocomposites. *Journal of Physical Chemistry C*. doi: 10.1021/jp807989b.

Yang, Y., Boom, R., Irion, B., van Heerden, D. J., Kuiper, P., & de Wit, H. (2012). Recycling of composite materials. *Chemical Engineering and Processing: Process Intensification*, *51*, 53–68. doi: 10.1016/j.cep.2011.09.007.

Zander, N. E. (2019). Recycled polymer feedstocks for material extrusion additive manufacturing [chapter]. In *Polymer-Based Additive Manufacturing: Recent Developments, Part 3 – Recycled Polymer Feedstocks for Material Extrusion Additive Manufacturing*. doi: 10.1021/bk-2019-1315.ch003.

Zander, N. E., Park, J. H., Boelter, Z. R., & Gillan, M. A. (2019). Recycled cellulose polypropylene composite feedstocks for material extrusion additive manufacturing. doi: 10.1021/acsomega.9b01564.

Zhang, J., Chevali, V. S., Wang, H., & Wang, C. H. (2020). Current status of carbon fibre and carbon fibre composites recycling. *Composites Part B: Engineering*, *193*(December 2019), 108053. doi: 10.1016/j.compositesb.2020.108053.

Zhang, X., Xu, Y., Wang, M., Liu, E., Zhao, N., Shi, C., Lin, D., Zhu, F., & He, C. (2020). A powder-metallurgy-based strategy toward three-dimensional graphene-like network for reinforcing copper matrix composites. *Nature Communications*. doi: 10.1038/s41467-020-16490-4.

Zhou, M. Y., Ren, L. B., Fan, L. L., Zhang, Y. W. X., Lu, T. H., Quan, G. F., & Gupta, M. (2020). Progress in research on hybrid metal matrix composites. *Journal of Alloys and Compounds*. doi: 10.1016/j.jallcom.2020.155274.

2 Preparation of Metal Matrix Composites by Solid-State Recycling from Waste Metal/Alloy Chips

Debasis Chaira
Department of Metallurgical and Materials Engineering,
National Institute of Technology Rourkela, Odisha, India

CONTENTS

2.1 Introduction 37
2.2 Solid-State Recycling Techniques 38
    2.2.1 Powder Metallurgy (PM) 38
    2.2.2 Hot Extrusion Method 42
    2.2.3 Friction Stir Processing (FSP) 44
    2.2.4 Equal Channel Angular Pressing (ECAP) 45
    2.2.5 High Pressure Torsion (HPT) 47
    2.2.6 Hot Press Forging Operation 48
2.3 Conclusions 50
Acknowledgment 54
References 54

2.1 INTRODUCTION

A large amount of metal/alloy chips are generated during the machining of metals and alloys. During the recycling of metal and alloy chips during the re-melting technique, several remarkable problems arise:

1. Relatively higher metal loss due to oxidation
2. Toxic gas generation from combustion of the oil emulsion adhering to the scrap chips
3. Relatively high-energy consumption and recycling cost
4. Harmful environmental impacts

DOI: 10.1201/9781003148760-2

Lazzaro and Atzori (1991) have shown the metal yield in conventional recycling of aluminum scraps is approximately 55%. Moreover, the high-energy requirement for melting aluminum alloys partly nullifies the advantages of recycling (Gabrielle et al., 2012). The melting of metals/alloys chips during recycling leads to changes in chemistry of the materials. Some innovative techniques are being developed to have a higher yield than conventional recycling processes of re-melting the waste product. The solid-state recycling process is one such method that can be used to overcome many disadvantages of the re-melting process. Solid-state recycling is considered a more suitable choice, especially for scrap chips or turning, and it can directly convert scraps into a bulk product with superior mechanical properties, bypassing melting. Most of the solid-state recycling techniques are based on severe plastic deformation (SPD) of the scrap to be recycled. These processes allow breakage of the oxide layer between the particle by imposing a large plastic strain, where temperature, pressure and process time allow solid bonding between the particles.

2.2 SOLID-STATE RECYCLING TECHNIQUES

Various solid-state recycling techniques are used to fabricate metal matrix composite from waste metal/alloy chips. These techniques are used to obtain a bulk structure with ultrafine grains and also modified properties. These can be obtained from a solid product where SPD is introduced. These techniques are also used to consolidate powder particles or a mixture of chips and powder. The various techniques are as follows:

1. Powder metallurgy (PM)
2. Hot extrusion (HE)
3. Equal channel angular processing (ECAP)
4. Friction stir processing (FSP)
5. Hot forging
6. High pressure torsion (HPT)

2.2.1 Powder Metallurgy (PM)

Powder metallurgy deals with the fabrication of useful engineering components from powder particles. PM is ideally suited for fabrication of small and intricate shape components from starting powders. It is economical for the fabrication of components from refractory materials where normal casting and metal forming routes are not feasible. There are three basic steps in powder metallurgy: powder production, powder compaction and sintering. Figure 2.1 shows a flow chart of the fabrication of metal matrix composite from waste metal/alloy chips.

The chips are initially cut into smaller sizes and then cleaned with acetone using an ultrasonic bath. Then, the grinding operation of chips is performed in milling devices to produce a powder. Various milling devices like ball mill, pulverisette and dual drive planetary mill and attritor mills are used, where chips

Preparation of Metal Matrix Composites

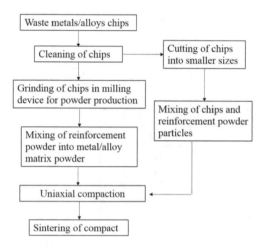

FIGURE 2.1 Flow chart of powder metallurgy process for fabrication of metal matrix composite from waste metal/alloy chips.

and steel/ceramic balls are used for grinding purposes. The size and shape of powder particles depend on various milling parameters: mill type, milling speed, ball to powder weight ratio, wet/dry milling and nature of materials. The details of the effect of milling parameters on the grinding operation are available elsewhere (Chaira & Karak, 2015; Suryanarayana, 2001). The metal/alloy powder is mixed with a reinforcement particle homogeneously by using mixing devices. Then, the powder mixture is compacted by uni-axial compaction. Figure 2.2 shows a schematic diagram of uni-axial die compaction used in powder metallurgy.

On the other hand, fine-size waste chips are mixed with reinforcement powder and then cold compacted. Finally, both of the compacts are sintered in a furnace at a high temperature in a vacuum/inert atmosphere, where metallurgical bonding between loose powder particles form to obtain a dense, solid product. Here, sintering is carried out in either a solid- or liquid-state sintering. A lower density is achieved in solid-state sintering than liquid-state sintering.

The composite fabricated by cold compaction and conventional sintering possess poor mechanical properties due to presence of porosity. Hence, such composites are fabricated by pressure-assisted sintering like hot iso-static pressing (HIP), hot pressing (HP) or spark plasma sintering (SPS) techniques, where simultaneous application of pressure and heat are used to achieve higher densification and better mechanical properties. The details of powder compaction and sintering techniques are available elsewhere (Chaira, 2021; German, 1994). Figure 2.3 shows the schematic diagram of HP used in powder metallurgy. In HP, a powder mixture is kept inside a graphite die and pressure is applied uni-axially. The whole assembly is kept inside a furnace, where the temperature can be monitored. The chamber can be heated in a vacuum or inert atmosphere can be

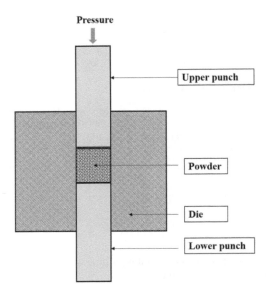

FIGURE 2.2 Schematic diagram of uni-axial die compaction used in powder metallurgy.

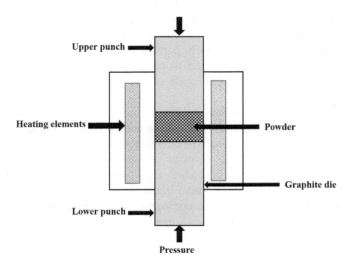

FIGURE 2.3 Schematic diagram of uni-axial HP used in in powder metallurgy.

maintained. The compact consolidated by uni-axial HP is not uniformly densified and uniform properties are not achieved since pressure is applied along a particular direction. Hence, HIP is adopted, where a deformable container is evacuated, degassed and is filled with powder. The consolidation of a powder-filled container

Preparation of Metal Matrix Composites 41

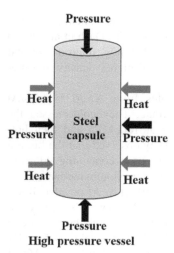

FIGURE 2.4 Schematic diagram of HIP process used in powder metallurgy.

is performed in a high-pressure vessel, where Ar or N_2 gas is purged to exert pressure and heat for densification. Figure 2.4 shows the schematic diagram of HIP set up and used for PM consolidation.

In SPS, high amperage pulse DC current is passed between powder particles in a graphite die through conductive electrodes. The combined application of pressure and high ampere current results in a consolidation of powder with close to theoretical density within a few minutes (5–10 minutes). SPS is best suited for consolidation of nanoparticles where the nanostructure can be retained after consolidation due to very fast consolidation. Awotunde et al. (2019) reviewed the influence of sintering methods on mechanical properties of Al nanocomposites reinforced with carbonaceous materials (CNTs, graphene, GNPs, etc.). They reported that SPS has emerged as a novel sintering technique because it restricts grain growth due to short sintering time and also preserves the integrity of the delicate structure of carbon allotrope reinforcement. The only limitation of the SPS technique is the cost factor. Viswanathan et al. (2006) also reviewed various consolidation techniques for nanocomposites. They highlighted several SPD and various sintering techniques that are used for the fabrication of nanocomposites. They also pointed out that traditional consolidation techniques like conventional pressureless sintering, HP and HIP have strong limitations of not being able to retain nanostructures during processing. Several researchers fabricated MMC by powder metallurgy using various consolidation techniques. Samal et al. (2013) fabricated Cu–graphite MMC by powder metallurgy through the conventional sintering and SPS process. They obtained higher density and improved mechanical properties of composites consolidated by SPS than conventional sintering, even when SPS (700°C) was performed at a lower temperature than conventional sintering (900°C). Khushbu Dash et al. (2012) fabricated Cu–Al_2O_3 MMC by powder metallurgy using conventional and SPS techniques. They achieved higher densification and better mechanical properties of composites fabricated by SPS than conventional. Khushbu Dash et al. (2013) also fabricated Al–Al_2O_3 MMC from Al powder and Al_2O_3 (micron size and nanosize) powder using the SPS technique and noticed higher density and mechanical properties of nanocomposites.

Canakci and Varol (2014) fabricated composites from AA7075 chips/Al powder–SiC powder via the powder metallurgy route. They cold-pressed the mixture of AA7075 chips/Al powder–SiC powder at 100 MPa and then hot-pressed at 500°C at 300 MPa pressure for 1 hour. They observed that the highest density of 2.73 g/cc could be achieved for the composite containing 90 wt.% AA7075 chips and 10 wt.% Al powder. They also found maximum hardness of 80 BHN for the composite containing 10 wt.% SiC and 10 wt.% Al powder. Kadir et al. (2017) fabricated composite from AA6061 chips–Al powder by cold compaction of the mixture followed by conventional sintering in tubular furnace at 552°C. They achieved maximum density of 2.47 g/cc, hardness value of 65.6 Hv and compressive strength of 307.7 MPa for a composite containing 100% AA6061 chip. They noticed that adding more Al powder into AA6061 chips degraded the mechanical properties. Shial et al. (2018) fabricated the Ti–TiC metal matrix composite via the powder metallurgy route. They initially milled the commercially pure (CP) Ti chips to produce powder in high-energy planetary mill and again milled Ti powder obtained from Ti chips and graphite powder mixture to synthesize Ti–TiC composite powder mixture. Emadi Shaibani and Ghambari (2011) prepared powders from gray cast iron chips by target jet milling and conventional ball milling. Afshari and Ghambari (2016) recycled bronze machining chips into bronze powder by jet milling and also subsequently compared with ball milling technique. Soufiani et al. (2010) fabricated Ti6Al4V powders from machining scraps by both planetary milling and shaker milling. Dikici et al. (2017) studied the effects of disc milling parameters on the physical properties and microstructural characteristics of Ti6Al4V powders. Verma et al. (2018) recycled extra low carbon and low-carbon steel chips via the powder metallurgy route. Aslan et al. (2019) conducted HP of bronze (CuSn10) chips and spheroidal graphite cast iron waste chips at 400°C for 20 minutes at 480, 640 and 820 MPa pressure. They achieved a porosity of 20%, hardness of 160 VHN and UTS of 325 MPa at a pressure of 820 MPa for the fabricated MMC.

2.2.2 Hot Extrusion Method

Direct hot extrusion is a very popular solid-state recycling technique where metal/alloy chips can be recycled directly without melting. Products of long shape and uniform cross section from light/ductile metal or alloy (Al/Cu) chips are manufactured by this technique. A high level of shear stress is required as the process is performed at a relatively low temperature. In this technique, initially, chips are cleaned, cut into small pieces and then compaction is carried out to form a billet of suitable size. Here, pure metal/alloy chips or a mixture of chips and reinforcing powder particles are used to fabricate composite. At this stage, a

Preparation of Metal Matrix Composites

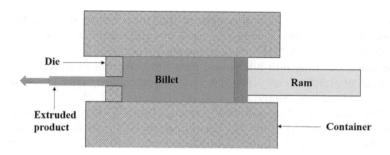

FIGURE 2.5 Schematic diagram of direct hot extrusion process.

billet is formed by mechanical interlocking of chips; formation of no metallurgical bonding takes place. Then the billet is preheated and finally extrusion is performed. Figure 2.5 shows a schematic diagram of a hot extrusion technique. The theoretical pressure for the extrusion process of fully dense materials is expressed as:

$$P = \sigma_o \ln(r), \qquad (2.1)$$

where P = pressure, σ_o = yield stress and r = extrusion ratio.

Shamsudin et al. (2016) reviewed the hot extrusion process as solid-state recycling of light metal and alloy chips. They reported that the final properties of the extruded product depend on several parameters:

i. Chip types and their morphologies: chips are generally produced by either milling or turning. Both types are suitable for the direct extrusion process but it affects the final properties.
ii. Method of compaction for fabrication of billet from chips: chips are either cold-compacted or hot-pressed. The temperature and pressure of compaction both affect the final properties of the extruded product.
iii. Purity and cleanliness of chips: affects the metallurgical bonding between chips after extrusion. Sometimes chips are cleaned and then annealed. It has been reported that annealed chips exhibit better densification.
iv. Extrusion ratio (r) and preheating temperature.

Łukasz Wzorek et al. (2017) fabricated a Cu–Al composite from a mixture of Al and Cu powder by cold compaction followed by hot extrusion. They cold-compacted the Al + Cu powder mixture and then extruded the compacted billet at 300, 350 and 400°C. They found a maximum UTS of 189 MPa and YS of 130 MPa when extrusion was performed at 300°C. Sherafat et al. (2009) recycled Al7075 alloy chips by fabricating an Al7075–Al composite by hot extrusion via powder metallurgy. They found that extrusion temperature should be lower than 500°C to obtain a good bonding between

Al7075 chip and Al. They also noticed that the strength of the composite increased but ductility decreased with increasing chip content. Hu et al. (2008) recycled an AZ91D magnesium alloy via the solid-state route using compaction of chips into billets and then hot extrusion of billets into rods. Mindivan et al. (2014) recycled pure magnesium chips by cold-press and hot-extrusion processes. Sugiyama et al. (2010) recycled an Al alloy and Cu turnings mixture by hot extrusion pressing. Initially, they prepared a billet from a mixture of Al alloy and Cu turning by multiple cold compactions. Then they preheated the billet and finally conducted hot extrusion to prepare extruded products of various shapes.

2.2.3 Friction Stir Processing (FSP)

In the friction stir extrusion process, a rotating tool is used to produce heat and plastic deformation during friction between the tool itself and workpiece (billet made from cold compaction of chips or powder) by compaction, stirring and extruding the material. The recycling process takes place in a unique way where significant savings of cost, labor and energy is possible compared to the conventional method. Figure 2.6 shows a schematic diagram of friction stir processing. The quality of the extruded product depends on the extrusion temperature and die rotating speed. It has been reported that at a very low die rotating speed, no extrusion is possible due to low heat input. On the other hand, high rotating speed and high strain lead to formation of swirl defects that affect the mechanical properties of the extruded product (Baffari et al., 2017). Harikishor Kumar et al. (2020) fabricated an Al alloy surface composite by friction stir processing an A356 cast alloy with reinforcement (red

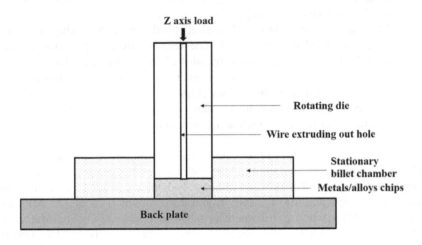

FIGURE 2.6 Schematic of friction stir processing setup.

mud + fly ash). They found that surface composite exhibited higher strength and wear resistance but ductility was reduced due to incorporation of reinforcement. Ghanbari et al. (2015) investigated the effect of the passes number on microstructural and mechanical properties of the Al2024/SiC composite produced by friction stir processing. Dolatkhah et al. (2012) studied the effects of process parameters on microstructural and mechanical properties of Al5052/SiC metal matrix composite fabricated via friction stir processing. Ma et al. (2008) reviewed friction stir processing technology and reported that FSP is an effective solid-state recycling technique that can provide localized modification and control of microstructure in the near surface layer of processed metallic components. They reported that intense plastic deformation and thermal exposure during friction stir processing leads to several phenomena: breakup of coarse dendritic, second phase precipitates and dissolution of precipitates which promotes closure of pores and generates fine, homogeneous and pore free wrought structures. Hsu et al. (2005) reported fabrication of in situ ultrafine grained Al–Al$_2$Cu metal matrix composite by FSP. Hsu et al. (2006) also investigated the effect of FSP on the reaction between Al and Ti in an Al–Ti sintered billet. Hosseinzadeh and Yapici (2018) fabricated an Al2024/SiC metal matrix composite by FSP. They observed that SiC particles enhanced the microhardness of FSPed samples up to 50% relative to the as-received condition. According to them, yield strength of the as-received samples improved about 2.5 times and the hardness of MMC increased to 130 VHN after four passes, compared to 90 VHN of the received Al204 sheet.

2.2.4 Equal Channel Angular Pressing (ECAP)

Equal channel angular pressing (ECAP) is another SPD technique that is extensively used to fabricate bulk nanostructure materials by shear deformation. The main advantages of this process is that exceptionally large, unidirectional and uniform deformation can be imparted to a bulk solid at a very low load. Figure 2.7 shows the schematic of the ECAP process. In this process, a billet is multipressed in a special die. The angle of two channel intersections is generally 90° but it may vary for difficult to deform materials. ECAP can be conducted at a higher temperature for hard-to-deform materials. The direction and number of bullet passes through channels are the most important parameters for microstructure and property modification. When the sample passes through the channel, shear stress is introduced. Although intense shear strain is introduced into the sample the cross-sectional area of the sample remains the same, allowing repetitive deformation. Wan et al. (2017) reviewed the ECAP process. ECAP is generally used for SPD of bulk materials but it can also be used to consolidate powders, chips or fabrications of a metal matrix composite.

Xu et al. (2009) fabricated an Al–Al$_2$O$_3$ MMC from powder mixture by back pressure ECAP which was carried out at 400°C at a pressure of 200 MPa. They

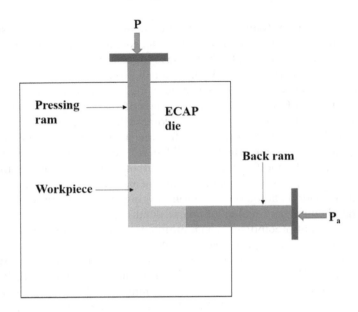

FIGURE 2.7 Schematic diagram of ECAP process.

obtained a fully dense Al–Al$_2$O$_3$ MMC with hardness of 2285 HV, UTS-740 MPa and 0.2% proof stress-460 MPa in compression and 1% plastic strain to fracture after four passes. Goussous et al. (2009) fabricated an Al–C MMC from a powder mixture by back pressure ECAP, which was carried out at 400°C at a pressure of 50 MPa. They achieved maximum density of 2.65 g/cc, hardness of 6.5 HV, UTS of 350 MPa in compression for Al–5 wt.% C MMC after eight passes. Qiong Xu et al. (2020) studied the enhancement of strength and ductility of SiC$_{(p)}$/AZ91 by rotary die (RD) ECAP process. The cast sample was processed by rotary-die equal channel angular pressing (RD-ECAP) at 250°C for three different passes (4P, 8P, 16P). They found that after 4P, 8P, 16P ECAP, the strength of the composites increased to 306, 288 and 285 MPa, respectively, and the ductility increased to 6.4, 7.3 and 8.2%, respectively. They also noticed that the strength of the composite was slightly decreased and the elongation was gradually increased with the pass number increasing from 4–16P. Aqeel Abbas et al. (2020) investigated the effects of ECAP processing on stir cast WS2/AZ91 Mg MMC. The composite was homogenized and then subjected to two-pass ECAP with 120° ECAP die. They found 22.143%, 44.735% and 92% increments in YTS, UTS and elongation in two passes of 1 wt.%WS2/AZ91 MMC as compared to homogenized monolithic AZ91 alloy as combined effects of ECAP and WS2.

Preparation of Metal Matrix Composites

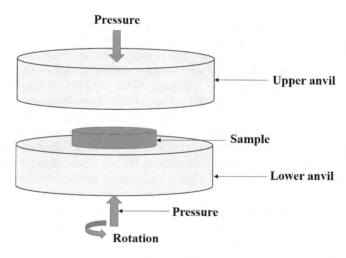

FIGURE 2.8 Schematic diagram of the HPT process.

2.2.5 HIGH PRESSURE TORSION (HPT)

In HPT, a specimen is held between two anvils and strained under the torsion of several GPa. Here, the upper anvil is fixed and a lower anvil rotates under hydrostatic pressure. A lower anvil performs torsional rotation with different revolutions per minute. During HPT, high plastic strains are generated. HPT is not only used for refinement of microstructures but also is used for consolidation of powders/chips. As a starting material, powders/chips or bulk solids can be used. HPT is one of the most powerful technique to prepare ultrafine-grained materials where non-homogeneous deformation with a large strain gradient occurs. Hence, a non-homogeneous microstructure is developed in materials processed by HPT. Figure 2.8 shows a schematic diagram of the HPT process. There are three basic limitations of HPT techniques:

1. The sample shape is in the form of a disc, which is not suitable for many industrial applications where thin sheets or wires are used.
2. The sample size is generally limited to 35 mm to maintain high compressive pressure.
3. The strain generated is non-uniform in proportion to the distance from the disc center.

Mohammad Khajouei-Nezhad et al. (2019) fabricated an Al7075/Al MMC from a mixture of 7075 Al alloy chips + pure Al powder mixtures by the HPT process of 1.2 GPa pressure using one and four turns. They achieved a density > 99% of RD, maximum UTS-500 MPa and elongation to failure is 7.9% for annealed chips after four turns. They found that the wt.% of chips and powder play an important role in determining mechanical properties. They observed that a higher amount of chips or powder showed poor mechanical properties along with inhomogeneous microstructure. The optimum properties and homogeneous microstructure were obtained at 50–50% of chips and powder content. Galiia Korznikova et al. (2020) fabricated a Cu–1 wt.% graphene laminated MMC by HPT at 6 GPa pressure (constrained and non-constrained HPT). They observed improvement of hardness from 1,410 to 1,850 HV at the center and 1,460–1,900 MPa at the edge of the sample after the first stage of HPT (constrained HPT). After the second stage of HPT, the hardness increased from 1,900 to 2,950 HV at the edge of the sample; however, both ductility and strength decreased after the second stage of HPT (non-constrained HPT). Bazarnik et al. (2019) fabricated Cu–SiC MMC by SPS followed by HPT. They observed the microhardness increased both after the addition of SiC and HPT processing of a SPSed sample. They also observed that UTS, YS and elongation decreased after the addition of SiC particles during SPS processing. However, UTS and YS increased and elongation decreased after HPT. Mohammad Jahedi et al. (2015) fabricated Cu–20 vol.% SiC MMC by a high-pressure double torsion process. They observed the homogeneity of a microstructure after HPT. F.X. Li et al. (2018) studied the microstructure and strength of CNT-reinforced Ti MMC fabricated by HPT. They observed a refinement of the grain size with increasing CNT contents, and the microhardness and tensile properties of the composites increased.

2.2.6 Hot Press Forging Operation

Hot press forging is a new alternative direct recycling approach which has fewer steps, lower energy consumption and is cost effective as it eliminates the two traditional intermediate processes of cold compacting and pre-heating. It is a deformation process where a workpiece is compressed between two dies of either impact load or hydraulic load to deform it. Figure 2.9 shows a schematic diagram of the forging operation. It can be classified as cold, warm or hot

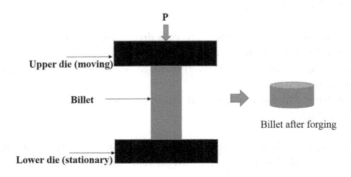

FIGURE 2.9 Schematic diagram of forging operation.

depending on the working temperature. Hot forging is performed above the recrystallization temperature (0.75 T_m) of a workpiece so that strain hardening will not occur. The high temperature process changes the grain structure of workpiece by breaking down into fine grain structure. Warm forging is conducted below recrystallization temperature but above room temperature (0.3 T_m). Cold forging is done at below recrystallization temperature of workpiece, often at or near room temperature. Forging can also be categorized as open or closed die forging depending on the type of forming operation. In open die forging, the workpiece is compressed between two flat platens or dies, thus allowing the metal to flow without any restriction in the sideward direction relative to the die surfaces.

Closed die forging, also called impression die forging, is performed in dies that have the impression that will be imparted to the workpiece through forming. In the intermediate stage, the initial billet deforms partially, giving a bulged shape. During the die full closure, impression is fully filled with a deformed billet and further moves out of the impression to form a flash.

In a multistage operation, separate die cavities are required for a shape change. In the initial stages, uniform distribution of properties and microstructures is seen. In the final stage, actual shape modification is observed. When drop forging is used, several blows of the hammer may be required for each step (R. Ganesh Narayanan).

Azlan Ahmad et al. (2017) fabricated the AA6061-based MMC by hot press forging a mixture of AA6061 chips and Al_2O_3 powder at 430, 480 and 530°C for 60, 30 and 120 minutes. They observed an increase in density and hardness of MMC with an increase in both temperature and holding time. Lajis et al. (2017) fabricated the MMC by hot press forging a AA6061 chips–Al_2O_3 powder (1, 2, 3 and 5 wt.%) mixture at 530°C and at a pressure of 35 tonnes for a period of 120 minutes. They found that the MMC containing 2 wt.% Al_2O_3 exhibited the highest UTS (315 MPa) and 20% elongation before failure. They also noticed that the further addition of Al_2O_3 leads to a reduction in strength of the composites.

Table 2.1 shows the summary of salient features on the processing of various solid-state recycling techniques. The table shows that in PM processing, initially a compact is made either from a mixture of metal/alloy scraps and powder or from a mixture of metal/alloy and reinforcement powder. Then, the green compact is consolidated at a high temperature either by pressureless or pressure-assisted sintering techniques. In hot extrusion, a billet is first made from chips by cold compaction and then it is preheated and finally hot extrusion of the preheated billet is carried out. In FSP, a billet is made from chips/powder and then the friction between the workpiece and tool extrudes the material. In the ECAP process, a billet is first made and then the workpiece is processed by using intense shear strain in a ECAP die. In HPT, the billet that made chips and powder is held between two anvils. The tortional stress of the lower anvil processes the workpiece. In high-pressure forging, the billet made from chips/

TABLE 2.1
Summary of Various Solid-State Recycling Techniques with Processing Features for Fabrication of MMC

Processing Method	Salient Processing Features
Powder metallurgy	i. Cold compaction of powder or a mixture of chips and powder followed by sintering ii. Consolidation by conventional pressureless sintering in furnace or by pressure-assisted sintering like HP, HIP and SPS
Direct hot extrusion	i. A billet is made by cold/hot compaction of powder or chips and powder mixture. ii. The preheated billet is deformed by shear force in a die where there is reduction in cross-sectional area.
Friction stir processing (FSP)	i. A billet is made from chips or powder by compaction. ii. A rotating tool is used to produce heat and plastic deformation during friction between the tool itself and workpiece (billet) by compaction, stirring and extruding the material.
Equal channel angular pressing (ECAP)	i. A billet is multiple pressed in a special die. ii. The sample is passed through the channel, shear stress is introduced. Although intense shear strain is introduced into the sample but cross sectional area of sample remains the same, allowing repetitive deformation.
High pressure torsion (HPT)	The specimen is held between two anvils. The torsional shear stress of several GPa generated by rotation of lower anvil consolidates the specimen/refine the microstructure.
Hot press forging operation	The workpiece is compressed between two dies via either impact or hydraulic loading.

powder is forged by either hydraulic or impact loading. Table 2.2 provides a summary of the advances of MMC fabricated by various solid-state recycling techniques.

2.3 CONCLUSIONS

This chapter discusses the importance and relevance of solid-state recycling techniques over conventional liquid recycling techniques. It provides basic principles and processing of fabrication of metal matrix composites by using solid-state recycling techniques. This chapter presents a comprehensive review and summary of recent research on the fabrication of MMC through solid-state recycling techniques. The main processing features of various MMC fabrication methods are also highlighted.

TABLE 2.2
Summary of Recent Advances of MMC Fabricated by Various Solid-State Recycling Techniques

Authors	Composite Materials System	Synthesis Techniques	Experimental Details	Properties Achieved
Canakci and Varol (2014)	AA7075 chips/ Al powder–SiC particles	Powder metallurgy using cold-pressing followed by HP	Cold-pressed at 100 MPa and then hot-pressed at 500°C at 300 MPa pressure for 1 h	Highest density of 2.73 g/cc for 90 wt.% AA7075 chips and 10 wt.% Al powder. Maximum hardness of 80 BHN for composite containing 10 wt.% SiC and 10 wt.% Al powder.
Kadir et al., (2017)	AA6061 chips–Al powder	Powder metallurgy using cold compaction followed by conventional sintering in furnace	Cold compaction followed by sintering in tubular furnace at 552°C	Maximum density of 2.47 g/cc, hardness value of 65.6 Hv and compressive strength of 307.7 MPa are achieved for composite containing 100% AA6061 chip.
Aslan et al. (2019)	Bronze (CuSn10) chips and spheroidal graphite cast iron chips	Powder metallurgy using HP	HP of bronze (CuSn10) chips and spheroidal graphite cast iron chips at 400°C for 20 min at 480, 640 and 820 MPa pressure	Porosity of 20%, hardness of 160 VHN and UTS of 325 MPa at a pressure of 820 MPa.
Łukasz Wzorek et al. 2017)	Cu + Al powder mixture	Direct hot extrusion	Cold compaction of powder mixture followed by hot extrusion at 300, 350 and 400°C.	Maximum UTS of 189 MPa and YS of 130 MPa for extrusion temperature of 300°C.
Sherafat et al. (2009)	Al7075 chips +Al composite	Direct hot extrusion	Cold pressed the Al7075 chips and Al powder and then extruded at 300–500°C	Maximum UTS of 250 MPa and hardness of 100 VHN for composite containing 40% Al powder when extrusion temp. 500°C. Increasing the chips wt. % in the products, the strength increases and ductility decreases.

(*Continued*)

TABLE 2.2 (Continued)
Summary of Recent Advances of MMC Fabricated by Various Solid-State Recycling Techniques

Authors	Composite Materials System	Synthesis Techniques	Experimental Details	Properties Achieved
Sugiyama et al. (2010)	A7075 aluminum alloy turnings + pure Cu composite	Direct hot extrusion	Cold compaction followed by direct hot extrusion	Not reported
Baffari et al. (2017)	Al–SiC MMC from AA2024 chips and SiC particles	Friction stir processing	Friction stir extrusion of AA2024 chips and SiC powder	Excessive amount of reinforcement ($p > 1\%$) causes the formation of big inter-granular conglomerate that affects deeply extrudes quality, causing crack opening and non-uniform mechanical properties.
Harikishor kumar et al. (2020)	Surface composite of A356 cast alloy reinforced with red mud and fly ash	Friction stir processing	Friction stir processing on surface of A356 cast alloy with red mud and fly ash	Surface composite exhibited higher strength and wear resistance but ductility reduced due to incorporation of reinforcement. Maximum UTS-290 MPa, YS-200 MPa, elongation-20%, hardness-85 VHN
Hosseinzadeh and Yapici (2018)	Al2024/SiC metal matrix composite	Friction stir processing	3 mm thick Al2024 sheets were used for FSP and reinforced with SiC particles of average size 4.5 μm	SiC particles enhanced the microhardness of FSPed samples up to 50% relative to the as-received condition. Yield strength of the as-received samples improved about 2.5 folds. Hardness of MMC rises to 130 VHN after 4 passes compared to 90 VHN received Al2024 sheet.
Xu et al. (2009)	Al–Al$_2$O$_3$ MMC	Back pressure ECAP	Back pressure ECAP of Al–Al$_2$O$_3$ powder mixture was	Fully dense Al–Al$_2$O$_3$ MMC with hardness of 2285 HV, UTS-740 MPa and 0.2% proof

(Continued)

Preparation of Metal Matrix Composites

TABLE 2.2 (Continued)
Summary of Recent Advances of MMC Fabricated by Various Solid-State Recycling Techniques

Authors	Composite Materials System	Synthesis Techniques	Experimental Details	Properties Achieved
			carried out at 400°C at a pressure of 200 MPa	stress-460 MPa in compression and 1% plastic strain to fracture.
Goussous et al. (2009)	Al–C MMC	Back pressure ECAP	Back pressure ECAP of Al–C powder mixture was carried out at 400°C at a pressure of 50 MPa	Maximum density of 2.65 g/cc, hardness of 6.5 HV, UTS of 350 MPa in compression were obtained for Al-5 wt.% C MMC after 8 passes.
Mohammad Khajouei-Nezhad et al. (2019)	MMC from 7075 Al alloy chips +pure Al powder mixtures	High pressure torsion	7075 Al alloy chips + pure Al powder mixtures was subjected to HPT of 1.2 GPa pressure with 1 and 4 turns	Density >99% of RD, maximum UTS-500 MPs and elongation to failure- 7.9% for annealed chips after 4 turns
Galiia Korznikova et al. (2020)	Cu–graphene laminated MMC	Two high pressure torsion processes (constrained and non-constrained HPT)	Cu–1 wt.% graphene laminated MMC by HPT at 6 GPa pressure	Improvement of hardness from 1,410 to 1,850 HV at the centre and 1,460–1,900 at the edge after first stage HPT. After second stage HPT, hardness increased from 1,900 to 2,950 HV at the edge; however, both ductility and strength decreased after second stage HPT.
Bazarnik et al. (2019)	Cu–SiC MMC	SPS + HPT	SPS of Cu–SiC (10, 20 wt.%) powder mixture at 950°C and a pressure of 150 MPa and then HPT of compact at 6 GPa pressure.	Microhardness increases both after addition of SiC and HPT processing of SPSed sample. UTS, YS and elongation decreases after addition of SiC particles during SPS processing. However, UTS and YS increases

(*Continued*)

TABLE 2.2 (Continued)
Summary of Recent Advances of MMC Fabricated by Various Solid-State Recycling Techniques

Authors	Composite Materials System	Synthesis Techniques	Experimental Details	Properties Achieved
				and elongation decreases after HPT.
Azlan Ahmad et al. (2017)	MMC from AA6061 Al chips–Al_2O_3 powder mixture	Hot press forging	Hot press forging of a mixture of AA6061 chips + Al_2O_3 powder at 430, 480 and 530°C for 60, 30 and 120 min.	Increase in density and hardness with increase in both temperature and holding time.
Lajis et al. (2017)	MMC from AA6061chips–Al_2O_3 powder mixture	Hot press forging	Hot press forging of AA6061 chips–Al_2O_3 (1, 2, 3 and 5 wt.%) mixture at 530°C at a pressure 35 tonnes for a period of 120 min	MMC containing 2 wt.% Al_2O_3 exhibits highest UTS (315 MPa) and 20% elongation before failure. Further addition of Al_2O_3 leads to reduction in strength.

ACKNOWLEDGMENT

The author is grateful to all previous authors for their work in the field of solid-state recycling techniques. The author is also grateful to the National Institute of Technology Rourkela, Odisha, India for providing infrastructural facilities. The author appreciates any suggestions/criticisms for further improvement of the article.

REFERENCES

Ab Kadir, Muhammad Irfan, Mustapa, Mohammad Sukri, Latif, Noradila Abdul, & Mahdi, Ahmed Sahib (2017). Microstructural analysis and mechanical properties of direct recycling aluminium chips AA6061/Al powder fabricated by uniaxial cold compaction technique. *Procedia Engineering, 1849*, 687–694.

Abbas, Aqeel, & Huang, Song-Jeng (2020). Investigation of severe plastic deformation effects on microstructure and mechanical properties of WS2/AZ91 magnesium metal matrix composites. *Materials Science & Engineering A, 780*, 139211. doi: 10.1016/j.msea.2020.139211.

Afshari, Elham, & Ghambari, Mohammad (2016). Characterization of pre-alloyed tin bronze powder prepared by recycling machining chips using jet milling. *Materials and Design, 103*, 201–208.

Ahmad, Azlan, Mohd Amri, Lajis, & Yusuf, Nur Kamilah (2017). On the role of processing parameters in producing recycled Aluminum AA6061 based metal matrix composite (MMC-Al$_R$) prepared using hot press forging (HPF) process. *Materials, 10*, 1098. doi:10.3390/ma10091098.

Aslan, A., Salur, E., Gunes, A., Sahin, O. S., Karadag, H. B., & Akdemir, A. (2019). The mechanical properties of composite materials recycled from waste metallic chips under different pressures. *International Journal of Environmental Science and Technology, 16*, 5259–5266. doi:10.1007/s13762-019-02317-3.

Awotundea, Mary A., Adegbenjoa, Adewale O., Obadelea, Babatunde A., Okoroa, Moses, Shongweb, Brendon M. & Olubambic, Peter A. (2019). Influence of sintering methods on the mechanical properties of aluminium nanocomposites reinforced with carbonaceous compounds: A review. *Journal of Materials Research Technology, 8*(2), 2432–2449. doi: 10.1016/j.jmrt.2019.01.026.

Baffari, Dario, Buffa, Gianluca, Campanella, Davide, & Fratini, Livan (2017). Al-SiC Metal matrix composite production through friction stir extrusion of aluminum chips. *Procedia Engineering, 207*, 419–424.

Bazarnik, P., Nosewicz, S., Romelczyk-Baishya, B., Chmielewski, M., Strojny Nędza, A., Maj, J., Huang, Y., Lewandowska, M., & Langdon, T. G. (2019). Effect of spark plasma sintering and high-pressure torsion on the microstructural and mechanical properties of a Cu–SiC composite. *Materials Science & Engineering A, 766*, 138350. doi:10.1016/j.msea.2019.138350.

Canakci, Aykut, & Varol, Temel (2014). Microstructure and properties of AA7075/Al–SiC composites fabricated using powder metallurgy and hot pressing. *Powder Technology, 268*, 72–79. doi:10.1016/j.powtec.2014.08.016.

Chaira, D., & Karak, S. K. (2015). Fabrication of nanostructured materials by mechanical milling. In M. Aliofkhazraei (Ed.), *Handbook of Mechanical Nanostructuring* (pp. 379–416). Wiley-VCH Verlag GmbH & Co. KGaA. doi:10.1002/9783527674947.ch16.

Chaira, Debasis (2021). *Powder Metallurgy Routes for Composite Materials Production, Reference Module in Materials Science and Materials Engineering*. Elsevier. doi:10.1016/B978-0-12-803581-8.11703-5.

Dash, K., Chaira, D., & Ray, B. C. (2013). Synthesis and characterization of aluminium–alumina micro- and nano-composites by spark plasma sintering. *Materials Research Bulletin, 48*, 2535–2542. doi:10.1016/j.materresbull.2013.03.014.

Dash, K., Ray, B. C., & Chaira, D. (2012). Synthesis and characterization of copper–alumina metal matrix composite by conventional and spark plasma sintering. *Journal of Alloys and Compounds, 516*, 78–84. doi:10.1016/j.jallcom.2011.11.136.

Dikici, Tuncay, & Sutcu, Mucahit (2017). Effects of disc milling parameters on the physical properties and microstructural characteristics of Ti6Al4V powders. *Journal of Alloys and Compounds, 723*, 395–400.

Dolatkhah, A., Golbabaei, P., Besharati Givi, M. K., & Molaiekiya, F. (2012). Investigating effects of process parameters on microstructural and mechanical properties of Al5052/SiC metal matrix composite fabricated via friction stir processing. *Materials and Design, 37*, 458–464. doi:10.1016/j.matdes.2011.09.035.

Emadi Shaibani, M., & Ghambari, M. (2011). Characterization and comparison of gray cast iron powder produced by target jet milling and high energy ball milling of machining scraps. *Powder Technology, 212*, 278–283.

Gabrielle, Gaustad, Elsa, Olivetti, & Kirchain, Randolph (2012). Improving aluminum recycling: A survey of sorting and impurity removal technologies. *Resources, Conservation and Recycling, 58*, 79–87.

German, R. M., (1994). *Powder metallurgy science*, second ed. Metal Powder Industrial Federation.

Ghanbari, D., Kasiri Asgharani, M., & Amini, K. (2015). Investigating the effect of passes number on microstructural and mechanical properties of the Al2024/SiC composite produced by friction stir processing. *Mechanika, 21*(6), 430–436. doi:10.5755/j01.mech.21.6.12227

Goussous, S., Xu, W., Wu, X., & Xia, K. (2009). Al–C nanocomposites consolidated by back pressure equal channel angular pressing. *Composites Science and Technology, 69*, 1997–2001. doi:10.1016/j.compscitech.2009.05.004.

Hosseinzadeh, Ali & Yapici, Guney Guven (2018). High temperature characteristics of Al2024/SiC metal matrix composite fabricated by friction stir processing. *Materials Science & Engineering A, 731*, 487–494. doi:10.1016/j.msea.2018.06.077.

Hsu, C. J., Chang, C. Y., Kao, P. W., Ho, N. J., & Chang, C. P. (2006). Al–Al3Ti nanocomposites produced in situ by friction stir processing. *Acta Materialia, 54*, 5241–5249.

Hsu, C. J., Kao, P. W., & Ho, N. J. (2005). Ultrafine-grained Al–Al2Cu composite produced in situ by friction stir processing. *Scripta Mater, 53*, 341–345.

Hu, Maoliang, Ji, Zesheng, Chen, Xiaoyu, & Zhang, Zhenkao (2008). Effect of chip size on mechanical property and microstructure of AZ91D magnesium alloy prepared by solid state recycling. *Materials Characterization, 59*, 385–389.

Jahedi, Mohammad, Paydar, Mohammad Hossein, & Knezevic, Marko (2015). Enhanced microstructural homogeneity in metal-matrix composites developed under high-pressure-double-torsion. *Materials Characterization, 104*, 92–100. doi:10.1016/j.matchar.2015.04.012.

Khajouei-Nezhad, Mohammad, Hossein Paydar, Mohammad, Mokarizadeh Haghighi Shirazi, Majid, & Gubicza, Jenő (2019). Microstructure and tensile behavior of Al7075/Al composites consolidated from machining chips using HPT: A way of solid-state recycling. *Metals and Materials International.* doi:10.1007/s12540-019-00428-7.

Korznikova, Galiia, Czeppe, Tomasz, Khalikova, Gulnara, Gunderov, Dmitry, Korznikova, Elena, Litynska-Dobrzynska, Lidia, & Szlezynger, Maciej (2020). Microstructure and mechanical properties of Cu-graphene composites produced by two high pressure torsion procedures. *Materials Characterization, 161*, 110122. doi:10.1016/j.matchar.2020.110122.

Kumar, Harikishor, Prasad, Rabindra, Kumar, Parshant, Tewari, S. P., & Singh, J. K. (2020). Mechanical and tribological characterization of industrial wastes reinforced aluminum alloy composites fabricated via friction stir processing. *Journal of Alloys and Compounds, 831*, 154832. doi:10.1016/j.jallcom.2020.154832

Lajis, M. A., Ahmad, A., Yusuf, N. K., Azami, A. H. & Wagiman, A. (2017). Mechanical properties of recycled aluminium chip reinforced with alumina (Al_2O_3) particle. *Materialwissenschaft und Werkstofftechnik*, 1–5. doi:10.1002/mawe.201600778.

Lazzaro, G., & Atzori C. (1991). Recycling of aluminium trimmings by conform processes. In *Light Metals* (pp. 1379–1384). Minerals, Metals & Materials SOC(TMS).

Li, F. X., Hao, P. D., Yi, J. H., Chen, Z., Prashanth, K. G., Maity, T., & Eckert, J. (2018). Microstructure and strength of nano-/ultrafine-grained carbon nanotube reinforced titanium composites processed by high-pressure torsion. *Materials Science & Engineering A, 722*, 122–128. doi:10.1016/j.msea.2018.03.007. http://www.iitg.ac.in/engfac/ganu/public_html/Metal%20forming%20processes_full.pdf

Ma, Z. Y. (2008). Friction stir processing technology: A review. *Metallurgical and Materials Transactions A, 642*, 39A. doi:10.1007/s11661-007-9459-0.

Mahboubi Soufiani, M. H., & Enayati, F. (2010). Karimzadeh, fabrication and characterization of nanostructured Ti6Al4V powder from machining scraps. *Advanced Powder Technology, 21*, 336–340.

Mindivana, H., Taskinb, N. & Kayalib, E. S. (2014). Recycling of pure magnesium chips by cold press and hot extrusion processes. *Acta Physica Polonica A, 125*, 429–431.

Samal, C. P., Parihar, J. S., & Chaira, D. (2013). The effect of milling and sintering techniques on mechanical properties of Cu–graphite metal matrix composite prepared by powder metallurgy route. *Journal of Alloys and Compounds, 569*, 95–101. doi:10.1016/j.jallcom.2013.03.122.

Shamsudin, S., Lajis, M. A. & Zhong, Z. W. (2016). Solid-state recycling of light metals: A review. *Advances in Mechanical Engineering, 8*(8), 1–23. doi:10.1177/1687814 016661921.

Sherafat, Z., Paydar, M. H., & Ebrahimi, R. (2009). Fabrication of Al7075/Al, two phase material, by recycling Al7075 alloy chips using powder metallurgy route. *Journal of Alloys and Compounds, 487*, 395–399.

Shial, S. R., Masanta, M., & Chaira, D. (2018). Recycling of waste Ti machining chips by planetary milling: Generation of Ti powder and development of in situ TiC reinforced Ti-TiC composite powder mixture. *Powder Technology, 329*, 232–240.

Soufiani, A.M., Enayati, M.H., & Karimzadeh, F. (2010). Fabrication and characterization of nanostructured Ti6Al4V powder from machining scraps. *Advanced Powder Technology, 21*, 336–340.

Sugiyama, S., Mera, T., & Yanagimoto, J. (2010). Recycling of minute metal scraps by semisolid processing: Manufacturing of design materials. *Transactions of Nonferrous Metals Society of China, 20*, 1567–1571.

Suryanarayana, C. (2001). Mechanical alloying and milling. *Progress in Materials Science, 46*, 1–184. doi:10.1016/S0079-6425(99)00010-9.

Verma, P., Saha, R., & Chaira, D. (2018). Waste steel scrap to nanostructured powder and superior compact through powder metallurgy: Powder generation, processing and characterization. *Powder Technology, 326*, 159–167.

Viswanathan, V., Laha, T., Balani, K., Agarwal, A., & Seal, S. (2006). Challenges and advances in nanocomposite processing techniques. *Materials Science and Engineering R, 54*, 121–285. doi:10.1016/j.mser.2006.11.002.

Wan, Bingbing, Chen, Weiping, Lu, Tiwen, Liu, Fangfang, Jiang, Zhenfei, & Mao, Mengdi (2017). Review of solid-state recycling of aluminum chips. *Resources, Conservation & Recycling, 125*, 37–47.

Wzorek, Łukasz, Wędrychowicz, Mateusz, Wiewióra, Marcel, Noga, Piotr, Skrzekut, Tomasz, Wzorek, Agata, & Łyp-Wrońska, Katarzyna (2017). Possibility of Al-Cu composite manufacturing from fine metal fractions by recycling process. *Metallurgy and Foundry Engineering, 3*(2), 117–124. doi:10.7494/mafe.2017.43.2.117.

Xu, Qiong, Ma, Aibin, Saleh, Bassiouny, Li, Yuhua, Yuan, Yuchun, Jiang, Jinghua, & Ni, Chaoying (2020). Enhancement of strength and ductility of SiCp/AZ91 composites by RD-ECAP processing. *Materials Science & Engineering A, 771*, 138579. doi:10.1016/j.msea.2019.138579.

Xu, W., Wu, X., Honma, T., Ringer, S. P., & Xia, K. (2009). Nanostructured Al–Al_2O_3 composite formed in situ during consolidation of ultrafine Al particles by back pressure equal channel angular pressing. *Acta Materialia, 57*, 4321–4330. doi:10.1016/j.actamat.2009.06.010.

3 A Comprehensive Study on the Recycled Aluminum Matrix Composites Reinforced with NiAl Intermetallics and TiB$_2$–TiC Ceramic Powders

H. Murat Enginsoy[1,2], Özgür Aslan[3], Fabio Gatamorta[4], Dhurata Katundi[1], and Emin Bayraktar[1]

[1]Mechanical and Manufacturing Engineering School, ISAE-Supmeca-PARIS, France
[2]Canakkale Onsekiz Mart University, Department of Industrial Engineering, Canakkale, Turkey
[3]Atilim University, Department of Mechanical Engineering, Ankara, Turkey
[4]University of Campinas-UNICAMP/FEM-Campinas/SP-Brazil

CONTENTS

3.1 Introduction 60
3.2 Experimental Conditions 61
    3.2.1 Composition Preparation 61
    3.2.2 Experimental Setup 61
    3.2.3 Finite Element Modeling 62
3.3 Results and Discussion 63
    3.3.1 Microstructure Analyses 63
    3.3.2 Quasi-Static Compression Test Results 65
    3.3.3 Low Velocity Impact Test Results 67
3.4 Conclusions 69

DOI: 10.1201/9781003148760-3

Acknowledgments.. 69
References.. 70

3.1 INTRODUCTION

Aluminum-based hybrid composites are favored in many different industrial applications where fuel consumption low/high mechanical performance is required in vehicles used in air-land and sea transportation, due to their many advantages (Enginsoy et al., 2019; Rajkovic et al., 2014; Sadoun & Fathy, 2019; Shehata et al., 2009; Wagih & Fathy, 2016). Although there are many reasons for this preference, high strength, lightness, easy-fast-low-cost controllable production features are among the most important reasons for preference (Enginsoy et al., 2020; Ferreira et al., 2016; Need et al., 2013).

In this study, the sinter and forging integrated method, which is a solid-state-based manufacturing technique that enables rapid production of high dimensional precision parts in industrial applications, is used. In addition, the recycled fresh scrap chips Al431 + 1050 material, which is used as a matrix component, has enabled the reduction of production costs and faster production (Abd-Elwaheda et al., 2020; Bayraktar et al., 2014; Ferreira et al., 2016; Zhang et al., 2019). As a result of our extensive research in literature, it has been determined that there are many different reinforcing elements that can be used in intermetallic hybrid composite materials (Choi et al., 2012; Ferreira et al., 2016; Gao et al., 2016; Hassan et al., 2009; Kurşun et al., 2016; Kwon et al., 2017; Liu et al., 2012; Shirvanimoghaddam et al., 2016a, 2016b). However, as a result of the evaluations we made on these studies, it was determined that the powders used as reinforcement elements were used on the same material in less type-limited proportions (Akbari et al., 2017; Chen et al., 2015; Lurie et al., 2016; Mazahery & Shabani, 2012; Park et al., 2017). In addition, it was observed that the powders produced in these studies did not have a detailed comparative analysis, both experimentally and numerically (Abdizadeh & Baharvandi, 2008; Abedini & Chen, 2014; Akbari, Baharvandi et al., 2013; Ghiţă & Popescu, 2012; Sharma et al., 2014; Shirvanimoghaddam et al., 2017). For this reason, hybrid composite materials have been produced with the aim of having high reinforcement powder types and different proportional use effects used within the scope of our own study. Thus, the materials we manufacture and chemically design have been able to gain multifunctional features that will allow them to work under different service conditions. The modeling of the new hybrid material, which was created by interacting with the matrix component of the composite material, of the reinforcement elements we used, was also performed using the finite element method.

Our next studies, which are in the main structure of our research, are discussed as follows: In Section 3.2, the technical infrastructures used in the preparation/production of hybrid composite material, experimental test studies, and the creation of the numerical model using the finite element method are given. In Section 3.3, microstructure formation studies obtained using SEM-EDS and mapping analysis, experimental results obtained from quasi-static compression-low velocity impact tests and finite element Johnson-Cook damage model (VUMAT in Abaqus®/Explicit software environment and FORTRAN® software). By preparing a subroutine application, the numerical results obtained from the nonlinear model study are discussed and explained comparatively. In the last section, a summary evaluation of the results obtained from these data is given.

3.2 EXPERIMENTAL CONDITIONS

3.2.1 COMPOSITION PREPARATION

Al431 + 1050 in the form of recycled fresh scrap chips were used as a matrix in hybrid composites developed within the scope of the common research project supported by the French aeronautic society. All of the chemical productions were taken from the VWR-French Branch/Chemical Company. The compositions of these matrix materials are given in Tables 3.1 and 3.2, respectively. The atomization process was applied on these materials before they were used as matrix components. Then it was subjected to high-energy milling process for approximately 150 minutes through a planetary ball mill with a ratio of 20:1. General composition of the specimen groups given in Table 3.3 were produced by the combined method of sinter and forging.

3.2.2 EXPERIMENTAL SETUP

"SEM" and "EDS" chemical analyses have been carried out to show the distribution of the reinforcements and interface relations with the matrix of these reinforcements. The density measurements of all composite samples produced

TABLE 3.1
Compositions of Al 431 (wt.%)

Al	Cu	Zn	Mg
B	1.5	5.5	2.5

TABLE 3.2
Compositions of AA 1050 (wt.%)

Al	Mn	Fe	Cu	Mg	Si	Zn	Ti
B	0.04	0.025	0.03	0.035	0.20	0.065	0.03

TABLE 3.3
Compositions of the Composite Specimens Prepared in Ten Groups (wt.%)

Specimen Group	Name	Al431 + AA1050	NiAl	TiB$_2$	TiC	Mo	Cu	CNTs	GNPs
F1	F1a	B	30	5	–	1	2	–	–
	F1b	B	30	10	–	1	2	–	–
	F1c	B	30	5	–	1	2	0.3	–
	F1d	B	30	15	–	1	2	–	–
	F1e	B	30	5	–	1	2	–	0.3
F2	F2a	B	30	–	5	1	2	–	–
	F2b	B	30	–	10	1	2	–	–
	F2c	B	30	–	5	1	2	0.3	–
	F2d	B	30	–	15	1	2	–	–
	F2e	B	30	–	5	1	2	–	0.3

were determined as 2.78 and 3.46 g/cm^3 by the Archimedes method. The destructive tests performed on the samples were repeated at least three times for each specimen group and the average values were considered. Quasi-static compression (using standard DIN 50106; using test device Zwick Z250) tests and low velocity impact tests have been carried out to compare their mechanical behavior of these composites.

3.2.3 Finite Element Modeling

The Johnson-Cook fracture model (via finite element method Abaqus®/Explicit and FORTRAN® software) was used for the phenomological damage model of the hybrid intermetallic composite material produced within the scope of the study (Bathe, 2008; Budiansky, 1965; Hashin, 1962; Ming, 2018; Wang, 2003). This model is given in Eq. (3.1).

$$D = \sum \frac{\Delta \varepsilon_p}{D_1(P^* + T^*)^{D_2}}, \quad (3.1)$$

where $\Delta\varepsilon_p$ is the equivalent plastic strain; D_1, D_2 is the material constant; P^* is the normalized hydrostatic pressure and T^* expresses the normalized maximum tensile hydrostatic pressure. The D value in Eq. (1) is equal to 0 and no plastic deformation occurred; $0 < D < 1$ damage started and progressed in the range, in the case of $D = 1$, it indicates that the building was completely damaged.

The Johnson-Cook constitutive model given in Eq. (2) was used in the simulation process in order to model the experimental studies as a whole with the finite element [15–18].

TABLE 3.4
Constitutive Parameters for Johnson-Cook Damage Model

Property	Parameter	Unit	Value
Density (Al)	ρ_1	g/cm^3	2.78
Density (NiAl)	ρ_2	g/cm^3	5.90
Density (TiC)	ρ_3	g/cm^3	4.92
Density (TiB$_2$)	ρ_4	g/cm^3	4.52
Shear modulus	G	MPa	3.92×10^4
Yield stress	A_1	MPa	3.15×10^2
Hardening constant	B_1	MPa	5.32×10^2
Hardening exponent	N_1	–	0.44
Strain rate constant	C_1	–	0.020
Thermal softening exponent	M_1	–	1
Melting temperature	T_M	K	1,120

$$\sigma_Y = \left(A_1 + B_1 \varepsilon_p^{N_1}\right)(1 + C_1 ln \dot{\varepsilon}_p^*)\left(1 - T_H^{M_1}\right), \qquad (3.2)$$

where σ_Y is the strain and temperature dependent yield strength, A_1 is material strength, ε_p is equivalent plastic strain, B_1 and N_1 are the strain hardening constants, C_1 is the strain rate relation, $\dot{\varepsilon}_p^*$ is the normalized equivalent plastic strain rate, T_H is the normalized temperature and M_1 is the temperature softening constant. The parameters used in the modeling process are given in Table 3.4. In addition, the other parameters in this table, the effects of the reinforcement powders included in the hybrid composite

It has been determined that the effects of the experimental conditions on the finite element model and the fact that the model results are compatible with the experimental data it directly affects. For this reason, Abaqus®/Explicit (together with the VUMAT subroutine application created in accordance with the experimental conditions in FORTRAN® software) was used (Ming, 2018).

3.3 RESULTS AND DISCUSSION

3.3.1 Microstructure Analyses

"SEM" and "EDS" chemical analyses have been carried out on the composite groups given in Table 3.3. General microstructures of the series of the specimen groups were given in Figure 3.1 known hereafter as F1 and F2, respectively.

As a general meaning, "EDS" chemical analyses for two groups were also presented in Figure 3.2. As for the distribution of the reinforcements in the matrix, more or less a homogeneous distribution even if some of the

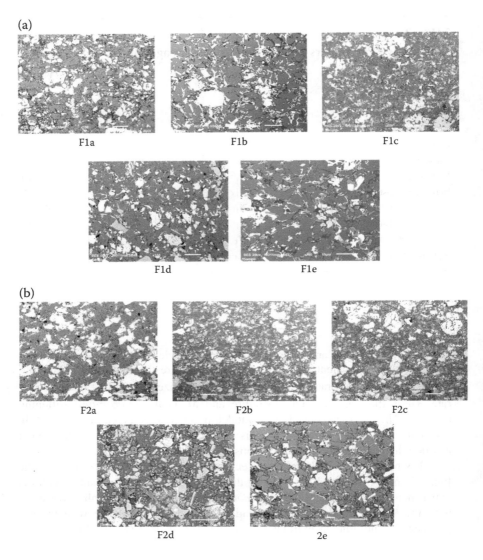

FIGURE 3.1 General microstructures of the series of F1 and F2 containing primary reinforcement, respectively.

agglomeration of the reinforcements in the matrix is due to short milling effect. As a typical observation at the interface between the intermetallics reinforcements and ceramic powder with matrix, one may find a strong cohesion at the interface due to a chemical bonding diffusion that occurred mutually between the matrix and reinforcements.

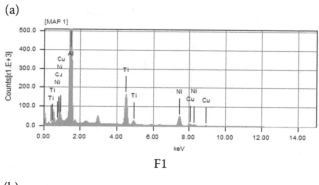

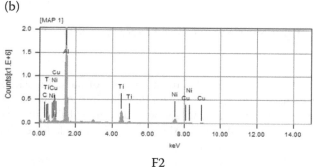

FIGURE 3.2 EDS analyses for the series of F1 and F2 in general meaning.

3.3.2 Quasi-Static Compression Test Results

The results obtained from the quasi-static compression tests performed on hybrid intermetallic composite samples were given in Figure 3.3 in detail with the finite element model results obtained in the F1 sample group. In the sample compositions given in Table 3.3, the comparison was made by taking four basic reinforcement powders as variables. In Figure 3.3a, the effects on the quasi-static compression behavior of TiB_2 reinforcement powders in F1a-b-d samples in the F1 sample group were given. It has been determined that increasing amounts of TiB_2 increase the compressive strength (in low deformation amount) of the composite sample. Likewise, a similar situation is observed in Figure 3.3b. It was determined that the sample with CNTs in its composition (F1c) has a higher compressive strength than the sample without any CNTs (F1a). It was observed that the composite sample reached a higher compression strength value under low deformation by adding GNPs at the same rate (F1e) instead of CNTs in the composition.

The use of high levels of TiB_2 and, additionally, the use of GNPs show that a higher amount of eutectic reactions occur within the microstructure and that a mutual diffusion mechanism works very well and tighter chemical bonds are

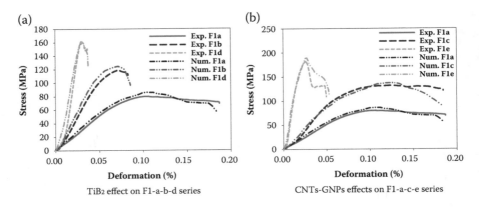

FIGURE 3.3 TiB$_2$-CNTs-GNPs effects on F1 specimen group.

formed. In this way, it has revealed the possibility of the composite structure to carry higher compressive stress with less deformation. As can be seen from all the graphs given in Figure 3.3, it has been determined that the results of the experimental and numerical model (Abaqus®/Explicit) made for each test sample show a good agreement with each other.

The results obtained from the quasi-static compression tests performed on hybrid intermetallic composite samples were given in Figure 3.4 in detail, together with the finite element model results obtained in the F2 sample group. In the sample compositions given in Table 3.3, the comparison was made by taking four basic reinforcement powders as variables. In Figure 3.4a, the effects on the quasi-static compression behavior of TiC reinforcement powders in the F2a-b-d samples in the F2 sample group were given. It has been determined that

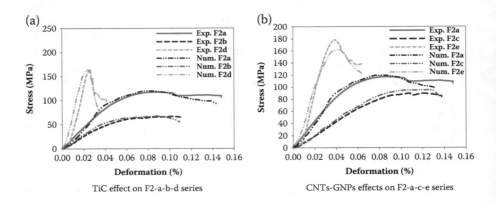

FIGURE 3.4 TiC-CNTs-GNPs effects on F2 specimen group.

increasing amounts of TiC do not always increase the compressive strength (with low deformation amount) of the composite sample. Quasi-static compressive strength occurred as F2d (15% TiC), F2a (5% TiC) and F2b (10% TiC), respectively, from high to low. Likewise, a similar situation is observed in Figure 3.4b. It was determined that the sample with CNTs (F2c) in its composition has almost the same compressive strength compared to the sample without any CNTs (F2a). It was observed that the composite sample reached a higher compressive strength values under low deformation by adding GNPs at the same rate (F2e) instead of CNTs in the composition. With the use of high levels of TiC and additionally the use of GNPs, the composite structure is less deformed and has the opportunity to carry a higher compressive stress. As can be seen from all the graphics given in Figure 3.4, it has been determined that the results of the experimental and numerical model (Abaqus®/Explicit) made for each test sample show a good agreement with each other.

3.3.3 Low Velocity Impact Test Results

The results obtained from the low velocity impact tests performed on hybrid intermetallic composite samples were given in Figure 3.5 in detail with the finite element model results obtained in the F1 sample group. In the sample compositions given in Table 3.3, four basic reinforcement powders were taken as variables and compared. In Figure 3.5a, the effects of TiB_2 reinforcement powders in the F1a-b-d samples in F1 sample group on low velocity impact behavior were given. There is no linear correlation between the increasing rates of TiB_2 and the impact strength of the composite specimen. Low velocity impact strength occurred as F1b (10% TiB_2), F1d (15% TiB_2) and F1a (5% TiB_2), respectively. When CNTs-GNPs effects were evaluated from the same point of view, it was determined that the sample (F1c) with CNTs in its composition in

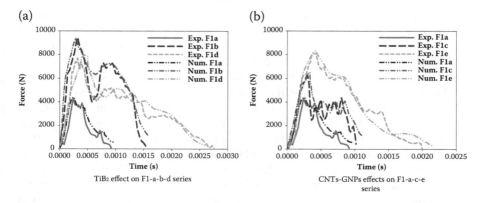

FIGURE 3.5 TiB_2-CNTs-GNPs effects on F1 specimen group.

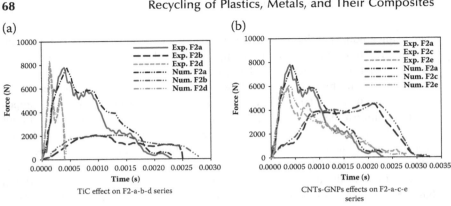

FIGURE 3.6 TiC-CNTs-GNPs effects on F2 specimen group.

Figure 3.5b had a higher impact resistance than the sample without any CNTs (F1a). It was observed that by adding only GNPs (F1e) into the composition instead of CNTs at the same rate, the composite sample reached higher impact strength values. The use of 10 wt.% TiB_2 and additionally the use of GNPs show that higher amounts of eutectic reactions occur in the microstructure and tighter chemical bonds are formed. In this way, it was revealed that there is a possibility that the composite structure has a higher impact capacity. As can be seen from all the graphics given in Figure 3.5, it has been determined that the results of the experimental and numerical model (Abaqus®/Explicit) made for each test sample show a good agreement with each other.

The results obtained from the low velocity impact tests performed on hybrid intermetallic composite samples were given in Figure 3.6 in detail, together with the finite element model results obtained in the F2 sample group. In the sample compositions given in Table 3.3, the comparison was made by taking four basic reinforcing powders as variables. In Figure 3.6a, the effects of TiC reinforcement powders in the F2a-b-d samples in the F2 sample group on low velocity impact behavior were given. There is no linear correlation between the effects of increasing TiC on the impact strength of the composite sample. Low velocity impact strength occurred as F2a (5% TiC), F2d (15% TiC) and F2b (10% TiC), respectively. Using the minimum amount of TiC in the present composition can be considered the most economical approach to achieve the highest impact resistance. When CNTs-GNPs effects were evaluated from the same point of view, it was determined that the sample without any CNTs (F2a) in Figure 3.6b had a higher impact resistance than the samples containing CNTs (F2c) and GNPs (F2e). By adding CNTs and GNPs to the composition, it was observed that the composite sample only had lower impact strength values over a longer period of time. Based on the data obtained from this experimental group, it was determined that only 5 wt.% of TiC would be sufficient for high impact resistance. In this respect, it has been determined that the use of only a minimum amount of TiC in order to create tighter chemical bonds in eutectic reactions occurring in the microstructure can provide higher impact resistance at less cost. As can be seen from all the graphics given in Figure 3.6, it has been determined that the results of the experimental and numerical model (Abaqus®/Explicit) made for each test sample show a good agreement with each other.

3.4 CONCLUSIONS

In the frame of the common research project, a comprehensive study has been carried out on the aluminum-based hybrid composites from fresh scrap chips of Al 431 + 1050 reinforced with basically NiAl intermetallics and TiB_2–TiC ceramic powders. For the production of these composites, a combined manufacturing method called sinter and forging was used as an efficient and economic method. The effects of the reinforcement elements have been evaluated on the mechanical properties of these composites.

As for the distribution of the reinforcements in the matrix, more or less a homogeneous distribution even if some of the agglomeration of the reinforcements in the matrix due to a short milling effect. As a typical observation at the interface between the intermetallics reinforcements and ceramic powder with matrix, one may find a strong cohesion at the interface due to a chemical bonding diffusion that occurred mutually between the matrix and reinforcements.

Additionally, a detailed study has also been carried out by means of FEM, finite element method. For this case, the Johnson-Cook damage model asw used. One may observe that the static compression test results have shown a very suitable approach in agreement with each other. As for the low velocity impact test, thanks to the subroutine application developed within the scope of the study, more compatible results were obtained. In this study, the experimental and numerical analysis techniques of recycled metal matrix composites can be applied to another type of material, polymer-based composites in a similar way. Thus, it can be suggested that a more comprehensive examination of the developed materials can be carried out (Aisyah et al., 2021; Alsubari et al., 2021; Asyraf et al., 2020; Ilyas et al., 2021, 2021b; Nurazzi et al., 2021; Omran et al., 2021; Sabaruddin et al., 2021).

The results obtained here are not limited because they have been carried out in the laboratory scale and they can give more practical results for the production of these composites in industrial scale and they should be a useful tool for the manufacturing facilities point of view.

ACKNOWLEDGMENTS

This work has been carried out at Supmeca/Paris-FRANCE Composite Research Laboratory. The authors acknowledge and appreciate Dr. G. ZAMBELIS from Airbus-Helicopter-Paris/FRANCE for supplying materials and for technical help. The authors acknowledge the technical staff of Supmeca/Paris, Mr. Christophe BEN BRAHIM and Mr. Abdelghani LARBI for mechanical tests and data acquisition system analyses and installation of electronic measurement devices.

REFERENCES

Abd-Elwaheda, M. S. & Meselh, A. F. (2020). Experimental investigation on the mechanical, structural and thermal properties of Cu–ZrO$_2$ nanocomposites hybridized by graphene nanoplatelets. *Ceramics International, 46*(7), 9198–9206. doi:10.1016/j.ceramint.2019.12.172.

Abdizadeh, H., & Baharvandi, H. (2008). Comparing the effect of processing temperature on microstructure and mechanical behavior of (ZrSiO$_4$ or TiB$_2$)/aluminum composites. *Materials Science and Engineering A, 498*, 53–58. doi:10.1016/j.msea.2008.07.009.

Abedini, A., & Chen, Z. T. (2014). A micromechanical model of particle-reinforced metal matrix composites considering particle size and damage. *Computational Materials Science, 85*, 200–205. doi:10.1016/j.commatsci.2014.01.012.

Aisyah, H. A., Paridah, M. T., Sapuan, S. M., Ilyas, R. A., Khalina, A., Nurazzi, N. M., Lee, S. H., & Lee, C. H. (2021). A comprehensive review on advanced sustainable woven natural fibre polymer composites. *Polymers, 13*(3), 471. doi:10.3390/polym13030471.

Akbari, M. K. et al., (2017). Al-TiB$_2$ micro/nanocomposites: Particle capture investigations, strengthening mechanisms and mathematical modelling of mechanical properties. *Materials Science and Engineering A, 682*, 98–106. doi:10.1016/j.msea.2016.11.034.

Akbari, M. K., Baharvandi, H., & Mirzaee, O. (2013). Fabrication of nano-sized Al$_2$O$_3$ reinforced casting aluminum composite focusing on preparation process of reinforcement powders and evaluation of its properties. *Composites Part B: Engineering, 55*, 426–432. doi:10.1016/j.msea.2008.07.009.

Alsubari, S., Zuhri, M. Y. M., Sapuan, S. M., Ishak, M. R., Ilyas, R. A., & Asyraf, M. R. M. (2021). Potential of natural fiber reinforced polymer composites in sandwich structures: A review on its mechanical properties. *Polymers, 13*(3), 423. doi:10.3390/polym13030423.

Asyraf, M. R. M., Rafidah, M., & Ishak, M. R., et al., (2020). Integration of TRIZ, morphological chart and ANP method for development of FRP composite portable fire extinguisher. *Polymer Composites, 41*, 2917.

Bathe, K. J. (2008). *Finite Element Method*. Wiley Online Library

Bayraktar, E., Ayari, F., Tan, M. J., Tosun Bayraktar, A., & Katundi, D. (2014). Manufacturing of aluminum matrix composites reinforced with iron-oxide nanoparticles: Microstructural and mechanical properties. *Metallurgical and Materials Transactions, 45B*(26), 352–362. doi:10.1007/s11663-013-9970-1.

Budiansky, B. (1965). On the elastic moduli of some heterogeneous materials. *Journal of the Mechanics and Physics of Solids, 13*(4), 223–227. doi:10.1016/0022-5096(65)90011-6.

Chen, L.-Y. et al., (2015). Processing and properties of magnesium containing a dense uniform dispersion of nanoparticles. *Nature, 528*(7583), 539–543. doi:10.1038/nature16445.

Choi, H., Konishi, H., & Li, X. (2012). Al2O3 nanoparticles induced simultaneous refinement and modification of primary and eutectic Si particles in hypereutectic Al-20Si alloy. *Materials Science and Engineering A, 541*, 159–165. doi:10.1016/j.msea.2012.01.131.

Clayton, J. D., & Tonge, A. L. (2015). A nonlinear anisotropic elastic–inelastic constitutive model for polycrystalline ceramics and minerals with application to boron carbide. *International Journal of Solids and Structures, 64*, 191–207. doi:10.1016/j.ijsolstr.2015.03.024.

Diyana, Z. N., Jumaidin, R., Selamat, Mohd Zulkefli, Ghazali, Ihwan, Julmohammad, Norliza, Huda, Nurul, & Ilyas, R. A. (2021). Physical properties of thermoplastic starch derived from natural resources and its blends: A review. *Polymers, 13* (9), 1–20. doi: 10.3390/polym13091396.

Enginsoy, H. M., Bayraktar, E., Katundi, D., Gatamorta, F., & Miskioglu, I. (2020). Comprehensive analysis and manufacture of recycled aluminum based hybrid metal matrix composites through the combined method; sintering and sintering + forging. *Composites Part B: Engineering, 194*, 108040. doi: 10.1016/j.compositesb.2020.108040.

Enginsoy, H. M., Gatamorta, F., Bayraktar, E., Robert, M. H., & Miskioglu, I. (2019). Numerical study of Al-Nb$_2$Al composites via associated procedure of powder metallurgy and thixoforming. *Composites Part B: Engineering, 162*, 397–410. doi: 10.1016/j.compositesb.2018.12.138.

Ferreira, L. F. P., Bayraktar, E., Miskioglu, I., & Katundi, D. (2016). Design of hybrid composites from scrap aluminium bronze chips. *SEM, Mechanics of Composite and Multi-functional Materials,. 15*, 131–138. doi: 10.1007/978-3-319-41766-0_15.

Ferreira, L. F. P., Bayraktar, E., Robert, M. H., & Miskioglu, I. (2016). Particles re-inforced scrap aluminum based composites by combined processing; sintering + thixoforming. *SEM, Mechanics of Composite and Multi-functional Materials, 7*, 145–152. doi: 10.1007/978-3-319-41766-0_17.

Ferreira, L. F. P., Bayraktar, E., Robert, M. H., & Miskioglu, I. (2016). Particles re-inforced scrap aluminium-based composites by combined processing sintering + thixoforming. *SEM, Mechanics of Composite and Multi-functional Materials, 17*, 145–152. doi: 10.1007/978-3-319-41766-0_17.

Gao, Y., Li, D., Zhang, W., Guo, Z., Yi, C., & Deng, Y. (2019). Constitutive modelling of the TiB2–B4C composite by experiments, simulation and neutral network. *International Journal of Impact Engineering, 132*, 103310. doi: 10.1016/j.ijimpeng.2019.05.024.

Gao, Y. B., Tang, T. G., Yi, C. H. et al., (2016). Study of static and dynamic behavior of TiB2–B4C composite. *Material Design, 92*, 814–822. doi: 10.1016/j.matdes.2015.12.123.

Ghiţă, C., Popescu, I. N. (2012). Experimental research and compaction behaviour modelling of aluminium based composites reinforced with silicon carbide particles. *Computational Materials Science, 64*, 136–140. doi: 10.1016/j.commatsci.2012.05.031.

Hashin, Z. (1962). The elastic moduli of heterogeneous materials. *Journal of Applied Mechanics, 29*(1), 143–150. doi: 10.1115/1.3636446.

Hassan, A. M. et al., (2009). Prediction of density, porosity and hardness in aluminum–copper-based composite materials using artificial neural network. *Journal of Materials Processing Technology, 209*(2), 894–899. doi: 10.1016/j.jmatprotec.2008.02.066.

Ilyas, R. A., Sapuan, S. M., Asyraf, M. R. M., Dayana, D. A. Z. N., Amelia, J. J. N., Rani, M. S. A., Faiz Norrrahim, Mohd Nor, et al., (2021). Polymer composites filled with metal derivatives: A review of flame retardants. *Polymers, 13* (11), 1701. doi: 10.3390/polym13111701.

Ilyas, R. A., Sapuan, S. M., Harussani, M. M., Hakimi, M. Y. A. Y., Haziq, M. Z. M., Atikah, M. S. N., Asyraf, M. R. M., et al., (2021). Polylactic acid (PLA) biocomposite: Processing, additive manufacturing and advanced applications. *Polymers, 13*(8), 1326. doi: 10.3390/polym13081326.

Johnson, G. R., & Holmquist, T. J. (1993). An improved computational constitutive model for brittle materials. In S. C. Schmidt, J. W. Shaner, G. A. Samara, & M. Ross, (Eds.), *High Pressure Science and Technology*. AIP Press. doi: 10.1063/1.46199.

Kurşun, A., Bayraktar, E., & Enginsoy, H. M. (2016). Experimental and numerical study of alumina reinforced aluminium matrix composites: Processing, microstructural aspects and properties. *Composites Part B: Engineering, 90*, 302–314. doi:10.1016/j.compositesb.2016.01.006.

Kwon, H. et al., (2017). Graphene oxide-reinforced aluminum alloy matrix composite materials fabricated by powder metallurgy. *Journal of Alloys and Compounds, 698*, 807–813. doi:10.1016/j.jallcom.2016.12.179.

Lankford, J. (1977). Compressive strength and microplasticity in polycrystalline alumina. *Journal of Materials Science, 12*(4), 791–796. doi:10.1007/BF00548172.

Liu, Z. Y. et al., (2012). Singly dispersed carbon nanotube/aluminum composites fabricated by powder metallurgy combined with friction stir processing. *Carbon, 50*(5), 1843–1852. doi:10.1016/j.carbon.2011.12.034.

Lou, J. F., Wang, Z., Hong, T. et al., (2009). Numerical study on penetration of semi-infinite aluminum-alloy targets by tungsten-alloy rod. *Chinese Journal of High Pressure Physics, 23*(1), 65–70. In Chinese.

Lurie, S. et al., (2016). Multiscale modelling of aluminium-based metal–matrix composites with oxide nanoinclusions. *Computational Materials Science, 116*, 62–73. doi:10.1016/j.commatsci.2015.12.034.

Meyers M. A. (1994). *Dynamic Behavior of Materials*. John Wiley & Sons.

Mazahery, A., & Shabani, M. O. 2012 Mechanical properties of A356 matrix composites reinforced with nano-SiC particles. *Strength of Materials, 44*(6), 686–692. doi:10.1007/s11223-012-9423-0.

Ming, L. (2018). *A Numerical Platform for the Identification of Dynamic Non-linear Constitutive Laws Using Multiple Impact Tests: Application to Metal Forming and Machining*. PhD Thesis, Institut National Polytechnique de Toulouse (INP Toulouse).

Need, R. F., Alexander, D. J., Field, R. D., Livescu, V., Papin, P., Swenson, C. A., & Mutnick, D. B. (2013). The effects of equal channel angular extrusion on the mechanical and electrical properties of alumina dispersion-strengthened copper alloys. *Journal of Material Sciences and Engineering, 565*, 450–458. doi:10.1016/j.msea.2012.12.007.

Nurazzi, N. M., Asyraf, M. R. M., Khalina, A., Abdullah, N., Aisyah, H. A., Rafiqah, S. A., Sabaruddin, F. A., Kamarudin, S. H., Norrrahim, M. N. F., Ilyas, R. A., & Sapuan, S. M. (2021). A Review on natural fiber reinforced polymer composite for bullet proof and ballistic applications. *Polymers, 13*(4), 646. doi:10.3390/polym13040646.

Omran, Abdoulhdi A.B., Mohammed, Abdulrahman A.B.A., Sapuan, S. M., Ilyas, R. A., Asyraf, M. R. M., Rahimian Koloor, Seyed S., & Petrů, Michal. (2021). Micro- and nanocellulose in polymer composite materials: A review. Polymers, *13*(2), 231. doi:10.3390/polym13020231.

Paliwal, B., Ramesh, K. T., & McCauley, J. W. (2006). Direct observation of the dynamic compressive failure of a transparent polycrystalline ceramic (AlON). *Journal of the American Ceramic Society, 89*(7), 2128–2133. doi:10.1111/j.1551-2916.2006.00965.x.

Paliwal, B., & Ramesh, K. T. (2008). An interacting micro-crack damage model for failure of brittle materials under compression. *Journal of the Mechanics and Physics of Solids, 56*(3), 896–923. doi:10.1016/j.jmps.2007.06.012.

Park, H. K., Jung, J., & Kim, H. S. (2017). Three-dimensional microstructure modeling of particulate composites using statistical synthetic structure and its thermo-mechanical finite element analysis. *Computational Materials Science, 126*, 265–271. doi:10.1016/j.commatsci.2016.09.033.

Rajkovic, V., Bozic, D., Stasic, J., Wang, H., & Jovanovic, M. T. (2014). Processing, characterization and properties of copper-based composites strengthened by low amount of alumina particles. *Powder Technology, 268*, 392–400. doi:10.1016/j.powtec.2014.08.051.

Sabaruddin, Fatimah A., Paridah, M. T., Sapuan, S. M., Ilyas, R. A., Lee, Seng H., Abdan, Khalina, Mazlan, Norkhairunnisa, Roseley, Adlin S.M., & Abdul Khalil, H. P. S. (2021). The effects of unbleached and bleached nanocellulose on the thermal and flammability of polypropylene-reinforced Kenaf core hybrid polymer bionanocomposites. *Polymers, 13*(1), 116. doi:10.3390/polym13010116.

Sadoun, A. M., & Fathy, A. (2019). Experimental study on tribological properties of Cu–Al$_2$O$_3$ nanocomposite hybridized by graphene nanoplatelets. *Ceramics International, 45*, 24784–24792. doi:10.1016/j.ceramint.2019.08.220.

Sharma, N. K. et al., (2014). Effective Young's Modulus of Ni–Al$_2$O$_3$ composites with particulate and interpenetrating phase structures: A multiscale analysis using object oriented finite element method. *Computational Materials Science, 82*, 320–324. doi:10.1016/j.commatsci.2013.10.005.

Shehata, F., Fathy, A., Abdelhameed, M., & Moustafa, S. F. (2009). Preparation and properties of Al$_2$O$_3$ nanoparticle reinforced copper matrix composites by in situ processing. *Material Design, 30*(7), 2756–2762. doi:10.1016/j.matdes.2008.10.005.

Shirvanimoghaddam, K. et al., (2016b). Boron carbide reinforced aluminium matrix composite: Physical, mechanical characterization and mathematical modelling. *Materials Science and Engineering A, 658*, 135–149. doi:10.1016/j.msea.2016.01.114.

Shirvanimoghaddam, K. et al., (2017). Cheetah skin structure: A new approach for carbonnano-patterning of carbon nanotubes. *Composites Part A: Applied Science and Manufacturing, 95*, 304–314. doi:10.1016/j.compositesa.2017.01.023.

Shirvanimoghaddam, K. et al., (2016a). Effect of B4C, TiB$_2$ and ZrSiO$_4$ ceramic particles on mechanical properties of aluminium matrix composites: Experimental investigation and predictive modelling. *Ceramics International, 42*(5), 6206–6220. doi:10.1016/j.ceramint.2015.12.181.

Shojaei, A., Li, G., Fish, J.. et al., (2014). Multi-scale constitutive modeling of ceramic matrix composite by continuum damage mechanics. *International Journal of Solids and Structures, 51*(23–24), 4068–4081. doi:10.1016/j.ijsolstr.2014.07.026.

Wagih, A., & Fathy, A. (2016). Experimental investigation and FE simulation of nanoindentation on Al–Al2O3 nanocomposites. *Advanced Powder Technology, 27*(2), 403–410. doi:10.1016/j.apt.2016.01.021.

Wang, X.-c. (2003). *Finite Element Method*. Tsinghua University Press.

Westerling, L., Lundberg, P., & Lundberg, B. (2001). Tungsten long-rod penetration into confined cylinders of boron carbide at and above ordnance velocities. *International Journal of Impact Engineering., 25*(7), 703–714. doi:10.1016/S0734-743X(00)00072-5.

Zhang, P., Zhang, L., Wei, D. B., Wu, P. F., Cao, J. W., Shijia, C. R., Qu, X. H., & Fu, K. X. (2019). Effect of graphite type on the contact plateaus and friction properties of copper- based friction material for high-speed railway train. *Wear, 432–433*, 202927. doi:10.1016/j.wear.2019.202927.

4 Recycling for a Sustainable World with Metal Matrix Composites

Uğur Aybarç[1] and M. Özgür Seydibeyoğlu[2,3]
[1]CMS Wheel Company
[2]Izmir Katip Çelebi University, Izmir, Turkey
[3]University of Maine, Advanced Structures and Composite Center, Maine, USA

CONTENTS

4.1 Introduction 76
4.2 What Is the Composite Material? 77
    4.2.1 The Classification of Composites for Matrix Material 78
        4.2.1.1 Ceramic Matrix Composites (CMCs) 78
        4.2.1.2 Metal Matrix Composites (MMCs) 78
        4.2.1.3 Polymer Matrix Composites (PMCs) 78
4.3 The Classification of Composites for Reinforcement Material 79
    4.3.1 Particle-Reinforced 79
    4.3.2 Fiber Reinforced 79
    4.3.3 Laminate Reinforced 79
4.4 Particle-Reinforced Composite Materials Overview 80
4.5 Production Methods of Composites 80
    4.5.1 LiquidState 80
        4.5.1.1 Stir Casting 80
        4.5.1.2 Compocasting 80
        4.5.1.3 Squeeze Casting 80
        4.5.1.4 Spray Deposition 82
        4.5.1.5 Infiltration Process 82
        4.5.1.6 In-situ Processing 82
        4.5.1.7 Ultrasonic Assisted Casting 83
    4.5.2 Solid State 83
        4.5.2.1 Powder Blending and Consolidation 83
        4.5.2.2 Diffusion Bonding 83

DOI: 10.1201/9781003148760-4

        4.5.2.3 Physical Vapor Deposition...83
4.6 Recycling of MMCs...84
    4.6.1 Recycling Techniques of MMCs...84
        4.6.1.1 Mechanical...84
        4.6.1.2 Chemical..85
        4.6.1.3 Thermal..85
    4.6.2 The Economic and Environmental Aspects of Recycle 86
References..88

4.1 INTRODUCTION

During the last century, material science has started to take on a vital importance for scientists. Because for many industries such as electronic, automotive and aerospace, and also daily use of properties, there is a need of nonconventional material to improve their mechanical endurance and care of environmental aspects. Conventional material cannot satisfy the expectations in some cases. For this reason, engineers working in the automotive sector have started to investigate nonconventional materials such as composites because composite materials have higher mechanical strength, longer useful life and also lighter weight than the well-known materials which have been used for years.

    Composite materials can be classified into two main groups. One of them is matrix materials and the other one is reinforcement materials. Matrix materials consist of metal, plastic and ceramic while reinforcement materials consist of fiber, particulate or whisker. Scientists have investigated different combinations of matrix and reinforcement materials. One of the most popular is metal matrix composite (MMC) with particle reinforcements. Additionally, carbides and oxides are generally used for reinforcement particles such as silicon carbide (SiC) and alumina. Meanwhile, different production methods have been tried by scientists, also. The stirring casting technique is widely used to produce MMCs because of its simplicity, flexibility and cost efficiency. But this method causes porosity inside the microstructure and negatively affects the final product in terms of mechanical properties. Some additional processes such as hot isostatic press, ultrasonic vibration or surface coating of particles have been investigated to improve the wetness between the particle and matrix phase. So, to choose the production technique is also of vital importance in the useful life of the final product.

    On the other hand, recycle, reuse and reproduce should be taken into consideration also, because the use of composite material instead of conventional material is not enough for environmental aspect. The environmental aspect of composites should be taken into consideration in many respects. Scraps obtained during the production system and also after the final product's useful life is finished should be recycled. But recycling of MMCs has not been taken into consideration in detail. Some researchers declared that the determination of separation technology and reproduction method of composite materials were important. In this case, two main options can be used. One of them is chemical recycling and the other one is re-melting-casting.

In this book chapter, all the fundamentals of metal processing with composite structures will be covered with a recycling outlook and detailed literature review will be presented. This review will shed light on circular economy and the economic use of MMC materials. A future outlook for the MMCs on circularity will be demonstrated.

4.2 WHAT IS THE COMPOSITE MATERIAL?

The "composite" expressed as a material system consists of more than two different materials. One of them is the continuous phase, called the matrix, and the others (non-continuous) are the additives dispersed in the matrix called reinforcement (Rajeswari et al., 2014; Surappa, 2003). The history of composite is back to ancient times and the first known examples of composites were used to build a house with a mixture of straw and mud by Egyptians and Mesopotamian settlers in 1500s BC (Nagavally, 2016). Over the years, people have produced and used composites in many areas, from the construction of buildings to the hunting and the construction of arrows and bows, to sustain their lives (Nagavally, 2016, Ngo, 2019; Rajeswari et al., 2014; Surappa, 2003). After the Second World War, the commercial use of composites started in military applications, especially aircraft industry (Botelho et al., 2006).

Nowadays, composites have become available in mass production conditions as commercial products used in many points of our daily lives. The most important reason for this is that composites have higher mechanical properties than each component (Garg et al., 2012; Kandpal et al., 2014; Ramnath et al., 2014; Reddy et al., 2007) and can be used in many production sectors. Additionally, the final products produced by using composite materials, in particular, have lower weights than those produced by using conventional materials. This situation not only provides significant advantages to the designers during the design of more creative unique products, but also decreases the negative effect of greenhouse gases, especially in the transportation sector. In the last years, the increasing use of passenger cars not only increases the environmental pollution but also decreases the amount of natural resources. Because of this, developed countries have started to make some regulations to keep control of the increasing pollution effects. This situation has prompted engineers working in the automotive and aerospace industry to search for new engineering materials that can replace conventional materials (Sozhamannan et al., 2012). According to the theoretical calculations, if the mass of the vehicle is decreased 10%, the fuel consumption would decrease about 5.5% (Milos et al., 2011). These expectations on material have led to detailed research and development efforts on composites (Dhanasekaran et al., 2016; Milos et al., 2011; Sozhamannan et al., 2012).

As given in Figure 4.1, composites can be classified in two main groups, one of them is matrix material and the other one is reinforcement material.

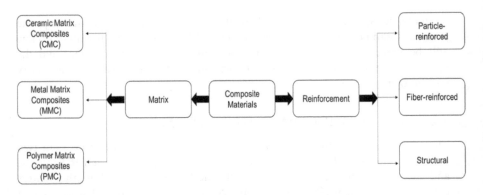

FIGURE 4.1 The classification of composites.

4.2.1 The Classification of Composites for Matrix Material

Composites are divided into three groups, ceramic, metal and polymer, according to the matrix material.

4.2.1.1 Ceramic Matrix Composites (CMCs)

Although ceramic materials have high stiffness and strength, they are unreliabile in terms of mechanical properties. Because of this disadvantage, CMCs have been developed (Cho et al., 2009). In this composite group, a matrix is a ceramic material and reinforcement can be ceramic fibers or fillers reinforced (Basutkar & Kolekar, 2015;Mathur & Kedar Bairwa, 2017). CMCs have higher toughness, stiffness, thermal shock resistance and hardness (Cho et al., 2009; Mathur & Kedar Bairwa, 2017). This kind of composite can be used in aerospace and automotive industries, medical implants and military applications.

4.2.1.2 Metal Matrix Composites (MMCs)

In this composite group, a matrix is a metal material. MMCs are the most widely used composite type and have begun to replace conventional materials used in many industries. MMCs have higher strength, elastic modulus, wear resistance, fatigue properties and lower weight than conventional materials (Rohatgi, 1993). Additionally, corrosion resistance is better at higher temperatures than for polymer matrix composites (Mathur & Kedar Bairwa, 2017). They can be used in automotive components (pump housings, brake calipers, gears, drive shafts, engine blocks, brake rotors, pistons, etc. (Mahajan & Aher, 2012) and in aircraft applications (Mathur & Kedar Bairwa, 2017).

4.2.1.3 Polymer Matrix Composites (PMCs)

In this composite group, a matrix is a polymer material (such as polyester, epoxy) and reinforcement can be thin-diameter fibers (such as boron, graphite and aramids) (Mathur & Kedar Bairwa, 2017). As known, polymer materials are not suitable for many applications because of their stiffness and strength. So, PMCs are developed to improve flexibility, toughness, adhesiveness, strength and conductivity. These properties strongly depend on a matrix and reinforcement materials (Mohan et al., 2017). They can be used in biomedical applications (medical implants and orthopedic devices, etc.), electrical applications (panels, insulators and connectors, etc.), marine applications (boat bodies and canoes, etc.) and in the automotive industry (body panels, drive shaft, doors, etc.) (Divya et al., 2016).

4.3 THE CLASSIFICATION OF COMPOSITES FOR REINFORCEMENT MATERIAL

Composites are divided into three groups, particle, fiber and structural, according to the reinforcement material.

4.3.1 Particle-Reinforced

In this material system, reinforcements are particle (inorganic, ceramic or metal) and distributed in a matrix phase as a second phase. The reinforcement particles work to improve the mechanical and physical properties of matrix material with the help of their mechanical and physical properties (Parameshwari et al., 2016). Generally, particles are ellipsoidal, spherical, irregular or polyhedral in shape and can be classified as dispersion-strengthened (small particles 60 μm) and large particle reinforced (Kanagaraj et al., 2019). Particle size, shape and amount affect the mechanical properties of composites (Ye et al., 2017).

4.3.2 Fiber Reinforced

In this material system, fibers are used as reinforcements and classified according to length (a long one is called a continuous fiber and a short one is called a discontinuous fiber). Additionally, it can be classified as natural and synthetic fiber (Rajak et al., 2019). The usage of fiber enhances the mechanical properties of the polymeric matrix (Prashanth et al., 2017).

4.3.3 Laminate Reinforced

Laminates are continuous fiber composites formed by stacking single fiber layers in different directions with high volumes from 60% to 70%. If the fibers in all layers are located in the same orientation, this is called a lamina. Otherwise, if the fibers in all layers are located at various angles, this is called a laminate (Campbell, 2010).

4.4 PARTICLE-REINFORCED COMPOSITE MATERIALS OVERVIEW

If papers published are taken into consideration, it can be seen that aluminum and its alloys have been widely used as matrix material (Aybarc et al., 2018; Garg et al., 2012, Ramnath et al., 2014). Additionally, different kinds of ceramic particles have been used as reinforcement material. Because of this, particle-reinforced aluminum MMCs are dealt with in the following pages. Additionally, the summary of publications of composites produced with aluminum matrix is given in Table 4.1.

4.5 PRODUCTION METHODS OF COMPOSITES

Production methods of composites consist of two main groups: liquid state and solid state processes.

4.5.1 LiquidState

The main characteristic of this group is that reinforcement materials are added into a molten matrix material. The most important criteria is wettability and behavior of the interface between reinforcement and matrix material. The techniques, given below, are known as liquid state processes (Bartolucci et al., 2011; Kandpal et al., 2014; Surappa, 2003).

4.5.1.1 Stir Casting

In this technique, a molten matrix metal and reinforcement material are mixed with the help of mechanical stirring. The stir casting method is commonly preferred because it is cost effective and is a simple process (Balasivanandha Prabu et al., 2006; Kandpal et al., 2014). But in this technique, porosity, homogeneity distribution of reinforcements and performance of wettability between reinforcements and the matrix material are problems to be overcome (Aqida et al., 2004; Priyadarshi & Sharma, 2016; Sozhamannan et al., 2012).

4.5.1.2 Compocasting

This technique is nearly the same with stir casting. The major differences is the matrix material is a semi-solid form when the reinforcement particles are added into it (Ray, 1993). In this technique, it obtained better wettability, lower porosity and also better reinforcement distributions (Saravanan et al., 2015).

4.5.1.3 Squeeze Casting

It can be said that this technique is combination of casting and forging processes. Liquid matrix metal with reinforcement is injected into a prepared die cavity with the help of a piston (Manickam & Manickam, 2014). This process is limited by a preformed shape mold, application cost is moderate and reinforcements can be severely damaged (Adat et al., 2015; Manickam & Manickam, 2014; Saravanan et al., 2015).

TABLE 4.1
Publications about Aluminum Matrix Composites

Researchers	Additive Materials	Particle Size	Ratio of Reinforcement	Production Method	References
T. Ozben, E. Kilickap, O. Cakir	SiC	30–60 μm	5, 10, 15	Pressure casting	Ozben et al. (2008)
M. Singla, D.D. Dwivedi, L. Singh, V. Chawla		~44 μm	5, 10, 15, 20, 25, 30	Stir casting	Singla et al. (2009)
S.B. Prabu, L. Karunamoorthy, S. Kathiresan, B. Mohan		60 μm	10	Stir casting	Balasivanandha Prabu et al. (2006)
K.L. Meena, Dr.A. Manna, Dr. S.S. Banwait, Dr. Jaswanti		~69–40–37 μm	5, 10, 15	Stir casting	Meena et al. (2013)
K. Karvanis, D. Fasnakis, A. Maropoulos, S. Papanikolaou		~44 μm	3, 6, 9, 12, 15	Centrifugal casting machine	Karvanis et al. (2016)
R. Agnihotri, S. Dagar		325 mesh	5, 10, 15	Mechanical stir casting	Agnihotri and Dagar (2017)
Md. H. Rahman, H.M. Mamun Al Rashed		53–74 μm	0, 5, 10, 20	Stir casting	Rahman et al. (2014)
K.K. Alaneme, M.O. Bodunrin	Al_2O_3	28 μm	6, 9, 15, 18	Stir casting	Alaneme and Bodunrin 2013
L. Singh, B. RamA. Singh		75–105–150 μm	3, 6, 9	Stir casting	Singh et al. (2013)
D. Sujan, Z. Oo, M.E. Rahman, M.A. Maleque, C.K. Tan		400 μm	5, 10, 15	Stir casting	Sujan et al. (2012)
S.A. Sajjadi, H.R. Ezatpour, H. Beygi		20 μm–50 nm	1, 3, 5, 10 (for μm) 1, 2, 3 (for nm)	Stir casting	Sajjadi et al. (2011)
S. Mula, P. Padhi, S.C. Panigrahi, S.K. Pabi S.Ghosh		10 nm	2	Ultrasonic chamber	Mula et al. (2009)
M. Kok		16–32–66 μm	7, 15, 23	Stir casting	Kok (2005)

(Continued)

TABLE 4.1 (Continued)
Publications about Aluminum Matrix Composites

Researchers	Additive Materials	Particle Size	Ratio of Reinforcement	Production Method	References
S.F. Bartolucci, J. Paras, M. A. Rafiee, J. Rafiee, S. Lee, D. Kapoor, N. Koratkar	Graphene		0,1	Hot isostatic pressing, and hot extrusion	Bartolucci et al. (2011)
M.M. Narwate, K.K. Mohandas	Graphene (aluminum with 10% fixed TiO2)		0.5, 0.75, 1.0	Stir casting	Narwate and Manjunath (2016)
P. Kumar, S. Aadithya, K. Dhilepan, N. Nikhil	Graphene (and SiC; fixed 5% wt.)		1, 3, 5	Ultrasonic assisted stir casting	Kumar et al. (2016)
B. S. Jagadish	Graphene		0.25, 0.5, 0.75, 1	Powder metallurgy	Jagadish (2015)

4.5.1.4 Spray Deposition

In this technique, molten metal is sprayed on a reinforcement (Kandpal et al., 2014). Additionally, this technique is used to produce a limited shape and large size product and its cost is expensive (Adat et al., 2015).

4.5.1.5 Infiltration Process

In this process, the filament-type reinforcement is normally used (Saravanan et al., 2015). This technique is similar to squeeze casting but the difference between the infiltration process and squeeze casting is that the applied pressure is provided by gas (Ray, 1993).

4.5.1.6 In-situ Processing

In this process, reinforcements such as TiC, NbB2 and TiB2 are created inside a molten matrix material by using their powder form. But before using this technique, it should be determined which reinforcement materials are wanted to be obtained in composites and designed before adding with the help of thermodynamic and kinetic studies (Ray, 1993). It is obtained through good reinforcement, matrix coordination and homogeneity of reinforcement particle distribution by using this technique, but it is expensive (Saravanan et al., 2015).

4.5.1.7 Ultrasonic Assisted Casting

The ultrasonic process is commonly used for nanoparticle dispersion (Liu et al., 2015). In this process, a determined sonotrode is submerged in a molten matrix material and reinforced material is added in it. Sonotrode creates cavitation and mixes matrix and reinforcement material together. With this technique, uniform distribution is obtained but the cost of the process is expensive (Liu et al., 2015; Saravanan et al., 2015).

4.5.2 Solid State

The main characteristic of this group is that a mutual diffusion bonding occurs between a matrix and reinforcement materials at a high process temperature and under pressure. The techniques, given below, are known as solid-state processes (Bartolucci et al., 2011; Kandpal et al., 2014; Surappa, 2003).

4.5.2.1 Powder Blending and Consolidation

In this technique, matrix and reinforcement materials are used in powder form and include a variety of process steps, such as sieving, blending, pressing, degassing and final consolidation (by forging, rolling, extrusion, etc.) (Ibrahim et al., 1991). The disadvantages of this process include the reinforcement can be fractured and the cost of the process is expensive (Bhandare & Sonawane, 2013).

4.5.2.2 Diffusion Bonding

This technique can be used for both similar and dissimilar metals. Matrix and reinforcement materials are used as foil or sheet forms. Inter-diffusion of atoms occurs between clean surfaces of metal and fiber reinforcement at an elevated temperature and applied pressure. The cost of the process is expensive (Kandpal et al., 2014; Saravanan et al., 2015).

4.5.2.3 Physical Vapor Deposition

In this technique, a workpiece surface is coated with a determined reinforced material atom by atom or molecule by molecule. To obtain composite products, vapor is produced in a protective atmosphere with the help of a high-power electron beam to deposit a determined metal on a reinforcement fiber (Buradagunta, 2015; Surappa, 2003). The cost of this process is moderate (Saravanan et al., 2015).

Saravanan et al. (2015) declared that the stir casting process is the cheapest one and applicable to large-quantity production. But in this process, porosity is of vital importance. Porosity decreases a material's service life. In order to eliminate porosity, additives are pre-heated at a suitable temperature and time. Thus, the angle between the molten aluminum and the reinforcement particles' contact surface is reduced and the wettability is increased. In addition to this application, the addition of 1% magnesium to the molten aluminum affects the wettability positively. In another application, the process of coating the surfaces of reinforcement particles with materials that can form intermetallic compounds by interacting with aluminum is also widely used (Urena et al., 2001).

4.6 RECYCLING OF MMCS

Recycling processes have gained importance, especially in recent years. In this section, the techniques used in the recycling of MMCs and the economic and environmental impacts of the recycling of composite materials are discussed.

4.6.1 Recycling Techniques of MMCs

Recycling processes are an importance topic, especially nowadays. This situation is important not only in terms of economic benefits, but also because of the decreasing natural resources, especially environmental pollution. For this reason, developed countries determine some policies for sustainable raw material use and environmental pollution reduction with many new regulations. The processing of a substance from ore requires both high costs and decreasing natural resources due to the use of energy and resources required for production. Environmental pollution is also negatively affected by flue gases, wastes, etc. during production. Therefore, recycling processes not only provide cost advantages compared to ore production, but also ensures that raw material is obtained in an environmentally sensitive manner if it is done under suitable conditions.

Two important criteria can be taken into consideration in the creation of recycling systems. The first one is the type of input materials to be used in recycling, and the other is the type of matrix material used in the production of composite material. The input material to be used in the recovery of composite materials can be divided into two groups. Waste products, burrs and chips formed during the production of composite materials can be classified in the first group. Composite materials, of which the useful life has expired, can be considered as the second group (Yang et al., 2012). In these groups, the second one is more important in terms of collecting all the products of which the useful life has expired. Because the recycling inputs to be obtained from the first group are where the production takes place, there will be no additional transportation and collection costs for their reproduction.

The type of matrix material plays an important role in the recovery process of composite materials. Metal and polymer matrix composites are easier to recover than ceramic matrix composites. Because of the characteristic of the matrix material, composites with ceramic matrix should be processed at very high temperatures (Yang et al., 2012). On the other hand, recycling process can be dived in three groups: mechanical, chemical or thermal.

4.6.1.1 Mechanical

The mechanical process is generally used for recycling all composites. In this process, retired composites or scraps are cut, ground and classified to improve the separation ability of the matrix (generally a thermoset resin) and fibers from the composite structures (Asmatulu et al., 2013; Oliveux et al., 2015). The hammer mill process's basic principle is to downsize the material by the impact and shear action until the parts pass through the predefined milling screen holes (Shuaib & Mativenga, 2016).

4.6.1.2 Chemical

The chemical recycling process can be divided two main categories. One of them is known as solvolysis, in which chemical dissolution reagents are used to depolymerize the matrix. During this process, the solvent or solvent mixture should be taken into consideration for the reaction rate (Dang et al., 2005; Yang et al., 2012). The other one is known as hydrolysis, in which water is used as a solvent to depolymerize the matrix (Wang et al., 2020).

4.6.1.3 Thermal

To decompose the resin, a high temperature is applied, between 300 and 1000°C. The quality of recovered fibers or filler materials depend on the process temperature (Yang et al., 2012). Additionally, the remelting method can also be considered in this category. During the reclamation of the composite, it can use insoluble fluxes, which have a smaller surface energy to overcome the free energy barrier of the composite system. On the other hand, the gravity technique, aided by blowing bubbles, can also be used to separate particles and matrix material (Bhuiyan et al., 2016).

These different processes can be used together for the recycling composite system. For the MMC system, mechanical and chemical processes can be used together. The squeeze casting method can be used for the mechanical process and fluxes containing sodium, potassium, NaCl and KCl can be used for the chemical process (Nishida, 2001). However, Yang et al. (2012) have created a comparative analysis of the recycling methods of different composites. Figure 4.2 shows the summary of this analysis.

Alternative recycling processes compared to the matrix material are clearly seen in Figure 4.2 and are not actually easy in terms of both investment and continuity of the process because the recycling process also has some problems. These problems can be listed as follows (Nagolska, 2007):

- the emergence of technically useless alloys containing components that are very difficult to separate by metallurgical methods;
- the occurrence of contaminations caused by reactions between the composite components, the remelted material obtained from the separation of the composite components or reactions occurring between the matrix and difficult to eliminate;
- determination of appropriate recycling process conditions that enable the separation of composite components.

Despite the difficulties in recycling processes, researchers are investigating methods and creating alternative recycling flow charts. Schuster et al. (1993)

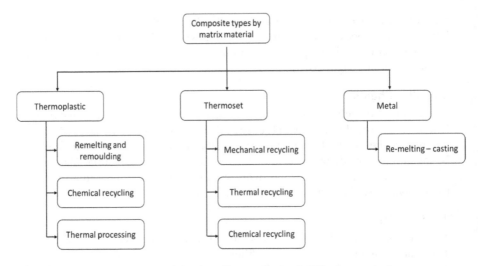

FIGURE 4.2 A comparison of the recycling methods of different composites.

investigated the reclamation and recycling of MMCs produced by SiC and Al_2O_3 reinforced. Ling and Gupta (Ling & Gupta, 2000) applied the disintegrated melt deposition technique to recycle aluminum matrix composites with SiC reinforcement.

The recycling of composite materials is of great importance for the efficient use of natural resources, disposal of waste and especially for the circular economy. However, in addition to all of these, measuring and determining the quality of composites that are recycled is another important issue. Within the scope of this subject, researchers have started to determine various methods. Shuaib and Mativenga (2016) analyzed the recycling of glass fiber reinforced plastic and applied three methods as fiber length measurement, sieving and furnace treatment to determine the quality analyses of recyclabes.

4.6.2 The Economic and Environmental Aspects of Recycle

One of the biggest causes of increasing environmental pollution today is the widespread use of passenger cars. This not only increases fossil fuel use, but also increases the emission of exhaust gases. Therefore, especially developed countries have started to make some new regulations in order to keep air pollution and exhaust gas emissions that cause greenhouse gas effect under control. The best example of this is that the European Union's determination of the CO_2 emissions for passenger cars as 130 g/km for 2015 and targeting this value as 95 g/km for 2021 (International Council on Clean Transportation). This prompts OEM (original equipment manufacturer) companies to produce new projects, in particular to reduce total vehicle weights. Conventional materials currently used are both insufficient to meet these targets and hinder vehicle weight reduction. Engineers and designers have begun to use composite materials as an alternative to conventional products and have shared many research results in the literature. However, most of the studies are directed towards increasing composite material production and mechanical properties. When this situation is considered in terms of the final products produced, it can be seen that the negative effects on the environment are reduced. But, another important issue is that after the useful life of the products produced with composite materials has expired, detailed studies have not been conducted on how to dispose of them and how they can be regained into the economy.

Yang et al. (2012) propose a chain of operations to recycle composites and bring them back to the economy. This chain includes the following topics:

- The availability of the composite scrap
- Collection and transport
- Reprocessing – recycling
- Market of the recycled products – recycles

Figure 4.3 shows that the availability of the composite scrap formed during production and composite materials that expired their useful life. After they collected, they should reprocess/recycle according the suitable recycling process in order to produce new composites to market.

In economic aspects, carbon fibers, glass fibers and resin can be given, for example. If carbon fiber, glass fiber and resin are reused, their economic value can amount approximately up to €97/kg (Urena et al., 2001), €0.25/kg and €2.25/kg (Cousins et al., 2019), respectively. In addition, while the production costs of recycled carbon fibers, which can be used as a secondary source, are quite high and

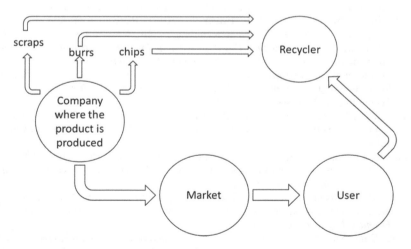

FIGURE 4.3 The availability of the composite scrap.

the energy consumption costs used for 1 kg production of it is 14 times the steel production cost (Anane-Fenin & Akinlabi, 2017).

As mentioned in the previous pages, recycling of composite materials is of great importance both in terms of the economy and environmental protection. For this reason, developed countries constantly express the importance of recycling processes not only for composites but also for conventional products. In addition, they emphasize that the weight of the passenger cars, which adversely affect the nature and cause greenhouse gas effects, should be reduced, especially in the transportation sector due to exhaust gases. In order to prevent these negative effects and keep them under control, they create various regulations and apply material sanctions if these regulations are not followed. According to new regulations, not only OEM companies but also countries have started to work to adapt. Because of this, linear economies which base on unidirectional material flow (production-consumption-waste) have begun to turn into a circular economy model which is based on production–consumption–reuse frameworks (Garcia-Bernabeu et al., 2020). Installing a new system and investing in infrastructure is not an easy task for companies and it is costly. Because of this, Tarverdi (2010) expressed that government aid and incentives should be.

REFERENCES

Adat, Ravikiran, Kulkarni, Sunil G., & Kulkarni, Sushma S. (2015). Manufacturing of particulate reinforced aluminum metal matrix composites using stir casting process. *International Journal of Current Engineering and Technology, 5*(4), 2808–2812.

Agnihotri, Rajesh, & Dagar, Santosh (2017). Mechanical properties of Al-SiC metal matrix composites fabricated by stir casting route. *Research in Medical & Engineering Sciences, 2*(5) 178–183.

Alaneme, Kenneth Kanayo, & Bodunrin, Michael (2013). Mechanical behaviour of alumina reinforced AA 6063 metal matrix composites developed by two step-stir casting process. *Acta Technica Corviniensis-Bulletin of Engineering, 6*(3), 105–110.

Anane-Fenin, Kwame, & Akinlabi, Esther T. (2017). Recycling of fibre reinforced composites: A review of current technologies. In *Proceedings of the DII-2017 Conference on Infrastructure Development and Investment Strategies for Africa: Infrastructure and Sustainable Development – Impact of Regulatory and Institutional Framework.* https://www.researchgate.net/publication/320536248_Recycling_of_Fibre_Reinforced_Composites_A_Review_of_Current_Technologies, Accessed 24.07.2020."

Aqida, S. N., Ghazali, Mohd Imran, & Hasim, J. (2004). Effects of porosity on mechanical properties of metal matrix composite: An Overview. *Journal Teknologi, 40*(A), 17–32.

Asmatulu, Eylem, Twomey, Janet & Overcash, Michael (2013). Recycling of fiber-reinforced composites and direct structural composite recycling concept. *Journal of Composite Materials, 48*(5), 593–608.

Aybarc, Uğur, Dispinar, Derya, & Seydibeyoglu, Mehmet Ozgur (2018). Aluminum metal matrix composites with SiC, Al_2O_3 and Graphene – review. *Archives of Foundry Engineering, 18*(2), 5–10.

Aybarc, Ugur, Yavuz, Hakan, Dispinar, Derya, & Seydibeyoglu, Mehmet Ozgur (2018). The use of stirring methods for the production of SiC-reinforced aluminum matrix composite and validation via simulation studies. *International Journal of Metalcasting, 13*, 190–200.

Balasivanandha Prabu, S., Karunamoorthy, L., Kathiresan, S., & Mohan, B. (2006). Influence of stirring speed and stirring time on distribution of particles in cast metal matrix composite. *Journal of Materials Processing Technology, 171*(2), 268–273.

Bartolucci, Stephen F., Paras, Joseph, Rafiee, M. A., Rafiee, J., Lee, Sabrina, Kapoor, Deepak, & Koratkar, Nikhil (2011). Graphene–aluminum nanocomposites. *Materials Science and Engineering: A, 528*(27), 7933–7937.

Basutkar, Ameya G., & Kolekar, Aniket (2015). A review on properties and applications of ceramic matrix composites. *International Journal pof Research and Scientific Innovation, II*(XII), 28–30.

Bhandare, Rajeshkumar Gangaram, & Sonawane, Parshuram S. (2013). Preparation of aluminium matrix composite by using stir casting method. *International Journal of Engineering and Advanced Technology, 3*(2), 61–65.

Bhuiyan, M. S. H., & Degischer, H. P. (2016). Metal matrix composites, recycling of. *Reference Module in Materials Science and Materials Engineering*, 1–4. doi: 10.1 016/B978-0-12-803581-8.03648-

Botelho, Edson Cocchieri, Silva, Rogerio Almeida, Pardini, Luiz Claudio, & Rezende, Mirable Cerquira (2006). A review on the development and properties of continuous fiber/epoxy/aluminum hybrid composites for aircraft structures. *Materials Research, 9*(3), 247–256.

Buradagunta, Ratna Sunil. (2015). Developing surface metal matrix composites: A comparative survey. *International Journal of Advanced in Materials Science and Engineering, 4*(3), 9–16.

Campbell, F. C. Structural Composite Materials. (2010). *Printed in the United States of America*, ASM International.

Cho, Johann, Boccaccini, Aldo R., & Shaffer, Milo S. P. (2009). Ceramic matrix composites containing carbon nanotubes. *Journal of Materials Science, 44*, 1934–1951.

Cousins, Dylan S., Suzuki, Yasuhito, Murray, Robynne E., Samaniuk, Joseph R., & Stebner, Aaron P. (2019). Recycling glass fiber thermoplastic composites from wind turbine blades. *Journal of Cleaner Production, 209*, 1252–1263.

Cshuster, David M., Skibo, Michael D., Bruski, Richard S., Provencher, Robert, & Riverin, Gaston (1993). The recycling and reclamation of metal-matrix composites. *Recycling Advanced Material, 45*, 26–30.

Dang, Weirong, Kubouchi, Masatoshi, Sembokuya, Hideki, & Tsuda, Ken (2005). Chemical recycling of glass fibre reinforced epoxy resin cured with amine using nitric acid. *Polymer, 46*(6), 1905–1912.

Dhanasekaran, Subramaniam, Sunilraj, S., Ramya, G., & Ravishankar, S. (2016). SiC and Al_2O_3 reinforced aluminum metal matrix composites for heavy vehicle clutch applications. *Transactions of the Indian Institute of Metals, 69*, 699–703.

Divya, H. V., Laxmana Naik, L., & Yogesha, B. (2016). Processing techniques of polymer matrix composites – a review. *International Journal of Engineering Research and General Science, 4*(3), 357–362.

Garcia-Bernabeu, Ana, Hilario-Caballero, Adolfo, Pla-Santamaria, Daid, & Salas-Molina, Francisco. (2020). A process oriented MCDM approach to construct a circular economy composite index. *Sustainability, 12*(618), 1–14.

Garg, Harish K., Verma, Ketan, Manna, Alakesh, & Kumar, Rajesh (2012). Hybrid metal matrix composites and further improvement in their machinability – A Review. *International Journal of Latest Research in Science and Technology, 1*(1), 36–44.

Ibrahim, I. A., Mohamed, F. A., & Lavernis, E. J. (1991). Particulate reinforced metal matrix composite – a review. *Journal of Materials Science, 26*, 1137–1156.

International Council on Clean Transportation. (2014). *EU CO$_2$Emission Standards for Passenger Cars and Light-Commercial Vehicles.* ICCT Web. https://theicct.org/publications/ldv-co2-stds-eu-2030-update-jan2019. Accessed 24.07.2020

Jagadish, Bhavanam. (2015). Synthesis and characterization of aluminium2024 and graphene metal matrix composites by powder metallurgy. *SSRG International Journal of Mechanical Engineering, 2*(7), 13–17.

Kanagaraj, A., Franciskennathamreth, C., Ajithkumar, M., Anandh, V., & Nagaraj, R. (2019). Development of particle reinforced composite by plastic and e-waste. *International Journal of Advanced Research, Ideas and Innovations in Technology, 5*(2), 280–285.

Kandpal, Bhaskar Chandar, Kuma, Jatinder, & Singh, Hari (2014). Production technologies of metal matrix composites: A review. *International Journal of Research in Mechanical Engineering & Technology, 4*(2), 27–32.

Karnik, Tarverdi (2010). Improving the mechanical recycling and reuse of mixed plastics and polymer composites. *Management, Recycling and Reuse of Waste Composites,* 281–302. doi:10.1533/9781845697662.3.281

Karvanis, K., Fasnakis, D., Maropoulos, A., & Papanikolaou, S. (2016). Production and mechanical properties of Al-SiC metal matrix composites. In *IOP Conference Series: Materials Science and Engineering 161.* IOP Publishing.

Kok, Metin. (2005). Production and mechanical properties of Al$_2$O$_3$ particle-reinforced 2024 aluminium alloy composites. *Journal of Materials Processing Technology, 161*(3), 381- 387.

Kumar, P., Aadithya, S., Dhilepan, K., Nikhil, N. (2016). Influence of nano reinforced particles on the mechanical properties of aluminum hybrid metal matrix composite fabricated by ultrasonic assisted stir casting. *ARPN Journal of Engineering and Applied Sciences, 11*(2), 1204–1210.

Ling, Pin Soon, & Gupta, Manoj (2000). Recycling of aluminium based metal matrix composite using disintegrated melt deposition technique. *Materials Science and Technology, 16,* 568–574.

Liu, Xiaoda, Jia, Shian, & Nastac, Laurentiu (2015). Ultrasonic cavitation-assisted molten metal processing of cast A356-nanocomposite. *International Journal of Metalcasting, 8*(3), 51–58.

Mahajan, Gokul V., & Aher, Vishnu S. (2012). Composite material: A review over current development and automotive application. *International Journal of Scientific and Research Publications, 2*(11), 1–5

Manickam, Dhanashekar, & Manickam, Senthil Kumar V. S. (2014). Squeeze casting of aluminium metal matrix composites – An overview. *Procedia Engineering, 97,* 412–420.

Mathur, Nitin Mukesh, Kedar Bairwa, Rajkumar (2017). A literature review on Composite material and scope of sugar cane bagasse. *International Journal of Engineering Development and Research, 5*(4), 125–133.

Meena, K. L., Manna, A., Banwait, S. S., & Jaswanti (2013). An analysis of mechanical properties of the developed Al/SiC-MMC's. *American Journal of Mechanical Engineering, 1*(1), 14–19.

Milos, Katica, Juric, Ivica & Skorput, Pero (2011). Aluminium-based composite materials in construction of transport means. *Science in Traffic and Transport, 23*(2), 87–96.

Mohan, N., Senthil, P., Vinodh, S., & Jayanth, N. (2017). A review on composite materials and process parameters optimisation for the fused deposition modelling process. *Virtual and Physical Prototyping, 12*(1), 47–59.

Mula, Suhrit, Padhi, P., Panigrahi, Sarat, Pabi, S. K., & Ghosh, Sudipto (2009). On structure and mechanical properties of ultrasonically cast Al–2%Al$_2$O$_3$ nanocomposite. *Materials Research Bulletin, 44*(5), 1154- 1160.

Nagavally, Rahul Reddy (2016). Composite materials – history, types, fabrication techniques, advantages and applications. *International Journal of Advances in Science Engineering and Technology, 5*(9), 82–87.

Nagolska, Dorota (2007). The quality of the recovered matrix material as an element of assessment of effectiveness of the composite recycling process. *Archives of Foundry Engineering, 7*(2), 163–168.

Narwate M., & Manjunath, K. N. Mohandas (2016). A study on mechanical and tribological properties of aluminum metal matrix composite reinforced with TiO_2 and graphene oxide. *International Journal, 4*(4), 729–732.

Ngo, Tri-Dung (2019). "Introduction to composite materials", Composite and Nanocomposite Materials From Knowledge to Industrial Applications. DOI: 10.5 772/intechopen.91285.

Nishida, Y. (2001). Recycling of metal matrix composites. *Advanced Engineering Materials, 3*(5), 315–317.

Oliveux, Geraldine, Dandy, Luke O., & Leeke, Gary A. (2015). Current status of recycling of fibre reinforced polymers: Review of technologies, reuse and resulting properties. *Progress in Materials Science, 72*, 61–99.

Ozben, Tamer, Kilickap, Erol, & Çakır, Orhan (2008). Investigation of mechanical and machinability properties of SiC particle reinforced Al-MMC. *Journal of Materials Processing Technology, 198*(1), 220–225.

Parameshwari, R., Raj, R. M., & Sundar, P. B. (2016). Experimental investigation of a particulate reinforced composite. *International Journal of Innovations in Engineering and Technology, 6*(4), 251–257.

Prashanth, S., Subbaya, K. M., Nithin, K., & Sachhidananda, S. (2017). Fiber reinforced composites – a review. *Journal of Material Science & Engineering, 6*(3), 1–6.

Priyadarshi, Devinder, & Sharma, Rajesh Kumar (2016). Porosity in aluminium matrix composite: Cause, effect and defence. *Material Science An Indian Journal, 14*(4), 119–129.

Rahman, Md. Habibur, Mamun Al & Rashed, H. M. (2014). Characterization of silicon carbide reinforced aluminum matrix composites. *Procedia Engineering, 90*, 103–109.

Rajak, Dipen Kumar, Pagar, Durgesh D., Menezes, Pradeep L., & Linul, Emanoil (2019). Fiber-reinforced polymer composites: Manufacturing, properties and applications. *Polymers, 11*(10), 1–37.

Rajeswari, Balakrishnan, Amirthagadeswaran, K. S., & Ramya, K. (2014). Microstructural studies of aluminium 7075-silicon carbide- alumina metal matrix composite. *Advanced Materials Research, 984–985*, 194–199.

Ramnath, B. Vijaya, Elanchezhian, C., Annamalai, R. M., Aravind, S., Sri Ananda Atreya, T., Vignesh, V., & Subramanian, C. (2014). Aluminium metal matrix composites – a review. *Reviews on Advanced Materials Science, 38*(5), 55–60.

Ray, Subrata. (1993). Review synthesis of cast metal matrix particulate composites. *Journal of Materials Science, 28*, 5397–5413.

Reddy, B. S. B., Das, Karabi, & Siddhartha, Das (2007). A review on the synthesis of in situ aluminum based composites by thermal, mechanical and mechanical–thermal activation of chemical reactions. *Journal of Materials Science, 42*, 9366–9378.

Rohatgi, Pradeep K. (1993). Metal-matrix composites. *Defence Science Journal, 43*(4), 323–349.

Sajjadi, Seyed Abdolkarim, Ezatpour, Hamit Reza, & Beygi, Hossein (2011). Microstructure and mechanical properties of $Al–Al_2O_3$ micro and nano composites fabricated by stir. *Materials Science and Engineering: A, 528*(29–30), 8765–8771.

Saravanan, Chinnusamy, Subramanian, K., Sivakumar, D. B., Sathyanandhan, M., & Narayanan, R. S. (2015). Fabrication of aluminium metal matrix composite – a review. *Journal of Chemical and Pharmaceutical Sciences Special Issue, 7*, 82–87.

Schuster, D.M., Skibo, M. D., Bruski, R. S., Provencher, R., & Riverin, G. (1993). There cycling and reclamation of metal-matrix composites. *Recycling Advanced Material, 45*, 26–30. https://doi.org/10.1007/BF03223214.

Shuaib, Norshah Aizat, & Mativenga, Paul Tarisai (2016). Effect of process parameters on mechanical recycling of glass fibre thermoset composites. *Procedia CIRP, 48*, 134–139.

Singh, Lakhvir, Baljinder, Ram, & Singh, Amandeep (2013). Optimization of process parameter for stir casted aluminium metal matrix composite using Taguchi method. *International Journal of Research in Engineering and Technology, 2*(08), 375–383.

Singla, Manoj, Dwivedi, D. Deepak, Singh, Lakhir, & Chawla, Vikas (2009). Development of aluminium based silicon carbide particulate metal matrix composite. *Journal of Minerals and Materials Characterization and Engineering, 8*(06), 455.

Sozhamannan, G. G., Balasivanandha Prabu, S., & Venkatagalapathy, V. S. K. (2012). Effect of processing paramters on metal matrix composites: Stir casting process. *Journal of Surface Engineered Materials and Advanced Technology, 2*, 11–15.

Sujan, D., Oo, Z., Rahman, M. E., Maleque, M. A., & Tan, C. K. (2012). Physio-mechanical properties of Aluminium metal matrix composites reinforced with Al2O3 and SiC. World Academy of Science, *Engineering and Technology, 6*, 288–291.

Surappa, Mirle Krishnegowda (2003). Aluminium matrix composites: Challenges and opportunities. *Sadhana, 28*(1&2), 319–334.

Tarverdi, K. (2009). Improving the mechanical recycling and reuse of mixed plastic sand polymer composites. *Management, Recycling and Reuse of Waste Composites*, 281–302. https://doi.org/10.1533/9781845697662.3.281.

Urena, Alejandro, Herrero, Pilar Rodrigo, Baldonedo, Juan Luis, & Gil, Luis (2001). Active coatings for SiC particles to reduce the degradation by liquid aluminium during processing of aluminium matrix composites: Study of interfacial reactions. *Journal of Microscopy, 201*(2), 122–136.

Wang, Peng Hao, & Zimmermann, Natalie (2020). Composite recycling techniques: A literature review. *Juniper Online Journal Material Science, 6*(1), 11–17. https://www.sciencedirect.com/science/article/pii/B9780128035818036481?via%3Dihub#!

Yang, Yongxiang, Boom, Rob, Irion, Brijan, Heeden, Derk-Jan van, Kuiper, Pieter, & de Wit, Hans (2012). Recycling of composite materials. *Chemical Engineering and Processing, 51*, 53–68.

Ye, Junjie, Chu, Chenchen, Zhai, Zhi, Wang, Yongkun, Shi, Baoquan & Qiu, Yuanying (2017). The interphase influences on the particle-reinforced composites with periodic particle configuration. *Applied Sciences, 7*(1), 1–13.

5 Properties of Recycled Metal Matrix Composites

A. Atiqah[1], N. Ismail[2], K.K. Lim[3], A. Jalar[1,2], M.A. Bakar[1], M.A. Maleque[4], R.A. Ilyas[5,6], and A.B.M. Supian[7]

[1]Institute of Microengineering and Nanoelectronics, Universiti Kebangsaan Malaysia, Selangor
[2]Deparment of Applied Physics, Faculty of Science and Technology, Universiti Kebangsaan Malaysia, Selangor
[3]Pusat Citra Universiti, Universiti Kebangsaan Malaysia, Selangor
[4]Department of Materials and Manufacturing, Kulliyah of Engineering, International Islamic University Malaysia, Kuala Lumpur
[5]School of Chemical and Energy Engineering, Faculty of Engineering, Universiti Teknologi Malaysia, Johor, Malaysia
[6]Centre for Advanced Composite Materials, Universiti Teknologi Malaysia, Johor, Malaysia
[7]Advanced Engineering Materials and Composites Research Centre, Department of Mechanical and Manufacturing Engineering, Universiti Putra Malaysia, Selangor, Malaysia

CONTENTS

5.1 Introduction 94
5.2 Fabrication of Recycled Metal Matric Composites 95
    5.2.1 Stir Casting 95
    5.2.2 Powder Metallurgy 96
5.3 Physical Properties of Recycled MMCs 96
5.4 Mechanical Properties of Recycled MMCs 99
5.5 Thermal Properties of Recycled MMCs 101

DOI: 10.1201/9781003148760-5

5.6 Summary 103
References 104

5.1 INTRODUCTION

The utilization of recycled materials has been trending for the unconcerned effective consumption of ecosystems and environmental pollution prevention. This can be implemented for machine elements and industrial products. In the manufacturing industry, the production of excessive metallic chips is coming from the machining of metals. The metallic chips are categorized as waste material from the industry.

A lot of scientists and researchers are working with recycling metal such as aluminium (Duflou et al., 2015; Paraskevas, Kellens, et al., 2015; van der Harst et al., 2016), bronze (Abdullah Aslan, Salur, et al., 2018; Delfino, 2014; Hosseini & Paydar, 2020; Karadag & Akdemir, 2019) and cast iron (Laila et al., 2014; Li et al., 2020). According to Karadağ (2012), the preferable method to recycle metallic chips is stir casting. The oxide layer that is coated on the surface of metallic chips is produced during the machining process. From this process, a dirt layer produced from metal cutting liquids led to electrical and heated conduction, which caused the electrical resistance oven, and the conduction oven is not functioning. The drawback of using this method is the gas emit dangerous for the human and environment also.

Previous work done by other researchers are using stir casting and sintering process for aluminium chips (Baffari et al., 2019; Christy et al., 2020; Karuppasamy et al., 2020), bronze chips (Abdullah Aslan, Salur, et al., 2018) and cast iron (Guluzade et al., 2013). Furthermore, other techniques available for production recycling metal matrix composites (MMCs) such as powder metallurgy/solid-state processing (Abd Rashid et al., 2014; Abdullahi & Abidoye, 2020; Bhadra et al., 2017; Bhouri & Mzali, 2020b; Joharudin et al., 2020; Mahdi et al., 2017), high-energy ball milling mixing and sintering (Ahmad et al., 2020; Mahdi et al., 2015), microwave sintering (Mahapatra et al., 2019; Mishra & Sharma, 2016) and spark/plasma sintering (Paraskevas et al., 2014; Paraskevas et al., 2016; Paraskevas, Vanmeensel, et al., 2015b). In this chapter, the properties of recycling metal matrix composites consisting of physical, mechanical and thermal properties will be covered for aluminium, bronze and cast iron. Moreover, the recycling process of metallic chips composites will be brief shortly, and the most selected fabrication of recycling, such as stir casting and powder metallurgy, will be presented in this chapter. A summary of the previous metal recycling process is shown in Table 5.1.

TABLE 5.1
Recycling Metal Process

Type of Recycling Metal	Metal Recycled Process	References
Al6061-T6 alloy	1. Dry milled by using high-speed machining to produce three different cycling chips 2. Cleaned by ultrasonic bath by using acetone solution 3. Dried at thermal oven at 60°C	(Khamis et al., 2015)
Al6061 alloy	1. Chips generated by dry machining	(Paraskevas, Vanmeensel et al., 2015a)
Spheroidal graphic cast iron	1. The bars were milled in a ball mill to produce the metallic chip 2. Metallic chip was sieved with 1–2 mm size	(Aslan et al., 2019)
Al 6061	1. Dry milled by using a high-speed milling machine 2. Cleaned in acetone solution using an ultrasonic bath 3. Milled using Planetary Ball Mill for 1 h, at 100, 150 and 200 rev/min with the ball/powder ration 10:1	(Fuziana et al., 2014)
Al 2017	1. 2017 aluminium alloy rods were machined using dry turning process 2. Mechanically milled through a high-energy ball milling process for 20 h, ratio ball to powder of 10:1, 250 rpm	(Bhouri & Mzali, 2020a)

5.2 FABRICATION OF RECYCLED METAL MATRIC COMPOSITES

5.2.1 STIR CASTING

The stir casting method is an inexpensive, simpler and preferable technology in molten phase recycling MMCs (Ramanathan et al., 2019; Kandpal et al., 2018; Kumar & Menghani, 2016). The stir casting process engages the liquid state fabrication method, which needs the embodiment of reinforcing particle into a molten form to produce the uniform distribution shown in Figure 5.1. The most important in stir casting is to maintain wettability, such as the close bonding between solid and liquid phase. Besides, there are many requirements in the stir casting method that should be considered in terms of good mechanical strength, lower porosity and reinforced material homogenous distribution.

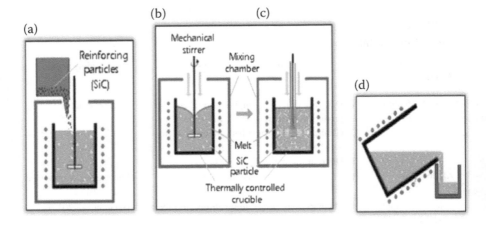

FIGURE 5.1 Stir casting method for recycled aluminium MMCs (Bulei et al., 2018).

5.2.2 Powder Metallurgy

Powder metallurgy or solid-phase process is an expensive method due to a higher cost of fundamental materials such as foil matrix or powder, etc. (Hosseini & Paydar, 2020; Singh et al., 2019; Meignanamoorthy & Ravichandran, 2020; Akhtar et al., 2018; Wakeel et al., 2017). Previous work that carried out the recycling MMCs has demonstrated an interesting study; Rojas-Díaz et al. (2020), for instance, recycled aluminium chips from the sawing process varied the grinding time. They conclude that grinding time (30–55 hours) is vital in processing the recycling aluminium chips from the sawing process. Figure 5.2 shows the common process of powder metallurgy for recycling MMCs.

5.3 PHYSICAL PROPERTIES OF RECYCLED MMCS

Most of the physical properties of recycled MMCs consider the density, apparent porosity, and very few water absorption tests were carried out. In another study, Bhouri and Mzali et al. (2020b) investigated recycled 2017 aluminium alloy matrix with a variation of graphite content using powder metallurgy. The recycled aluminium chips were obtained from 2017 aluminium alloy rod by using a dry turning process. The final dimension of an aluminium chip has a width less than 1 mm, a length between 20 and 75 mm and a thickness of 0.25 mm. The composite samples' sintered densities were decreased with decreasing graphite content due to the lower density of graphite compared to 2017 aluminium alloy. The other factor that influences the density of this recycled MMCs is that graphite particles reduce the contact area between the composites. This will provide poor bonding or network; and thus, will reduce the sintered density properties of the composites. The summary of physical properties of recycling MMCs is listed in Table 5.2.

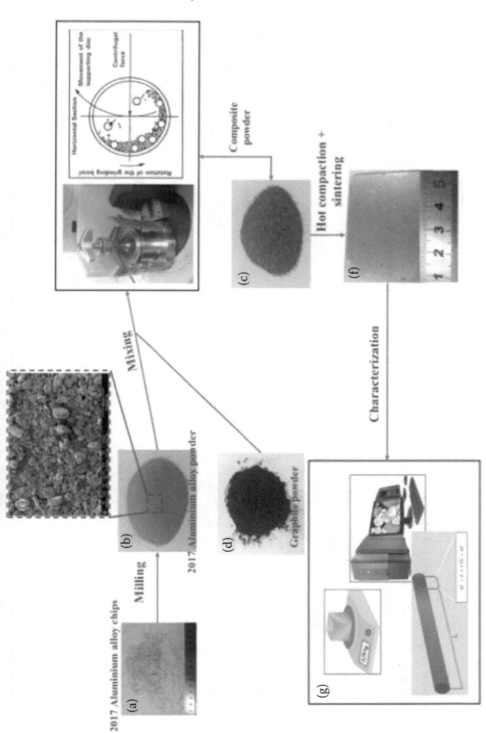

FIGURE 5.2 Fabrication of 2017 aluminium/graphite composites through powder metallurgy (Bhouri & Mzali, 2020a).

TABLE 5.2
Physical Properties of Recycled MMCs

Recycled Metal	Reinforcing Particle	Treatment	Composite Fabrication	Density (g/cm³)	Apparent Porosity (%)	Water Absorption (%)	Hardness Properties	Refereces
Recycled 2017 aluminium	Rice husk	Chemical treatment	Powder metallurgy	2.3–2.5	5–10	0.25–9	50–70 Hv	(Bhouri & Mzali, 2020b)
Recycled aluminium chip AA7075	Rice husk	–	Powder Metallurgy	2.3–2.5	5–16	2–8	50–70 Hv	(Joharudin et al., n.d.)
Recycled aluminum cans	Silicon Carbide	–	Powder Metallurgy	2.5–2.7	–	–	30–60 HRB	(Abdullahi & Abidoye, 2020)
Aluminium sawing chips	NaCl (60–70)	–		0.83–0.98	62–69	–	–	(Thalmaier et al., 2019)
Aluminium AA6061	Aluminium powder	–	Cold compaction solid-state sintering	2.3–2.5	2–12	–	40–70 Hv	(Ab Kadir et al., 2017)
Aluminium AA6061	Aluminium powder	–	Spark plasma sintering	2.7	–	–	713.2 ± 28.4 (surface) 661.2 ± 35.3 Hv	(Paraskevas, Vanmeensel et al., 2015a)
AA7075 aluminium alloy	–	–	Dry Turning Process–Mechanically milled through the ball mill process	2.85–3.34 g/cm³	–	–	–	(Bhouri & Mzali, 2020a
2017 aluminium powder	–	–		2,200–2,500 kg/m³	11–18%	–	–	(Enginsoy et al., 2020)
Aluminium 6061	4% Alumina	–	Powder metallurgy	2.19–2.45 kg/m³	–	–	60-150 Hv	(Fuziana et al., 2014)

The work by Joharudin et al. (2020) investigated the physical properties of recycled aluminium chip AA075 with untreated rice husk ash (URHA) and treated rice husk ash (TRHA) with variation 0%, 2.5%, 5.0%, 7.5%, 10.0% and 12.5%. The URHA and TRHA of rice husk content were chemical treatment consisted of SiO_2, C, K_2O, P_2O_5, CaO, MgO, SO_3, Cl, FeO_3 and Al_2O_3 compositions. The recycled aluminium chop AA075 was produced by a high-speed computer numerical control (CNC) machine. Properties such as micro-hardness, water absorption, apparent porosity and density of the MMCs were evaluated in this study. When increasing rice husk content, the higher the micro-hardness at the value of 53.59 Hv to 65.93 Hv, the water absorption and apparent porosity of the MMCs. However, the density of MMC was increased up to 5wt.% and then reduced when increasing the mass fraction of rice husk is due to the weak bonding of the reinforced particles. The treated rice husk content MMC was slightly higher with untreated density, apparent porosity, water uptake and microhardness properties.

5.4 MECHANICAL PROPERTIES OF RECYCLED MMCS

Recycling MMCs have the potential to meet a wide variety of mechanical properties. The mechanical properties of recycled MMCs depend on the proper selection of high temperature of reinforcement with an appropriate metal matrix, lightweight and a high-temperature recycling MMCs can be manufactured. The study by Lajis et al. (2017) investigated the effect of alumina's varying content from 1 to 5wt.% with a recycled aluminium chip that obtained from milling AA6061-T6. The addition of 1 to until 2wt.% alumina increases the ultimate tensile strength. This is because alumina avoids the tendency of dislocation in the recycled chip matrix composites – conversely, the higher amount of alumina from 3 to 5wt.% the ultimate tensile strength is decreased is probably initiated weak bonding in the material.

Enginsoy et al. (2020) carried out experimental and optimization on the hybridization of recycled metal AA7075 aluminium alloy with copper to produce new aluminium-copper MMCs. Two variation techniques to fabricate those composites with sintering and combined sintering and forging methods. Moreover, three different ZnO constituents, Nb_2Al and SiC were substituted with recycled metal AA7075/copper composites. Mechanical testing such as dynamic, compression and wear test showed that the recycled metal AA7075/copper composites implemented the combination of sintering and forging improved the mechanical properties compared to the composites with the sintering process only.

In another study by Gatamorta et al. (2020) employed the combination of sintering and forging for recycled AA431 aluminium mix with TiC (5 and 10wt.%), Mo and Cu (4 and 4wt.%) as reinforcement for aluminium matrix composites. The mechanical properties such as microhardness, static compression, three-point bending and impact drop weight showed that the higher content of 10 wt.% of TiC showed improved mechanical properties due to the combined method of sintered and forged increased the toughness and ductility for these aluminium metal composites.

Abdullahi and Abidoye (2020) investigated recycled aluminium can matrix with a variation of silicon carbides particulates from 0 to 10wt.% as the reinforcement through the stir casting method. Besides, two variations of stir temperature, such as 700°C and 750°C were carried out to observe the mechanical properties of recycled aluminium composites. From this research, the researchers found that the higher content of silicon carbide improved the tensile strength by 90%, impact strength by 119%, flexural strength by 53% and hardness by 61% compared to the unreinforced composites. Moreover, the higher stirring temperature at 750°C also enhances the mechanical properties of aluminium metal composites.

The investigation by Joharudin et al. (2020) studied the effect of rice husk's burning temperature as reinforcement with recycled aluminium chip AA7075 to produce aluminium metal composites. The different rice husk silica content from 2.5 until 12.5wt.% were prepared and burned at 700°C and calcination burned at 1000°C. Based on the investigation, the increasing rice husk silica content improved the hardness of metal matric composites. The optimum temperature of rice husk ash at 700°C showed better hardness properties compared to other temperatures.

Krishnan et al. (2019) investigated the scarp aluminium alloy with two different reinforcements consisting of alumina and spent alumina catalyst through the stir-squeeze casting method. The scrap aluminium alloy/alumina exhibited abrasive wear loss (0.11 mg for the finest abrasive), ultimate compressive strength (UCS) (312 MPa) and second-highest ultimate tensile strength (UTS) (125 MPa) compared to scrap aluminium with spent alumina catalyst composites. The use of scrap aluminium alloy can be employed due to high-performance aluminium MMCs and low-cost matrix material rather than virgin $AlSi_7Mg$.

Kamble and Kulkarni (2019) carried out study aluminium scrap with waste sugarcane bagasse ash (SCBA) with varying content from 0 to 0.25wt.% to cast aluminium MMCs. The role of waste bagasse ash is to improve the mechanical properties of aluminium composites. The hardness with a high content of waste sugarcane bagasse ash at 2.5wt.% exhibited a higher hardness value at 55.37 HB. The morphology of 2.5% SCBA confirms this; the morphology of samples was embedded and uniformly distributed reinforcement particles.

Aslan et al. (2018) investigated the use of spheroid cast iron and bronze chip to produce MMCs by varying the temperature of 350–450°C. These waste materials have good mechanical properties for journal bearing and produce clean surfaces due to machining without cooling fluids and good tribological properties. The hardness of composites that have a higher content of spheroid cast iron exhibited higher hardness from the study. Conversely, the higher compressive strength when the processing temperature is lower, the composites result in good bonding between chips and relatively higher strength and hardness.

Kadir et al. (2017) investigated the variation of aluminum powder content with aluminium AA6061 via cold compaction solid-state sintering suitable for fabrication direct recycling aluminium chips. In this research, the increasing amount of aluminium powder exhibited the lower mechanical properties of AA6061/aluminium powder. The non-uniform and randomly distributed microstructural showed that the weak bonding between the aluminium AA6061 recycling and aluminium powder proved that the weak bonding existed with the higher amount of aluminium powder.

Badarulzaman et al. (2014) carried out a study on recycling aluminium AA6061 with Sn matrix via the direct conversion method. The difference in volume of Sn was investigated and the mechanical properties such as yield strength (YS) and ultimate tensile strength (UTS) of recycling Al6061/Sn composites. In this study, the YS and UTS trend increases from 0 until 20 vol.% of Sn, though the curve starts decreasing because the aluminium properties are stronger and have a higher sintering temperature than Sn (232°C). Moreover, other factors also influenced matrix-reinforcement bonding, such as the amount of matrix/reinforcement, form and size of reinforcing shape, the sintering temperature and the cold pressing parameter (Gronostajski et al., 2000; Gronostajski & Matuszak, 1999).

Paraskevas et al. (2015a) discovered the use of spark plasma sintering by mixing aluminium atomized powder and machining chips. The innovative techniques that used meltless or solid-state recycling methods was developed for aluminium metal composites. The same type of aluminium exhibited the homogenous properties from this study even though it consisted of two different phases. The mechanical properties of Al powder with Al6061 scraps such as hardness and compression properties were approximate to 0.5% to the reference 6061 alloy ingot. The aluminium powder acts as a binder and matrix that promotes a good bonding for the alloy scrap and eliminating residual porosity. Besides, the merit of using plasma sintering can produce the finer form Al scrap with Al powder, desorb the entrapped gas, activate metallic surfaces of powder particles, easily fracture and disperse the surface oxides of the Al chip which results in the fully dense near-net sintered products. The mechanical properties of recycling aluminium MMCs were presented in Table 5.3.

5.5 THERMAL PROPERTIES OF RECYCLED MMCS

Based on the literature available, a few works carried out thermal properties of recycling MMCs. For instance, Bhouri et al. (2020a) investigated the thermoelastic properties of recycled aluminium alloy/graphite composites with variation of graphite content 5 and 10wt.% through the powder metallurgy process. The increase of graphite particles at 10wt.% reduced the coefficient of thermal expansion (CTE) composites. The composites with a lower CTE are good candidates for electronic applications. These composites can replace the existing material with higher CTE, which also improved the thermal management of electronics devices.

TABLE 5.3
Mechanical Properties of Recycled Aluminium Metal Composites

Type of Recycled Aluminium	Reinforcement	Fabrication Method	Mechanical Properties	References
Aluminium Chip AA7075	Rice Husk Silica (2.5–12.5 wt.%)	Powder Metallurgy	Hardness (50–70 Hv)	(Joharudin et al., n.d.)
Scarp aluminium alloy wheel	Alumina	Squeeze Casting	Hardness (Brinell)-58.475 Abrasive wear (ES20)-0.42 UTS (MPa)-125 UCS (MPa)-312	(Krishnan et al., 2019)
Scrap aluminium wheel	Spent alumina catalyst	Squeeze Casting	Hardness (Brinell)-55.95 Abrasive wear (ES20)-0.72 UTS (MPa)-82 UCS (MPa)-274	(Krishnan et al., 2019)
Scrap aluminium	Waste sugar cane bagasse ash	Stir casting	Hardness (Brinell)-42.23–55.7	(Kamble & Kulkarni, 2019)
Aluminium chip	Aluminium powder	Cold compaction solid-state recycling	Hardness-65.6 Hvcompression strength (307.7 MPa)	(Ab Kadir et al., 2017)
Aluminium	Sn powder	A solid-state direct conversion method	Yield strength-3 Paultimate tensile strength-8.3 Pa	(Badarulzaman et al. 2014)
Aluminium 6061	Aluminium	Powder metallurgy	Yield strength-120.6 ± 3.6 MPa compressive strength-228–280, 6 MPa The strain hardening exponent (n)-0.4 ± 0.02 strength coefficient (K)-103.7 MPa	(Paraskevas, Vanmeensel, et al., 2015a)
Aluminium 6061	4% Alumina	Powder metallurgy	Hardness-60–150 Hv	(Fuziana et al. 2014)

Kim et al. (2017) used waste metal sludge from steel mill plants as thermal conductive filler in composites. The properties of waste metal sludge with higher thermal conductivity and electrical resistivity can easily reach a lower cost.

TABLE 5.4
Thermal Properties of Recycled MMCs

Recycled Material	Reinforcement	Thermal Properties	References
2017 aluminium alloy	Graphite content (5 and 10 wt.%)	Coefficient of thermal expansion (CTE)20.0–22.5 × 10^{-6}/°C	(Bhouri & Mzali, 2020a)
Aluminium chips	(60–70) NaCl	Thermal conductivity-0.84 W/mKSpecific heat-137 J/kg K3thermal diffusivity-3.1 (m^2/s)	(Thalmaier et al., 2019)
Bronze chip/ Spheroid cast iron	–	Compressive strength (MPa)- 480–820brinell hardness	(Abdullah Aslan, Güneş, et al., 2018)
Waste metal sludge	–	Thermal conductivity-0.93 Wm^{-1} K^{-1}	(Kim et al., 2017)

Besides, the waste metal sludge doesn't need special treatment like other conducting material such as BN that is difficult to handle due to intrinsic repulsion force and chemical inertness (Eichler & Lesniak, 2008). The silane coupling agent (APTES) was employed to improve the interfacial affinity with the polymer. The thermal conductivity of 60wt.% of waste metal sludge composites was improved at 0.93 Wm^{-1} K^{-1} after surface modification with APTES. The potential of using waste metal sludge from metal still plants as thermally conductive filler promotes environmental preservation. Thermal properties of recycled MMCs were shown in Table 5.4.

5.6 SUMMARY

In this chapter, summarizing the works of literature and the recycling of metal to produce MMCs show that although the aluminium alloy used most recycling methods through the milling process, and sieved to obtain the uniform particle size of metallic chips. Then, better mechanical properties showed that powder metallurgy techniques that consisted of a mixing, compaction and sintering process are the most preferable to fabricate MMCs, which employed the meltless reaction that can avoid the dangerous chemical reaction such as stir casting. The benefit of recycling metal such as aluminum, cast iron and bronze from the manufacturing industry can mitigate environmental protection and improved energy consumption.

The authors would like to sincerely acknowledge the collaborating research institute, the Institute of Microengineering and Nanoelectronics IMEN, Faculty of Science and Technology (FST), Pusat Citra Universiti and International Islamic University Malaysia (IIUM) for publishing this chapter.

REFERENCES

Ab Kadir, Muhammad Irfan, Mustapa, Mohammad Sukri, Latif, Noradila Abdul, & Mahdi, Ahmed Sahib (2017). Microstructural analysis and mechanical properties of direct recycling aluminium chips AA6061/Al powder fabricated by uniaxial cold compaction technique. *Procedia Engineering, 184*, 687–694.

Abd Rashid, Mohd Warikh, Yacob, Fariza Fuziana, Lajis, Mohd Amri, Mohd Asyadi'Azam Mohd Abid, Mohamad, Effendi, & Ito, Teruaki (2014). 2301 a review: The potential of powder metallurgy in recycling aluminum chips (Al 6061&Al 7075). In *The Proceedings of Design & Systems Conference 24th Design Engineering Systems Division JSME Conference Japan Society of Mechanical Engineers* [No.14–27], 17–19th September 2014 at: Tokushima, Japan. The Japan Society of Mechanical Engineers.

Abdullahi, I., & Abidoye, J. K. (2020). Development and characterisation of recycled aluminum silicon carbide composites. *Nigerian Journal of Technology, 39*(2), 433–441.

Ahmad, Azlan, Lajis, Mohd Amri, Yusuf, Nur Kamilah, & Ab Rahim, Syaiful Nizam (2020). Statistical optimization by the response surface methodology of direct recycled aluminum-alumina metal matrix composite (MMC-AlR) employing the metal forming process. *Processes, 8*(7), 805.

Akhtar, Shagil, Saad, Mohammad, Misbah, Mohd Rasikh, & Sati, Manish Chandra (2018). Recent advancements in powder metallurgy: A review. *Materials Today: Proceedings 5*(9), 18649–18655.

Aslan, A., Salur, E., Gunes, A., Sahin, O. S., Karadag, H. B., & Akdemir, A. (2019). The mechanical properties of composite materials recycled from waste metallic chips under different pressures. *International Journal of Environmental Science and Technology 16*(9), 5259–5266. 10.1007/s13762-019-02317-3.

Aslan, Abdullah, Güneş, Aydın, Salur, Emin, Şahin, Ömer Sinan, Karadağ, Hakan Burak, & Akdemir, Ahmet (2018). Mechanical properties and microstructure of composites produced by recycling metal chips. *International Journal of Minerals, Metallurgy, and Materials, 25*(9), 1070–1079.

Aslan, Abdullah, Salur, Emin, Güneş, Aydın, Şahin, Ömer Sinan, Karadağ, Hakan Burak, & Akdemir, Ahmet (2018). Production and mechanical characterization of prismatic shape machine element by recycling of bronze and cast-iron chips. *Journal of the Faculty of Engineering and Architecture of Gazi, 33*(3), 1013–1027.

Badarulzaman, Nur Azam, Karim, Siti Rodiah, & Lajis, Mohd Amri (2014). Fabrication of Al-Sn composites from direct recycling aluminium alloy 6061. *Applied Mechanics and Materials, 465*, 1003–1007.

Baffari, Dario, Reynolds, Anthony P, Masnata, Attilio, Fratini, Livan, & Ingarao, Giuseppe (2019). Friction stir extrusion to recycle aluminum alloys scraps: Energy efficiency characterization. *Journal of Manufacturing Processes, 43*, 63–69.

Bhadra, J., Al-Thani, N., & Abdulkareem, A. (2017). Recycling of Polymer-Polymer Composites. *Micro and Nano Fibrillar Composites (MFCs and NFCs) from Polymer Blends*. Elsevier. 10.1016/B978-0-08-101991-7.00011-X.

Bhouri, Mariem, & Mzali, Foued (2020a). Analysis of thermo-elastic and physical properties of recycled 2017 aluminium alloy/Gp composites: Thermal management application. *Materials Research Express, 7*(2), 26546.

Bhouri, Mariem, & Mzali, Foued (2020b). Analysis of thermo-elastic and physical properties of recycled 2017 aluminium alloy/Gp composites: Thermal management application. *Materials Research Express, 7*(2). 10.1088/2053-1591/ab5eeb.

Bulei, C., Todor, M. P., & Kiss, I. (2018). Metal matrix composites processing techniques using recycled aluminium alloy. *IOP Conference Series: Materials Science and Engineering, 393*(1). 10.1088/1757-899X/393/1/012089.

Christy, John Victor, Arunachalam, Ramanathan, Mourad, Abdel-Hamid I., Krishnan, Pradeep Kumar, Piya, Sujan, & Al-Maharbi, Majid (2020). Processing, properties, and microstructure of recycled aluminum alloy composites produced through an optimized stir and squeeze casting processes. *Journal of Manufacturing Processes, 59*, 287–301.

Delfino, Davide (2014). Bronze recycling during the bronze age: Some consideration about two metallurgical regions. *Antrope, 1*, 120–143.

Duflou, Joost R, Tekkaya, A Erman, Haase, Matthias, Welo, Torgeir, Vanmeensel, Kim, Kellens, Karel, Dewulf, Wim, & Paraskevas, Dimos (2015). Environmental assessment of solid state recycling routes for aluminium alloys: Can solid state processes significantly reduce the environmental impact of aluminium recycling? *CIRP Annals, 64*(1), 37–40.

Eichler, Jens, & Lesniak, Christoph (2008). Boron nitride (BN) and BN composites for high-temperature applications. *Journal of the European Ceramic Society, 28*(5), 1105–1109.

Enginsoy, H. M., Bayraktar, Emin, Katundi, D., Gatamorta, F., & Miskioglu, I. (2020). Comprehensive analysis and manufacture of recycled aluminum based hybrid metal matrix composites through the combined method; sintering and sintering + forging. *Composites Part B: Engineering, 108040.*

Fuziana, Y. F., Warikh, A. R. M., Lajis, M. A., Azam, M. A., & Muhammad, N. S. (2014). Recycling aluminium (Al 6061) chip through powder metallurgy route. *Materials Research Innovations, 18*(sup6), S6–354.

Gatamorta, F., Miskioglu, Ibrahim, Bayraktar, E., & Melo, M. L. N. M. (2020). Recycling of aluminium-431 by high energy milling reinforced with TiC-Mo-Cu for new composites in connection applications. In *Mechanics of Composite and Multi-Functional Materials* (Vol. 5, pp. 41–46). Springer.

Gronostajski, J., Marciniak, H., & Matuszak, A. (2000). New methods of aluminium and aluminium-alloy chips recycling. *Journal of Materials Processing Technology, 106*(1–3), 34–39.

Gronostajski, J., & Matuszak, A. (1999). The recycling of metals by plastic deformation: An example of recycling of aluminium and its alloys chips. *Journal of Materials Processing Technology, 92*, 35–41.

Guluzade, Reshad, Avcı, Ahmet, Demirci, M Turan, & Erkendirci, Ö Faruk (2013). Fracture toughness of recycled AISI 1040 steel chip reinforced AlMg1SiCu aluminum chip composites. *Materials & Design (1980–2015), 52*, 345–352.

Harst, Eugenie van der, Potting, José, & Kroeze, Carolien (2016). Comparison of different methods to include recycling in LCAs of aluminium cans and disposable polystyrene cups. *Waste Management, 48*, 565–583.

Hosseini, Majid, & Paydar, Mohammad Hossein (2020). Fabrication of phosphor bronze/Al two-phase material by recycling phosphor bronze chips using hot extrusion process and investigation of their microstructural and mechanical properties. *International Journal of Minerals, Metallurgy and Materials, 27*(6), 809–817.

Joharudin, Nurul Farahin Mohd, Latif, Noradila Abdul, Mustapa, Mohammad Sukri, Azam, Nur, & Badarulzaman, Muhammad Faisal Mahmod. n.d. Effect of burning temperature on rice husk silica as reinforcement of recycled aluminium chip AA7075. *Materialwissenschaft und Werkstofftechnik, 50*(3), 283–288.

Joharudin, Nurul Farahin Mohd, Noradila Abdul Latif, Mohammad Sukri Mustapa, & Badarulzaman, Nur Azam (2020). Effects of untreated and treated rice husk ash

(RHA) on physical properties of recycled aluminium chip AA7075. *International Journal of Integrated Engineering, 12*(1), 132–137.

Kamble, Ajinkya, & Kulkarni, S. G. (2019). Microstructural examination of bagasse ash reinforced waste aluminium alloy matrix composite. In *AIP Conference Proceedings, 2105*, 20011. AIP Publishing LLC.

Kandpal, Bhaskar Chandra, Kumar, Jatinder, & Singh, Hari (2018). Manufacturing and technological challenges in stir casting of metal matrix composites – a review. *Materials Today: Proceedings, 5*(1), 5–10.

Karadağ, H. B. (2012). *Production and Mechanical Properties of Steel/Bronze Chips Composite*. PhD Thesis, Selçuk University, Department of Mechanical Engineering, Konya.

Karadag, Hakan Burak, & Akdemir, Ahmet (2019). Production and mechanical properties of bronze/steel chips composite materials produced by direct recycling. *Hittite Journal of Science & Engineering, 6*(2), 83–90.

Karuppasamy, R., Barik, Debabrata, Sivaram, N. M., & Dennison, Milon Selvam (2020). Investigation on the effect of aluminium foam made of A413 aluminium alloy through stir casting and infiltration techniques. *International Journal of Materials Engineering Innovation, 11*(1), 34–50.

Khamis, S. S., Lajis, M. A., & Albert, R. A. O. (2015). A sustainable direct recycling of aluminum chip (AA6061) in hot press forging employing response surface methodology. *Procedia CIRP, 26*, 477–481.

Kim, Kiho, Lee, Gaehang, Yoo, Youngjae, & Kim, Jooheon (2017). Recycling of metal sludge wastes for thermal conductive filler via sintering and surface modification. *Journal of Alloys and Compounds, 694*, 1011–1018.

Krishnan, Pradeep Kumar, Christy, John Victor, Arunachalam, Ramanathan, Mourad, Abdel-Hamid I, Muraliraja, Rajaraman, Al-Maharbi, Majid, Murali, Venkatraman, & Chandra, Majumder Manik (2019). Production of aluminum alloy-based metal matrix composites using scrap aluminum alloy and waste materials: Influence on microstructure and mechanical properties. *Journal of Alloys and Compounds, 784*, 1047–1061.

Kumar, Bharat, & Menghani, Jyoti V (2016). Aluminium-based metal matrix composites by stir casting: A literature review. *International Journal of Materials Engineering Innovation, 7*(1), 1–14.

Laila, Assayidatul, Nanko, Makoto, & Takeda, Masatoshi (2014). Upgrade recycling of cast iron scrap chips towards β-FeSi$_2$ thermoelectric materials. *Materials, 7*(9), 6304–6316.

Lajis, M. A., Ahmad, A., Yusuf, N. K., Azami, A. H., & Wagiman, A. (2017). Mechanical properties of recycled aluminium chip reinforced with alumina (Al2O3) particle: Mechanische Eigenschaften von Mit Aluminiumoxid (Al2O3) Verstärkten Recycelten Aluminiumspänen. *Materialwissenschaft Und Werkstofftechnik, 48*(3–4), 306–310.

Li, Ping, Li, Xinran, & Li, Fengjun (2020). A novel recycling and reuse method of iron scraps from machining process. *Journal of Cleaner Production*. doi:10.1016/j.jclepro.2020.121732.

Mahapatra, Rajendra Prasad, Srikant, Satya Sai, Rao, Raghupatruni Bhima, & Mohanty, Bijayananda (2019). Recovery of basic valuable metals and alloys from E-waste using microwave heating followed by leaching and cementation process. *Sādhanā, 44*(10), 209.

Mahdi, Ahmed Sahib, Mustapa, Mohammad Sukri, Lajis, Mohd Amri, & Abd Rashid, Mohd Warikh (2015). Effect of holding time on mechanical properties of recycling aluminium alloy AA6061 through ball mill process. *International Journal of Mechanical Engineering and Technology (IJMET), 6*(9), 133–142.

Mahdi, Ahmed Sahib, Mohammad Sukri Mustapa, Mahmod Abd Hakim Mohamad, Abdul Latif M. Tobi, Muhammad Irfan Ab Kadir, & Samsi, Mohd Arif (2017). Dry sliding wear behavior of the reinforced by graphite particle and heat treated of recycled aluminum AA6061 based MMC fabricated by powder metallurgy method. *Key Engineering Materials, 740*, 9–16. Trans Tech Publ.

Meignanamoorthy, M., & Ravichandran, M. (2020). Synthesis of metal matrix composites via powder metallurgy route: A review. *Mechanics and Mechanical Engineering 22*(1), 65–76.

Mishra, Radha Raman, & Sharma, Apurbba Kumar (2016). A review of research trends in microwave processing of metal-based materials and opportunities in microwave metal casting. *Critical Reviews in Solid State and Materials Sciences, 41*(3), 217–255.

Paraskevas, Dimos, Dadbakhsh, Sasan, Vleugels, Jef, Vanmeensel, Kim, Dewulf, Wim, & Duflou, Joost R (2016). Solid state recycling of pure mg and az31 mg machining chips via spark plasma sintering. *Materials & Design, 109*, 520–529.

Paraskevas, Dimos, Kellens, Karel, Dewulf, Wim, & Duflou, Joost R (2015). Environmental modelling of aluminium recycling: A life cycle assessment tool for sustainable metal management. *Journal of Cleaner Production, 105*, 357–370.

Paraskevas, Dimos, Vanmeensel, Kim, Vleugels, Jef, Dewulf, Wim, Deng, Yelin, & Duflou, Joost R (2014). Spark plasma sintering as a solid-state recycling technique: The case of aluminum alloy scrap consolidation. *Materials, 7*(8), 5664–5687.

Paraskevas, Dimos, Vanmeensel, Kim, Vleugels, Jef, Dewulf, Wim, & Duflou, Joost (2015a). The use of spark plasma sintering to fabricate a two-phase material from blended aluminium alloy scrap and gas atomized powder. *Procedia Cirp, 26*, 455–460.

Paraskevas, Dimos, Vanmeensel, Kim, Vleugels, Jef, Dewulf, Wim, & Duflou, Joost R (2015b). Solid state recycling of aluminium sheet scrap by means of spark plasma sintering. *Key Engineering Materials, 639*, 493–498. Trans Tech Publ.

Ramanathan, Arunachalam, Krishnan, Pradeep Kumar, & Muraliraja, Rajaraman (2019). A review on the production of metal matrix composites through stir casting–furnace design, properties, challenges, and research opportunities. *Journal of Manufacturing Processes, 42*, 213–245.

Rojas-Díaz, L. M., Verano-Jiménez, L. E., Muñoz-García, E., Esguerra-Arce, J., & Esguerra-Arce, A. (2020). Production and characterization of aluminum powder derived from mechanical saw chips and its processing through powder metallurgy. *Powder Technology, 360*, 301–311.

Singh, Narinder, Singh, Rupinder, Ahuja, I. P. S., Farina, Ilenia, & Fraternali, Fernando (2019). Metal matrix composite from recycled materials by using additive manufacturing assisted investment casting. *Composite Structures, 207*, 129–135. 10.1016/j.compstruct.2018.09.072.

Thalmaier, György, Sechel, Niculina Argentina, & Vida-Simiti, Ioan (2019). Heat transfer enhancement of paraffin phase change composite material using recycled aluminum sawing chips. *JOM, 71*(3), 1049–1055.

Wakeel, Saif, Saleem, Mustafa, & Nemat, Ammar (2017). A review on the mechanical properties of aluminium based metal matrix composite via powder metallurgy. *International Journal of Mechanical and Production Engineering, 5*(4), 2092–2320.

6 Morphology of Recycled Metal Composites

V. Anandakrishnan and S. Sathish
Department of Production Engineering, National Institute of Technology, Tamil Nadu, India

CONTENTS

6.1 Introduction 109
6.2 Microstructure Overview 110
6.3 Recycled Metal Composites with Aluminium Matrix 112
6.4 Recycled Metal Composites with Magnesium Matrix 121
6.5 Recycled Metal Composites with Other Matrix Materials 122
6.6 Summary 123
References 124

6.1 INTRODUCTION

Recycling of materials has a significant role in energy conservation, which is the most crucial factor that has a broader impact on environmental issues. When a new product or a component is manufactured, it needs an abundant energy source, which may be in the form of material, power and human resource. Notably, the developments in the automotive, aerospace and defense sectors resulted in the utilization of composite materials in a higher amount. Particularly, the utilization of metal matrix composites is profoundly expanded. The volume of metal matrix composite production is found to have increased nearly 8,000 tons in 2019, whereas it had been nearly 5,500 tons in 2012 (Ajay Kumar et al., 2020). This shows the demand and significant contribution of metal matrix composites in global applications. There is always left-out scrap during the production of any part/product, and it is unavoidable. This, along with the drastic increase in the utilization of metal matrix composites, has been a direct result in the dumping of composite scraps. Also, after the specified service period, the developed composites are also considered waste or scrap. Thus, recycling a metal matrix composite has become a mandate in the present scenario, which will undoubtedly provide a solution for the above issues. The phrase "recycling of metal matrix composite" seems to be simple, but in reality, it is more complicated as it is made with physically distinct materials. It is quite challenging to separate physically distinct and dispersed reinforcement materials from the matrix material, and it also needs specialized processes, whereas recycling the metal matrix components or scrap through remelting or any other technique is quite possible and very similar to the production processes that are used to synthesize newer composites. Although metal matrix composites are produced with reinforcements in various forms, such as particulate, long fiber and short fiber or whisker, the recycling is almost possible only for particulate reinforced composites. Hence, the present chapter mostly focuses on the recycling of particulate reinforced metal composites. In the advent of recycling, there arise questions on how far the morphology and strength of recycled materials will compete with the virgin composite materials. Mostly, the production of recycled composite evolves, using scrap matrix material with the addition of available reinforcements, using scrap matrix and recycled reinforcements and by recycling the virgin composite materials alone. Due to the variants in the production technique of recycled composites, there arises a necessity to evaluate the performance of the material in terms of metallurgy, mechanical, chemical and tribological behaviors. The successive topics primarily discuss the exploration of the metallurgy of recycled metal matrix composites. Also, it gives an insight into the metallurgy of the recycled composites in comparison with the effect of processing routes and different reinforcements.

6.2 MICROSTRUCTURE OVERVIEW

It is well known that the strength of pure metals is mostly lower when compared with alloys or composites. The enhancement in the strength is mainly attained by the induced variations in the microstructure due to the addition of metals, nonmetals or ceramics. Figure 6.1 shows the schematic representation of the possible outcome of material A (metal) with the addition of another material B (metal/non-metal). The typical illustration (Figure 6.1) displays with the casting technique and similar outcomes may also be attained with other techniques like powder metallurgy, friction stirring, diffusion bonding and so on. When material B is added in material A (may be called as matrix/ base material), upon its solubility, it results in the formation of either an alloy or composite. An alloy is a material that resulted from the incorporation of an added material into the matrix material to a specified extent either by substitution or insertion, referred to as substitutional and interstitial alloys, respectively. If the added material has a chemical reaction with the matrix material, then it results in the intermetallic formation, whereas if the added material does not undergo any change and retains it as a distinct phase, then it exhibits the formation of composite material. In composite materials, the distribution of added material (called reinforcements) has a dominant effect on the material's strength. The added reinforcements may be dispersed in any of the below-mentioned patterns in the matrix material. Figure 6.2a–d shows the possible ways of the distribution of reinforcements in the matrix material.

Morphology of Recycled Metal Composites

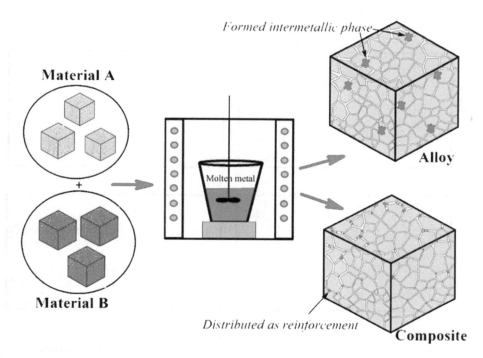

FIGURE 6.1 Typical schematic illustration showing the possible outcomes of a material added with another material (matrix/base).

a. Dispersion in the grain boundaries.
b. Dispersion in the matrix grains.
c. Dispersion in both grain boundaries and matrix grains.
d. Uniform dispersion in a consecutive way of matrix and reinforcements.

The distribution of reinforcements in a matrix material highly depends on the processing routes and its operating conditions. If the composite is processed through suitable processing techniques using optimized operating parameters, the uniform distribution of reinforcements in the matrix is attained (Figure 6.3a) or else it leads to the clustered distribution of reinforcements (Figure 6.3b). In some instances, the added reinforcements may result in the formation of a reaction-formed particle due to the induced chemical reaction of reinforcements with the matrix material. A typical schematic illustration of the formation of reaction particles in the matrix material is displayed in Figure 6.3c. There is more chance in the formation of reaction induced particles in recycled metal composites as it has an extended temperature and exposure time.

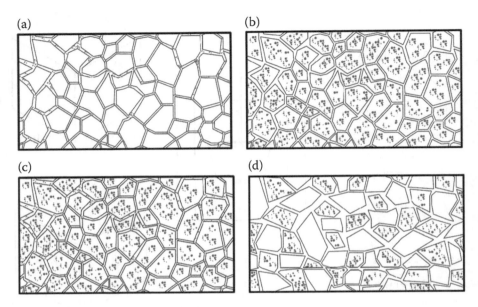

FIGURE 6.2 Typical illustration of the possible ways of the distribution of reinforcement particles in the matrix (base) material: (a) reinforcement at grain boundaries, (b) reinforcement inside matrix grain, (c) reinforcement in both matrix and grain boundary and (d) distribution of reinforcement in a consecutive sequence.

6.3 RECYCLED METAL COMPOSITES WITH ALUMINIUM MATRIX

The morphology of the recycled metal matrix composites made with aluminium as matrix material under different processing techniques is discussed further. The most common, economical and reliable processing technique is casting, and it has a significant contribution in recycled composites, too. Such utilization of the casting technique in the development of recycled metal composite with scrap 8011 aluminium alloy sheets through an in-situ reaction technique attempted by the authors in their earlier work gives an insight into the production of recycled composites with scrap alloys. The inclusion of halide salts at the time of re-melting the alloy sheets with appropriate holding time induced the endothermic reaction among the halide salts results in the in-situ formation of TiB_2 particles (Selvan et al., 2019). Figure 6.4a shows the in-situ formed TiB_2 particles, which is uniformly distributed over the 8011 matrices. The magnified view of the composite material (Figure 6.4b) displays the formation of hexagonal-shaped TiB_2 particles with a clear interface around the in-situ formed particles. Also, the broken in-situ formed TiB_2 particles (*pointed with arrow*) is observed in the magnified view that shows the disintegration of reacted particles during the

Morphology of Recycled Metal Composites

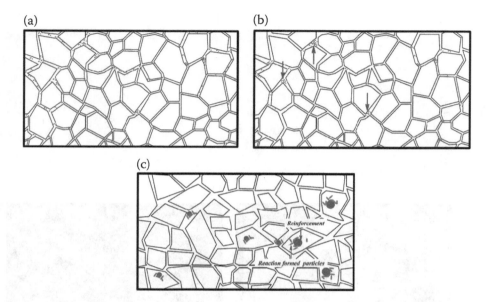

FIGURE 6.3 Typical illustrations of microstructure of composite material: (a) uniform distribution of reinforcement, (b) clustered distribution of reinforcement and (c) reaction formed particles around the reinforcement.

stirring process. Figure 6.4c–e shows the elemental analysis carried out on the recycled composite that displays the elements of aluminium, titanium and boron, respectively. Figure 6.4f shows the merged view of the elemental map of recycled composite. Similarly, the recycled composites reinforced with in-situ ZrB_2 particles were also synthesized using the scrap 8011 aluminium alloy sheets (Muthamizh Selvan & Anandakrishnan, 2019; Selvan, Anandakrishnan et al., 2018; Selvan, Munivenkatappan et al., 2018). The microstructural observation shows the in-situ formed ZrB_2 with good interfacial bonding. In another work, the aluminium matrix composite with scrap aluminium 6061, composites reinforced with 20 vol.% alumina was successfully recycled through casting four times. At the same time, the primary composite was processed through direct chill casting. The recycled composite showed the formation of spinel $MgAl_2O_4$ on the surface of alumina particles due to the reaction of magnesium with alumina. The volume percentage of the reinforced alumina is found to be slightly varied under recycling (i.e., the volume percentage of alumina particles; Primary – 20%, Recycled 1× – 20.2%, 2× – 20.5%, 3× – 19.4%, and 4× – 20.6%) and it is found to be within limits, which shows the feasibility of recycling of metal matrix composites.

The fluidity and microstructure of the scrap AA6063-10 vol.% B_4C composite obtained in cast and extruded conditions during its recycling is analyzed (Shi et al., 2012). The scrap aluminium composites were melted in an electric furnace

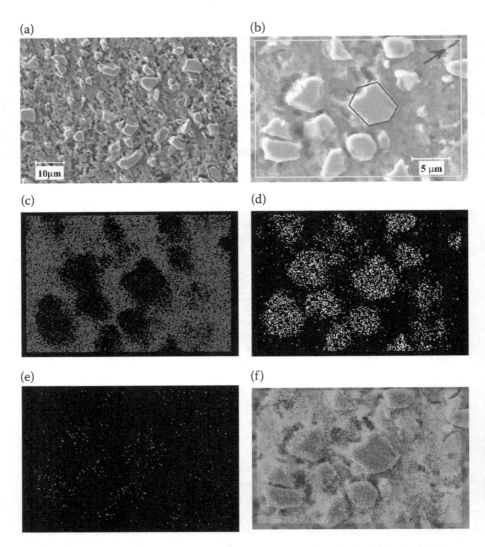

FIGURE 6.4 Microstructures of recycled 8011 aluminium composite reinforced with in-situ TiB$_2$ particles: (a) SEM image of composite, (b) magnified view, (c, d, e) elements of aluminium, titanium and boron from EDS map analysis and (f) merged view of all elements in SEM images.

for 8.5 hours to analyze its variations in microstructure and fluidity. The microstructure of the received scrap composites exhibits the presence of boron carbide along with in-situ formed Al$_3$BC, TiB$_2$ and AlB$_2$ (rarely observed) around the boron carbide particles. In the as received scrap composites, the boron carbide particle clusters were observed in cast conditions, whereas in extruded conditions, the clusters were found to be less dense compared to cast condition, as the clusters were broken down during material deformation. The most common defect of oxide film formation in aluminium is observed in both the cast and extruded conditions. In the cast condition, the reinforcement and in-situ formed phases were enclosed within the film to form a network, whereas in the extruded condition, the network was broken down to tiny discontinuous slices due to the material deformation. The remelting and holding of the scrap aluminium composite (i.e., in recycled composite) exhibited an increased particle size and increased quantity of in-situ formed phases Al_3BC and TiB_2 along with distinct formation of AlB_2 and oxide film in both received cast and extruded composite. This shows that the remelting and increased holding time significantly influenced the in-situ reaction of boron carbide with the matrix material. Even after remelting and holding, the in-situ formed phase is observed around boron carbide particles within the oxide film network in the recycled composites of cast condition. Similar observations of the in-situ formed phase are perceived in the recycled composites of extruded condition. More uniformly distributed in-situ formed phases away from boron carbide are observed as the extruded scrap composite shows the relatively dispersed reinforcement.

Similar experimentation of the microstructure and fluidity analysis is done on the recycled AA1100 aluminium composite with 16 vol.% B_4C for 2.5 hours (Shi et al., 2012). The microstructural analysis on the as-received cast and rolled scrap composite exhibits the in-situ formed Al_3BC, TiB_2 and less amount of AlB_2. The presence of needle-shaped TiB_2 is observed due to the higher percentage of titanium in the matrix material (i.e., 1.5 wt.%). The in-situ formed phases were found to be dispersed in the rolled sheet, whereas it is accumulated closer to boron carbide particles in cast scrap composite. Dense clusters of formed phases were observed in the cast scrap condition, whereas it is dispersed in the rolled composite due to severe deformation. After recycling, the increased amount and increased particle size of in-situ formed phases of Al_3BC and TiB_2 is observed along with the distinct formation of AlB_2.

Figure 6.5 shows the schematic illustration of a typical microstructure of AA1100-16 vol.% B_4C composite. The formation of a TiB_2 protective layer around the boron carbide particles is observed in a cast composite, as shown in Figure 6.5a. In the received scrap composite under rolled condition, this protective layer is being partially damaged and peeled off from the boron carbide particle (Figure 6.5b), and thereby promotes the reaction of boron carbide with aluminium. Hence, the induced in-situ formed Al_3BC and AlB_2 particles are agglomerated around the boron carbide particle (as shown in Figure 6.5c) due to remelting with increased holding time, whereas in the remelting of cast composite, the formed protective layer of TiB_2 around boron carbide limits the further in-situ reactions of the boron carbide particle with aluminium and prevents the agglomeration.

The attempt on the development of recycled composite with scrap aluminium beverage cans was successful. The aluminium matrix composite was developed with the addition of pumice particulates through stir casting (Dagwa & Adama,

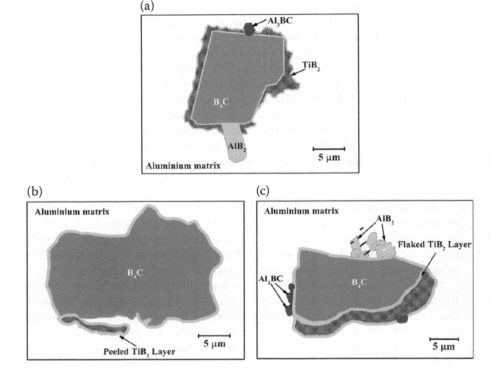

FIGURE 6.5 Typical illustration to explain the micro mechanism of protective layer around the boron carbide reinforcement (a) as received scrap composite under cast condition, (b) as received scrap composite under rolled condition and (c) recycled composite from the rolled condition.

2018). The 35 μm particle size of pumice was used as the reinforcement with the weight percentages of 2, 4, 6 and 8. The morphology of pumice particles is found to be a porous lamella amorphous structure. The microstructural analysis of the composite shows uniformly distributed pumice particles with good interfacial bonding. Likewise, the recycled hybrid composite is made with the addition of magnesium oxide and baggage ash through stir casting (Shashi Prakash & Garima, 2020). In the process of recycling the waste bagasse (sugarcane waste), it was dried, milled to a powder and burnt to attain the ash. The chemical composition of the bagasse ash majorly consists of SiO_2, Al_2O_3, Fe_2O_3, CaO, MgO, K_2O and Na_2O. The microstructural studies exhibited a uniform distribution of magnesium oxide and baggage ash in the aluminium matrix.

The recycled aluminium composite is made with the reinforcement of nano copper oxide and recycled spent alumina catalyst through the double stir casting technique (Gayathri & Elansezhian, 2020). Along with LM25 aluminium matrix,

Morphology of Recycled Metal Composites 117

5% of scrap aluminium is added to encourage waste reduction. The microstructural analysis on the recycled composite exhibits a perfect mixture of spherical ball-shaped copper oxide particles and unstructured white grains of spent alumina catalyst in the matrix along with its uniform dispersion. There is no observation of defects like pores, cracks, voids and agglomerations, which is evident of the excellent bonding of reinforcement with matrix. Generally, the addition of spent alumina catalyst induces pores, but it was eliminated in the recycled composite by the insertion of nano copper oxides.

Similarly, the recycled aluminium composite was prepared by using aluminium 1050 alloy strips with the addition of 38 μm silicon carbide particles through disintegrated melt deposition (Ling & Gupta, 2000). Then the disintegrated composite was hot extruded and subjected to metallurgical analysis. Subsequently, the composite was recycled once, and twice in a similar way to study the variations in the microstructure. There is no significant observation of macropores and macro segregations from the macrostructural analysis of composites. The microstructural analysis on the new and recycled composite exposed the uniform distribution and good interface integrity of silicon carbide with the aluminium matrix, as shown in Figure 6.6. Also, the irregular-shaped micro pores are observed closer to the silicon carbide particles due to its poor wettability, whereas the porosity of a recycled composite is observed to be less compared to a new composite due to the improved wettability with extended temperature exposure and breakage of the cluster during extrusion. The recycled composite exhibits a reduction in the silicon carbide particle size as it is subjected to axial stress and shear stress during extrusion that led to a silicon carbide particle fracture. Except for the reduction in silicon carbide particle size, there is no significant variations in the morphology of recycled composites compared to

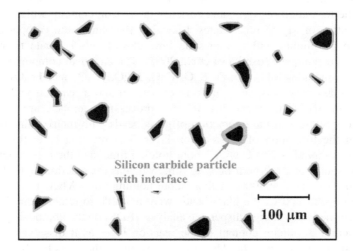

FIGURE 6.6 Typical illustration to explain the interface formation around silicon carbide reinforcement in the recycled aluminium composite.

virgin composites. There is a slight reduction in the quantity of silicon carbide in the recycled composite; however, it is observed within the deviation limits, i.e., 12.9 wt.% for a new composite and 12.2 wt.% for a twice-recycled composite.

Another attempt in the development of recycled composite through displacement reaction method is made successful. The alumina reinforced aluminium composite is made using the scrap aluminium alloy with the insertion of silicon dioxide rods (Yoshikawa et al., 2003). The insertion of silicon dioxide rod in the melt induces a reaction with aluminium that results in the formation of alumina. The microstructural analysis on the composite exhibits two different regions of bright and dark, which shows the aluminium and alumina phase, respectively. The homogenous distribution of alumina particles is observed in both edge and central areas of the composite. Further, there is no significant presence of MgSiO, MgAlSiO and MgO in the recycled composite, which shows the incorporation of magnesium with ceramic phase. The elemental analysis on the ceramic phase exhibits a higher magnesium content, which confirms the above incorporation.

An additional promising technique used to synthesize the recycled composite is the infiltration technique. The process of infiltration casting is effectively used to produce the recycled composite using beverage cans in addition to silicon and magnesium to accomplish the matrix Al–18Si–15Mg (Escalera-Lozano et al., 2007). The hybrid composite was produced with 75 μm silicon carbide, and 90 μm fly ash in the volume ratio of 30:20 to accomplish the Al–SiC–$MgAl_2So_4$ composite. The morphological analysis of the hybrid composite exposed the in-situ formed Mg_2Si and spinel phase $MgAl_2So_4$ during the infiltration process. The morphology of composite with the spinel phase of round in shape (shape of fly ash) as it results from the reaction of silicon dioxide and mullite (available in fly ash) with matrix aluminium and added magnesium. The above reaction avoids the formation of Al_4C_3 and enriches the silicon content, which results in the formation of Mg_2Si precipitates. Likewise, the aluminium composite using the scrap aluminium with 1 mm soda lime glass beads is made through the infiltration technique (Yoshikawa et al., 2005). The chemical composition of the glass bead includes SiO_2, Na_2O, K_2O, CaO, MgO, Al_2O_3 and Fe_2O_3. The reinforced glass beads display a uniform dispersion and a reaction with the aluminium matrix. The variations in the processing temperature and time profoundly influenced the interaction of glass beads with matrix aluminium resulted in the formation of the reaction layer. The reaction layer thickness of 50 μm is observed at 700°C with 1-hour holding time, and the layer thickness is found to double at the 2-hour holding time, whereas the reaction layer thickness of 70 μm is observed at 800°C with a 1-hour holding time. When the temperature goes up beyond 800°C, the glass beads were not able to retain their shape and were deformed. The energy dispersive analysis (EDS) on the composite explored the reduction of silicon content in the reaction layer as it dissolves with aluminium to form a precipitate. Also, the Fe content in the glass beads resulted in the formation of precipitates.

Similarly, the recycled aluminium composite is produced through infiltration technique using the turned chips of AA2024-45 vol.% SiC composite and with the addition of silicon carbide particles in the volume of 67% and 30% (Tan et al., 2020). The turned composite chips were milled to a maximum of 30 hours that resulted in chip powders (referred to as composite particles) of 900, 650 and 300 μm for 6, 12 and 18 hours, respectively. After 18 hours, the milled powders exhibited the equiaxed morphology along with the alumina formation on its surface. The composites made with three different-sized chips exhibited a uniform distribution of silicon carbide particles without any agglomeration and defects. Even though the silicon carbide has a poor wettability, good interfacial bonding is observed in the composites. The possible reason for this bonding is the formation of alumina and large dimensions of composite particles. Also, the formation of white precipitates is observed in the composite particles and matrix. The transmission electron microscopic analysis on the recycled composite with 300 μm composite particles displays the uniform distribution of silicon carbide along with a clear interface. The observed diffraction pattern in the aluminium and composite particles resembles the aluminium diffraction pattern, whereas the diffraction pattern in the interface is observed to be an amorphous diffraction pattern. The EDS analysis on the interface shows the reduction in atomic percentage for aluminium and an increase in atomic percentage for oxygen. The observation in the variation of diffraction patterns and the elemental analysis confirms the amorphous alumina.

The recycled aluminium composite is successfully made with the chips of aluminium and chips of tin through cold forging followed by sintering (Badarulzaman et al., 2014). The composites with different volume percentages of tin, namely 6%, 20% and 40% were made in the process sequence of milling, compaction and sintering. The microstructure of the composite shows the clear interface boundary between aluminium and tin. The existence of voids is observed to be less in the composite made of 40 vol.% tin compared to 6 vol.% tin composite. Similarly, aluminium copper matrix recycled composite is made with the recycled aluminium 7075 alloys through sintering and sinter plus forging process (Enginsoy et al., 2020). With an equal weight ratio, the aluminium alloy and pure copper were taken as a matrix with the considered reinforcement of NB_2Al (10 wt.%), SiC (10 wt.%) and ZnO (15 wt.% and 30 wt.%). Besides, the nature of wettability is improved with nano aluminium and zinc of 5 wt.% and 2 wt.%, respectively. The microstructural analysis of the recycled composite showed the presence of added reinforcements, NB_2Al, SiC and ZnO, along with the formation of $CuAl_2$. The addition of nano aluminium enhanced the formation of eutectic structure in reaction with copper. The homogenous distribution of particles, along with good interfacial bonding, is observed. The enforcement of sic particles into the matrix grains was observed in the forged composites due to the action of forging. The increased density of composites after the forging process clearly expressed the virtual elimination of porosity.

The aluminium composite is reinforced with AISI 1040 steel chips in the matrix AlMglSiCu aluminium chips through cold pressing and sintering

(Guluzade et al., 2013). The composites were made in three different weight percentages of steel chips, namely 20%, 30% and 40%. The microstructural analysis of all three composites exhibited the layers of steel and aluminium with a definite chip border interface. Also, few pores and voids in the chip border and the oxidation is observed in the aluminium chips. The EDS on the aluminium composite shows the reduction in the percentage of aluminium with an increased weight percentage of steel chips, which shows the diffusion of aluminium atoms into the iron crystal. Similarly, the aluminium composite reinforced with tungsten particles is produced through compaction and hot extrusion using pure aluminium chips and AlMg$_2$ alloy chips (Gronostajski et al., 1998). The aluminium chips were milled along with tungsten powder, then compacted and hot extruded. The microstructure of the composite (Al–0.7 wt.% W and AlMg$_2$–1.2 wt.% W) exhibits few micro pores with the distribution of tungsten particles. The process of compaction induced an excellent mechanical bonding of tungsten with aluminium chips and further extrusion results in the denser composite. A similar finding was observed with the aluminium AlCu$_4$ composite reinforced with tungsten particle through similar processing technique. Likewise, the alumina reinforced aluminium composite was made with aluminium 6061 alloy chips through the optimized cold pressing followed by hot extrusion. The microstructural analysis on the composite exhibits a uniform distribution of alumina particles with microstructure refinement.

The effect of process parameters on the recycling of scrap aluminium in synthesizing the alumina reinforced aluminium composite was analyzed through the hot-pressing technique (Ahmad et al., 2017). The alumina particles of 2 wt.% is added with the aluminium chips got from the process milling as received 6061 aluminium alloy. The hot pressing of the mixture was performed at different temperatures (i.e., 430, 480 and 530°C) and time (i.e., 60, 90 and 120 minutes) at a load of 350 kN. The microstructural analysis of the developed composites showed that the increase in holding time and temperature reduces the average grain diameter of the composite. The maximum average grain diameter of 60.07 μm is observed for the composite prepared for 60 minutes at 430°C. At the same time, the minimum average grain diameter of 28.33 μm is observed for the composite prepared for 120 minutes at 530°C.

Friction stir extrusion is a noticeable technique that is used to synthesize the recycled composite in solid-state condition. With the help of friction stir extrusion, the chips of aluminium 2024 alloy were transformed into a recycled composite with the addition of 2-micron silicon carbide particles (Baffari et al., 2017). The analysis on the chip morphology exhibits the average dimensions of 5 mm length, 2 mm width and 0.2 mm thick. The increase in the percentage of silicon carbide exhibited deep visual cracks on the composite surface due to the conglomeration of silicon carbide particles on the grain boundary as an incoherent cluster. The microstructural analysis on the friction extruded recycled alloy and composite, displayed a fine equiaxial grains due to the dynamic recrystallization and the uniform dispersion of silicon carbide particles at grain boundaries, respectively.

6.4 RECYCLED METAL COMPOSITES WITH MAGNESIUM MATRIX

The morphology of the recycled metal matrix composites made with magnesium as a matrix material under different processing techniques is discussed here. Only a few attempts were made on the development of magnesium-based recycled composites. One such magnesium composite is made through friction stir processing using AZ91 alloy as a matrix with the recycled chopped carbon fibers (size of 1 mm) as reinforcement (Afrinaldi et al., 2018). The microstructure of the cast magnesium alloy exhibits the $Mg_{17}Al_{12}$ eutectic β phase, and the friction stirred alloy shows the refined grains with the breakup of eutectic β phases along with onion rings. The composite was friction stirred with two different pin tools, one of threaded pin tool and other is of the 3-flat pin tool. The microstructure of the recycled carbon fiber magnesium composite stirred with the threaded pin tool is analyzed in transverse and longitudinal cross sections. The transverse cross section exhibits the ring-like structure with refined grains with a size of 12 μm and uniform distribution of carbon fibers. The microstructure displays very fine carbon fibers and short fibers of less than 20 μm which shows that the fibers are broken down by the plastic deformation during stirring. The longitudinal cross-section of the composite processed with threaded tool exhibits the defects of large holes in the upper region and the uniform distribution of carbon fibers in the matrix. Similar findings were observed in the microstructure of composite processed with 3-flat pin tool in both transverse and longitudinal cross sections. The defect density and hole size is smaller in the composite processed with a 3-flat pin tool compared to threaded pin composite.

The AZ31 magnesium composite with in-situ reinforced Mg_2Si and MgO is produced through solid-state processing (i.e., compaction followed by hot extrusion) through the addition of scrap silicon dioxide glass fiber (Kondoh & Luangvaranunt, 2003). The fine silicon dioxide glass powders (16.8 μm) are reinforced at different mass percentages, namely 0%, 2%, 4%, 6% and 8%. The microstructure of the composite reinforced with 2 mass % of silicon dioxide exhibits the relatively uniform distribution of in-situ formed dispersoids Mg_2Si and MgO particles along with few clusters at grain boundaries. The particle size of the in-situ formed dispersoids is found to be similar to the glass powder particle size without any growth or coarsening. At the same time, the refinement of grains is observed in the matrix material nearly 10–15 μm due to the dynamic recrystallization during the process of extrusion. Similarly, the composite was synthesized through the repeated plastic working (RPW) method. The microstructure of RPW composite reinforced with 2 mass % of silicon dioxide exhibits more uniformly distributed in-situ formed Mg_2Si and MgO particles without any pores and clusters. The particle size of the in-situ formed dispersoids is reduced to 3.4 μm from 16.8 μm due to the fracture of reinforcements during repeated deformation. Also, the grain size of the magnesium matrix is significantly reduced to around 3 μm from 8 μm due to repeated deformation.

The magnesium composite is produced using pure magnesium with the addition of 38 μm silicon carbide particles through disintegrated melt deposition (Soon & Gupta, 2002). Then the disintegrated composite is hot extruded and subjected to metallurgical analysis. Subsequently, the composite was recycled once or twice in a similar way to study the variations in the microstructure. The microstructural analysis on the new and recycled composite exposed the uniform distribution and good interface integrity of silicon carbide with the aluminium matrix. Also, irregular-shaped micro pores closer to the silicon carbide particles are observed due to its poor wettability, whereas the breakage of clusters and reduction in porosity is observed in the recycled composite compared to new composite due to the improved wettability with extended temperature exposure and extrusion. Also, a reduction in the quantity of silicon carbide is observed in the recycled composite, i.e., 12.7 wt.% for a new composite and 11 wt.% for a twice-recycled composite.

6.5 RECYCLED METAL COMPOSITES WITH OTHER MATRIX MATERIALS

Some of the recycled metal matrix composites produced with ferrous, titanium and nickel as matrix material are discussed in further. The scrap tungsten carbide reinforced ferrous composite was produced through the centrifugal casting technique (Song & Wang, 2012). The ferrous scrap composite with 75% volume fraction of tungsten carbide is made as a ring in the dimensions of 234 mm outer diameter, 110 mm inner diameter with 78 mm length. The recycled composite exhibits a two-layered structure; one is the outer layer of more than 15 mm width, reinforced with tungsten carbide particles, and the other is the inner end of the Fe–C matrix. The microstructure of the recycled composite displays the tungsten carbide particles with an average size of 35 μm and its uniform distribution in Fe–C matrix. Also, the crystallites of the bone-like structure are observed with the matrix and the results of EDS confirm tungsten carbide with the presence of tungsten, iron and carbon elements. The bone-like structure is the precipitate that results from the reaction of supersaturated tungsten with matrix alloy. The Fe–C matrix alloy in the inner layer of composite shows the bainite matrix with the presence of graphite and carbides. The average particle size of the tungsten carbide is found to be reduced when compared with the primary composite. Also, the volume fraction of tungsten carbide is slightly reduced from 75% to 70% in the recycled composite. The chemical composition of the recycled composite and primary composite is found to be almost same.

The titanium matrix composite reinforced with titanium carbide particles is produced through powder metallurgy with titanium chips (Jaya Teja et al., 2020). The composite is made at a different compaction pressure and sintered at 1200°C for 2 hours. The microstructural analysis of the composite shows significant variations in the microstructure. The increase in compaction pressure resulted in decreased pores, and the composite microstructure is found to be relatively fine at a higher compaction pressure. The reason is increased pressure; there is less chance of inter-particle gap, which results in the adequate bonding of particles while sintering.

The nickel-based composite reinforced with alumina particles is produced through cold isostatic pressing followed by sintering with recycled nickel powder processed from scrap ferrous (Karayannis & Moutsatsou, 2012). The composite is produced at a different weight percentage of alumina particles. The microstructure of 20 wt.% and 30 wt.% of alumina-reinforced composite displays the metal and ceramic constituents in lighter and darker phase. The atomic diffusion among the nickel particles resulted in the densification of the matrix and bonding with reinforcement. The bonding of alumina with nickel particles is observed to be less at a higher percentage addition of alumina. Also, the agglomeration of alumina particles and the interfacial porosity is observed.

The metal matrix composite with CuSn10 bronze chips as matrix material and GGG40 cast iron chips as reinforcement are made through hot isostatic pressing (Aslan et al., 2018). The composite is produced with four different weight percentages of GGG40 cast iron chips, namely 10, 20, 30 and 40 wt.% at different temperature and pressure. The composite microstructure exhibits an excellent penetration of bronze chips into the narrow zones of cast iron chips. The proper fusion of the matrix with reinforcement can be obtained at appropriate pressure and temperature. The inappropriate selection of parameters results in a pore formation at the chip border due to the restriction of plastic deformation. The secondary phase spherical graphite in the cast iron chips is found to be elongated in composites, and it may have occurred during the machining process. The cast iron chip retains its morphology without any deformation even after cold-pressing and sintering.

6.6 SUMMARY

Metal matrix composites based on lightweight materials acquire more application to the automotive and aerospace industries, which results in the dump of scrap lightweight materials and metal matrix composites. The need for recycling and reclamation of metal matrix composites are being resurrected. The present research on the development of recycled metal composites is found to be very meager. This is the right time to move or focus the research endeavors in the accomplishment of recycled metal composites. The actual research accomplishments on the recycled composites has proven the ability to recycle and reclaim the metal matrix composites. As is evident that the morphology of material relies on the processing routes, utmost care is required during the selection of appropriate processing technique during the synthesis of recycled composite. Some of the processing techniques that are identified as successful methods in the development of recycled composites are stir casting, infiltration, disintegrated melt deposition, powder metallurgy, hot-pressing and friction stir processing. Some of the critical observations in the morphological analysis of the recycled metal composites are listed below:

- The recycled composites didn't show any significant deviations in the chemical compositions compared to the primary/virgin composite material.
- Relatively more uniform distribution of reinforcement is observed in the recycled metal composites.
- Enhancement in the surface integrity between the matrix and reinforcement is observed in the recycled composites due to the improved wettability by extended exposure to temperature and increased holding time.
- The recycled metal composites exhibited a lower porosity level compared to the primary/virgin composite materials.
- The recycled composite also exhibits in-situ formed (reaction formed) particles in the matrix material owing to the extended exposure to temperature and increased holding time.
- In some instances, the in-situ formed particles induced the agglomerations.
- In some instances, the formation of a protective interface layer is also observed between the matrix reinforcement.
- The volume or weight percentage of the reinforcement are found to be slightly varied in the recycled metal composites. But the deviations are found to be within limits, which promotes the feasibility of recycling of metal matrix composites.
- The refinement of the matrix grain is observed, and in some instances the refinement of reinforcement is also observed, i.e., reduction in the particle size of reinforcement.
- In some instance, the formation of intermetallic phases are observed in the recycled composites, which is not being displayed in the primary/virgin composite material.

REFERENCES

Afrinaldi, Angga, Kakiuchi, Toshifumi, Nakagawa, Shohei, Moritomi, Hiroshi, Kumabe, Kazuhiro, Nakai, Asami, Ohtani, Akio, Mizutani, Yoshiki, & Uematsu, Yoshihiko (2018). Fabrication of recycled carbon fiber reinforced magnesium alloy composite by friction stir processing using 3-flat pin tool and its fatigue properties. *Materials Transactions, 59*(3), 475–481. doi:10.2320/matertrans.M2017297.

Ahmad, Azlan, Lajis, Mohd Amri & Yusuf, Nur Kamilah (2017). On the role of processing parameters in producing recycled aluminum AA6061 based metal matrix composite (MMC-AlR) prepared using hot press forging (HPF) process. *Materials, 10*(9). doi:10.3390/ma10091098.

Ajay Kumar, P., Rohatgi, Pradeep & Weiss, David (2020). 50 years of foundry-produced metal matrix composites and future opportunities. *International Journal of Metalcasting, 14*(2), 291–317. doi:10.1007/s40962-019-00375-4.

Aslan, Abdullah, Güneş, Aydın, Salur, Emin, Şahin, Ömer Sinan, Karadağ, Hakan Burak, & Akdemir, Ahmet (2018). Mechanical properties and microstructure of composites produced by recycling metal chips. *International Journal of Minerals, Metallurgy and Materials, 25*(9), 1070–1079. doi:10.1007/s12613-018-1658-8.

Badarulzaman, N. A., Karim, S. R., & Lajis, M. A. (2014). Fabrication of Al-Sn composites from direct recycling aluminium alloy 6061. *Applied Mechanics and Materials, 465-466*, 1003–1007. doi:10.4028/www.scientific.net/AMM.465-466.1003.

Baffari, Dario, Buffa, Gianluca, Campanella, Davide, & Fratini, Livan (2017). AI-SiC metal matrix composite production through friction stir extrusion ofaluminum chips. *Procedia Engineering, 207*, 419–424. doi: 10.1016/j.proeng.2017.10.798.

Dagwa, I. M., & Adama, K. K. (2018). Property evaluation of pumice particulate-reinforcement in recycled beverage cans for Al-MMCs manufacture. *Journal of King Saud University – Engineering Sciences, 30*(1), 61–67. doi: 10.1016/j.jksues.2015.12.006.

Enginsoy, H. M., Bayraktar, E., Katundi, D., Gatamort, F., & Miskioglud, I. (2020). Comprehensive analysis and manufacture of recycled aluminum based hybrid metal matrix composites through the combined method; sintering and sintering + forging. *Composites Part B, 194*, 108040. doi: 10.1016/j.compositesb.2020.108040.

Escalera-Lozano, R., Gutiérrez, C. A., Pech-Canul, M. A., & Pech-Canul, M. I. (2007). Corrosion characteristics of hybrid Al/SiCp/MgAl$_2$O$_4$ composites fabricated with fly ash and recycled aluminum. *Materials Characterization, 58*(10), 953–960. doi: 1 0.1016/j.matchar.2006.09.012.

Gayathri, J., & Elansezhian, R. (2020). Influence of dual reinforcement (nano CuO + reused spent alumina catalyst) on microstructure and mechanical properties of aluminium metal matrix composite. *Journal of Alloys and Compounds, 829*. doi: 1 0.1016/j.jallcom.2020.154538.

Gronostajski, J. Z., Kaczmar, J. W., Marciniak, H., & Matuszak, A. (1998). Production of composites from Al and AlMg$_2$ alloy chips. *Journal of Materials Processing Technology, 300*(3–4), 37–41. doi: 10.1016/s0924-0136(97)00390-7.

Guluzade, Reshad, Avci, Ahmet, Demirci, M. Turan, & Erkendirci, Ö Faruk (2013). Fracture toughness of recycled AISI 1040 steel chip reinforced AlMg1SiCu aluminum chip composites. *Materials and Design, 52*, 345–352. doi: 10.1016/j.matdes.2013.05.025.

Jaya Teja, Pilli, Ranjan Shial, Satya, Chaira, Debasis, & Masanta, Manoj (2020). Development and characterization of Ti–TiC composites by powder metallurgy route using recycled machined Ti chips. *Materials Today: Proceedings, 26*(2020), 3292–3296. doi: 10.1016/j.matpr.2020.02.467.

Karayannis, V. G., & Moutsatsou, A. K. (2012). Synthesis and characterization of nickel-alumina composites from recycled nickel powder. *Advances in Materials Science and Engineering*, 1–9. doi: 10.1155/2012/395612.

Kondoh, Katsuyoshi, & Luangvaranunt, Tachai (2003). New process to fabricate magnesium composites using SiO$_2$ glass scraps. *Materials Transactions, 44*(12), 2468–2474. doi: 10.2320/matertrans.44.2468.

Ling, P. S., & Gupta, M. (2000). Recycling of aluminium based metal matrix composite using disintegrated melt deposition technique. *Materials Science and Technology, 16*(5), 568–574. doi: 10.1179/026708300101508090.

Muthamizh Selvan, B. M., & Anandakrishnan, V. (2019). Investigations on corrosion behaviour of AA 8011-ZrB 2 in situ metal matrix composites. *Lecture Notes in Mechanical Engineering, 2*, 335–342. doi: 10.1007/978-981-13-6374-0_39.

Selvan, B.M. Muthamizh, Anandakrishnan, V., Muthukannan Duraiselvam, Venkatraman, R., & Sathish, S. (2018). Multi objective optimization of wear behaviour of in situ AA8011-ZrB2 metal matrix composites by using Taguchi-Grey analysis. *Materials Science Forum, 928*, 162–167. doi: 10.4028/www.scientific.net/msf.928.162.

Selvan, Bellamballi Munivenkatappan Muthamizh, Anandakrishnan, Veeramani, Duraiselvam, Muthukannan, & Sundarameenakshi, Sivaraj (2019). Wear testing of in situ cast AA8011-TiB2 metal matrix composites. *Materials Testing, 61*(8), 779–786.

Selvan, Muthamizh, Munivenkatappan, Bellamballi, & Veeramani, Anandakrishnan (2018).

Investigation of tribological behavior of AA8011–ZrB$_2$ in-situ cast-metal-matrix composites. *Materials and Technology, 52*(4), 451–457. doi: 10.17222/mit.2017.046.

Shashi Prakash, Dwivedi, & Garima, Dwivedi (2020). Utilization of recycled hazardous waste bagasse as reinforcement to develop green composite material. *World Journal of Engineering, 17*(3), 399–406. doi: 10.1108/WJE-03-2019-0069.

Shi, C. J., Z. Zhang, & Chen, X. G. (2012). Microstructure and fluidity evolution during recycling of Al-B 4C metal matrix composites. *Journal of Composite Materials, 46*(6), 641–652. doi: 10.1177/0021998311420310.

Song, Yanpei, & Wang, Huigai (2012). High speed sliding wear behavior of recycled wcp-reinforced ferrous matrix composites fabricated by centrifugal cast. *Wear, 279–277,* 105–110.

Soon, L. P., & Gupta, M. (2002). Synthesis and recyclability of a magnesium based composite using an innovative disintegrated melt deposition technique. *Materials Science and Technology, 18*(1), 92–98. doi: 10.1179/026708301125000276.

Tan, Xin, Bin Zhang, Kai Liu, Xiaobo Yan, Jun Han, Xin Liu, Wenshu Yang, et al. (2020). Microstructure and mechanical property of the 2024Al matrix hybrid composite reinforced with recycled SiCp/2024Al composite particles. *Journal of Alloys and Compounds, 815.* doi: 10.1016/j.jallcom.2019.152330.

Yoshikawa, Noboru, Nakano, Yuuya, Sato, Kentarou, & Taniguchi, Shoji (2005). Fabrication of composite materials using al scrap and wasted glass. *Materials Transactions, 46*(12), 2582–2585. doi: 10.2320/matertrans.46.2582.

Yoshikawa, Noboru, Yamaguchi, Hideaki, Kitahara, Gaku, & Taniguchi, Shoji (2003). Utilization of Al scrap for fabrication of Al(Alloy)/Al$_2$O$_3$ composite material. *Materials Transactions, 44*(7), 1271–1275. doi: 10.2320/matertrans.44.1271.

7 Performance of Natural Fiber Reinforced Recycled Thermoplastic Polymer Composites Under Aging Conditions

M. Chandrasekar[1], T. Senthil Muthu Kumar[2], K. Senthilkumar[3], Sabarish Radoor[4], R.A. Ilyas[5,6], S.M. Sapuan[7,8], J. Naveen[9], and Suchart Siengchin[4]

[1]School of Aeronautical Sciences, Hindustan Institute of Technology and Science, Chennai, Tamil Nadu, India
[2]Department of Mechanical Engineering, Kalasalingam Academy of Research and Education, Krishnankoil, Tamil Nadu, India
[3]Center of Innovation in Design and Engineering for Manufacturing (CoI-DEM), King Mongkut's University of Technology North Bangkok, Bangkok, Thailand
[4]Department of Materials and Production Engineering, The Sirindhorn International Thai-German Graduate School of Engineering (TGGS), King Mongkut's University of Technology North Bangkok, Bangkok, Thailand
[5]School of Chemical and Energy Engineering, Faculty of Engineering, Universiti Teknologi Malaysia, Johor, Malaysia
[6]Centre for Advanced Composite Materials, Universiti Teknologi Malaysia, Johor, Malaysia
[7]Laboratory of Biocomposite Technology, Institute of Tropical Forestry and Forest Products (INTROP), Universiti Putra Malaysia, Selangor, Malaysia

[8]Department of Mechanical and Manufacturing Engineering, Faculty of Engineering, Universiti Putra Malaysia, Selangor, Malaysia
[9]School of Mechanical Engineering, Vellore Institute of Technology, Vellore, India

CONTENTS

7.1 Introduction 128
7.2 Thermal Properties 130
7.3 Mechanical Properties 131
7.4 Conclusion 135
References 136

7.1 INTRODUCTION

Many researchers are actively working on natural fiber reinforced thermoplastic composites due to their benefits: lightweight, less expensive, biodegradability and good mechanical properties (Atiqah et al., 2019; Chandrasekar et al., 2017, 2019, 2020; Krishnasamy et al., 2019; Krishnasamy, Muthukumar, et al., 2019; Muthu Kumar, et al., 2019; Senthilkumar et al., 2015, 2017, 2019; Thiagamani et al., 2019). Among the various influencing parameters, the fiber-matrix interface plays a significant role in fiber reinforced composites (Aisyah et al., 2021; Alsubari et al., 2021; Ayu et al., 2020; Ilyas et al., 2021; Nurazzi et al., 2021; Omran et al., 2021; Sabaruddin et al., 2020; Sapuan et al., 2020). For instance, to transfer the load from the fiber reinforcement to the matrix and determining the mechanical properties of the composites. Thus, the internal crack proliferation and stress transfer could be reduced and the mechanical performance of the composite structures is intended to enhance (Li et al., 2017; Ma et al., 2019; Zhu et al., 2004).

Due to the ecological and environmental needs, recycling and reuse of thermoplastic have gained huge attention (Thiagamani et al., 2019a). Furthermore, to tackle the problems with one-use plastics, several countries have banned using one-use plastics in some applications such as water bottles, teacups, carry bags, utensils, straws, etc. and have replaced them with reusable plastics (Senthil Muthu Kumar et al., 2019). Waste comprising recyclable materials, such as plastic, might be utilized as a crude material in numerous applications. The utilization of reused materials can reduce the waste generation, crude material utilization and energy required for producing the plastic materials (Kumar et al., 2021). Plastics are almost used in all fields and when they are thrown out, they pose a serious threat to the environment and the living organisms. These wastes account for a major part of the solid wastes and can have a wide change in their nature (Al-Maadeed &Labidi, 2014). The principle reason behind opting for the reuse and recycling of plastics is due to their slower degradation time in a landfill and affect the ecosystem. Further, it is essential to reprocess the used plastics to obtain the necessary useful properties. The utilization of natural or bio-based fibers/fillers is a significant method to reduce plastic waste. Various organic and inorganic fillers are used with polymers to improve their mechanical and physical properties of recycled plastics (Muthu Kumar et al., 2019).

Recycling plastic materials has seen a faster extension internationally in recent decades. Advancements in the blending and reprocessing of recyclable plastics have paved the way for new chances to redirect most plastic waste from landfills. Recently, enactments related to the natural effects of solid waste have urged the industry and researchers to discover innovative approaches to use the extensive measure of reused plastics and bio-based materials (Xu et al., 2015). The growing consumer market has also prompted significant issues on material accessibility and sustainability. In this specific circumstance, the utilization of sustainable materials such as lignocellulosic fillers with thermoplastic polymers as of late has expanded. Because of the huge amount of plastic waste created every day, the utilization of waste plastic for various applications has become a phenomenal choice (Silva et al., 2017).

Polymer composites can be recycled in many ways, such as mechanical, thermal and solvent-based. Mechanical-based recycling is the most common method implemented in the recycling of polymers and their composites. The process involves the grinding and milling of the polymers and is finally used as reinforcement material in the new composite products. The thermal process involves the controlled combustion of the organic component of the composite material. Solvent-based recycling involves the chemical decomposition of the polymer composites with solvents or supercritical fluids (Otheguy et al., 2009). Figure 7.1 below shows the various recycling methods employed for the polymer composites.

The recycling of polymers and their composites did not achieve economic viability due to the absence of appropriate markets for products manufactured from recycled material compared to their virgin materials. This may be due to the inferior functional properties of these materials. Further, the use of chemicals for

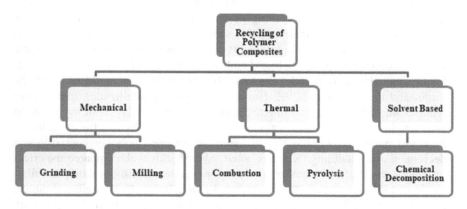

FIGURE 7.1 Various recycling methods in polymer composites.

recycling can contribute to environmental risk, and the process involved in converting the chemical mixtures to usable products is also complex.

Natural fiber reinforced polymer composites (NFRPCs) have been used in almost all engineering applications. The high accessibility, ease and relative strengthening activity of the various regular fibers made them inevitable for polymer composites, mostly with medium to low strength requirements. The creation of the composites from the recycled polymers and characteristic fibers can bring about the material of minimal effort and improved properties. Polyethylene and other thermoplastics are utilized in large quantities for nonstructural applications requiring low strength. For example, bundling, while at the same time thermosetting polymers, for example, epoxy-based composites, are utilized for high-performance applications such as sports equipment, automotive, aerospace structures, etc., and semi-structural applications (Moreno et al., 2018b). However, it is well known that thermoplastics are easier to recycle than thermoset polymers. Due to this reason, there has been considerable interest in the recycling of thermoplastic-based composites. This article aims to present an overview of the recycled thermoplastic polymer composites and their performance under accelerated aging conditions. The variations in the mechanical and thermal properties and the mechanism behind the degradation and response of the composites with recycled polymers to different aging conditions have been discussed.

7.2 THERMAL PROPERTIES

A thorough understanding of weathering behavior of composite is essential as it is related to the durability of the lignocellulosic fiber and the polymer employed. Many researchers have investigated the weathering characteristic of polymer composites. The effect of weathering on the performance of recycled and virgin forms of wood-polymer (polypropylene, PP, and high-density polypropylene, HDPE) composites were investigated in a study (Adhikary et al., 2010). The thermal analysis showed that both controlled and weathered PP samples exhibited double endothermic melting peaks, whereas a single endothermic peak was noted for the unweathered HDPE sample. The melting and crystallization enthalpies of the exposed wood flour/PP composites were higher than that of the control sample. On the other hand, both the melting enthalpy and the crystallization enthalpy were found to be lower for the composites reinforced with the virgin and the recycled HDPE. However, the peak crystallization temperature remains unchanged after weathering. On comparing the crystallinity of virgin and recycled polymer, it was observed that weathering increases the crystallinity of virgin polymers (PP and HDPE). But, for recycled polymer composites, a decrement in crystallinity (Xc) was noted. Abu-Sharkh et al. compared the effect of natural and artificial weathering conditions on the stability of date palm fiber reinforced PP composites (Abu-Sharkh & Hamid, 2004). They employed maleic anhydride grafted PP (MAPP) as the compatibilizer for the composites. After six months of natural weathering, a sudden drop in melting point was noted. This indicates that MAPP is more susceptible to oxidation during weathering than pure PP. Meanwhile, uncompatibilized samples displayed higher stability than pure PP. The lignocellulosic fibers in the composite material were believed to be responsible for slowing down the degradation. When exposed to artificial weathering, the PP polymer displayed less stability than the composite with and without a compatibilizer. This could be due to the depletion of a significant amount of stabilizers during the fabrication process. The compatibilized sample displayed higher stability towards artificial weathering, which could be because radiation frequency is more intense in natural weathering conditions than in the accelerated weathering chamber. Silva et al. monitored the effect of natural weathering on wood flour/recycled PP/ethylene vinyl acetate-based composites (Silva et al., 2017). Although natural weathering has no major impact on the melting point of composite, it significantly impacted the enthalpy values and degree of crystallinity of composite. After aging, the melting enthalpy and crystallinity of the composites increased from 30.8% to 37.5% and 45.13% to 54.83%, respectively. This is due to aging-induced polymer chain scission, which results in larger amounts of smaller crystalline phases. These observations were confirmed by DSC and TGA analysis. After aging, a new endothermic peak was observed, indicating the secondary recrystallization of polymeric segments. Furthermore, with the increase in the aging time, a prominent reduction in the mass loss was observed in the composite.

Long-term soil aging is reported to have an adverse effect on the thermal stability of the composite. The addition of mixed particle-size fiber could improve the thermal stability of such composites. A recent study showed that the introduction of mixed particle-size fibers improved interfacial adhesion and facilitated heat transfer from matrix to fiber. Tian et al. subjected PP-wood composite which is reinforced by waste-printed circuit board powders (WPCBP) to UV radiation for 15 days and evaluated its thermal and physical properties (Tian et al., 2019). After accelerated aging, thermal parameters (crystallization peak temperature, melting peak temperature and crystalline fraction) of PP composite declined due to aging-induced degradation of the polymer. However, on incorporating wood fiber and WPCBP, the thermal properties of composites increase. This is ascribed to the fact that the WF fillers and WPCBP resists the degradation of PP macromolecule chains.

7.3 MECHANICAL PROPERTIES

The natural fiber reinforced thermoplastic composites are used in automobile, marine, civil structures, sporting goods, electronics and biomedicals due to their environmental and economic reasons (Thiagamani et al., 2019b). Nevertheless, the application of NFRPC in outdoor applications is limited due to their degradation under aging or weathering effects such as ultraviolet (UV) radiation, corrosive environments and high temperature. Thus, the performance of the NFRPC under mechanical loads are found to be affected due to the deterioration of the fiber-matrix interfacial adhesion (Chen et al., 2018; Downey & Drzal,

2016; Tran et al., 2018; Senthilkumar et al., 2018; Vercher et al., 2020; Wu et al., 2016).

Moreover, the fiber-matrix interfacial bonding is easily affected by the moisture content. Hence, the natural fiber reinforced composites are subjected to (i) hydrolysis, (ii) differential swelling and (iii) residual stress relaxation mechanisms due to the prolonged moisture environments (Azwa et al., 2013; Brito et al., 2020). The interfacial bonding characteristics are usually determined using two approaches, namely interfacial shear strength (IFSS) and inter-laminar shear strength (ILSS). The former characteristic can be analyzed by employing a micromechanical test and a short beam test to determine the latter.

Methacanon et al. (2010) indicated that the wet natural fibers showed higher tensile strength and elongation than the dry ones. It was ascribed to (i) the higher order of crystallinity and (ii) the crystalline orientation of fibers due to the fiber wetting. Since the fibers observed the moisture content, they could be acted as lubricants; besides, the fibers can slide one over the other during stretching. Hence, the elongation of the wet natural fibers was found to be higher than the dry ones. Moreno et al. (2018a) evaluated tensile properties of the recycled low-density polyethylene (LDPE)/pinewood waste composites under accelerated aging conditions (Figure 7.2).

The tensile strength of reinforced composites after aging was insignificant compared to the virgin matrix because the pinewood waste observed the ultraviolet radiation during the evaluation time and helped to retain the tensile strength of the composites. However, the virgin matrix exhibited severe loss due to the degradation effects. Regarding the elongation at break, both fiber reinforced composites and the virgin matrix exhibited similar results after 4,032 hours of aging while the LDPE was severely suffered. In the case of Young's modulus, there was no significant difference observed between composite samples and the LDPE matrix. However, before aging, the fiber reinforced composites showed higher results than the matrix due to the reinforcing effects of pinewood waste.

In another work, Silva et al. investigated the natural aging on mechanical properties of wood flour/recycled PP composites, as shown in Figure 7.3 (Silva et al., 2017). The tensile modulus of wood floor composites declined post-aging; however, the stiffness was better than the pure matrix. It was ascribed to the greater stability of wood floor-filled composites. In the case of flexural strength, the wood floor-filled composites were found to decrease with aging. It was attributed to the increased swelling behavior of the wood floor due to humidity differences. The aged specimens also displayed a decreasing trend in the case of the flexural modulus.

Inácio et al. (2018) compared the mechanical properties of unaged and aged combinations of recycled PP/ethylene-propylene-diene monomer (EPDM)/talc/bamboo fiber/MAPP composites. These composites were prepared by varying the individual wt.% of (i) PP/EPDM/talc, (ii) bamboo fiber and (iii) compatibilizers, respectively. When comparing the aged specimens with the unaged specimens, the tensile modulus showed no significant differences for the

Performance of Natural Fiber

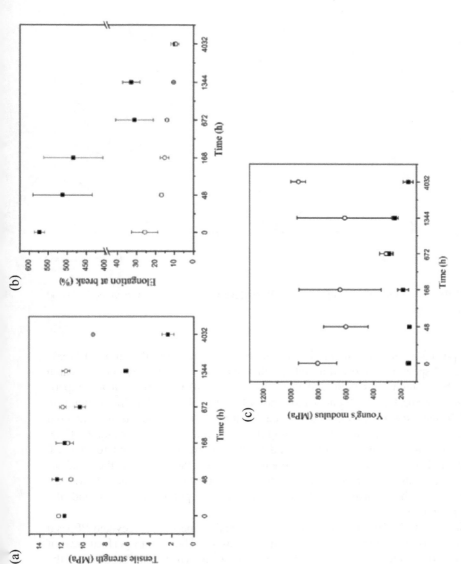

FIGURE 7.2 (a) Tensile strength, (b) elongation at break and (c) Young's modulus of low-density polyethylene (LDPE)/pinewood waste composites (reused with permission, License number 4870270564290).

*(■): LDPE; (○): LDPE/pinewood waste composites.

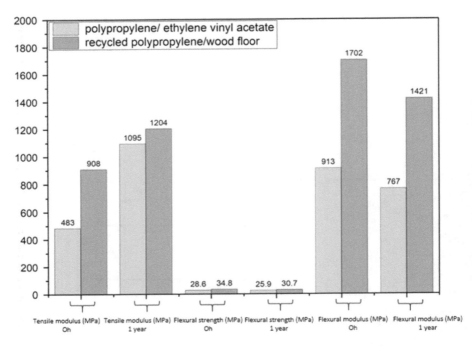

FIGURE 7.3 Mechanical properties of PP/ethylene vinyl acetate and recycled PP/wood floor composites.

composites with any compatibilizer (Figure 7.4). Regarding the strain at break, the aged specimens showed a higher value than the unaged specimens in which the fiber content was increased from 20 wt.% to 40 wt.% without a compatibilizer. Flexural strength under three-point bending displayed a similar trend, in which aged specimens exhibited better flexural strength than the unweathered specimens; it was ascribed to annealing that occurred due to the evaluation time. Likewise, the flexural modulus of all aged specimens was found to increase compared to the unweathered specimens. However, the energy at the break of aged materials showed a decreasing trend than the non-aged ones. It was attributed to the weakening of the bonds in the polymeric chain and the compatibilizer.

Joseph et al. (1999) examined the mechanical properties of recycled PP/sisal fiber reinforced composites under aging conditions. The overall performance of the biocomposites was found to be higher than the virgin matrix. Though the sisal fiber provided stiffness to the recycled PP composites, their mechanical properties were decreased due to the degradation of fiber-matrix bonding. Regarding the elongation at break, the virgin matrix outperformed compared to the recycled PP matrix; it was ascribed to their high brittleness and rigidity. The elongation at break for the fiber reinforced composites was decreased after

Performance of Natural Fiber

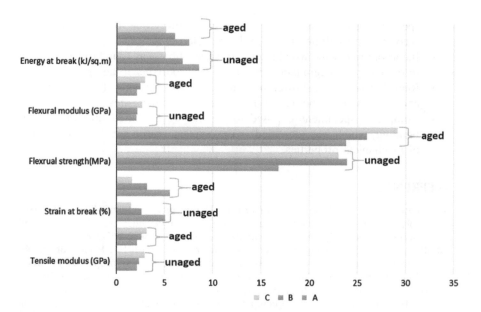

FIGURE 7.4 Mechanical properties of recycled PP composites (reused with permission, License number 5027171278538).

*A: (PP/EPDM/talc)/ (bamboo fiber)/(compatibilizer) = 80:20:0.

B: (PP/EPDM/talc)/ (bamboo fiber)/(compatibilizer) = 70:30:0.

C: (PP/EPDM/talc)/ (bamboo fiber)/(compatibilizer) = 60:40:0.

exposure to ultraviolet radiation of 170 hours. This reduction was attributed to the extensive chain scission of the virgin matrix. However, the weathering also induced the formation of fiber-matrix de-bonding due to the formation of microcracks on (i) the surfaces of the virgin matrix and (ii) the recycled PP matrix composites. Consequently, the composites were found to fail under lower loads. Also, the humidity played a significant role in deteriorating the interfacial bonding strengths of recycled PP/sisal fiber reinforced composites.

7.4 CONCLUSION

Following are the conclusions from the literature review of the works published on the aging effects of the natural fiber reinforced recycled thermoplastic composites:

- Composites with the recycled polymers displayed changes in crystallinity, endothermic peaks and melting point due to the aging. Aged composite

specimens showed a slightly higher crystallinity due to the formation of secondary crystalline phases in the recycled polymer.
- Incorporation of the wood filler and natural fiber was found to be effective in reducing the degradation of the recycled polymer to aging.
- Recycled polymers showed a decline in tensile strength, flexural strength and stiffness than the natural fiber reinforced composites.
- Composite specimens exposed to weathering exhibited fiber-matrix debonding, which was identified as the primary reason for the deterioration of mechanical properties.

REFERENCES

Abu-Sharkh, B. F., & Hamid, Halim (2004). Degradation study of date palm fiber/polypropylene composites in natural and artificial weathering: Mechanical and thermal analysis. *Polymer Degradation and Stability, 85*(3), 967–973.

Adhikary, Kamal Babu, Pang, Shusheng, & Staiger, Mark P. (2010). Effects of the accelerated freeze-thaw cycling on physical and mechanical properties of wood flour-recycled thermoplastic composites. *Polymer Composites, 31*(2), 185–194.

Aisyah, H. A., Paridah, M. T., Sapuan, S. M., Ilyas, R. A., Khalina, A., Nurazzi, N. M., Lee, S. H., & Lee, C. H. (2021). A comprehensive review on advanced sustainable woven natural fiber polymer composites. *Polymers, 13*(3). doi:10.3390/polym13030471.

Al-Maadeed, M. A., & Labidi, S. (2014). Recycled polymers in natural fiber-reinforced polymer composites. In *Natural Fibre Composites* (pp. 103–114). Elsevier.

Alsubari, S., Zuhri, M. Y. M., Sapuan, S. M., Ishak, M. R., Ilyas, R. A., & Asyraf, M. R. M. (2021). Potential of natural fiber reinforced polymer composites in sandwich structures: A review on its mechanical properties. *Polymers, 13*(3), 423. doi:10.3390/polym13030423.

Atiqah, A., Chandrasekar, M., Senthil Muthu Kumar, T., Senthilkumar, K., & Ansari, M. N. M. (2019). *Characterization and Interface of Natural and Synthetic Hybrid Composites. Reference Module in Materials Science and Materials Engineering.* Elsevier. doi:10.1016/b978-0-12-803581-8.10805-7.

Ayu, Rafiqah S., Khalina, Abdan, Saffian Harmaen, Ahmad, Zaman, Khairul, Isma, Tawakkal, Liu, Qiuyun, Ilyas, R. A., & Lee, Ching Hao (2020). Characterization study of empty fruit bunch (EFB) fibers reinforcement in poly(butylene) succinate (PBS)/starch/glycerol composite sheet. *Polymers, 12*(7), 1571. doi:10.3390/polym12071571.

Azwa, Z. N., Yousif, Belal, Manalo, A. C., & Karunasena, W. (2013). A review on the degradability of polymeric composites based on natural fibers. *Materials & Design, 47*, 424–442. doi:10.1016/j.matdes.2012.11.025.

Brito, C. B. G., Sales, R. C. M., & Donadon, M. V. (2020). Effects of temperature and moisture on the fracture behaviour of composite adhesive joints. *International Journal of Adhesion and Adhesives.* doi:10.1016/j.ijadhadh.2020.102607.

Chandrasekar, M., Ishak, M. R., Sapuan, S. M., Leman, Z., & Jawaid, M. (2017). A review on the characterisation of natural fibers and their composites after alkali treatment and water absorption. *Plastics, Rubber and Composites.* doi:10.1080/14658011.2017.1298550.

Chandrasekar, M., Shahroze, R. M., Ishak, M. R., Saba, N., Jawaid, M., Senthilkumar, K., Kumar, T Senthil Muthu, & Siengchin, Suchart (2019). Flax and sugar palm reinforced epoxy composites: Effect of hybridization on physical, mechanical, morphological and dynamic mechanical properties. *Materials Research Express, 6*(10), 105331. doi:10.1088/2053-1591/ab382c.

Chandrasekar, M., Siva, I., Senthil Muthu Kumar, T., Senthilkumar, K., Siengchin, Suchart, & Rajini, N. (2020). Influence of fiber inter-ply orientation on the mechanical and free vibration properties of banana fiber reinforced polyester composite laminates. *Journal of Polymers and the Environment,* 1–12. doi:10.1007/s1 0924-020-01814-8

Chen, Chin-Hsing, Chiang, Chin-Lung, Chen, Wei-Jen, & Shen, Ming-Yuan (2018). The effect of MBS toughening for mechanical properties of wood-plastic composites under environmental ageing. *Polymers and Polymer Composites, 26*(1), 45–58.

Downey, Markus A., & Drzal, Lawrence T. (2016). Toughening of carbon fiber-reinforced epoxy polymer composites utilizing fiber surface treatment and sizing. *Composites Part A: Applied Science and Manufacturing, 90,* 687–698.

Ilyas, R. A., Sapuan, S. M., Atikah, M. S. N., Asyraf, M. R. M., Ayu Rafiqah, S., Aisyah, H. A., Nurazzi, N Mohd, & Norrrahim, M. N. F. (2021). Effect of hydrolysis time on the morphological, physical, chemical, and thermal behavior of sugar palm nanocrystalline cellulose (*Arenga Pinnata* (Wurmb.) Merr). *Textile Research Journal, 91*(1–2), 152–167. doi:10.1177/0040517520932393.

Inácio, André L.N., Nonato, Renato C., & Bonse, Baltus C. (2018). Mechanical and thermal behavior of aged composites of recycled PP/EPDM/Talc reinforced with bamboo fiber. *Polymer Testing, 72,* 357–363. doi:10.1016/j.polymertesting.2018.10.035.

Joseph, P. V., Joseph, Kuruvilla & Thomas, Sabu (1999). Effect of processing variables on the mechanical properties of sisal-fiber-reinforced polypropylene composites. *Composites Science and Technology.* doi:10.1016/S0266-3538(99)00024-X.

Krishnasamy, Senthilkumar, Muthukumar, Chandrasekar, Nagarajan, Rajini, Thiagamani, Senthil Muthu Kumar, Saba, Naheed, Jawaid, Mohammad, Siengchin, Suchart, & Ayrilmis, Nadir (2019). Effect of fiber loading and Ca(OH) 2 treatment on thermal, mechanical, and physical properties of pineapple leaf fiber/polyester reinforced composites. *Materials Research Express, 6*(8), 085545. doi:10.1088/2053-1591/ab2702.

Krishnasamy, Senthilkumar, Kumar Thiagamani, Senthil Muthu, Kumar, Chandrasekar Muthu, Nagarajan, Rajini, R. M. Shahroze, Siengchin, Suchart, Ismail, Sikiru O. & M. P. Indira Devi (2019). Recent advances in thermal properties of hybrid cellulosic fiber reinforced polymer composites. *International Journal of Biological Macromolecules, 141,* 1–13. doi:10.1016/j.ijbiomac.2019.08.231.

Krishnasamy, Senthilkumar, Thiagamani, Senthil Muthu Kumar, Muthukumar, Chandrasekar, Tengsuthiwat, Jiratti, Nagarajan, Rajini, Siengchin, Suchart, & Ismail, Sikiru O. (2019). Effects of stacking sequences on static, dynamic mechanical and thermal properties of completely biodegradable green epoxy hybrid composites. *Materials Research Express, 6*(10), 105351. doi:10.1088/2053-1591/ab3ec7.

Kumar, T Senthil Muthu, Rajini, N., Alavudeen, A., Siengchin, Suchart, Varada Rajulu, A., & Ayrilmis, Nadir (2021). Development and analysis of completely biodegradable cellulose/banana peel powder composite films. *Journal of Natural Fibers, 18*(1), 151–160. doi:10.1080/15440478.2019.1612811.

Li, Ran, Gu, Yizhuo, Zhang, Gaolong, Yang, Zhongjia, Li, Min & Zhang, Zuoguang (2017). Radiation shielding property of structural polymer composite: Continuous basalt fiber reinforced epoxy matrix composite containing erbium oxide. *Composites Science and Technology, 143,* 67–74. doi:10.1016/j.compscitech.2017.03.002.

Ma, Renliang, Li, Weiwei, Huang, Momo, Feng, Ming, & Liu, Xiaojing (2019). The reinforcing effects of dendritic short carbon fibers for rigid polyurethane composites. *Composites Science and Technology, 170,* 128–134.

Methacanon, P., Weerawatsophon, U., Sumransin, N., Prahsarn, C., & Bergado, D. T. (2010). Properties and potential application of the selected natural fibers as limited life geotextiles. *Carbohydrate Polymers, 82*(4), 1090–1096.

Moreno, D. D. P., Hirayama, D., & Saron, C (2018a). Accelerated aging of pine wood waste/recycled LDPE composite. *Polymer Degradation and Stability, 149*, 39–44. doi: 10.1016/j.polymdegradstab.2018.01.014.

Moreno, D. D. P., Hirayama, D., & Saron, C. (2018b). Accelerated aging of pine wood waste/recycled LDPE composite. *Polymer Degradation and Stability, 149*, 39–44.

Muthu Kumar, Senthil T., Yorseng, Krittirash, Siengchin, Suchart, Ayrilmis, Nadir, & Rajulu, Varada A. (2019). Mechanical and thermal properties of spent coffee bean filler/poly(3-hydroxybutyrate-Co-3-hydroxyvalerate) biocomposites: Effect of recycling. *Process Safety and Environmental Protection, 124*, 187–195. doi: 10.1016/j.psep.2019.02.008.

Nurazzi, N. M., Asyraf, M. R. M., Khalina, A., Abdullah, N., Aisyah, H. A., Ayu Rafiqah, S., Sabaruddin, F. A., et al., (2021). A review on natural fiber reinforced polymer composite for bullet proof and ballistic applications. *Polymers, 13*(4), 646. doi: 1 0.3390/polym13040646.

Omran, Abdoulhdi A. Borhana, Mohammed, Abdulrahman A. B. A., Sapuan, S. M., Ilyas, R. A., Asyraf, M. R. M., Koloor, Seyed Saeid Rahimian, & Petrů, Michal (2021). Micro- and nanocellulose in polymer composite materials: A review. *Polymers, 13*(2), 231. doi: 10.3390/polym13020231.

Otheguy, M. E., Gibson, A. G., Findon, E., Cripps, R. M., Ochoa Mendoza, A., & Aguinaco Castro, M. T. (2009). Recycling of end-of-life thermoplastic composite boats. *Plastics, Rubber and Composites, 38*(9–10), 406–411.

Sabaruddin, Fatimah Athiyah, Paridah, M. T., Sapuan, S. M., Ilyas, R. A., Seng Hua Lee, Khalina Abdan, Norkhairunnisa Mazlan, Adlin Sabrina Muhammad Roseley, & Abdul Khalil, H. P. S. (2020). The effects of unbleached and bleached nanocellulose on the thermal and flammability of polypropylene-reinforced kenaf core hybrid polymer bionanocomposites. *Polymers, 13*(1), 116. doi: 10.3390/polym13010116.

Sapuan, S. M., Aulia, H. S., Ilyas, R. A., Atiqah, A., Dele-Afolabi, T. T., Nurazzi, M. N., Supian, A. B. M., & Atikah, M. S. N. (2020). Mechanical properties of longitudinal basalt/woven-glass-fiber-reinforced unsaturated polyester-resin hybrid composites. *Polymers, 12*(10). doi: 10.3390/polym12102211.

Senthil Muthu Kumar, T., Senthilkumar, K., Chandrasekar, M., Rajini, N., Suchart Siengchin, & Varada Rajulu, A. (2019). Characterization, thermal and dynamic mechanical properties of poly(propylene carbonate) lignocellulosic cocos nucifera shell particulate biocomposites. *Materials Research Express, 6*(9). doi: 10.1088/2 053-1591/ab2f08.

Senthilkumar, K., Chandrasekar, M., Rajini, N., Suchart Siengchin, & Rajulu, Varada (2019). Characterization, thermal and dynamic mechanical properties of poly (propylene carbonate) lignocellulosic cocos nucifera shell particulate biocomposites. *Materials Research Express, 6*(9), 96426.

Senthilkumar, K., Saba, N., Rajini, N., Chandrasekar, M., Jawaid, M., Siengchin, Suchart, & Alotman, Othman Y. (2018). Mechanical properties evaluation of sisal fiber reinforced polymer composites: A review. *Construction and Building Materials, 174*, 713–729. doi: 10.1016/j.conbuildmat.2018.04.143.

Senthilkumar, K., Siva, I., Rajini, N., & Jeyaraj, P. (2015). Effect of fiber length and weight percentage on mechanical properties of short sisal/polyester composite. *International Journal of Computer Aided Engineering and Technology, 7*(1), 60–71.

Senthilkumar, Krishnasamy, Siva, Irulappasamy, Hameed Sultan, Mohamed Thariq, Rajini, Nagarajan, Siengchin, Suchart, Jawaid, Mohamad, & Hamdan, Ahmad (2017). Static and dynamic properties of sisal fiber polyester composites – effect of interlaminar fiber orientation. *BioResources*. doi: 10.15376/biores.12.4.7819-7833.

Silva, Caroline Barbosa da, Martins, Andrea Bercini, Catto, André Luis, & Campomanes Santana, Ruth Marlene (2017). Effect of natural ageing on the properties of recycled polypropylene/ethylene vinyl acetate/wood flour composites. *Matéria (Rio de Janeiro), 22*(2).

Thiagamani, Senthil Muthu Kumar, Krishnasamy, Senthilkumar, Muthukumar, Chandrasekar, Tengsuthiwat, Jiratti, Nagarajan, Rajini, Siengchin, Suchart, & Ismail, Sikiru O. (2019). Investigation into mechanical, absorption and swelling behaviour of hemp/sisal fiber reinforced bioepoxy hybrid composites: Effects of stacking sequences. *International Journal of Biological Macromolecules, 140*, 637–646. doi: 10.1016/j.ijbiomac.2019.08.166.

Thiagamani, Senthil Muthu Kumar, Krishnasamy, Senthilkumar, & Siengchin, Suchart (2019a). Challenges of biodegradable polymers: An environmental perspective. *Applied Science and Engineering Progress, 12*(3), 149.

Thiagamani, Senthil Muthu Kumar, Krishnasamy, Senthilkumar, & Siengchin, Suchart (2019b). Challenges of biodegradable polymers: An environmental perspective. *Applied Science and Engineering Progress*. doi: 10.14416/j.asep.2019.03.002.

Tian, Shenghui, Luo, Yuanfang, Chen, Jizun, He, Hui, Chen, Yong, & Ling, Zhang (2019). A comprehensive study on the accelerated weathering properties of polypropylene – wood composites with non-metallic materials of waste-printed circuit board powders. *Materials, 12*(6), 876.

Tran, Phuong, Quynh Thuy Nguyen, & Lau, K. T. (2018). Fire performance of polymer-based composites for maritime infrastructure. *Composites Part B: Engineering, 155*, 31–48.

Vercher, Jose, Fombuena, Vicent, Diaz, Arturo, & Soriano, Maria (2020). Influence of fiber and matrix characteristics on properties and durability of wood–plastic composites in outdoor applications. *Journal of Thermoplastic Composite Materials, 33*(4), 477–500.

Wu, Guangshun, Ma, Lichun, Liu, Li, Wang, Yuwei, Xie, Fei, Zhong, Zhengxiang, Zhao, Min, Jiang, Bo, & Huang, Yudong (2016). Interface enhancement of carbon fiber reinforced methylphenylsilicone resin composites modified with silanized carbon nanotubes. *Materials & Design, 89*, 1343–1349.

Xu, Changyan, Xing, Cheng, Pan, Hui, Matuana, Laurent M. & Zhou, Handong (2015). Hygrothermal aging properties of wood plastic composites made of recycled high density polypropylene as affected by inorganic pigments. *Polymer Engineering & Science, 55*(9), 2127–2132.

Zhu, Jiang, Peng, Haiqing, Rodriguez-Macias, Fernando, Margrave, John L., Khabashesku, Valery N., Imam, Ashraf M., Lozano, Karen, & Barrera, Enrique V. (2004). Reinforcing epoxy polymer composites through covalent integration of functionalized nanotubes. *Advanced Functional Materials, 14*(7), 643–648.

8 Physical and Mechanical Properties of Recycled Metal Matrix Composites

Pradeepkumar Krishnan[1] and Ramanathan Arunachalam[2]
[1]Mechanical and Industrial Engineering Department, National University of Science and Technology, Muscat, Sultanate of Oman
[2]Mechanical and Industrial Engineering Department, Sultan Qaboos University, Muscat, Sultanate of Oman

CONTENTS

8.1 Introduction 141
8.2 MMCs Production Using Recycled Materials 144
    8.2.1 Direct-Chill Method 144
    8.2.2 Centrifugal Method 146
    8.2.3 Stir-Squeeze Casting Method 146
8.3 Separation Techniques Used for Separating the Reinforcements from the Matrix Materials 147
    8.3.1 Simple Remelting 149
    8.3.2 Electrolysis 150
    8.3.3 Aqueous Solution 150
    8.3.4 Supergravity 151
    8.3.5 Nozzle Filtering 152
    8.3.6 Salt Fluxing 153
8.4 Mechanical Properties of the Recycled MMCs 155
8.5 Conclusions 158
Acknowledgment 159
References 159

8.1 INTRODUCTION

Metals are a natural resource found in the Earth's crust, and their compositions vary between various locations, which contribute to spatial variations in

DOI: 10.1201/9781003148760-8

the concentration around them. The metal accumulation/distribution in the environment is determined by the specific material's characteristics and environmental influences (Jaishankar et al., 2014). The destruction of virgin materials is a significant cause of global habitat loss. Rapid consumption levels of raw materials such as iron (Fe), aluminum (Al), copper (Cu), magnesium (Mg) and zinc (Zn) are already having a significant effect on the world's environment. Heavy metals are becoming more critical because of their non-degradable nature and the pile-up in nature that causes dangerous environmental disasters (Bahrami et al., 2015). Disposing of used or waste materials has become one of the many serious challenges worldwide. Recycling is the only possible way to protect the environment and reduce the demand for virgin materials like metals. Recycling involves removing foreign particles from the waste stream to be used as raw materials to manufacture new products. Many environmental advantages can be gained from waste reduction and recycling strategies. Ecological benefits include reducing greenhouse emissions, reducing the release of toxins and conservation of energy. In recent years, a great deal of effort has been made to preserve the Earth's ecosystem (Bahrami et al., 2015). Many manufacturing companies and researchers have focused on finding new materials and recycling existing materials considered waste in nature. Due to the environmental aspect, there has been a growing interest in recycling metal matrix composites. Goñi et al. (2003) researched a low-cost composite metal matrix to enhance automobile pistons' thermal efficiency, automotive clutch discs and brake discs for railway trains. The technique of an in situ reaction is used to produce the composite. The physical and mechanical properties obtained from experiments carried out under service conditions are also favorable, and thus the industrialization of the components is feasible. The Scopus search (dated 6th Nov 2020) in "Title-Abstract-Keywords" for the terms: "metal AND matrix AND composites" resulted in 42012 documents refined using the term "recycling," resulting in 841 documents. These results were further refined using the term "aluminum" followed by "stir AND casting," resulting in 454 and 103 papers, respectively. The pie chart illustrated in Figure 8.1 shows the distribution of these 103 documents. It is self-evident that the number of literature on the recycling of Al MMCs, including review papers, is very scarce (only two in total). Among these, Sun et al. (2020) is the only recent literature that discusses the separation technique for separating the reinforcement from the matrix of scrap Al MMCs which is the focus of this chapter. However, in Sun et al. (2020), a few literature sources are cited that discuss separation techniques used in recycling MMCs. Al MMCs are in use for the last decade, and very soon, the market may be flooded with scrapped Al MMCs. Although the Al MMCs could be directly remelted and reused, these composites depend on the type of reinforcements used and may suffer from deteriorated mechanical properties due to the formation of interfacial reaction compounds (Sun et al., 2020). Because of the low mechanical properties of directly remelted MMCs, developing appropriate separation

Recycled Metal Matrix Composites

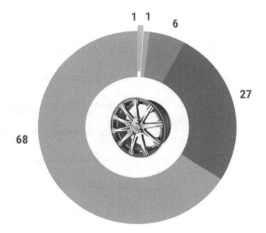

- Recycling of Al MMCs
- Review papers on recycling of Al MMCs
- Al MMCs produced from recycled (scrap / waste) Matrix materials
- Al MMCs produced from recycled (waste) reinforcement materials
- Others not directly relevant to recycling

FIGURE 8.1 Scopus search: Distribution of literature related to recycling of Al MMCs.

techniques becomes vital. In the pie chart (Figure 8.1), the Al MMCs produced from recycled matrix and reinforcements refer to waste or scrap materials being used as a matrix (such as beverage cans, car alloy wheels and others) and reinforcement (agriculture residues such as rice husk, bagasse, industry wastes such as glass, spent alumina catalyst and others), respectively, for producing the Al MMCs.

Figure 8.2 shows the schematic representation of potential matrix materials from waste metals/alloys that can be recycled and includes aluminum, magnesium, copper and ferrous alloys. Among these, aluminum wastes are one of the best recycled or reused materials. The energy costs involved in its processing can be reduced up to 95% by recycling Al scrap, as Al's production from its ore is both expensive and energy-consuming (Yamagiwa et al., 2003).

Although the recycled MMCs offer reasonable physical, mechanical strength and surface hardness, recycling these materials are challenging because they are sensitive to reinforcement, the mode of manufacture and the history of recycled materials. The properties are also influenced by the recycled MMCs parameters such as remelting temperature, holding time and the number of recycling (Fan et al., 2003). Significant amounts of energy consumption happen in traditional recycling methods, such as melting, which negatively impacts nature. Moreover, the price of MMCs is two to three times higher than the price of the matrix, and if the waste is not recycled, the cost of the product becomes higher and may not be viable commercially. A variety of eco-friendly recycling processes and low-

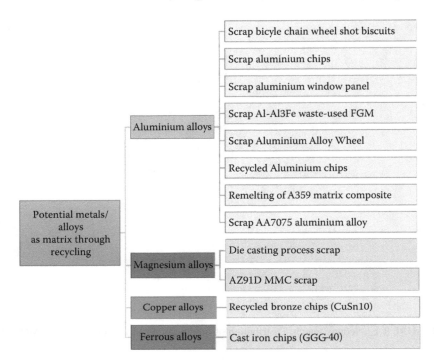

FIGURE 8.2 Schematic representation of potential metals/alloys that can be recycled as a matrix material for producing MMCs.

cost recycling technologies are therefore significant instead of traditional methods that possess low efficiency, high cost and, in particular, environmental risks (Aslan et al., 2019).

8.2 MMCS PRODUCTION USING RECYCLED MATERIALS

The recycling of MMCs refers to the processing of MMCs directly, while the reclamation of MMCs is the separation of the individual components. A few of the standard methods used for producing MMCs from recycled materials are discussed in this section. Metallic chips are recycled through hot deformation, extrusion, high-pressure torsion and sintering under pressure techniques. However, the most widely used method of recycling is the melting and casting processes.

8.2.1 DIRECT-CHILL METHOD

Direct-chill casting (DC) is a primary industrial semi-continuous casting method for partially finished products. The partially finished products are further processed by either rolling, forging or extrusion. The DC casting mold has holes arranged along its bottom end, through which jets of water flow, forming an

Recycled Metal Matrix Composites

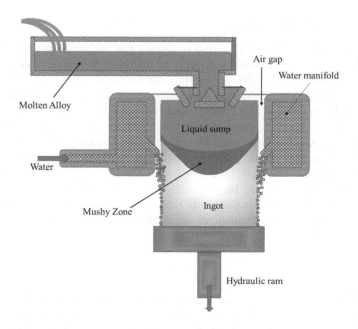

FIGURE 8.3 Schematic diagram of direct-chill casting method.

outer surface like a mold for direct chilling and solidification (Al-Helal et al., 2020). A simple schematic diagram of the direct-chill casting method is shown in Figure 8.3.

In this method, the mixture of reinforcements and alloy (such as cast log ends, extrusion butts and light cuttings) will undergo significant forms of recycling until it enters the final product as the reinforcing particles are combined with the alloy, as shown in Figure 8.4 (Schuster et al., 1993).

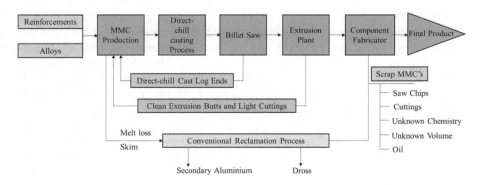

FIGURE 8.4 Production of MMC from recycled materials using the direct-chill method.

8.2.2 CENTRIFUGAL METHOD

Recycling of MMCs can be carried out using the vacuum centrifugal casting method. The centrifugal casting process is a casting technique used to manufacture thin-walled cylinders by pouring molten metal into a rapidly spinning cylindrical mold. The centrifugal rotational force exerts pressure on the molten metal. Under the Ar gas atmosphere, waste master alloys are heated to a molten state. After the alloy is melted, the chamber lids are closed, and a pump evacuates the vacuum chamber. The melt is kept under vacuum conditions to extract the gases dissolved in the melting alloy. This method will prevent harmful material oxidation. The melt is then preheated. The die oriented on the horizontal axis is selected according to the desired component shape. The centrifugal force serves to disperse molten metal in the mold at pressures reaching several times the force of gravity. The centrifugal force effect can be determined through the following equation:

$$G = 2DN^2,$$

where D is the diameter of the casting tube and N is the rate of revolution. Yamagiwa et al. produced a functionally graded material (FGM) by the centrifugal process using Al and Fe wastes (Yamagiwa et al., 2003). This process is illustrated graphically in Figure 8.5.

8.2.3 STIR-SQUEEZE CASTING METHOD

The stir-squeeze casting method is the most commonly used method for producing MMCs. Researchers have recently made an MMC material using scrap aluminum alloy wheels (SAAWs). They used the stir-squeeze casting technique to process MMC due to its cost-effectiveness, simplicity and versatility. Figure 8.6 displays the schematic diagram for the production of scrap aluminum alloy wheel (matrix) based composite material reinforced with agro-waste rice husk ash (RHA) generated as a by-product in the rice mill industry and spent alumina catalyst (SAC) generated in the oil and gas refineries. Excellent physical and mechanical properties such as tensile strength of 115.6 MPa, compression strength of 210.5 MPa and hardness of 86 HRB have been achieved using RHA and SAC as reinforcement materials and SAAW as matrix material (Dwivedi et al., 2020). The feasibility of using chemically cleaned scrap aluminum alloy wheels (SAAWs) as matrix material and spent aluminum catalysts (SACs) from oil refineries as reinforcement material was investigated and compared to virgin alloys and reinforcements. Optimum mechanical properties and tensile strength of 125 MPa, compression strength of 312 MPa and hardness of 57 BHN were obtained (Krishnan et al., 2019).

Recycled Metal Matrix Composites

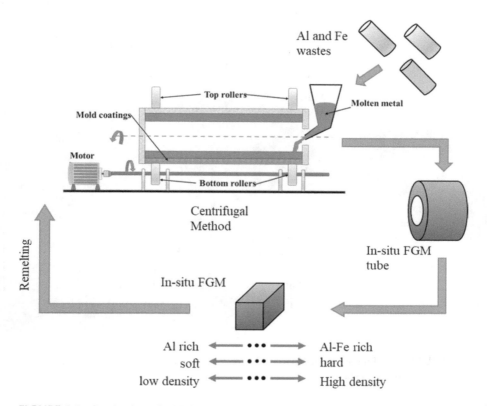

FIGURE 8.5 Production of Al-FGM using the centrifugal method.

8.3 SEPARATION TECHNIQUES USED FOR SEPARATING THE REINFORCEMENTS FROM THE MATRIX MATERIALS

MMCs consist of a matrix metal and reinforcement with a crystal structure and interfaces. The mixing entropy in the production of MMCs has a more negligible effect because MMCs are not atomic-level mixtures. Hence, the reinforcements can be isolated from the matrix, so the separation of the reinforcement from the composite is feasible. Several techniques have been developed and can be classified as shown in Figure 8.7, according to the principles employed to separate the constituents.

As more individuals and businesses become aware of environmental impacts, demand for environmentally friendly and sustainable products increases. Many products today can be recycled or reclaimed. Recycled items are products made from scrap or waste or discarded parts/products. For their original purpose or an alternative purpose, reclaimed objects can be reused in their new form. Specific modern separation techniques like salt or other fluxing methods can be implemented to recover metals and alloys. Rotary salt furnace technology proved helpful for this

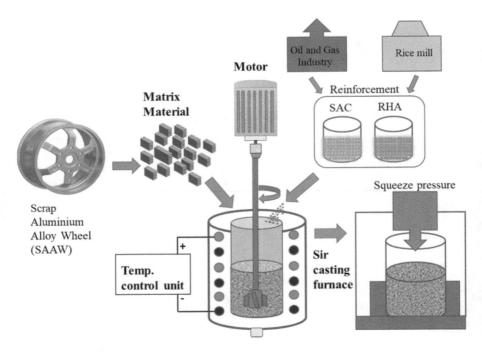

FIGURE 8.6 Production of MMCs using recycled scrap aluminium alloy using the stir-squeeze casting method.

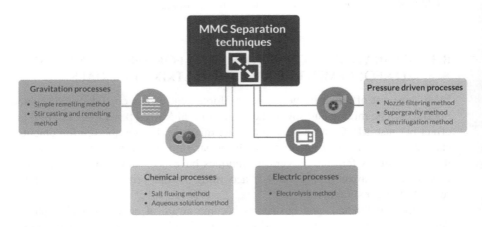

FIGURE 8.7 Classification of techniques used in the separation of MMC constituents.

Recycled Metal Matrix Composites

process. Through this process, anyone can recover over 60% of metals. There are various separation methods applied to separate matrix alloys and reinforcements from MMCs, such as the simple remelting separation method (Sharma & Ahn 2018), electrorefining in an ionic liquid separation method, electrolysis separation method (Kamavaram et al., 2005), aqueous solution separation method (Li et al., 2019), salt fluxing separation method (Schuster et al., 1993), supergravity centrifugal separation method (Sun et al., 2020) and nozzle filtering separation method (Mizumoto et al., 2007). These methods are discussed in the following sections.

8.3.1 Simple Remelting

Figure 8.8 demonstrates a simple and effective separation process of reinforcement particles from the matrix (Sharma & Ahn 2018). The scrap obtained during MMC production is remelted in an electric furnace. The molten slurry is then manually stirred to remove the oxide layer entrapped into it. The molten metal is then carefully poured into molds so that only the low-density melt would fill up the mold, while the high-density reinforcement would be separated from the melt. The equation can determine the percentage recovery of matrix alloy from MMC:

$$\% \text{ recovery of matrix alloy} = \frac{W_{RA}\,(g)}{W_C(g) - W_{Z(g)}} \times 100,$$

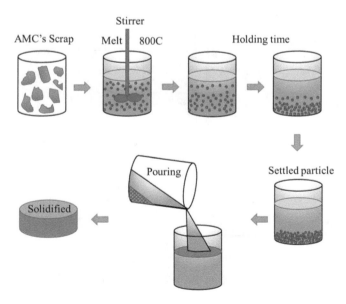

FIGURE 8.8 Separation of matrix alloy and reinforcement using a simple remelting method.

where
  W_{RA} = weight of recovered matrix alloy
  W_C = weight of the composite
  W_Z = weight of the reinforcement in the composite.

8.3.2 Electrolysis

Figure 8.9 (Kamavaram et al., 2005) shows the scheme for recycling the composite metal matrix using the electrolysis process in low-temperature ionic liquids. In this process, the composite metal matrix is used as an anode, copper is used as a cathode and pure metal is used as reference electrodes. The reference electrode was used to test the anode and cathode electrode potentials independently.

8.3.3 Aqueous Solution

The extraction of hydrogen from seawater is beneficial because seawater is the most abundant natural resource. The following reaction occurs when any metals come into contact with H_2O. As shown in Figure 8.10a, when the mass eutectic alloy sample sinks in an aqueous solution (Li et al., 2019), the formation of hydrogen gas bubbles around the sample surface can be easily observed. After some time, the mass eutectic alloy sample gets corroded and the hydrogen gas

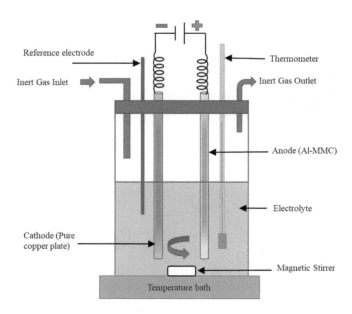

FIGURE 8.9 Separation of matrix alloy and reinforcement using the electrolysis method.

Recycled Metal Matrix Composites

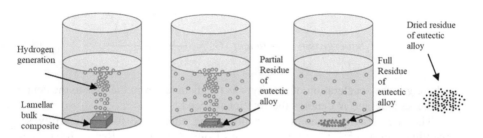

FIGURE 8.10 Extraction method for recycling the valuable metals using an aqueous solution.

bubble continues in the solution, as shown in Figure 8.10b. Ultimately in Figure 8.10c, the bulk sample gets wholly corroded, and after drying, the residues can be seen in Figure 8.10d.

8.3.4 Supergravity

Supergravity centrifugal separation is another efficient way to recycle MMCs. In this process, the centrifugal force is provided by the centrifugal apparatus shown in Figure 8.11. It consists primarily of a furnace and a counterweight that is symmetrically mounted on the centrifugal rotor. The furnace also has a cylindrical crucible with an isothermal resistance wire heated region. The furnace and counterweight shifts from vertical to horizontal as soon as the centrifugal rotor started to function. In the following equation, the gravity coefficient of the supergravity field is defined as the ratio of centrifugal acceleration to gravity's standard acceleration (Wang et al., 2018):

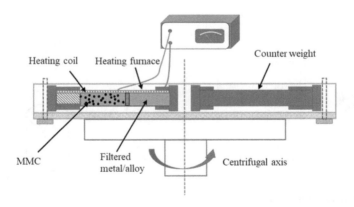

FIGURE 8.11 Schematic diagram of supergravity centrifugal separation.

$$\text{Supergravity cooeficient } (G) = \frac{\sqrt{g^2 + \left(\frac{N^2\pi^2 r}{900}\right)^2}}{g}, \qquad (8.1)$$

where g is the acceleration of gravity, r is the distance from the axis to the center of the sample and N denotes the rotational speed.

The supergravity technique is a promising alternative for successfully separating reinforcements and matrix alloys from MMC waste. Research has been conducted on the isolation of silicon carbide particles from Al MMC aluminum by a supergravity centrifugal force separation process, as shown in Figure 8.12. Through this process, high-purity aluminum alloy from waste MMC and SiC particles were further separated by dissolving the residue in the acids (Sun et al., 2020).

8.3.5 Nozzle Filtering

The other efficient process of reinforcement separation from MMC is the nozzle filtering method. In this method, the composite materials are melted using an electrical resistance furnace. At a completely molten state, Ar gas pressure forces the molten MMC to flow through the nozzle. Possible residues may be present, and these residues can be disintegrated by pickling in hydrochloric acid. Figure 8.13 shows the schematic illustration of the nozzle filtration method. Recently, the researchers have shown a keen interest in applying this separation method. The nozzle size effect on the nozzle filtering method has been studied by Mizumoto et al. (2007). A low-pressure infiltration process produced Al-Cu alloy reinforced with SiC composites, remelted and separated using the nozzle filtering method. The authors observed that due to the interfacial reaction between reinforcement and matrix alloy at high temperatures, the surface of the isolated reinforcement particles was rough.

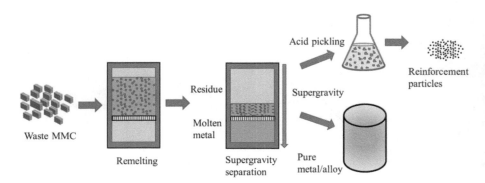

FIGURE 8.12 Process chart of recovering reinforcements and metals/alloys from waste MMC by supergravity process.

Recycled Metal Matrix Composites

8.3.6 Salt Fluxing

There is still commercial value for metal-matrix composite scrap that cannot be recycled into functional composites. The salt fluxing separation approach is the other successful process of reinforcement separation from MMCs. In order to recover the alloy matrix, reinforcement particles may be extracted from the MMCs by common salt or other fluxing techniques. This method incorporates solid salt mixtures or injecting reactive gases directly into the molten MMCs for effective dewetting of ceramic particles from the alloy matrix. This process is schematically displayed in Figure 8.14.

Separation of the matrix and reinforcement from the metal matrix composite (MMC) can be accomplished by adding a particular salt flux mix (Ravi et al., 2007). During the fabrication of MMCs, the minimum work, W, required is given by

$$W = (\gamma_{rm} - \gamma_{ra})dA, \tag{8.2}$$

where

γ_{rm} is the interface energy between reinforcement and matrix metal
γ_{ra} is the surface energy of the reinforcement surrounded by air

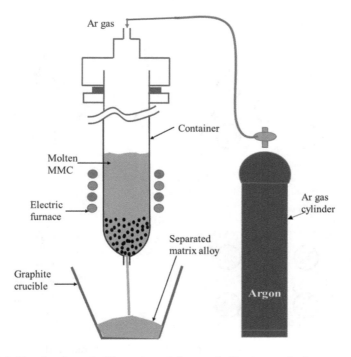

FIGURE 8.13 A schematic illustration of the nozzle filtering method.

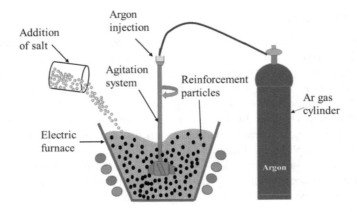

FIGURE 8.14 A schematic illustration of the salt fluxing separation method.

dA is the total interface area between matrix metal and reinforcement.

The separation takes place naturally when the interface free energy (ΔG_1) is negative, as shown in Figure 8.15a and b.

$$\Delta G_1 = \gamma_{ra} dA_1 + \gamma_{ma} dA_2 - \gamma_{rm} dA, \tag{8.3}$$

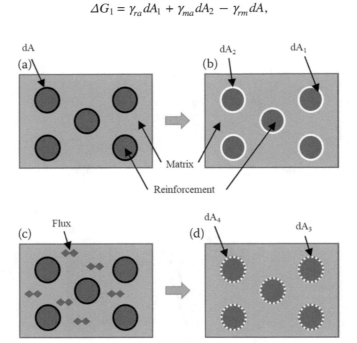

FIGURE 8.15 Surface area at the interface: (a) before separation, (b) after separation (before flux addition), (c) before separation and (d) after separation (after flux addition).

Recycled Metal Matrix Composites

where

γ_{ma} is the surface energy of the matrix surrounded by air.

dA_1 and dA_2 are the surface areas of reinforcement and matrix metal at the interface.

The free energy interface between reinforcement and matrix metal when salt fluxes are added to the separation process is shown in Figure 8.15c and d.

$$\Delta G_2 = \gamma_{rf} dA_3 + \gamma_{mf} dA_4 - \gamma_{rm} dA, \qquad (8.4)$$

where

γ_{rf} is the energy interface between the flux and the reinforcement

γ_{mf} is the energy interface between the flux and the matrix

dA_3 is the area interface between flux and the reinforcement

dA_4 is the area interface between flux and the matrix.

Assuming

$$dA = dA_3 = dA_4$$

since the difference between these three interface areas is very small,

$$\Delta G_3 = (\gamma_{rf} + \gamma_{mf} - \gamma_{rm}) dA \qquad (8.5)$$

According to the previous equation, when $\Delta G_3 < 0$ and the salt fluxes have negligible solubility in the metal matrix, then the separation of the matrix metal and the reinforcement of the composite takes place naturally. Therefore, adding salt flux results in complete isolation of matrix material and reinforcement from MMC scrap with a metal yield of 84%. The separated aluminum matrix can potentially be reusable.

8.4 MECHANICAL PROPERTIES OF THE RECYCLED MMCS

It is necessary to discuss the physical and mechanical properties of waste material to understand the factors that make it a resource. The mechanical properties of recycled MMCs are shown in Table 8.1. The table summarizes various techniques used in the production of MMCs using recycled materials and their applications. MMCs can be recycled into new products using several processing methods, and their applications are also presented in the table. The reuse of MMCs scrap from post-industrial waste is finding various applications such as automotive, packaging, architecture, electrical transmission and energy generation applications, military vehicles and aircraft applications. Based on the literature examined, there is a strong potential to enhance the reused metal matrix composites' material behavior by adding appropriate reinforcement to the metal matrix composite (Mohd Joharudin et al., 2019). Some waste materials are

TABLE 8.1
Recent Methods of Developing MMCs Using Recycled Materials and Their Mechanical and Physical Properties

MMCs	Processing Technique	Applications/Properties	References
Wrought alloy (6061) Scrap/Al_2O_3	Direct Chill (DC) casting	Chemical analysis reveals that no substance has contaminated the aluminum alloy. Suitable for applications that are both economical and ecologically demanding. Tensile strength: 367 MPa	(Schuster et al., 1993)
A380Al/SiC(p) Shot biscuit, overflow well of bicycle	Die casting	Bicycle chain wheel	(Lin et al., 1998)
Aluminum and bronze granulated chips AlCuAl12	Cold compaction and hot extrusion	Bearing materials, good frictional properties. Tensile strength: 109.4 MPa, Compression strength: 58.9 MPa	(Chmura & Gronostajski, 2000)
Scrap waste-used Al-Al_3Fe FGM	Centrifugal method	FGM with fine Al_3Fe particles is needed for commercial use, which has more ductility and greater strength. Tensile strength: 94 MPa	(Yamagiwa et al., 2003)
Scrap Aluminium Alloy chips from (AA6061/ Al_2O_3)	Cold press and hot extrusion	The mechanical properties of recycled composite materials are good or even improved. Tensile strength: 172 MPa	(Fogagnolo et al., 2003)
Recycled bronze chips (CuSn10)/ spheroidal cast iron chips (GGG40)	Hot isostatic pressing	Relatively clean surfaces lead to machining without coolant; optimal mechanical performance for journal bearing; strong tribological characteristics. Compression strength: 514 MPa, Hardness: 150 HRB	(Abdullah Aslan et al., 2018)
Recycled A380Al/ SiC(p) Composites (shot biscuits, runners, and overflow wells of bicycle chain wheel)	Partially closed stir casting	In the recycling of composites, very little porosity and inclusion have been found. Tensile strength: 258 MPa	(Lin et al., 1998)
Recycled aluminum chip/Al_2O_3 composite	Hotpress forging	When made from recycled aluminum using a hot press, it has the highest ultimate tensile strength and elongation capacity. Tensile strength: 318 MPa.	(Lajis et al., 2017)

(Continued)

TABLE 8.1 (Continued)
Recent Methods of Developing MMCs Using Recycled Materials and Their Mechanical and Physical Properties

MMCs	Processing Technique	Applications/Properties	References
Scrap AA7075 aluminium alloy/ Nb$_2$Al and SiC particles (hybrid)	Sintering and forging	The developed recycled composite can be used as an optional material for electric motor components, for instance, brushes, where wear and strength improvements are required. Tensile strength: 210 MPa, Microhardness (HV$_{0.3}$): 398	(Enginsoy et al., 2020)
Al/TiB$_2$	High pressure die casting, multi pouring sand casting	Pistons, train brake discs, and clutch discs. Tensile strength: 230 MPa, Hardness: 89 HB	(Goñi et al., 2003)
Al Scrap alloy (Window panel) + SiO$_2$	Ar atmosphere electric resistance furnace	850 MPa (Compressive fracture stress), Hardness: 400 GPa (Hv)	(Yoshikawa et al., 2003)
Scrap Aluminium Alloy Wheel (SAAW) + Ash from Rice Husk + Spent Catalyst	Stir-squeeze casting method	Tensile strength: 115.6 MPa, Compressive strength: 210.5 MPa, Hardness: 86 HRB	(Dwivedi et al., 2020)
Remelting of A359 matrix composite	Simple remelting	Tensile strength: 220 MPa	(Klasik et al., 2016)

generated during the manufacture of certain industrial MMCs such as brake disks, bicycle frames, motorcycle sprockets, seamless automobile shaft tubes, chain wheels, bicycle chain wheels, motorcycle sprockets, bearings, electric motor brushes, golf club heads and electrical equipment racks. For example, the die casting of motorcycles components produces plenty of shot biscuits and waste material in primary runners, primary and overflow wells (Schuster et al., 1993). The composite materials are generally classified as heterogeneous, complex and sometimes anisotropic materials. Their properties are affected by various variables, including the form of reinforcement, volume fraction, geometry, distribution, matrix/reinforcement interface and manufacturing processes (Haghshenas, 2016).

Fair mechanical and physical properties, including tensile strength, compressive strength and hardness of recycled MMCs, have been obtained. Yoshikawa et al. (2003) synthesized aluminum alloy synthetic scrap window frame melted and reacted with SiO_2. The produced composite showed higher strength and hardness. Recycling of metal matrix composites is carried out through the powder metallurgical process where the scrap MMCs are ball milled and then sintered. Alternatively, Fogagnolo et al. (2003) has employed a novel, cost-effective and straightforward method for recycling metal matrix composite through the cold press and hot extrusion. This method has achieved a comparatively higher mechanical strength, such as tensile and hardness, than the primary composite. Lin et al. (1998) introduced the novel metal matrix composite melting and degassing equipment. They mixed old material such as a composite bicycle chain wheel and new material of these composites in different weight ratios. After recycling for the third time, very little porosity and inclusion in composites recycled have been found. From tensile strength tests, tensile strength and elongation are the same for new materials and composites. Schuster et al. (1993) attempted to establish a sustainable approach for processing and reclamation of metal matrix composites, which is crucial to manufacturing these advanced materials. For the recycling process, recycling of both extrusion MMC scraps and foundry alloy scraps are chosen in their study. Both MMC scraps are recycled multiple times, and their mechanical properties have been examined. The MMC extruded alloy scrap was recycled into the direct chill cast method, and the resulting billet was checked to see whether the composite properties were degraded by repeated recycling. As a result, the author identified the produced composite is mostly not affected by the number of times the material being recycled. Chmura and Gronostajski (2000) investigated the processing and mechanical properties of bearing materials developed by recycling bronze aluminum chips through cold compaction and hot extrusion. The author established that the reinforcement and particle size phase content are key factors to enhance physical and mechanical properties. Better tensile strength and the worst compression strength are measured when the bronze chips are smaller in particle size. The author also concluded that, by using waste materials, composite bearing materials can be made. Abdullah Aslan et al. (2018) examined porous metal matrix composites' physical and mechanical characteristics consisting of spheroidal cast iron chips as reinforcement agent and bronze pieces as matrix agent. Hot isostatic pressing process was used to produce the composite under different temperature and pressure conditions. The maximum hardness and compressive strength obtained is mainly based on the nature of the matrix alloy. The research found that these types of metal chips could be recycled repeatedly, resulting in MMCs with strong mechanical properties.

8.5 CONCLUSIONS

Over the past decades, the mineral ore content of mined metal deposits has declined. Reducing ores' metal content leads to increased expenditure on raw material, overload, energy, emissions, etc., which should not be ignored as it has a significant impact on our ecosystem. This chapter discussed the feasibility of metal/alloy/MMC waste recycling systems. Recycling means reducing production and energy costs. Furthermore, to build a better ecosystem, the reused composite materials' physical characteristics and mechanical properties were presented. It can be concluded that using scrap materials as a matrix material for low-cost, high-performance metal matrix composite materials is a viable choice. Recycled alloys that may have similar physical and mechanical characteristics to virgin alloys may be obtained.

- Optimal tensile and compressive strength can be achieved by scrap metals/alloys. Besides, metals may be recycled without the unnecessary deterioration of their mechanical properties.
- Products such as aluminum chips, bicycle chain wheels, scrap aluminum window panels, scrap alloy wheels, bronze chips, and cast iron chips can be produced and recycled.
- Recyclability processes include remelting, holding and casting, direct chill casting, stir casting, cold compaction and hot extrusion, direct solid and semi-solid processing methods, hot isostatic pressing, powder metallurgy, thixoforming and rheocasting.
- Stir casting technology is a relatively simple and the most economical way to process recycled composites.
- For the recycling of MMCs, approximately 50–77% of the matrix metal can be retrieved from composite scraps through the matrix metal separation technique. Some of the successful separation methods are the equimolar mixture of the salt flux method and the ionic liquid processing method's electrolysis.

ACKNOWLEDGMENT

The first authors thank the management of the National University of Science and Technology (NUST). The second author thanks Sultan Qaboos University (SQU), Muscat, Sultanate of Oman, for support and encouragement in research activities.

REFERENCES

Al-Helal, Kawther, Patel, Jayesh B., Scamans, Geoff M., & Fan, Zhongyun (2020). Melt conditioned direct chill (MC-DC) casting and extrusion of AA5754 aluminium alloy formulated from recycled taint tabor scrap. *Materials, 13*(12). doi:10.3390/ma13122711.

Aslan, A., Salur, E., Gunes, A., Sahin, O. S., Karadag, H. B., & Akdemir, A. (2019). The mechanical properties of composite materials recycled from waste metallic chips under different pressures. *International Journal of Environmental Science and Technology, 16*(9), 5259–5266. doi:10.1007/s13762-019-02317-3.

Aslan, Abdullah, Güneş, Aydın, Salur, Emin, Şahin, Ömer Sinan, Karadağ, Hakan Burak, & Akdemir, Ahmet (2018). Mechanical properties and microstructure of composites produced by recycling metal chips. *International Journal of Minerals, Metallurgy and Materials, 25*(9), 1070–1079. doi:10.1007/s12613-018-1658-8.

Bahrami, A., Soltani, N., Pech-Canul, M. I., & Gutierrez, C. A. (2015). Development of metal-matrix composites from industrial/agricultural waste materials and their derivatives. *Critical Reviews in Environmental Science and Technology, 46*(2), 143–208. doi:10.1080/10643389.2015.1077067.

Chmura, W., & Gronostajski, J. (2000). Mechanical and tribological properties of aluminium-base composites produced by the recycling of chips. *Journal of Materials Processing Technology, 106*(1–3), 23–27. doi:10.1016/S0924-0136(00)00632-4.

Dwivedi, Shashi Prakash, Sharma, Pardeep, & Saxena, Ambuj (2020). Utilization of waste spent alumina catalyst and agro-waste rice husk ash as reinforcement materials with scrap aluminium alloy wheel matrix. *Proceedings of the Institution of Mechanical Engineers, Part E: Journal of Process Mechanical Engineering.* doi:10.1177/0954408920930634.

Enginsoy, H. M., Bayraktar, E., Katundi, D., Gatamorta, F., & Miskioglu, I. (2020). Comprehensive analysis and manufacture of recycled aluminum based hybrid metal matrix composites through the combined method; sintering and sintering + forging. *Composites Part B: Engineering, 194*, 108040. doi:10.1016/j.compositesb.2020.108040.

Fan, Tongxiang, Zhang, Di, Yang, Guang, Shibayanagi, Toshiya, Naka, Masaaki, Sakata, Takao, & Mori, Hirotaro (2003). Chemical reaction of SiCp/Al composites during multiple remelting. *Composites Part A: Applied Science and Manufacturing, 34*(3), 291–299. doi:10.1016/S1359-835X(03)00029-0.

Fogagnolo, J. B., Ruiz-Navas, E. M., Simón, M. A., & Martinez, M. A. (2003). Recycling of aluminium alloy and aluminium matrix composite chips by pressing and hot extrusion. *Journal of Materials Processing Technology, 143–144*(1), 792–795. doi:10.1016/S0924-0136(03)00380-7.

Goñi, J., Egizabal, P., Coleto, J., Mitxelena, I., Leunda, I., & Guridi, J. R. (2003). High performance automotive and railway components made from novel competitive aluminium composites. *Materials Science and Technology, 19*(7), 931–934. doi:10.1179/026708303225004413.

Haghshenas, M. (2016). Metal–matrix composites. *Reference Module in Materials Science and Materials Engineering*, 1–28. doi:10.1016/b978-0-12-803581-8.03950-3.

Jaishankar, Monisha, Tseten, Tenzin, Anbalagan, Naresh, Mathew, Blessy B., & Beeregowda, Krishnamurthy N. (2014). Toxicity, mechanism and health effects of some heavy metals. *Interdisciplinary Toxicology, 7*(2), 60–72. doi:10.2478/intox-2014-0009.

Kamavaram, V., Mantha, D., & Reddy, R. G. (2005). Recycling of aluminum metal matrix composite using ionic liquids: Effect of process variables on current efficiency and deposit characteristics. *Electrochimica Acta, 50*(16–17), 3286–3295. doi:10.1016/j.electacta.2004.12.002.

Klasik, A., Maj, M., Pietrzak, K., Wojciechowski, A., & Sobczak, J. (2016). Fatigue life and microstructure after multiple remelting of A359 matrix composites reinforced with SiC particles. *Archives of Metallurgy and Materials, 61*(4), 2123–2128. doi:10.1515/amm-2016-0340.

Krishnan, P. K., Christy, J. V., Arunachalam, R., Mourad, A.-H. I., Muraliraja, R., Al-Maharbi, M., Murali, V., & Chandra, M. M. (2019). Production of aluminum alloy-based metal matrix composites using scrap aluminum alloy and waste materials: influence on microstructure and mechanical properties. *Journal of Alloys and Compounds, 784.* doi:10.1016/j.jallcom.2019.01.115.

Lajis, M. A., Ahmad, A., Yusuf, N. K., Azami, A. H., & Wagiman, A. (2017). Mechanical properties of recycled aluminium chip reinforced with alumina (Al_2O_3) particle: mechanische eigenschaften von mit aluminiumoxid (Al_2O_3) Verstärkten Recycelten

Aluminiumspänen. *Materialwissenschaft Und Werkstofftechnik, 48*(3), 306–310. doi: 10.1002/mawe.201600778.

Li, Song Lin, Song, Jenn Ming, & Uan, Jun Yen (2019). Mg–Mg2X (X=Cu, Sn) eutectic alloy for the Mg2X nano-lamellar compounds to catalyze hydrolysis reaction for H2 generation and the recycling of pure X metals from the reaction wastes. *Journal of Alloys and Compounds, 772*, 489–498. doi: 10.1016/j.jallcom.2018.09.154.

Lin, C. B., Leu, C. E., & Lee, E. C. (1998). Manufacturing and recycling of A380Al/ SIC(p) composites. *Journal of Materials Engineering and Performance, 7*(2), 239–246. doi: 10.1361/105994998770347972.

Mizumoto, Masayuki, Ohgai, Takeshi, & Kagawa, Akio (2007). Novel separation technique of particle reinforced metal matrix composites by fused deposition method. *Materials Science Forum, 539–543*, 1028–1032. doi: 10.4028/www.scientific.net/msf.539-543.1028.

Mohd Joharudin, N. F., Abdul Latif, N., Mustapa, M. S., Mansor, M. N., Siswanto, W. A., Murugesan, J., & Yusof, F. (2019). Effect of amorphous silica by rice husk ash on physical properties and microstructures of recycled aluminium chip AA7075. *Materialwissenschaft Und Werkstofftechnik, 50*(3), 283–288. doi: 10.1002/mawe.201800229.

Ravi, K. R., Pillai, R. M., Pai, B. C., & Chakraborty, M. (2007). Separation of matrix alloy and reinforcement from aluminum metal matrix composites scrap by salt flux addition. *Bulletin of Materials Science, 30*(4): 393–398. doi: 10.1007/s12034-007-0063-0.

Schuster, David M., Skibo, Michael D., Bruski, Richard S., Provencher, Robert, & Riverin, Gaston (1993). The recycling and reclamation of metal-matrix composites. *SAE Technical Papers*, 26–30. doi: 10.4271/930182.

Sharma, Ashutosh, & Ahn, Byungmin (2018). Recycling of aluminum alloy from al-cu metal matrix composite reinforced with SiC particulates. *Korean Journal of Materials Research 28*(12), 691–695. doi: 10.3740/MRSK.2018.28.12.691.

Sun, Ningjie, Wang, Zhe, Guo, Lei, Wang, Lu & Guo, Zhancheng (2020). Efficient separation of reinforcements and matrix alloy from aluminum matrix composites by supergravity technology. *Journal of Alloys and Compounds, 843*. doi: 10.1016/j.jallcom.2020.155814.

Wang, Zhe, Gao, Jintao, Shi, Anjun, Meng, Long, & Guo, Zhancheng (2018). Recovery of zinc from galvanizing dross by a method of supergravity separation. *Journal of Alloys and Compounds, 735*, 1997–2006. doi: 10.1016/j.jallcom.2017.11.385.

Yamagiwa, Kazuhisa, Watanabe, Yoshimi, Fukui, Yasuyoshi, & Kapranos, Plato (2003). Novel recycling system of aluminum and iron wastes-in-situ Al-Al$_3$Fe functionally graded material manufactured by a centrifugal method. *Materials Transactions, 44*(12), 2461–2467. doi: 10.2320/matertrans.44.2461.

Yoshikawa, Noboru, Yamaguchi, Hideaki, Kitahara, Gaku & Taniguchi, Shoji (2003). Utilization of Al scrap for fabrication of Al(Alloy)/Al2O3 composite material. *Materials Transactions, 44*(7), 1271–1275. doi: 10.2320/matertrans.44.1271.

9 Thermal Properties of Recycled Polymer Composites

Marwah Rayung[1], Min Min Aung[1,2,3], and Hiroshi Uyama[4]

[1]Institute of Tropical Forestry and Forest Products, Universiti Putra Malaysia, Selangor, Malaysia
[2]Department of Chemistry, Faculty of Science, Universiti Putra Malaysia, Selangor, Malaysia
[3]Chemistry Division, Centre of Foundation Studies for Agricultural Science, Universiti Putra Malaysia, Selangor, Malaysia
[4]Department of Applied Chemistry, Graduate School of Engineering, Osaka University, Japan

CONTENTS

9.1 Introduction 163
9.2 Thermal Analysis for Polymeric Materials 165
9.3 Case Studies of Recycled Polymer Composites 166
    9.3.1 Recycled Polyethylene Terephthalate Composites 166
    9.3.2 Recycled Polyethylene Composites 167
    9.3.3 Recycled Polyvinyl Chloride Composites 169
    9.3.4 Recycled Polypropylene Composites 171
    9.3.5 Recycled Polystyrene Composites 171
9.4 Summary and Future Outlook 174
References 176

9.1 INTRODUCTION

The growing consumptions of polymeric materials globally show the dependency of human beings towards these materials. As their production increases, so does the amount of waste generated from this industry. The problems arise when those wastes cannot be naturally degraded; they take thousands of years to decompose. Since most of these wastes are resistant to biological degradation, they are usually disposed of using conventional methods such as landfill and incineration (Kiełbasa & Korenko, 2019). However, they are not a viable option in the long run as landfill takes up valuable spaces while incineration gives off toxic gaseous that could affect humans and the environment. Much effort had been made to find the proper system for managing polymeric wastes. Recycling is one of the suggested strategies for tackling this problem. In general, recycling is the process by which waste material is collected, processed and converted into new products. Implementing this strategy would minimize the quantity of waste sent to landfills and incinerators, while at the same time helping to protect the environment and natural resources (Grigore, 2017).

In general, polymers can be classified as thermoplastic polymers and thermosetting polymers. These two polymers can be distinguished by the way they respond to being heated. Upon heating, thermoplastics polymers will melt, soften and can be re-formed back when they are cooled down. On the other hand, when thermosetting polymers are being heated, they harden and cannot be re-formed. Thermoplastics polymers are easier to recycle as they can be melted and re-formed. As for thermosetting polymers, they are much more difficult to recycle, and usually, some of them are grounded and being used as fillers for another process, or in some cases, they are broken down into their chemical building blocks that can be reused (Delvere et al., 2019). In the polymer waste stream, types of wastes produced worldwide includes polyethylene terephthalate (PET), low-density polyethylene (LDPE), high-density polyethylene (HDPE), polyvinyl chloride (PVC), polypropylene (PP) and polystyrene (PS) at which HDPE and LDPE forms the largest fraction (Basha et al., 2020). These polymeric wastes would be beneficial if they can be reused, repurposed and reprocessed with no or little reduction in their properties.

Nowadays, there are four types of commonly known recycling methods for polymers: primary, secondary, tertiary and quaternary recycling. The primary and secondary involve the thermo-mechanical process of mixing and remolding of the polymers. They differ in the sense that the primary recycling is for uncontaminated materials, while secondary recycling is for post-consumer waste. So the secondary required extra steps for washing in the process. The tertiary involves a chemical recycling method at which the polymers undergo depolymerization method to produce the original monomers or oligomers of the materials. Further, quaternary recycling is for the energy recovery process based on the combustion of the polymeric waste (Veloso & Clodoaldo, 2019). However, other than high-cost production, the main barriers to polymer recycling is the weakening of their properties after being reprocessed (Hugo et al., 2011). One of the approaches being used to alleviate this problem and upgrading the quality of the recycled polymer is by the incorporation of fillers. A wide range of fillers can be used for this purpose such as natural fibers, agricultural industrial residue, mineral-based materials and so on. These fillers can be classified into two classes: organic and inorganic. The recycled polymer itself can act both as filler and matrix in the system. In this case, the combination of polymers and fillers is called a polymer composite.

In essence, polymer composites is a mixture of two or more materials with significantly different physical and chemical properties but they often complementary to each other (Wahit et al., 2012). The aim of polymer composite

Recycled Polymer Composites

development is to produce a product with performance characteristics that combine the positive attributes of each component. The constituent materials can be assigned as a matrix and filler or reinforcement component. The properties of the polymer composites are influenced by a number of factors including the type of composite system, choice of materials and their composition, phase compatibility and methods of processing (Kim et al., 2011). The same concept can be applied to recycled polymer composites. Therefore, understanding the fundamental properties and characteristics of the materials is an important basis before going through the whole production of these materials. In the processing of recycled polymer composites, there are several properties to look up such as chemical, electrical, mechanical and thermal analysis. In particular, knowledge about thermal properties is very important as it can be an indicator about the quality of the materials. The thermal characteristics depend on their molecular orientation, materials being used, methods of processing and thermal history. Some of them are very sensitive to temperature changes and these properties affect their overall performance. This chapter illustrates a few case studies on thermal behaviors involving different types of recycled polymer composite systems utilizing various forms of material as fillers.

9.2 THERMAL ANALYSIS FOR POLYMERIC MATERIALS

Thermal analysis is a powerful tool that can be used for inspection, process optimization and component failure analysis. Various thermal analyses have been used to study the thermal behavior of polymeric materials such as thermogravimetric analysis (TGA), differential scanning calorimetry (DSC), dynamic mechanical calorimetry (DMA), thermomechanical analysis (TMA), dilatometry (DIL), dielectric analysis (DEA), heat flow meters (HFM), laser flash technique (LFA) and others. Thermal analysis at a glance can provide information regarding the thermal transition, specific heat capacity, glass transition temperature, melting temperature, thermal conductivity, etc. Table 9.1 shows the thermal analysis techniques that are being used to study the thermal behavior of polymeric materials.

In the case of recycled polymer composites, one of the most frequently used thermal characterization methods is TGA. For TGA, the main information from this analysis is the thermal stability and compositional analysis such as matrix and filler content. Another important analysis is DSC, which can be used to determine the heat capacity and various structural transition temperatures of polymeric materials. DMA is used for measuring the viscoelasticity of the materials, glass transition and modulus values. The following section will discuss the thermal behavior of different types of recycled polymer composite systems. To date, most of the reported thermal analysis for recycled polymer composites is based on TGA and DSC methods of characterization, and some studies report on DMA analysis.

TABLE 9.1
Thermal Analysis Techniques for Particular Properties

Properties or Application	Instrument
Specific heat capacity	DSC
Glass transition temperatures	DSC, TMA, DMA, DEA, LFA
Degree of crystallinity	DSC
Melting point, crystallization behavior, enthalpies	DSC
Degree of crystallinity	DSC
Thermal transition	DSC, DIL, TMA, LFA
Thermal stability, thermal decomposition, depolymerization and degradation	TGA
Compositional analysis	DSC, TGA
Thermal expansion coefficient	DIL, DMA, TMA
Filler content	TGA
Thermal conductivity	HFM, LFA
Viscoelastic behavior	DMA, TMA
Mechanical damping	DMA

9.3 CASE STUDIES OF RECYCLED POLYMER COMPOSITES

9.3.1 Recycled Polyethylene Terephthalate Composites

Polyethylene terephthalate (PET) is one of the most commonly used thermoplastic polymers in packaging industries, monopolizing the bottles market for beverages. It is abbreviated by the resin identification for recycling as code #1. PET is a hard, lightweight, robust and dimensionally stable material. It can withstand high temperatures while maintaining its structural properties. PET can be highly transparent and colorless, but thicker sections are usually opaque and off-white. Depending on its processing and thermal history, PET can exist both as an amorphous and as a semi-crystalline polymer and it can be easily recycled at high temperature. The use of recycled PET (RPET) in place of the virgin PET has been significantly rose up due to its potential recyclability properties (Nisticò, 2020). The chemical structure and thermal properties of the PET is presented in Figure 9.1. T_g and T_m represent the glass transition temperature and melting point of the polymer.

PET can be recycled through mechanical, chemical, and energy recovery approaches. Since PET is a thermoplastic polymers, its properties is hugely depending on the reprocessing condition (Nonato & Bonse, 2016). Specific

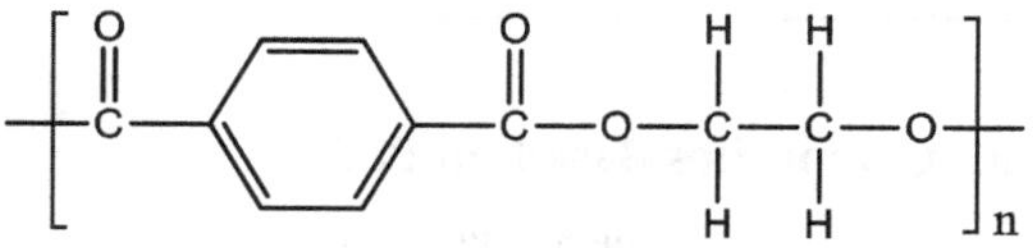

Recycled Polymer Composites

PET	Thermal properties	
	T_g (°C)	T_m (°C)
Amorphous	60-84	-
Semi-crystalline	68-80	255-265

FIGURE 9.1 Chemical structure and thermal properties of PET.

property enhancement has been done by several techniques such as incorporation of fillers, blending with other polymers and surface modification through physical or chemical treatments. The fillers used vary based on organic/inorganic and natural/synthetic fillers. Studies on RPET blended with other polymers have been carried out with other polymers such as PP, PE, PBS and so on. In some cases, the RPET was used as the filler in the composite system. Numerous systems have been developed for recycled PET with different reinforcement materials, as shown in Table 9.2.

Based on the literature study, the thermal properties of the RPET composites differ from one system to another depending on several factors such as the type, size and concentration of the filler. A study on RPET/Mg(OH)$_2$ nanocomposites for fire retardant applications shows that the thermal stability improved after addition of the nanofillers compare to the control. The thermal properties further increase with an increasing concentration of the Mg(OH)$_2$ (Naguib, 2018).

9.3.2 Recycled Polyethylene Composites

Polyethylene (PE) is one of the major waste products worldwide. PE is used in a variety of applications ranging from packaging, consumer goods, automotive, piping, textile, etc. PE can be classified by its density and branching. Low-density polyethylene (LDPE) and linear low-density polyethylene (LLDPE) belong to the branch version of PE. LDPE is softer and more flexible compared to HDPE and thus it is widely used in the production of plastics bags. On the other hand, HDPE is harder and has better resistance to heat. LDPE possesses high transparency due to the amorphous nature, while HDPE has low transparency because of the increased level of crystallinity. In the resin identification code for recycling, HDPE represents code #2, while LDPE is classified as code #4. Figure 9.2 shows the chemical structure and thermal properties of PE (Table 9.3).

TABLE 9.2
Recycled Polymer Composites Based on PET

Polymer Composites	Thermal Properties	References
RPET/PEN-MWCNT	DSC-No significant changes in the thermal propertiesDMA	(Yesil, 2013)
RPET/PEN-Carbon/Glass fiber	Thermal stability increases with the addition of PEN and short fiber	(Karsli et al., 2013)
RPET-HBPET	DSC-Thermal stability increases with the addition of HBPET	(Saeed et al., 2014)
RPET-Newspaper fiber	Thermal stability increases with the addition of the newspaper fiber	(Ardekani et al., 2014)
RPET/HDPE-EGMA-Rice husk flour	Thermal stability increases with the addition of rice husk flour	(Chen et al., 2014)
RPET-Sisal fiber-glycerol-castor oil	No significant changes in the thermal properties	(Passos et al., 2014)
RPET-Kenaf-POM	No significant changes in the thermal properties	(Dan-mallam et al., 2014)
RPET-ZnO nanocomposites	Thermal stability increases with the addition of diacid-capped zinc	(Mallakpour & Javadpour, 2016)
RPET/PP	Thermal stability increases with the addition of RPET	(Nonato & Bonse, 2016)
RPET-MWCNT-ZnO	Thermal stability increases with the addition of MWCNT-ZnO	(Mallakpour & Behranvand, 2017)
RPET-PBAT-Wollastonite	Thermal stability increases with the addition of wollastonite	(Chuayjuljit et al., 2017)
RPET-Mg(OH)$_2$ nanocomposites	Thermal stability increases with the addition of Mg(OH)$_2$	(Naguib, 2018)
RPET-PEN-Mica platelets	Thermal stability increases with the addition of PEN and mica platelets	(Mebarki & David, 2018)
RPET-Glass fiber	Thermal stability increases with the addition of PEN and glass fiber	(Mebarki & David, 2018)
RPET-MWCNT	Thermal stability increases with the addition of MWCNT	(Chowreddy et al., 2018)
RPET-Clay nanocomposites	Thermal stability decreases with increasing clay content	(Reddy et al., 2018)
RPET/TPU-MWCNT	Thermal stability increases with the addition of MWCNT	(Fang et al., 2018)
RPET-Biochar	Thermal stability increases with the addition of biochar	(Idrees et al., 2018)
RPET-Epoxy-coated kenaf fiber	Thermal stability decreases with the addition of kenaf fiber	(Owen et al., 2018)
RPET-GNP	Thermal stability decreases with increasing GNP content	(Yang et al., 2019)
RPET-CATAS nanocomposites	Thermal stability increases with the addition of CATAS	(Dominici et al., 2020)

(Continued)

TABLE 9.2 (Continued)
Recycled Polymer Composites Based on PET

Polymer Composites	Thermal Properties	References
Rubber-RPET-Wood fiber	–	(Cosnita et al., 2014)
Rubber-RPET-Halloysite nanotubes	Thermal stability increases with the addition of halloysite nanotubes	(Nabil & Ismail, 2015)
UPR-RPET	Thermal stability increases with the addition of RPET	(Dehas et al., 2018)
Epoxy resin-RPET-CaCO$_3$	Thermal stability increases with the addition of RPET and CaCO$_3$	(Nguyen et al., 2020)

EGMA – ethylene-glycidyl methacrylate; PBAT – poly(butylene adipate-co-terephthalate); CATAS – Calcium terephthalate anhydrous salts; MWCNT – multi-walled carbon nanotubes; NR – natural rubber; HBPET – Hyperbranched polyester; POM – Polyoxymethylene; UPR – unsaturated polyester resin; PEN – poly(ethylene 2,6-naphthalate); PBS – poly(butylene succinate).

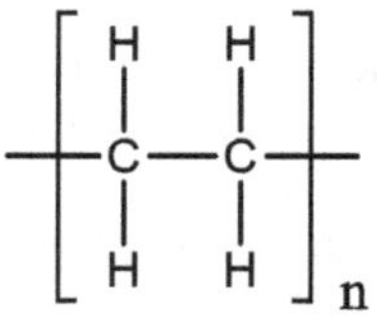

PE	Thermal properties	
	T_g (°C)	T_m (°C)
LDPE	-125	105-115
HDPE	-120	120-140

FIGURE 9.2 Chemical structure and thermal properties of PE.

9.3.3 Recycled Polyvinyl Chloride Composites

Polyvinyl chloride (PVC) is a thermoplastic polymer used in wire and cable insulation, pipes, packaging, clothing, automotive and others. PVC is usually produced either in a rigid or flexible form. The flexible type contains plasticizers and it often referred to as plasticized PVC. Furthermore, PVC is sensitive to thermal changes and has quite a small range of processing temperatures. For this reason, additives that can stabilize the materials at higher temperatures are added to the material during production. PVC is fully recyclable and it is identified as recycling code #3. PVC can be recycled several times without a decrease in its

TABLE 9.3
Recycled Polymer Composites Based on PE

Polymer Composites	Thermal Properties	References
RLDPE-Corn husk fibers	Thermal stability enhanced with the addition of corn husk fibers	(Youssef et al., 2015)
RLDPE-Sorghum bran	Thermal stability enhanced with the addition of sorghum bran	(Ogah, 2016)
RLDPE-Chitosan	Thermal stability reduced with the addition of chitosan	(Lim et al., 2016)
RLDPE-Al$_2$O$_3$	Thermal stability enhanced with increasing filler content	(Khanam et al., 2016)
RLDPE-Pine wood waste	Thermal stability reduced after the addition of pine wood fiber	(David et al., 2017)
RLDPE-Wheat straw powder	Thermal stability enhanced with the addition of wheat straw powder	(Touati et al., 2018)
RLDPE/PP-Diss fiber	No significant changes in thermal stability	(Touati et al., 2018)
RLDPE-Clay powder	Thermal stability enhanced with increasing amount of clay powder	(Onuegbu et al., 2019)
RLDPE-Napier grass fiber	No significant changes in thermal stability	(Grillo & Saron 2020)
RHDPE-Wood fiber-PAH	Thermal stability increased with the addition PAH	(Hanif et al., 2015)
RHDPE-Palm kernel shell	Thermal stability reduced with increasing amount of filler	(Husna et al., 2016)
RHDPE-Wood waste	Addition of wood waste reduced thermal stability of the composites	(Mürsit Tufan et al., 2016)
RHDPE-Hazelnut husk	Thermal stability reduced with addition of hazelnut husk	(M Tufan & Ayrilmis 2016)
RHDPE-Wood sawdust	–	(Brostow et al., 2016)
RHDPE-Wood flour-Nanographene	Addition of nanographene increase the thermal stability of the composite	(Beigloo et al., 2017)
RHDPE-Oil palm fibers	Thermal stability increased with filler addition	(Medupin et al., 2017)
RHDPE-Banana fiber-Fly ash	Increase in thermal stability	(Satapathy & Kothapalli, 2018)
RHDPE-Jute fiber-Fly ash-MAPE	Addition of jute fiber increases thermal stability	(Satapathy, 2018)
RHDPE-Cashew nutshell powder	No significant changes in thermal stability	(Gomes et al., 2018)
RHDPE-Sugarcane bagasse	Irradiated samples show improved thermal stability	(El-zayat et al., 2019)
RHDPE-Rice hulls-MAPE	Increased thermal stability after addition of filler	(Orji & McDonald, 2020)
RLLDPE-Date palm seed powder	Decreased thermal stability with increase in filler loading	(Appu et al., 2019)

(Continued)

TABLE 9.3 (Continued)
Recycled Polymer Composites Based on PE

Polymer Composites	Thermal Properties	References
RLLDPE-Palm rachis fiber-Nanosilica	No significant changes in thermal stability	(Abdel et al., 2020)
RPE-Rice husk-GPE compatibilizers	Rice husk increases the thermal stability of the composites. The compatibilizers improve the interfacial interaction between the RPE and rice husk and thus increase the thermal stability.	(Hong et al., 2016)
RPE-Gypsum-MAPE	–	(I. Bilici et al., 2019)
RPE-Colemanite	Increase in thermal stability after addition of colemanite	(İ. Bilici, 2019)
RPE-Wheat straw fibers	Size of wheat straw fibers have no significant effect on the thermal stability	(Zhang et al., 2020)

GPE – graft copolymers of polyethylene; MAPE – maleic anhydride-grafted polyethylene; PAH – phthalic anhydride.

chain length (VinylPlus 2016). Figure 9.3 displays the chemical structure and thermal properties of PVC (Table 9.4).

9.3.4 RECYCLED POLYPROPYLENE COMPOSITES

Polypropylene (PP) is a thermoplastics polymer that belongs to the polyolefin family and is one of the top three widely used polymers alongside LDPE and HDPE. PP found applications in packaging (flexible and rigid form), consumer goods, automotive parts, fibers and fabrics, medical applications and more. PP is fully recyclable and its resin identification code is #5. The chemical structure and thermal properties of PP is depicted in Figure 9.4 (Table 9.5).

9.3.5 RECYCLED POLYSTYRENE COMPOSITES

Polystyrene (PS) is a versatile plastics that has been used in a wide variety of consumer product applications and commercial packaging. It is a naturally transparent thermoplastic polymer. Depending on the intended use, PS usually made into solid or rigid foamed materials. Solid PS is commonly used in electronic appliances, automobile parts, medical tools, food container and more. The foamed version of PS or also known as expanded polystyrene or extruded polystyrene normally used in appliances insulation, protective packaging, containers, disposable tableware and others. The most popular product of polystyrene is styrofoam, produced in the 1941 and it is one of the major wastes

TABLE 9.4
Recycled Polymer Composites Based on PVC

Polymer Composites	Thermal Properties	References
RPVC-Rice husk	–	(Ramle et al., 2013)
Natural rubber-RPVC	Thermal stability improves with increasing RPVC content	(Norazlina et al., 2015)
RPVC-RPET powder	–	(Atar et al., 2017)
RPVC-Waste printed circuit board	Thermal stability improves with the addition of filler	(Das et al., 2019)
RPVC-Fly ash	Thermal properties improves with the addition of fly ash	(Gohatre et al., 2020)

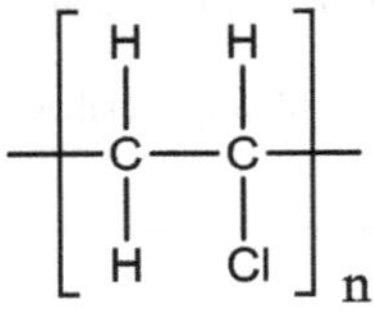

PVC	Thermal properties	
	T_g (°C)	T_m (°C)
Rigid	80	170-210
Flexible	59	170-210

FIGURE 9.3 Chemical structure and thermal properties of PVC.

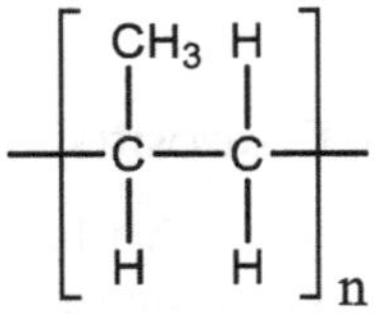

Polymer	Thermal properties	
	T_g (°C)	T_m (°C)
PP	-20	160-165

FIGURE 9.4 Chemical structure and thermal properties of PP.

TABLE 9.5
Recycled Polymer Composites Based on PP

Polymer Composites	Thermal Properties	References
RPP-Date palm wood flour-Glass fiber	No significant changes	(AlMaadeed et al., 2012)
RPP-Coconut shell powder-SDS	Thermal stability improves with addition of SDS as coupling agent	(Chun et al., 2013)
RPP-Fly ash-Palmitic acid	Thermal stability improves with the addition of palmitic acid	(Sengupta et al., 2013)
RPP-Sisal fiber	Thermal stability improves with the addition of sisal fiber	(Gupta et al., 2014)
RPP-OPEFB-Glass fiber	Incorporation of glass fiber increases the thermal stability	(Rivai et al., 2014)
RPP-Wood flour-Glass fiber	Glass fiber improves the thermal stability	(Al-Maadeed et al., 2014)
RPP-Kenaf fiber- MAPP	Treated fibers improves the thermal stability	(Muhammad Remanul Islam et al., 2014)
RPP-Chicken feathers	Thermal stability improves by the addition of chicken feathers	(Amieva et al., 2015)
RPP-Glass fiber-Ultrasound treated OPEFB	Thermal stability increases with treated OPEFB	(Muhammad R. Islam et al., 2015)
RPP-Waste wood particle-carbon ash	Thermal stability increases with the addition of carbon ash	(K. W. Chen et al., 2016)
RPP-Wheat straw fibers	Thermal stability decreases with the addition of wheat straw fibers	(Yu et al., 2016)
RPP-Kapok husk-stearic acid	Thermal stability of composites with kapok husk modified with stearic acid improved	(Chun et al., 2016)
RPP-MCC	Thermal stability reduces with the addition of MCC	(Samat et al., 2017)
RPP-Rice straw lignin-MAPP	–	(Karina et al., 2017)
RPP-Old newspaper fibers	Deinking of the old newspaper decreases thermal stability	(Zhang et al., 2018)
RPP-Buckwheat husk	–	(Salikhov et al., 2018)
RPP-Peanut shell powder	Irradiation improves thermal stability of the composites	(Zaaba et al., 2018; Zaaba & Ismail, 2018)
RPP-Wood flour-maleated soybean oil	Addition of maleated soybean oil increases the thermal stability	(M Poletto, 2019)
RPP- Newsprint fiber-based cellulose	No significant changes to virgin PP	(Bogataj et al., 2019)
RPP-Kenaf fiber-Graphite	Thermal stability improved with addition of graphite	(Ramli et al., 2020)

(Continued)

TABLE 9.5 (Continued)
Recycled Polymer Composites Based on PP

Polymer Composites	Thermal Properties	References
RPP-Cotton swab waste-Rice husk	Pyrolyzed rice husk improve thermal stability	(Moreno et al., 2020)
RPP-Bagasse fiber-Ionic liquid treated lignin	–	(Younesi-Kordkheili & Pizzi, 2020)
RPP-Waste printed circuit boards	Addition of waste printed circuit board increased thermal stability	(Grigorescu et al., 2020)

SDS – sodium dodecyl sulphate; OPEFB – oil palm empty fruit bunch; MAPP – maleic anhydride grafted polypropylene; MCC – microcrystalline cellulose

produced worldwide. In the recycling identification code, PS is denoted by code #6. The chemical structure and thermal properties (Sastri, 2010) of PS is presented in Figure 9.5 (Table 9.6).

9.4 SUMMARY AND FUTURE OUTLOOK

Recycling is considered an important approach for the waste management system of polymeric products. In the manufacturing of recycled polymer composites, knowledge about their thermal properties is essential as they directly influence the mechanical, electrical and chemical properties of the materials.

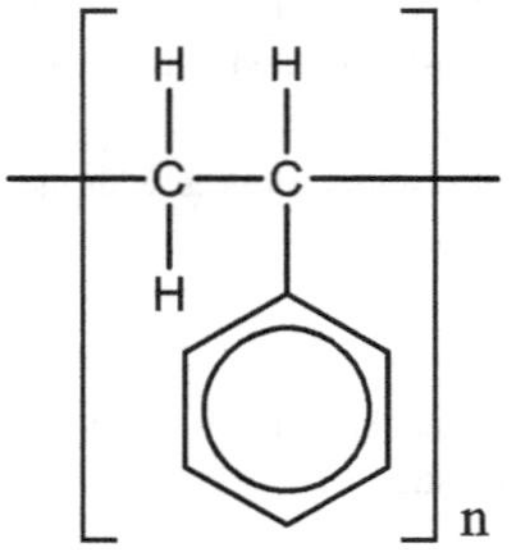

PS	Thermal properties	
	T_g (°C)	T_m (°C)
Solid	100	270
Foamed	90-95	270

FIGURE 9.5 Chemical structure and thermal properties of PS.

TABLE 9.6
Recycled Polymer Composites Based on PS

Polymer Composites	Thermal Properties	References
RPS-Wood flour	Improve thermal stability compare to virgin PS	(Z. Yang et al., 2010)
RPS-Maleated lignin	Improve thermal stability compare to virgin PS	(Lisperguer et al., 2013)
RPS-Curaua fibers	Addition of curaua fibers improve thermal stability of the composites	(Borsoi et al., 2013)
RPS-Eucalyptus nitens lignin	Thermal stability increased with the addition of lignin	(Pérez-Guerrero et al., 2014)
RPS-Wood flour-SMA	SMA improve thermal stability by improving interfacial adhesion between RPS and wood flour	(Matheus Poletto, 2016)
RPS-Wood flour-SCF	Improve thermal properties with the addition of wood flour and SCF	(Pramoda et al., 2018)
RPS-Durian husk fiber	Thermal stability reduced with the addition of durian husk fiber	(Koay et al., 2018)
RPS-Rice straw fiber-Magnesium hydroxide	Addition of magnesium hydroxide as a fire retardant improved the thermal stability of the composites	(Eskander et al., 2018)
RPS/virgin PS-Eggshell powder	Thermal stability improve due to the addition of RPS	(Hayeemasae et al., 2019)
RPS-Rice husk-Polyester polyol	Thermal stability reduced with the addition of polyester polyol	(Sandoval et al., 2020)
RPS-Coconut shell-MAPS	Thermal stability reduced with the addition of coconut shell fiber	(Ling et al., 2020)
RPS-Coffee slag	-	(Kuan et al., 2020)

MAPS – maleated polystyrene; SMA – styrene-co-maleic anhydride; SCF – short carbon fibers

Therefore, thermal analysis is important to be conducted before the recycled composites can be used for a particular product application. This information could be useful especially during the processing time and for quality control purposes. The present chapter has been focusing on the thermal properties of the waste coming from the thermoplastics polymers. The existing study shows that the thermal properties of recycled polymer composites are correlated to the materials used, especially the filler. Various types of fillers have been investigated from synthetic to natural fillers.

REFERENCES

Abdel, Wagih, Sadik, Alim, Ghaffar, Abdel, & El, Maghraby (2020). Impact of hybrid nanosilica and nanoclay on the properties of palm rachis-reinforced recycled linear low-density polyethylene composites. *Journal of Thermoplastic Composite Materials*, 1–20. doi: 10.1177/0892705720944213

Al-Maadeed, M. A., Shabana, Yasser M., & Noorunnisa Khanam, P. (2014). Processing, characterization and modeling of recycled polypropylene/glass fibre/wood flour composites. *Materials and Design, 58*, 374–380.

AlMaadeed, Mariam A, Kahraman, Ramazan, Khanam, P N, & Madi, Nabil (2012). Date palm wood flour/glass fibre reinforced hybrid composites of recycled polypropylene: mechanical and thermal properties. *Materials and Design, 42*, 289–294.

Amieva, E. Jiménez Cervantes, Velasco-Santos, C., Martínez-Hernández, A. L., Rivera-Armenta, J. L., Mendoza-Martínez, A. M., & Castaño, V. M. (2015). Composites from chicken feathers quill and recycled polypropylene. *Journal of Composite Materials, 49*(3), 275–283.

Appu, Sreekumar P., Ashwaq, Omar, Al-Harthi, Mamdouh, & Umar, Yunusa (2019). Fabrication and characterization of composites from recycled polyethylene and date palm seed powder. *Journal of Thermoplastic Composite Materials*, 1–12. doi: 10.1177/0892705719843960

Ardekani, Sara Madadi, Dehghani, Alireza, Al-Maadeed, Mariam A., Wahit, Mat Uzir, & Hassan, Azman (2014). Mechanical and thermal properties of recycled poly(ethylene terephthalate) reinforced newspaper fiber composites. *Fibers and Polymers, 15*(7), 1531–1538.

Atar, Ilkay, Karakus, Kadir, Basboga, Ibrahim Halil, Bozkurt, Fatma, & Mengeloglu, Fatih (2017). Investigation of effect of recycled PET powders on recycled polyvinyl chloride. *Journal of Engineering, 20*(2), 81–88.

Basha, Inayath Shaik, Ali, M. R., Al-Dulaijan, S U, & Maslehuddin, M. (2020). Mechanical and thermal properties of lightweight recycled plastic aggregate concrete. *Journal of Building Engineering, 32*, 101710.

Beigloo, J. G., Eslam, H. K., Hemmasi, A. H., Bayzar, B., & Ghasemi, I. (2017). Effect of nanographene on physical, mechanical, and thermal properties and morphology of nanocomposite made of recycled high density polyethylene and wood. *BioResources, 12*, 1382–1394.

Bilici, Ibrahim (2019). The effect of colemanite on thermal properties of recycled polyethylene. *Acta Physica Polonica A, 135*(5), 922–924.

Bilici, Ibrahim, Deniz, Celal U., & Oz, Baytullah (2019). Thermal and mechanical characterization of composite produced from recycled PE and flue gas desulfurization gypsum. *Journal of Composite Materials, 53*(23), 3325–3333.

Bogataj, Vesna, Fajs, Peter, Peñalva, Carolina, Omahen, Marko, Čop, Matjaž, & Henttonen, Ari (2019). Utilization of recycled polypropylene, cellulose and newsprint fibres for production of green composites. *Detritus, 7*, 36–43.

Borsoi, Cleide, Scienza, Lisete C., & Zattera, Ademir J. (2013). Characterization of composites based on recycled expanded polystyrene reinforced with curaua fibers. *Journal of Applied Polymer Science, 128*(1), 653–659.

Brostow, Witold, Datashvili, Tea, Jiang, Peter, & Miller, Harrison (2016). Recycled HDPE reinforced with sol–gel silica modified wood sawdust. *European Polymer Journal, 76*, 28–39.

Chen, Kuo Wei, Hu, Ta Hsiang, Perng, Yuan Shing, & Chen, Ching Shun (2016). Effect of carbon ash content on the thermal and combustion properties of waste wood particle/recycled polypropylene composites. *MATEC Web of Conferences, 67*, 1–6.

Chen, Ruey Shan, Hafizuddin, Mohd, Ghani, A. B., Ahmad, Sahrim, Salleh, Mohd Nazry, & Tarawneh, Mou A. (2014). Rice husk flour biocomposites based on recycled high-density polyethylene/polyethylene terephthalate blend: effect of high filler loading on physical, mechanical and thermal properties. *Journal of Composite Materials*, 1–14. doi:10.1177/0021998314533361

Chowreddy, Ravindra Reddy, Varhaug, Nord Katrin, & Rapp, Florian (2018). Recycled polyethylene terephthalate/carbon nanotube composites with improved processability and performance. *Journal of Materials Science, 53*, 7017–7029.

Chuayjuljit, Saowaroj, Chaiwutthinan, Phasawat, Raksaksri, Laksamon, & Boonmahitthisud, Anyaporn (2017). Effects of poly (butylene adipate-co -terephthalate) and ultrafined wollastonite on the physical properties and crystallization of recycled poly (ethylene terephthalate). *Journal of Vinyl and Additive Technology, 23*(2), 106–116.

Chun, Koay Seong, Husseinsyah, Salmah, & Azizi, Fatin Nasihah (2013). Characterization and properties of recycled polypropylene/coconut shell powder composites: effect of sodium dodecyl sulfate modification. *Polymer - Plastics Technology and Engineering, 52*, 287–294.

Chun, Koay Seong, Husseinsyah, Salmah, & Yeng, Chan Ming (2016). Green composites from kapok husk and recycled polypropylene: Processing torque, tensile, thermal, and morphological properties. *Journal of Thermoplastic Composite Materials, 29*(11), 1517–1535.

Cosnita, Mihaela, Cazan, Cristina, & Duta, Anca (2014). Interfaces and mechanical properties of recycled rubber–polyethylene terephthalate–wood composites. *Journal of Composite Materials, 48*(6), 683–694.

Dan-Mallam, Yakubu, Abdullah, Mohamad Zaki, Sri, Puteri, & Megat, Melor (2014). Mechanical properties of recycled kenaf/polyethylene terephthalate (PET) fiber reinforced polyoxymethylene (POM) hybrid composite Yakubu. *Journal of Applied Polymer Science*, 1–7. doi:10.1002/app.39831

Das, Rajesha K, Gohatre, Omdeo K., & Biswal, Manoranjan (2019). Influence of non-metallic parts of waste printed circuit boards on the properties of plasticised polyvinyl chloride recycled from the waste wire. *Waste Management & Research*, 1–9. doi:10.1177/0734242X19836725

David, Diego, Moreno, Pinzón, & Saron, Clodoaldo (2017). Low-density polyethylene waste/recycled wood composites. *Composite Structures, 176*, 1152–1157.

Dehas, Wided, Guessoum, Melia, Douibi, Abdelmalek, Jofre-Reche, Jose Antonio, & Martin-Martinez, Jose Miguel (2018). Thermal, mechanical, and viscoelastic properties of recycled poly(ethylene terephthalate) fiber- reinforced unsaturated polyester composites wided. *Polymer Composites*, 1–12. doi:10.1002/pc.24119

Delvere, Ieva, Iltina, Marija, Shanbayev, Maxat, Abildayeva, Aray, Kuzhamberdieva, Svetlana, & Blumberga, Dagnija (2019). Evaluation of polymer matrix composite waste recycling methods. *Environmental and Climate Technologies, 23*(1), 168–187.

Dominici, Franco, Sarasini, Fabrizio, Luzi, Francesca, Torre, Luigi & Puglia, Debora (2020). Thermomechanical and morphological properties of poly(ethylene terephthalate)/anhydrous calcium terephthalate nanocomposites. *Polymers,12*, 276.

El-Zayat, Mai M., Abdel-Hakim, A., & Mohamed, Maysa A. (2019). Effect of gamma radiation on the physico mechanical properties of recycled HDPE/modified sugarcane bagasse composite. *Journal of Macromolecular Science, Part A, 56*(2), 127–135.

Eskander, Samir B., Tawfik, Magda E., & Tawfic, Medhat L. (2018). Mechanical, flammability and thermal degradation characteristics of rice straw fiber-recycled polystyrene foam hard wood composites incorporating fire retardants. *Journal of Thermal Analysis and Calorimetry, 132*(2), 1115–1124.

Fang, Changqing, Yang, Rong, Zhang, Zisen, Zhou, Xing, Lei, Wanqing, Cheng, Youliang, Zhang, Wei, & Wang, Dong (2018). Effect of multi-walled carbon nanotubes on the physical properties and crystallisation of recycled PET/TPU composites. *RSC Advances, 8*, 8920–8928.

Giraldi, A. L. F. De M., Bartoli, J. R., Velasco, J. I., & Mei, L. H. I. (2005). Glass fibre recycled poly (ethylene terephthalate) composites: mechanical and thermal properties. *Polymer Testing, 24*, 507–512.

Gohatre, Omdeo Kishorrao, Biswal, Manoranjan, Mohanty, Smita, & Nayak, Sanjay K. (2020). Study on thermal, mechanical and morphological properties of recycled poly(vinyl chloride)/fly ash composites. *Polymer International, 69*, 552–563.

Gomes, Victor N. C., Carvalho, Amanda G., Furukava, Marciano, Medeiros, Eliton S., Morais, Dayanne D. S., & Colombo, Ciliana R. (2018). Characterization of wood plastic composite based on HDPE and cashew nutshells processed in a thermokinetic mixer. *Polymer Composites*, 1–12. doi:10.1002/pc.24257

Grigore, Mădălina Elena (2017). Methods of recycling, properties and applications of recycled thermoplastic polymers. *Recycling, 2*(24), 1–11.

Grigorescu, Ramona Marina, Ghioca, Paul, Iancu, Lorena, David, Madalina Elena, Andrei, Elena Ramona, Filipescu, Mircea Ioan, Ion, Rodica Mariana et al (2020). Development of thermoplastic composites based on recycled polypropylene and waste printed circuit boards. *Waste Management, 118*, 391–401.

Grillo, Camila Cezar, & Saron, Clodoaldo (2020). Wood-plastic from pennisetum purpureum fibers and recycled low-density polyethylene. *Journal of Natural Fibers*, DOI: 10.1080/15440478.2020.1764436.

Gupta, Arun Kumar, Biswal, Manoranjan, Mohanty, S., & Nayak, S. K. (2014). Mechanical and thermal degradation behavior of sisal fiber (sf) reinforced recycled polypropylene (RPP) composites. *Fibers and Polymers, 15*(5), 994–1003.

Hanif, M. P. M., Supri, A. G., & Zainuddin, F. (2015). Effect of phthalic anhydride on tensile properties and thermal stability of recycled high density polyethylene/wood fiber composites. *Jurnal Teknologi, 10*, 97–101.

Hayeemasae, Nabil, Song, Lim Wey, & Ismail, Hanafi (2019). Sustainable use of eggshell powder in the composite based on recycled polystyrene and virgin polystyrene mixture. *International Journal of Polymer Analysis and Characterization, 24*(3), 266–275.

Hong, Haoqun, Li, Xuesong, Liu, Hao, Zhang, Haiyan, He, Hui, Xu, Huanxiang, & Jia, Demin (2016). Transform rice husk and recycled polyethylene into high performance composites: using a novel compatibilizer to infiltratively enhance the interfacial interactions. *Progress in Rubber, Plastics and Recycling Technology, 32*(4), 253–268.

Hugo, A. M., Scelsi, L., Hodzic, A., Jones, F. R., & Dwyer-Joyce, R. (2011). Development of recycled polymer composites for structural applications. *Plastics, Rubber and Composites, 40*(6–7), 317–323.

Husna, A Nadiatul, Lim, B. Y., Salmah, H., & Voon, C. H. (2016). Palm kernel shell filled recycled high-density polyethylene: effect of filler loading. *Materials Science Forum, 857*, 191–195.

Idrees, Mohanad, Jeelani, Shaik, & Rangari, Vijaya K. (2018). 3D printed sustainable biochar-recycled PET composite. *ACS Sustainable Chemistry and Engineering, 6*(11), 1–8.

Islam, Muhammad R., Rivai, Makson, Gupta, Arun, & Beg, Mohammad Dalour H. (2015). Characterization of ultrasound-treated oil palm empty fruit bunch-glass fiber-recycled polypropylene hybrid composites. *Journal of Polymer Engineering, 35*(2), 135–143.

Islam, Muhammad Remanul, Beg, Mohammad Dalour H., & Gupta, Arun (2014). Characterization of alkali-treated kenaf fibre-reinforced recycled polypropylene composites. *Journal of Thermoplastic Composite Materials, 27*(7), 909–932.

Karina, Myrtha, Syampurwadi, Anung, Satoto, Rahmat, Irmawati, Yuyun, & Puspitasari, Tita (2017). Physical and mechanical properties of recycled polypropylene composites reinforced with rice straw lignin. *BioResources 12*(3), 5801–5811. doi:10.15376/biores.12.3.5801-5811.

Karsli, Nevin Gamze, Yesil, Sertan, & Aytac, Ayse (2013). Effect of short fiber reinforcement on the properties of recycled poly(ethylene terephthalate)/poly(ethylene naphthalate) blends. *Materials and Design, 46*, 867–872.

Khanam, P. Noorunnisa, Al-Maadeed, M. A., & Mrlik, Miroslav (2016). improved flexible, controlled dielectric constant material from recycled LDPE polymer composites. *Journal of Materials Science: Materials in Electronics, 27*(8), 8848–8855.

Kiełbasa, Paweł, & Korenko, Maroš (2019). Physical and chemical properties of waste from pet bottles washing as a component of solid fuels. *Energies, 12*, 1–17.

Kim, Hee Soo, Lee, Byoung Ho, Lee, Sena, Kim, Hyun Joong, & Dorgan, John R. (2011). Enhanced interfacial adhesion, mechanical, and thermal properties of natural flour-filled biodegradable polymer bio-composites. *Journal of Thermal Analysis and Calorimetry, 104*, 331–338.

Koay, Seong Chun, Subramanian, Varnesh, Chan, Ming Yeng, Pang, Ming Meng, Tsai, Kim Yeow, & Cheah, Kean How (2018). Preparation and characterization of wood plastic composite made up of durian husk fiber and recycled polystyrene foam. *MATEC Web of Conferences, 152*, 1–7.

Kuan, Hsu Chiang, Chiang, Chin Lung, Shen, Ming Yuan, & Kuan, Chen Feng (2020). The study on coffee slag/recycled polystyrene circulation materials and its application on blinds. *Modern Physics Letters B, 34*(7–9), 1–5.

Lim, B. Y., Voon, C. H., Salmah, H., & Nordin, H. (2016). Chitosan filled recycled low density polyethylene composite: melt flow behavior and thermal degradation properties. In *The 2nd International Conference on Functional Materials and Metallurgy (ICoFM 2016)*, 1–7.

Ling, Sing Li, Koay, Seong Chun, Chan, Ming Yeng, Tshai, Kim Yeow, Chantara, Thevy Ratnam, & Pang, Ming Meng (2020). Wood plastic composites produced from postconsumer recycled polystyrene and coconut shell: effect of coupling agent and processing aid on tensile, thermal, and morphological properties. *Polymer Engineering and Science, 60*(1), 202–210.

Lisperguer, Justo, Nuñez, Christian, & Perez-Guerrero, Patricio (2013). Structure and thermal properties of maleated lignin-recycled polystyrene composites. *Journal of the Chilean Chemical Society, 58*(4), 1937–1940.

Mallakpour, Shadpour, & Behranvand, Vajiheh (2017). Recycled PET/MWCNT-ZnO quantum dot nanocomposites: adsorption of Cd(II) ion, morphology, thermal and electrical conductivity properties shadpour. *Chemical Engineering Journal, 313*, 873–881.

Mallakpour, Shadpour, & Javadpour, Mashal (2016). The thermal, optical, flame retardant, and morphological consequence of embedding diacid-capped zno into the recycled PET matrix. *Journal of Applied Polymer Science, 133*(20), 43433.

Mebarki, Fouzia, & David, Eric (2018). Dielectric characterization of thermally aged recycled polyethylene terephthalate and polyethylene naphthalate reinforced with inorganic fillers. *Polymer Engineering and Science, 58*(5), 701–712.

Medupin, R. O., Abubakre, O. K., Abdulkareem, A. S., Muriana, R. A., Kariim, I., & Bada, S. O. (2017). Thermal and physico-mechanical stability of recycled high density polyethylene reinforced with oil palm fibres. *Engineering Science and Technology, an International Journal, 20*, 1623–1631.

Mondadori, N. M. L., Nunes, R. C. R., Zattera, A. J., Oliveira, R. V. B., & Canto, L. B. (2008). Relationship between processing method and microstructural and mechanical properties of poly(ethylene terephthalate)/short glass fiber composites. *Journal of Applied Polymer Science, 109*, 3266–3274.

Moreno, Diego David Pinzón, de Camargo, Rayane Veloso, Luiz, Denise dos Santos, Branco, Lívia Teresinha Pimentel, Grillo, Camila Cezar, & Saron, Clodoaldo (2021). composites of recycled polypropylene from cotton swab waste with pyrolyzed rice husk. *Journal of Polymers and the Environment, 29*, 350-362.

Nabil, H, & Ismail, H. (2015). Preparation and properties of recycled poly (ethylene terephthalate) powder/halloysite nanotubes hybrid-filled natural rubber composites. *Journal of Thermoplastic Composite Materials, 28*(3), 415–430.

Naguib, Hamdy M. (2018). Environmental-friendly recycled polyester/Mg(OH)2 nanocomposite: fire-retardancy and thermal stability. *Polymer Testing, 72*, 308–314.

Nguyen, Dang Mao, Vu, Thi Nhung, Mai, Thi, Nguyen, Loan, Nguyen, Trinh Duy, Thuc, Chi NHan Ha, Bui, Quoc Bao, Colin, Julien, & Perre, Patrick (2020). Synergistic influences of stearic acid coating and recycled pet microfibers on the enhanced properties of composite materials. *Materials, 13*, 1–16.

Nisticò, Roberto (2020). Polyethylene terephthalate (PET) in the packaging industry. *Polymer Testing 90*, 1–18.

Nonato, Renato Carajelescov, & Bonse, Baltus Cornelius (2016). A study of PP/PET composites: factorial design, mechanical and thermal properties. *Polymer Testing, 56*, 167–173.

Norazlina, H., Firdaus, R. M., & Hafizuddin, W. M. (2015). Enhanced properties from mixing natural rubber with recycled polyvinyl chloride by melt blending. *Journal of Mechanical Engineering and Sciences, 8*, 1440–1447.

Ogah, Anselm Ogah (2016). Characterization of sorghum bran / recycled low density polyethylene for the manufacturing of polymer composites. *Journal of Polymers and the Environment, 25*, 533–543.

Onuegbu, G. C., Onyedika, G. O., & Ogwuegbu, M. O. C. (2019). *Mechanical and Thermal Behaviour of Clay Filled Recycled Low Density Polyethylene*. Springer International Publishing.

Orji, B. O., & McDonald, A. G. (2020). Evaluation of the mechanical, thermal and rheological properties of recycled polyolefins. *Materials 13*, 667.

Owen, M. M., Ishiaku, U. S., Danladi, A., Dauda, B. M., & Romli, A. Z. (2018). The effect of surface coating and fibre loading on thermo-mechanical properties of recycled polyethylene terephthalate (RPET)/epoxy-coated kenaf fibre composites. In *National Symposium on Polymeric Materials 2017 (NSPM)*, 1–14.

Passos, Rachel, Santos, De Oliveira, Castro, Daniele Oliveira, & Coll, Adhemar (2014). Processing and thermal properties of composites based on recycled pet, sisal fibers, and renewable plasticizers. *Journal of Applied Polymer, 131*(12), 1–13.

Pérez-Guerrero, Patricio, Lisperguer, Justo, Navarrete, José, & Rodrigue, Denis (2014). Effect of modified eucalyptus nitens lignin on the morphology and thermo-mechanical properties of recycled polystyrene. *BioResources, 9*(4), 6514–6526.

Poletto, M. (2019). Maleated soybean oil as coupling agent in recycled polypropylene/wood flour composites: mechanical, thermal, and morphological properties. *Journal of Thermoplastic Composite Materials, 32*(8), 1056–1067.

Poletto, Matheus (2016). Effects of the coupling agent structure on the adhesion of recycled polystyrene wood flour composites: thermal degradation kinetics and thermodynamics parameters. *Journal of Composite Materials, 50*(23), 3291–3299.

Pramoda, K. P., Lim, Q. F., & Chen, Shilin (2018). Synergistic effects of fillers on recycled polystyrene composites. *Polymer Bulletin, 75*(3), 1185–1195.

Ramle, M. S., Romli, A. Z., & Abidin, M. H. (2013). Tensile properties of aminosilane treated rice husk/recycled PVC composite. *Advanced Materials Research, 812*, 151–156.

Ramli, Ros Azlinawati, Zulkifli, Muhammad Syafiq, & Rabat, Nurul Ekmi (2020). Effect of graphite on mechanical, thermal and morphological properties of kenaf recycle polypropylene wood plastic composites. *Materials Science Forum, 981* MSF, 144–149. doi: 10.4028/www.scientific.net/MSF.981.144.

Reddy, Ravindra, Katrin, Chowreddy, Varhaug, Nord, & Rapp, Florian (2018). Recycled poly (ethylene terephthalate)/clay nanocomposites: Rheology, thermal and mechanical properties. *Journal of Polymers and the Environment, 27*, 37–49.

Rezaeian, I., Jafari, S. H., Zahedi, P., & Nouri, S. (2009). An investigation on the rheology, morphology, thermal and mechanical properties of recycled poly(ethylene terephthalate) reinforced with modified short glass fibers. *Polymer Composites, 30*(7), 993–999.

Rios, Cornier H., Sundaram, P. A., & Celorie, J. T. (2007). Effect of recycling on material properties of glass-filled polyethylene terephthalate. *Journal of Polymers and the Environment, 15*, 51–56.

Rivai, M., Gupta, A., Islam, M. R., & Beg, M. D. H. (2014). Characterization of oil palm empty fruit bunch and glass fibre reinforced recycled polypropylene hybrid composites. *Fibers and Polymers, 15*(7), 1523–1530.

Saeed, Haroon A. M., Eltahir, Yassir A., Xia, Yumin, & Yimin, Wang (2014). Properties of recycled poly(ethylene terephthalate)(pet)/hyperbranched polyester (hbpet) composite fibers. *The Journal of the Textile Institute, 106*(6), 601–610.

Salikhov, R. B., Bazunova, M. V., Bazunova, A. A., Salikhov, T. R., & Zakharov, V. P. (2018). Study of thermal properties of biodegradable composite materials based on recycled polypropylene. *Letters on Materials, 8*(4), 485–488. doi: 10.22226/2410-3535-2018-4-485-488.

Samat, N., Lazim, N. H.M., Motsidi, S. N.R., & Azlina, H. N. (2017). Performance properties of irradiated recycled polypropylene as a compatibilizer in recycled polypropylene/microcrystalline cellulose composites. *Materials Science Forum, 894* MSF, 62–65. doi: 10.4028/www.scientific.net/MSF.894.62.

Sandoval, N. A., Cruz, A. P., & Murillo, E. A. (2020). Evaluation of a hyperbranched polyester polyol as plasticizing agent for composites of recycled polystyrene and rice husk. *Journal of Physics: Conference Series, 1587*, 1–8.

Sastri, Vinny R. (2010). Commodity thermoplastics: Polyvinyl chloride, polyolefins and polystyrene. In *Plastics in Medical Devices*, 73–119. William Andrew Applied Science Publisher. doi: 10.1016/b978-0-8155-2027-6.10006-6.

Satapathy, Sukanya (2018). Development of value-added composites from recycled high-density polyethylene, jute fiber and flyash cenospheres: mechanical, dynamic mechanical and thermal properties. *International Journal of Plastics Technology, 22*, 386–405.

Satapathy, Sukanya, & Kothapalli, Raju V. S. (2018). Mechanical, dynamic mechanical and thermal properties of banana fiber/recycled high density polyethylene biocomposites filled with flyash cenospheres. *Journal of Polymers and the Environment, 26*, 200–213.

Sengupta, Shubhalakshmi, Ray, Dipa, & Mukhopadhyay, Aniruddha (2013). Sustainable materials: value-added composites from recycled polypropylene and fly ash using a green coupling agent. *ACS Sustainable Chemistry and Engineering, 1*(6), 574–584.

Sharma, Amit Kumar, & Mahanwar, Prakash A. (2010). Effect of particle size of fly ash on recycled poly(ethylene terephthalate)/fly ash composites. *International Journal of Plastics Technology, 14*, 53–64.

Touati, Zohra, Boulahia, Hakima, Belhaneche, Naima, & Valérie, Bensemra (2018). Modification of diss fibers for biocomposites based on recycled low-density polyethylene and polypropylene blends. *Waste and Biomass Valorization, 10*, 2365–2378.

Tufan, M, & Ayrilmis, Naidr (2016). Potential use of hazelnut husk in recycled high-density polyethylene composites. *BioResources, 11*, 7476–7489.

Tufan, Mürsit, Akbas, Selcuk, Yurdakul, Sema, Gulec, Turker, & Eryilmaz, Hasan (2016). Effects of different filler types on decay resistance and thermal, physical, and mechanical properties of recycled high - density polyethylene composites. *Iran Polymer Journal, 25*, 615–622.

Veloso, Rayane, & Clodoaldo, De Camargo (2019). Mechanical–chemical recycling of low-density polyethylene waste with polypropylene. *Journal of Polymers and the Environment, 28*, 794–802.

VinylPlus (2016). PVC Recycling Technologies. *Brussels Belgium.* https://www.vinylplus.eu/

Wahit, Mat Uzir, Akos, Noel Ibrahim, & Laftah, Waham Ashaier (2012). Influence of natural fibers on the mechanical properties and biodegradation of poly(lactic acid) and poly(μ-caprolactone) composites: a review. *Polymer Composites, 33*(7), 1045–1053.

Yang, Bin, Chen, Jin, Su, Li-fen, Miao, Ji-bin, Chen, Peng, Qian, Jia-sheng, Xia, Ru & Shi, You (2019). Melt crystallization and thermal properties of graphene platelets (gnps) modified recycled polyethylene terephthalate (rpet) composites: the filler network analysis. *Polymer Testing, 77*, 1–15.

Yang, Zhe, Peng, Hongdan, Wang, Weizhi, & Liu, Tianxi (2010). Thermal and mechanical properties of wood flour–polystyrene blends from postconsumer plastic waste. *Journal of Applied Polymer Science, 116*(5), 2658–2667.

Yesil, Sertan (2013). Effect of carbon nanotube reinforcement on the properties of the recycled poly (ethylene terephthalate) poly(ethylene naphthalate)(r-pet/pen) blends containing functional elastomers. *Materials and Design, 52*, 693–705.

Younesi-Kordkheili, Hamed, & Pizzi, Antonio (2020). Ionic liquid- modified lignin as a bio- coupling agent for natural fiber- recycled polypropylene composites. *Composites Part B: Engineering, 181*(May), 107587. doi:10.1016/j.compositesb.2019.107587.

Youssef, Ahmed M, Ahmed El-Gendy, & Kamel, Samir (2015). Evaluation of corn husk fibers reinforced recycled low density polyethylene composites. *Materials Chemistry and Physics, 152*, 26–33.

Yu, Min, He, Chunxia, Huang, Runzhou, Liu, Junjun, & Lu, Derong (2016). Accelerated weathering of recycled polypropylene packaging bag composites reinforced with wheat straw fibers. *Forest Products Journal, 66*(7–8), 485–494.

Zaaba, Nor Fasihah, & Ismail, Hanafi (2018). Comparative study of irradiated and non-irradiated recycled polypropylene/peanut shell powder composites under the effects of natural weathering degradation. *BioResources 13*(1), 487–505.

Zaaba, Nor Fasihah, Ismail, Hanafi, & Jaafar, Mariatti (2018). Recycled polypropylene/peanut shell powder (RPP/PSP) composites: property comparison before and after electron beam irradiation. *Polymer Composites, 39*(9), 3048–3056.

Zhang, Xiaolin, Bo, Xiangfeng, Cong, Longkang, Wei, Liqing, & McDonald, Armando G. (2018). Characteristics of undeinked, alkaline deinked, and neutral deinked old newspaper fibers reinforced recycled polypropylene composites. *Polymer Composites, 39*(10), 3537–3544.

Zhang, Xiaolin, Wang, Zhe, Cong, Longkang, Nie, Sunjian, & Li, Jia (2020). Effects of fiber content and size on the mechanical properties of wheat straw/recycled polyethylene composites. *Journal of Polymers and the Environment, 28*, 1833–1840.

10 Thermal Properties of Recycled Polymer Composites

Havva Hande Cebeci[1], Korkut Açıkalın[2], and Aysel Kantürk Figen[1]*
[1]Department of Chemical Engineering, Yıldız Technical University, Istanbul, Turkey
[2]Department of Energy Systems Engineering, Yalova University, Yalova, Turkey

CONTENTS

10.1 Overview of Recycled Polymer Composites 185
10.2 Fundamentals of Thermal Analysis Techniques 188
10.3 Thermal Properties of Composites 189
10.4 Factors Influencing the Thermal Properties of Recycled Polymer Composites 190
10.5 Conclusion Remarks 193
References 193

10.1 OVERVIEW OF RECYCLED POLYMER COMPOSITES

Extensive use of polymers in daily life has led to increased related waste, giving rise to environmental problems. Nowadays, plastic recycling has become an important issue due to ecological and environmental requirements. Plastic wastes consisting of various polymer chains constitute the most significant part of solid wastes in the world. Since polymer recycling's main difficulty is its degradation, it is recycled by strengthening its physical and mechanical properties using a reinforcing component.

These structures with high strength are called composite materials. The purpose of incorporating these structures into a polymer matrix is to improve their properties. Phenolic, epoxy-polyester resins, glass, aramid, and carbon are among the most used synthetic materials as reinforcing components (Scelsi et al., 2011). Nanofillers are primarily used to improve the critical properties (mechanical, thermal, rheological and electrical) of waste polymers to fabricate commercial products suitable for markets (Danesh et al., 2012; Scelsi et al., 2011). Composites, including recycled polymer, will have a wide range of

DOI: 10.1201/9781003148760-10

applications in different industries when fabricated with desired key properties. Therefore, many researchers have focused on developing an economic fabrication process of recycled polymer composites with appropriate properties.

Used fiberglass is less costly as a reinforcement with recycled polypropylene (PP) in wooden floors. Because of their high resistance and low weight, glass fiber reinforced composites are preferred in the automotive, aerospace and construction sectors (Singha & Thakur, 2008). The increasing environmental awareness in the world has started the green chemistry and legal regulation establishments accordingly. Glass fiber and other synthetic polymer fibers, which are the new generation reinforcements for polymer materials, have many advantages. These fibers' main benefits are their low cost, renewability, biodegradability, low specific gravity, abundancy, high specific strength, hardness and the existence of commercial application areas. Several types of crops, such as palm oil, banana, bamboo, pineapple leaf, wheat straw, sisal, kenaf, abaca and coconut, are used in the production of vegetable fibers (Ajorloo 2020; Narayanan et al., 2011; Scelsi et al., 2011). Natural fibers are favored over glasses or carbon when it contributes to minimizing noise in vehicle interior materials. In most cases, natural fibers are more commercially viable than synthetic fibers. However, one of the disadvantages of natural fiber reinforced polymeric composites is their hydrophilic structure. To improve fiber-matrix adherence and eliminate durability loss, the fiber surface must be subjected to comprehensive physical and chemical treatment (Dhakal et al., 2007).

Recycled polymer reinforced by natural fiber can be used for non-structural parts in the automotive industry, such as car roof liners, door panels and covers (Fuqua et al., 2012). The biggest problem of natural fiber reinforced polymer composites is the poor adhesion of natural fibers to polymers. A composite based on recycled polypropylene reinforced with secondary fiber has been developed to ensure the fiber/matrix interface (Kumar et al., 2011).

In one study, waste paper-based secondary fiber has been used to reinforce engineering plastics such as polypropylene. Newspaper fibers processed with sodium silicate and magnesium chloride are used as reinforcements in natural and acrylonitrile rubbers (Mekap & Palsule, 2012; Nashar et al., 2004). The most popular spherical fillers are calcium carbonate, clay, glass beads, carbon black and alumina trihydrate. Since it is readily available and cheap, calcium carbonate is the most widely used filler.

Plate-like fillings are more effective than sphere fillers as reinforcement. In unprocessed materials, it increases the modulus, while decreasing the curvature and the cost of the material. The coated filler effect on the matrix interface varies depending on the polymer type. The coated filler improved impact resistance in polypropylene (PP) homopolymer but decreased hardness in PP and high-density polyethylene. Mica works better only with fibrous materials. It must be treated with silane or mixed with maleic anhydride-altered polymers for better interaction with plastics. In industry, silane reinforcement is not preferred. Binding agents, such as titanate or silane coatings, maleic anhydride and acrylic acid, are necessary for fiber-matrix bonding. It has been stated in the studies that the addition of small amounts of mica to glass fiber reinforced polyolefins increased the module and decreased the cost. Due to the increase in module is micas of favorable effects on fiber-matrix adherence (Tolinski, 2009). According to industrial and academic knowledge, glass fiber reinforcement is mostly used to progress recycled thermoplastics to long-permanent items. The strength, hardness, fracture toughness and heat resistance of the plastic can be increased through this reinforcement. There are studies about coated fillers such as mica and calcium carbonate that can be used instead of glass fiber or with glass fiber (Hugo et al., 2011).

Because of its high strength and low weight, using glass fiber as reinforcement with a recycled PP wooden floor is more cost effective. Recycled PP hybrid composites containing palm wood flour and glass fiber have also been studied by researchers. According to the tensile analysis, they investigated that adding grinded wood to recycled PP improved the tensible features (AlMaadeed et al., 2012; Saiful Islam et al., 2012; Zhang et al., 2013). Banana fiber is combined with a binding agent to polylactic acid chains, increasing the compatibility and stability of polylactic acid. Pineapple leaf fiber and recycled polylactic acid or polybutylene succinate chains are combined in environmentally friendly reinforced composites. As a result, these structures can optimize fiber-polymer matrix suitability while also increasing the physical strength of the materials (Shih & Huang, 2011). The Sri-Lanka project has produced low-cost and environmentally friendly construction structures using recycled low-density polyethylene plastic composites and banana fibers made from agricultural waste. No fiber treatments or compatibilizers were used to reduce the cost. Thus, environmental problems have been reduced, and a new source of income for the people of the region was provided (Bolduc et al., 2018).

Rice straw may be used to strengthen recycled polystyrene to enhance material's mechanical features. Thus, it can produce hardwood-polymer composites for various practical applications. There are many advantages of using agricultural rice straw fibers in the composite formulation. The most important of these is their light weight, good qualification, cheap charge, yearly renewability, mechanical features, low energy consumption and ecological friendly (Mohsen, 2013). Hardwood composite panels were produced by recycling swelled polystyrene foam waste and make strong it with rice straw. The rice straw was not processed in any way and, just dried to decrease recycling cost. The results indicated that boosting the bonding agent ingredients increased the tensile strength property and decreased the water absorption by over 45% in the final products.

It also indicated that value-added hardwood composites could be created from wastes. These hardwood products can replace natural wood in many everyday applications. They have performed mechanical characterization, dimensional stability and good resistance (Tawfik et al., 2017). It can be utilized in picnic tables, park benches and landscaping timbers.

In another research, rice husk filling fibers with different loading contents were loaded into recycled HDPE composites. Maleic acid polyethylene (MAPE) was added as a bonding agent. The tensile and flexural features increased with the increase of the rice husk loading, but the composites' impact strength decreased (Tong et al., 2014).

In another study, recycled bamboo fibers were added to polypropylene resin since bamboo fibers can increase impact strength, tensile strength and compression strength for PP resin. It has a lower cost, and it is an industry-friendly process compared to those reinforced with synthetic fibers. Recycling agricultural waste such as bamboo can prevent damage to the environment (Mosawi et al., 2015).

The by-product from the construction industry, paper industry and coating factory was used as fillers to produce recycled polymer composites. A composite was created as a sample using wooden fiber as a filling material. Stretchable features and effect durability were evolved. Compared with the reference substance, the liquid waste from the paper sectors reduced the tensile strength and modulus and stone dust and rock wool from the building industry, its impact modulus remained the same, while the powder coating waste from the coating factory increased porosity and decreased density (Malekani et al., 2014).

10.2 FUNDAMENTALS OF THERMAL ANALYSIS TECHNIQUES

The thermal analysis techniques, which are subjected to modern developments, play an important role in material characterization, especially for polymers, polymeric composites and recycled polymers. Generally, thermal analysis is applied to measure the change of physical and chemical properties of the materials against temperature and time under controlled conditions. Nowadays, thermal analysis techniques provide helpful information for material design, reaction kinetics, reactor design, process optimization and energy analysis. Moreover, supplying the recycled polymers instead of wood, fiber and synthetic polymers for the effective recycling process, correlation of chemical structure and physical properties with thermal analysis results is necessary. Thermogravimetry (TG), differential scanning calorimetry (DSC), differential thermal analysis (DTA), dynamic thermal analysis (DMA) and thermomechanical analysis (TMA) have been extensively employed for polymer characterization such as thermal transitions, crystallinity degree, characteristic temperatures, residue amount, the activation energy of decomposition, oxidative degradation and stability (Dobkowski, 2006). Besides these, viscoelastic properties of polymers are measured for characterization of mechanical behavior by using the DMA technique where minor cyclic stress is applied, and the response stress is recorded (Krishnasamy et al., 2019).

The essential physical properties of a polymer can be classified under three groups as (A) phase transitions, volumetric properties, calorimetric properties, electrical/optical/magnetic properties, mechanical and acoustic properties; (B) properties determining mass transfer in polymers, rheological properties of polymer melts, chemical, thermal and UV stability; and (C) ultimate mechanical and electrical properties such as creep, failure, toughness, hardness, friction, wear,

TABLE 10.1
Commonly Used Thermal Analysis Techniques to Measure the Key Properties of Polymer Materials

Key-Properties	TA Method
T_g-value	DSC, DMA
T_m-value	DSC
H_f-value	DSC
Thermal stability	TG, DTG
Moisture sensitivity	TG, DTG

Source: Dobkowski, 2006 and Groenewoud, 2001.

yield strength, aging effects, tracking and dielectric strength. It can be seen that there are so many variables to identify a polymeric system. Groenewoud defined the "key properties" as the T_g-value (glass transition temperature), the T_m-value (melting temperature), the H_f-value (heat of fusion), the thermal stability and the moisture sensitivity instead of calculating all the properties mentioned in groups A, B and C (Groenewoud, 2001). Standard thermal analysis techniques used to measure polymer materials' key properties are listed in Table 10.1.

These measured key properties depend on program parameters or factors such as sample mass, heating or cooling rate, crucible material, carrier (or reactive) gas, gas flow velocity and reference substance. A series of standardized methods were developed and reported by ISO, ASTM, DIN and others to improve repeatability and reproducibility of experimental results (Liu & Li, 2009).

10.3 THERMAL PROPERTIES OF COMPOSITES

Recycled polymer composites, consisting of recycled polymeric matrix and filler as a reinforcement, have been used increasingly in automotive, packaging, construction, etc., areas in the last years regarding the control of plastic pollution (Espert et al., 2003). Formulation, compatibilization, modification, characterization and fabrication of recycled polymer composites have become very attractive, and a lot of research has been conducted in this field. Mechanical and thermal shortcomings in polymers occur during recycling, and reinforcement must be handled by blending or through polymerization. Different types of blend, filler or additives such as sepiolite (Farshchi & Ostad, 2020), wood dust (Chawla et al., 2020), Jute fabrics (Rokbi et al., 2020), fly ash (Gohatre et al., 2020), organoclay Cloisite 15A (Honorato et al., 2020), leaf fiber (Dehghani et al., 2013), pyrolyzed rice husk (Moreno et al., 2020), etc. can be used to enhance the properties of polymers.

Generally, it's significant to explore the thermal properties of recycled polymer before fabrication of composites containing recycled polymer to optimize the temperature range for composites' processing. Waste fly ash from thermal power plants was mixed with recycled poly(vinyl chloride) (rPVC) recovered from waste electrical and electronic products.

The thermal properties of r-PVC were observed to improve by incorporating fly ash as revealed from the DSC and TGA studies. DSC analyses showed that the fly ash's presence did enhance the T_g value of r-PVC. The efficient dipole-dipole interaction between the fly ash and the rPVC matrix was discovered based on the thermal degradation profiles of rPVC and rPVC/fly ash composite (Gohatre et al., 2020).

After the fabrication, performing thermal analyses is still required to check the composites' key thermal properties. In a double cylinder grinder, sustainable polymer composites were made from a combination of waste recovered rubber and wood flour (WF). Before and after the vulcanization, the thermal properties of composites were investigated by DSC and TGA techniques, respectively. The obtained thermal analysis results showed that the addition of wood flour to the waste reclaimed rubber shifted the T_g temperature from ~15°C to 10°C and decreased composites' thermal stability. It was found that the residual amount has been reduced when the amount of WF was increased. The inference that the amount of volatile components was increased. The authors suggested using WF-filled reclaimed rubber composites for insulation applications (Phiri et al., 2020).

10.4 FACTORS INFLUENCING THE THERMAL PROPERTIES OF RECYCLED POLYMER COMPOSITES

Many researchers have focused on investigating the main factors affecting the thermal properties of recycled polymer composites, including properties of polymer, filler and formulation of composites.

Several kinds of fillers have been incorporated to improve the properties of various composites. A comparative investigation on applications underlined that fiber reinforced polymer composites demonstrate better mechanical and thermal properties than metals and alloys (Nagaprasad et al., 2020). In addition to this, natural fibers are the alternative reinforcements against glass and carbon fibers due to their superior properties such as acoustic absorption, good thermal features, increased energy recovery, high hardness, non-corrosive structure, decreased tool abrasion and decreased dermal and respiratory irritation (Eskander & Tawfik, 2018). It is necessary to explore a few case studies to understand the effect of various fillers on composite thermal features. The fillers contact the polymer chains during processing and play a key role in reinforcing. Plasticizer content is also critical for enhancing the compatibility of filler particles in the polymer matrix (Gohatre et al., 2020). The importance of chemical treatment was also underlined for composite preparation ensuring compatibility between surface and reinforcement materials. The material cost can be decreased by effectively reducing the waste from pineapple leaves and recycled chopsticks and reusable. Hybrid fibers are then produced with polylactic acid (PLA) and polybutylene succinate (PBS) after chemically treating pineapple leaf fiber and recycled disposable chopstick fiber with an alkaline solution and triethoxyvinlysilane, which is a silane binding agent. SEM analysis results showed the best properties and limited water absorption for 30% chemically modified reinforcement material.

The tensile strengths of PBS and PLA were developed, and the heat deflection temperature was increased, respectively, by 33.6% and 75% (Rocha & Rosa, 2019). Starch gum (SG) was used as a binding agent for raw or recycled polymer matrix composites, reinforced with wood flour. Aqueous starch solutions containing 3% and 5% by weight of coupling agent were used, and a coating was allowed to form on the reinforcement. The comparison was made with another binding agent, maleic anhydride grafted (MAPP). The starch gum coating produced low pore space in the recycled polymer and provided a more compatible layer on the reinforcement (Danesh et al., 2012). Table 10.2 lists the recycled polymer matrixes with suitable filler types used for enhancing thermal properties.

To build hybrid composite recycled polypropylene (RPP) from palm wood flour/glass fiber, separate weights of the two reinforcements were utilized. According to the differential thermal calorimeter results, it was observed that there was no difference between the two in the case of T_m. Still, the T_c values of composites and recycled polypropylene changed marginally. The high amount of

TABLE 10.2
Suitable Filler Types Used for Enhancing Thermal Properties of the Recycled Polymer Matrix

Filler	Polymer	Enhanced Properties	References
Date palm seed	Vinyl ester	Thermal stability	(Nagaprasad et al., 2020)
Fly Ash	Polypropylene	Melting, crystallization crystallinity temperature	(Ajorloo et al., 2020)
Fly Ash	Poly(vinyl chloride)	Thermal stability	(Gohatre et al., 2020)
Kapok fibers	Polyethylene	Melting temperature and degree of crystallinity	(Macedo et al., 2020)
Organoclay Cloisite 15A	Polyethylene terephthalate	Glass transition temperature thermal stability composite degradation temperature	(Honorato et al., 2020)
Wollastonite	Polypropylene	Crystallization and melting behavior	(Ding et al., 2020)
Graphene oxide nanoflakes	Poly(ethylene terephthalate)	Thermal stability	(Bayat et al., 2019)

Source: Dobkowski, 2006 and Groenewoud, 2001.

wood flour in composites reduced the crystallinity (%), whereas glass fiber's inclusion increased the crystallinity (%). It was concluded that the use of a certain amount of wood flour as a filler for glasses fiber reinforced hybrid composites will decrease the material cost without significantly damaging the properties (Shih et al., 2014).

The composition is the other important factor affecting the thermal properties of composites. The different content of fillers has been shown to provide a different level of enhancement. Melt mixing and compression is used to create recycled high-density polyethylene (RHDPE) composites and natural fibers. The effects of fiber (wood and pulp), binder form and concentration on composite properties were studied. The study also discovered that composites had lower crystallization peak temperatures and a broader crystalline temperature spectrum than pure RHDPE, as well as higher thermal stability (Lei et al., 2007). Using epoxy adhesives, recycled poly (ethylene terephthalate) (PET), ground rubber rubber (GTR) and graphene oxide (GO), the adhesive formulations were arranged by varying the amounts of the ingredients given above. It indicated that the addition of the adhesive (GO) increased the initial decomposition temperature, the maximum decomposition temperature and the thermal stability in terms of the amount of epoxy residual coal. Due to the presence of these particles in the epoxy matrix, the movements of the system sample became difficult and the formation of molecules was prevented, research indicated (Bayat et al., 2019). It has been focused on the formulation of composite materials from thermoplastic polymer waste, especially for structural applications. The effect of filler types and amount of additive in the compound on the mechanical and thermal properties were investigated. When small amounts of mica were added to glass fiber reinforced mixtures, it caused a significant change in tensile strength and modulus (Hugo et al., 2011). Polypropylene/rice husk (non-pyrolyzed/NRH and pyrolyzed/PRH) composites were prepared from cotton swab waste and agro-industrial waste extrusion with different contents. The PP/RH composites prepared by extrusion range from 2 to 20 wt.% of non-pyrolyzed (NRH) and pyrolyzed (PRH) rice husk. Thermal analyses showed that the final decomposition temperature was shifted to a higher value in PP/NRH composites. It reached 483°C in the composites containing NRH fillers at 10 wt.% and 20 wt.% (Moreno et al., 2020).

The effects of fibers and binding agents on the composite produced by melt bonding with recycled high-density polyethylene (HDPE) and composites based on natural fibers such as pinewood flour and pulp were investigated. Without binder agents, HDPE and pulp composites have a slightly higher modulus, while HDPE and pine composites have been found to have low tensile strength and impact strength. For the blend derived with carboxylated polyethylene, the mechanical properties of HDPE and pulp composites reached maximum value. For the mixture derived from titanium, RHDPE improved the mechanic features of the pulp system however decreased the major mechanic features of pine. The mechanic features of fiber-reinforced recycled HDPE were determined for two thermal degradation stages for untreated HDPE/fibre composites. Binding agents has a minimum effect on the thermal attitude of composite. Thermal degradation occurred at a lower temperature than fiber-reinforced recycled HDPE due to the fiber structure (Kishchynskyi et al., 2016).

10.5 CONCLUSION REMARKS

Composites of recycled polymer reinforced with fillers were successfully fabricated through the hand layup process, spray-up process, vacuum bag molding process, resin transfer molding process, vacuum infusion process and compression molding process, etc. The stability, durability and flammability characteristics of recycled polymer composites should be examined. Concerning the thermal properties, the filler looks to play a significant role in enhancing recycled polymer. Besides this, modification methods applied to fillers led to more desirable thermal stabilities and thermomechanical properties. TG, DTA, DSC and DMA are the best methods to determine polymer-filler compatibility and enhancement of thermal properties.

REFERENCES

Ajorloo, Mojtaba, Ghodrat, Maryam, & Kang, Won-Hee (2020). Incorporation of recycled polypropylene and fly ash in polypropylene-based composites for automotive applications. *Journal of Polymers and the Environment*. doi:10.1007/s1 0924-020-01961-y.

AlMaadeed, Mariam A., Kahraman, Ramazan, & Khanam, Noorunnisa P. et al. (2012). Date palm wood flour/glass fibre reinforced hybrid composites of recycled polypropylene: mechanical and thermal properties. *Materials & Design*, 42, 289–294.

Bachtiar, Dandi, Sapuan, Salit, & Hamdan, Mohamad M. (2008). The effect of alkaline treatment on tensile properties of sugar palm fibre reinforced epoxy composites. *Materials & Design*, 29(7), 1285–1290.

Bayat, Sahar, Jazani, Omid Moini, Molla-Abbasi, Payam, Jouyandeh, Maryam, & Saeb, Mohammad Reza (2019). Thin films of epoxy adhesives containing recycled polymers and graphene oxide nanoflakes for metal/polymer composite interface. *Progress in Organic Coatings*, 136, 105201.

Bolduc, Sean, Jung, Kyungmin, & Venkata, Pramathanath et al. (2018). banana fiber/low-density polyethylene recycled composites for third world eco-friendly construction applications – waste for life project Sri Lanka. *Journal of Reinforced Plastics and Composites*, 37(21), 1322–1331.

Chawla, Kapil, Singh, Janspreet, & Singh, Rupinder (2020). On recyclability of thermosetting polymer and wood dust as reinforcement in secondary recycled ABS for nonstructural engineering applications. *Journal of Thermoplastic Composite Materials*. doi:10.1177/0892705720925135.

Cheung, Hoi-yan, Ho, Mei-po, & Lau, Kintak et al. (2009). Natural fibre-reinforced composites for bioengineering and environmental engineering applications. *Composites Part B: Engineering*, 40(7), 655–663.

Danesh, Mohammad A., ZiaeiTabari, Hassan, & Hosseinpourpia, Reza et al. (2012). Investigation of the morphological and thermal properties of waste newsprint/ recycled polypropylene/ nanoclay composite. *BioResources*, 7(1), 936–945.

Dehghani, Alireza, Ardekani, Madadi S., & Al-Maadeed, Mariam A. et al. (2013). Mechanical and thermal properties of date palm leaf fiber reinforced recycled poly (ethylene terephthalate) composites. *Materials & Design, 52*, 841–848.

Dhakal, Hom N., Zhang, Z. Y., & Richardson, Mel O.W. (2007). Effect of water absorption on the mechanical properties of hemp fibre reinforced unsaturated polyester composites. *Composites Science and Technology, 67*(7–8), 1674–1683.

Ding, Qian, Fu, Hao, & Hua, Chaoran et al. (2020). Effect of β-nucleating agent on crystallization of wollastonite-filled recycled polypropylene composites. *Journal of Thermal Analysis and Calorimetry*. doi: 10.1007/s10973-020-09769-7

Dobkowski, Zbigniew. (2006). Thermal analysis techniques for characterization of polymer materials. *Polymer Degradation and Stability, 91*(3), 488–493.

Eskander, Samir B., & Tawfik Magda E. (2018). Impacts of gamma irradiation on the properties of hardwood composite based on rice straw and recycled polystyrene foam wastes. *Polymer Composites, 40*(18). doi: 10.1002/pc.25036

Espert, Ana, Camacho, Walker, & Karlson, Sigbritt (2003). Thermal and thermomechanical properties of biocomposites made from modified recycled cellulose and recycled polypropylene. *Journal of Applied Polymer Science, 89*(9), 2353–2360.

Farshchi, Negin & Ostad Yalda K. (2020). Sepiolite as a nanofiller to improve mechanical and thermal behavior of recycled high-density polyethylene. *Progress in Rubber, Plastics and Recycling Technology, 36*(3), 185–195.

Fuqua, Micheal A., Huo, Shanshan, & Ulven, Chad A. (2012). Natural fiber reinforced composites. *Polymer Reviews, 52*(3), 259–320.

Gohatre, Omdeo K., Biswal, Manoranjan, & Mohanty, Simita et al. (2020). Study on thermal, mechanical and morphological properties of recycled poly(vinyl chloride)/ fly ash composites. *Polymer International, 69*(6), 552–563.

Groenewoud, W. M. (2001). Chemical structure/physical properties correlations. In *Characterisation of Polymers by Thermal Analysis* (pp. 230–281). Elsevier.

Honorato, Luciana R., Rodrigues, Patrícia de F., & Silva, Adriana dos A. et al. (2020). Synergistic effects of organoclay cloisite 15A on recycled polyethylene terephthalate. *Journal of Materials Research and Technology, 9*(6), 13087–13096.

Hugo, Annie M., Scelsi, Lino, & Jones, Frank R. et al. (2011). Development of recycled polymer composites for structural applications. Plastics, Rubber and Composites, 40(6-7), 317–323.

Kishchynskyi, Sergii, Nagaychuk, Vasyl, & Bezuglyi, Artem (2016). Improving quality and durability of bitumen and asphalt concrete by modification using recycled polyethylene based polymer composition. *Procedia Engineering, 143*, 119–127.

Kumar, Vinay, Tyagi, Lalit, & Sinha, Shishir (2011). Wood flour–reinforced plastic composites: a review. *Reviews in Chemical Engineering, 27*, 253–264.

Krishnasamy, Senthilkumar, Thiagamani, Senthil M. K., & Kumar, Chandresekar M. et al. (December 2019), Recent advances in thermal properties of hybrid cellulosic fiber reinforced polymer composites. *International Journal of Biological Macromolecules, 141*, 1–13.

Lei, Yong, Wu, Qinglin, & Yao, Fei et al. (2007), Preparation and properties of recycled HDPE/natural fiber composites. *Composites Part A: Applied Science and Manufacturing, 38*(7), 1664–1674.

Liu, Zhenhai & Li, Hongfei (2009). Application to polymer characterization by thermal analysis. *Frontiers of Chemistry in China, 4*(4), 352–359.

Macedo, Murilo J. P., Giovanna, Silva S., & Feitor, Michele C. et al. (2020). Composites from recycled polyethylene and plasma treated kapok fibers. *Cellulose, 27*(4), 2115–2134.

Malekani, Mehdi, Bazyar, Behzad, & Talaiepour, Mohammad et al. (2014). Influence of maleic-anhydride-polypropylene (MAPP) on the physical properties of polypropylene/sawdust fir flour composite. *Journal of Applied Environmental and Biological Sciences, 4*(3), 340–343.

Mekap, Dibyaranjan & Palsule, Sanjay (2012). Secondary fiber/recycled polypropylene composites. *Asian Journal of Research in Chemistry, 5*(5), 655–658.

Mohsen, Riham A. (2013). *Investigation of Reinforced Polystyrene Foam Waste with Natural or Synthetic Fibers*. The American University in Cairo.

Moreno, Diego D. P., Camargo, Rayane V., & Luiz, Denise dos S., et al. (2020). Composites of recycled polypropylene from cotton swab waste with pyrolyzed rice husk. *Journal of Polymers and the Environment, 29*, 350–362.

Mosawi, Al A., Abdulsada, Shaymaa A., & Hashim, Abbass (2015). Mechanical properties of recycled bamboo fibers reinforced composite. *European Journal of Advances in Engineering and Technology, 2*(4), 20–22.

Nagaprasad, Nagaraj, Stalin, Balasubramaniam, & Vignesh, Venkataraman et al. (2020). Effect of cellulosic filler loading on mechanical and thermal properties of date palm seed/vinyl ester composites. *International Journal of Biological Macromolecules, 147*, 53–66.

Narayanan, Venkateshwaran, ElayaPerumal, Ayyasamy, & Alavudeen, Aziz et al. (August 2011). Mechanical and water absorption behaviour of banana/sisal reinforced hybrid composites. *Materials & Design, 32*(7), 4017–4021.

Nashar, Doaa E.E., Messieh, A. E., & Basta, A. H.. (2004). Newsprint paper waste as a fiber reinforcement in rubber composites. *Journal of Applied Polymer Science, 91*(5), 3410–3420.

Phiri, Mohau J., Phiri, Mapoloko M., & Mpitso, Khotso et al. (2020). Curing, thermal and mechanical properties of waste tyre derived reclaimed rubber–wood flour composites. *Materials Today Communications, 25*, 101204.

Rocha, Daniel B., & Rosa, Derval D. (2019). Coupling effect of starch coated fibers for recycled polymer/wood composites. *Composites Part B: Engineering, 172*, 1–8.

Rokbi, Mansour, Khaldoune, Abderaouf, & Sanjay, M. R. et al. P (2020). Effect of processing parameters on tensile properties of recycled polypropylene based composites reinforced with jute fabrics. *International Journal of Lightweight Materials and Manufacture, 3*(2), 144–149.

Saiful Islam, Md., Hamdan, Sinin, & Rezaur, Rahman Md. et al. (2012), The effect of alkali pretreatment on mechanical and morphological properties of tropical wood polymer composites. *Materials & Design, 33*, 419–424.

Scelsi, Lino, Hodzic, Alma, & Soutis, Constantinos et al. (2011). A review on composite materials based on recycled thermoplastics and glass fibres. *Plastics, Rubber and Composites, 40*(1), 1–10.

Tolinski, Micheal (2009). *Additives For Polyolefins: Getting the Most Out of Polypropylene, Polyethylene and TPO*. Elsevier Ltd.

Shih, Yeng F., Chang, Wen C., & Liu, Wei C. et al. (2014). Pineapple leaf/recycled disposable chopstick hybrid fiber-reinforced biodegradable composites. *Journal of the Taiwan Institute of Chemical Engineers, 45*(4), 2039–2046.

Shih, Yeng F. & Huang, Chien C. (2011). Polylactic acid (PLA)/banana fiber (BF) biodegradable green composites. *Journal of Polymer Research, 18*(6), 2335–2340.

Singha, Amar & Thakur, Vijay K. (October 2008). Mechanical properties of natural fibre reinforced polymer composites. *Bulletin of Materials Science, 31*(5), 791–799.

Tawfik, Magda, Eskander, Samir, & Nawwar, Gelal A. M. (2017). Hard wood-composites made of rice straw and recycled polystyrene foam wastes. *Journal of Applied Polymer Science, 134*(18). doi: 10.1002/APP.44770.

Tong, Jia Y., Royan, Nishata R. R., & Ng, Christopher et al. (2014). Study of the mechanical and morphology properties of recycled HDPE composite using rice husk filler. *Advances in Materials Science and Engineering, 2014*, 1–6.

Zhang, Xiuju, Yang, Huajun, & Lin, Zhidan et al. (2013). polypropylene hybrid composites filled by wood flour and short glass fiber: effect of compatibilizer on structure and properties. *Journal of Thermoplastic Composite Materials, 26*(1), 16–29.

11 Flame Retardancy of Recycled Polymer Composites

Maryam Jouyandeh[1], Henri Vahabi[1], Fouad Laoutid[2], Navid Rabiee[3], and Mohammad Reza Saeb[1,4]

[1]Université de Lorraine, CentraleSupélec, Metz, France
[2]Laboratory of Polymeric and Composite Materials, Materia Nova Research Center, Mons, Belgium
[3]Department of Physics, Sharif University of Technology, Tehran, Iran
[4]Department of Polymer Technology, Faculty of Chemistry, Gdańsk University of Technology, Gdańsk, Poland

CONTENTS

11.1 Introduction.. 197
11.2 Recycling of Flame Retardant Polymer Composites 199
        11.2.1 Recycling of Flame Retardant Textiles 200
        11.2.2 Recycling of Waste Electrical and Electronic Equipment..... 200
11.3 Development of Flame-Retardant Materials from Recycled Plastics... 205
11.4 Ultimate Properties of the Recycled Flame Retardant Polymer Composites... 211
11.4 Conclusion 212
References.. 214

11.1 INTRODUCTION

The use of flame retardants is a classical approach for reducing the fire hazards of polymer materials (Jouyandeh et al., 2019; Lu & Hamerton, 2002; Pearce, 2012; Saeb et al., 2017; Vahabi et al., 2018). Massive annual production of additive-contained polymer composites consists of about 35–40% of thermoplastics, which are generally reinforced by glass, carbon-based particles and fibers and about 60% of the reinforced thermoset composites are used as flame retardant materials (Marosi et al., 2017; Paran et al., 2019; Seidi et al., 2020). Along with the undeniable benefits of flame retardant polymer composites, utilization of these multicomponent polymer composites is associated with serious sustainability and environmental safety issues. The large amounts of daily use of flame retardant polymeric products also create an unwanted outcome, which is the disposal of wastes (Zadeh et al., 2017). Using petroleum-based polymers and disposal of their wastes deplete the fossil fuel reserves. In addition, the use of materials of different nature and quantity in the flame retardant polymer composites make difficult their reprocessing and recycling. Large quantities of end-of-life flame retardant polymer composite wastes are being produced yearly, which increases the volume of landfilled waste due to the difficulty of their recycling procedures. Moreover, since all flame retardant polymer composites are not exposed to flame during the application, they are added to plastic wastes at the end of their life (Hadavand 2020; Jouyandeh et al., 2020; Tikhani et al., 2020). With the expansion of plastic wastes, in turn, the danger of fire risk has dramatically increased, endangering our daily lives (Bayat et al., 2019; Ebrahimi Jahromi et al., 2017; Irvine et al., 2000; Kobes et al., 2009; Shamsi et al., 2019). Infusing various resources and chemicals into the flame retardant products may offend the environment rather than protect it.

Typically, flame retardants may contain various chemical compositions such as halogens (bromine and chlorine), phosphorus, nitrogen, aluminum, magnesium, boron, antimony, molybdenum or recently developed nano-fillers (Vahabi et al., 2018, 2021). Notwithstanding, some of the chemicals present in flame retardant products can degrade by exposure to some natural phenomena including microbial activity, light, oxidation or hydration. However, the others such as commercial flame retardant cannot easily degrade; instead, they can be found in a large quantity in the air (Saito et al., 2007), surface water (Rodil et al., 2012), drinking water (Bacaloni et al., 2007), wastewater (Meyer & Bester, 2004) and dust (Stapleton et al., 2009). It is not surprising to say that flame retardant chemicals have also been found in fish species (Sundkvist et al., 2010) and human breast milk (LeBel & Williams, 1986), which can cause asthma and cancer. By launching stricter environmental legislations, industries are being encouraged/forced to make recycled flame retardant polymeric materials and valorize the recycled polymers. Increasing the amount of recycled flame retardant polymers and limiting the disposing of flame retardant wastes are essential challenges associated with choosing proper strategies to reduce CO_2 emissions, as well as controlling the release of toxins and the migration of flame retardants into the environment.

Generally, recycled flame retardant polymer composites contain different additives, nanoparticles and other waste polymers (Delgado et al., 2007). The use of dissimilar materials in the flame retardant polymer composite makes their recycling difficult. The concept of sustainable development applied in this field implies that fire retardants should have little impact on health and the environment during their life cycle, recycling and disposal. Accordingly, in this chapter we attempt to point out the difficulty in separating the flame retardants from the polymers in the recycling process, approaches of flame retardant detection and various recycling methods of flame retardant polymer composite wastes. In addition, the development and use of recycled flame retardant polymer composites integrate them into a new material is discussed in the next sections. Finally, the effect of utilization of recycled flame retardant polymer composites on the flame retardancy performance and other composite properties were investigated.

11.2 RECYCLING OF FLAME RETARDANT POLYMER COMPOSITES

Recycling is the main alternative to the undesired disposal of flame retardant products- a more sustainable and affordable approach rather than producing new flame retardant materials. Waste management hierarchy is shown in Figure 11.1 (Kindgom, 2002). The direct disposal of flame retardant materials is in the bottom of the hierarchy as the least desirable technique. The recycling and reuse of flame retardant products are on the upper stages of waste management hierarchy as supportive ways to reduce the environmental threats.

General steps to reuse the recycled polymer composites consist of (i) reproduction and reuse of polymer composite waste for applications where products with similar specifications to the original product are prepared; (ii) reuse of polymer composite wastes in products that require lower performance;

FIGURE 11.1 Waste management hierarchy denoting possible ways for recycling of wastes (Yasin et al., 2016). It is inspired by Maslow's hierarchy of needs, such that from the bottom of the hierarchy to the top we have least desirable treatments being gradually advanced with highly efficient ones.

(iii) depolymerization or devulcanization followed by the reuse in organic polymers and recovery/reuse of additives; (iv) pyrolysis, cracking and partial recovery of materials; and (v) combustion with energy recovery (Reijnders, 2014; Shamsi et al., 2019; Vahabi et al., 2019). The recovery process of fillers stoutly linked to the polymers is easier than the recovery of suspended fillers in the flame retardant systems (Monopoli et al., 2010; Wang et al., 2011). Therefore, complete recovery does not occur in practice, such that the plastic recycling rate above 10% is rarely reported (Reijnders, 2011). More complexities are associated with polymer blend composites in which two or more types of polymers are used, or the simple polymer composites in which two or more kinds of flame retardants are utilized. The recycling of a simple flame retardant polymer system does not necessarily provide support, either scientifically or technically, for the recycling of complex flame retardant polymer composites from a real waste stream. Nevertheless, there might be some recycling options to be explored by which one would afford to recycle such systems.

11.2.1 RECYCLING OF FLAME RETARDANT TEXTILES

The flame retardant textiles are utilized in many daily used goods and devices such as baby clothing, pushchairs, carpets and car seats. The large volume of the used flame retardant textiles in return generates a massive amount of wastes, leading to toxic chemical procedures and high CO_2 emissions released into the water, air and soil. The life cycle of flame retardant textiles is shown in Figure 11.2.

There is lack of a sustainable approach for recycling flame retardant textile wastes. Landfill and incineration of flame retardant textiles brings the consequence of incorporation of various toxic components into the environment. In this sense, researchers have tried to find sustainable recycling approach for discarding flame retardant textile wastes. Vecchiato et al. (2018) used an enzyme-based approach for the recycling of waste Rayon fibers. They recovered glucose and phosphorus additives with no change in their chemical structures. By complete hydrolysis of the Rayon fibers by cellulose preparation, a high yield of ca. 98% of the glucose and >99% of the flame-retardant pigment recovery was reached. Yasin et al. (2016) used mechanical technique for recycling the flame retardant treated cotton curtains. They reported that the higher CO_2 emissions and higher energy consumption were obtained during flame retardant cotton curtain production phase by chemical recycling compared to mechanical recycling.

11.2.2 RECYCLING OF WASTE ELECTRICAL AND ELECTRONIC EQUIPMENT

About 2.7 million tons of waste electrical and electronic equipment (WEEE) containing flame retardants are annually discarded. These wastes contain about 18 wt.% bromine and 31 wt.% phosphorus-based flame retardants that cause the environmental or human health risks (Peeters et al., 2014). The recycling rate of

Recycled Polymer Composites

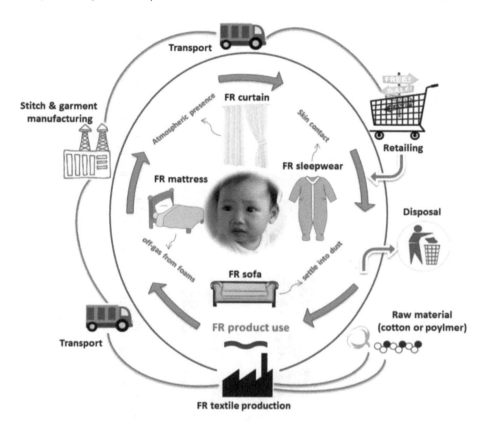

FIGURE 11.2 The life cycle of flame retardant textiles (Yasin et al., 2016). "Attention! Baby on Board"; it can be seen that such a complex cycle made this sweet baby worried about the future. The borderline separating childhood territory from the real adult world, which is typically hazardous and unreliable to a larger extent of baby's imagination, is easily permeable and quite delicate to the threats.

flame retardant WEEE plastics is still very low, e.g., only 12 wt.% of WEEE plastics are recycled in Europe (Biddle, 2012; Wang et al., 2012). The complexity of the WEEE plastics in which up to 300 different types of plastics are involved is the main reason for such a low recycling efficiency. Peeters et al. (2014) reported a size reduction based recycling of the plastics containing flame retardant from acrylonitrile butadiene styrene (ABS)/polycarbonate (PC) end-of-life liquid crystal display (LCD) TVs in which the product fraction contained more than 18 wt.% other plastics (Figure 11.3). In addition, as shown in Figure 11.3, they indicated that the overall recovery for flame retardant ABS/PC can be performed between 55–71% by using a rotary table.

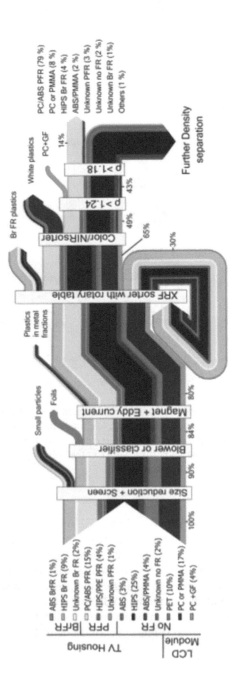

FIGURE 11.3 A detailed map of separation process of LCD TV plastics based on size screening followed by recognition and weighing of flame retardant plastics (Peeters et al., 2014). Even to this level it looks elaborate. However, the mysterious part of this plant seems to be how the composition and size could mutually affect separation efficiency and reliability.

In a number of studies, the recyclability of flame retardant polymers from real waste plastic mixtures, mainly from the WEEE, has been investigated (Dias et al., 2018; Gu et al., 2016; Işıldar et al., 2019; Reijnders, 2011). The process of mechanical recycling of WEEE is increasingly expected by regulators and demanded by original equipment manufacturers. Albeit mechanical recycling is the most economical recycling method used worldwide, it seems technically challenging for recycling of WEEE plastics (Imai et al., 2003). Considering the fact that 12% of WEEE plastics require the use of flame retardants to ensure a sufficient level of fire safety, there is a need for comparative test data on the recyclability of different flammable plastics (Imai et al., 2003).

Wagner et al. (2020) used failure mode and effect analysis (FMEA) as a powerful tool for the risk assessment related to the WEEE plastic recycling. Their results showed that the availability of information on the origin of WEEE plastics and processing history can help identify high-risk sources such as undesirable kinds of plastics, glass, metals, wood, rubber and foam as well as additives like talc, calcium carbonate or brominated flame retardants (Figure 11.4). Therefore, significant potential is expected from the implementation of innovative clustering strategies to improve the removal of hazardous materials and reduce the complexity of material inlet flows to increase the performance and the quality of recycled plastics.

Achieving success in recycling depends on many factors such as the type of polymer matrix (thermoplastic or thermoset) and the type of reinforcing agent that is fiber, glass, carbon, etc. In a mechanical recycling process, a solvent/nonsolvent system can be used to separate and recycle the polymers in a very easy way. Separation and recycling of polymers by this method seems to be technically possible and commercially significant. However, during the mechanical recycling process, including extrusion and injection molding, the plastic material is exposed to high temperatures up to 250°C (Ragaert et al., 2017). After exposure of plastic wastes to such a harsh condition, the recycled product should to a large degree present properties comparable to those of the virgin material to guarantee required application. The recycled products should also retain their physical properties, remain with the same color as the virgin material, have the same level of fire safety, and must not contain toxic by-products such as dioxins and furans. It goes without emphasis that such a high level of expectation is the reason why the efficiency of flame retardant polymer recycling is quite low in practice.

In the case of polyesters such as polylactic acid (PLA) and polyethylene terephthalate (PET), in addition to the mechanical recycling, recycling of the polymer raw material is also possible, because the polymer can be hydrolyzed to lactic acid (Anderson, 2017; Chauliac et al., 2020) or ethylene glycol and terephthalic acid, respectively. Thermal degradation of the polymeric materials through thermal cracking or pyrolysis in the absence of oxygen and in a neutral atmosphere results in valuable products (Balat, 2008; Maher & Bressler, 2007). During pyrolysis at elevated temperatures, depending on the polymer, the

Grading scale 1 2 3 4 5 6 7 8 9 10	TV Housings	Washing Machines	Fridges	Fridge Drawers	FTIR	XRF	MCA(CV)	TGA	DSC	Density	MFI	Spectrophotometer	GC, MS or ICP with MS	Mechanical Properties	Aesthetics	Legal Compliance	Rheological Properties	No testing	FTIR+XRF+MCA (CV)	All techniques
-Impurities																				
Plastic types	9	7	7	3	2	10	9	6	3	6	10	10	9	9	8	2	6	810	162	162
Glass, concrete, ceramic	10	10	10	4	9	10	3	10	10	8	10	10	10	9	6	1	2	900	270	270
Metals	10	10	10	2	9	2	2	10	10	6	10	6	10	9	6	1	2	900	180	180
Wood, Paper	5	5	5	3	5	10	2	10	6	10	10	10	6	9	8	2	5	450	90	90
Rubber	7	8	8	3	2	10	2	10	4	6	10	10	6	9	7	1	4	720	144	144
Foam	7	7	10	2	2	10	2	10	7	7	10	10	6	8	6	2	3	800	160	160
-Additives																				
Talc	3	10	9	9	4	10	10	3	10	6	10	10	2	8	3	1	4	800	320	160
CaCO₃	3	10	9	9	4	10	10	3	10	6	10	10	2	8	3	1	4	800	320	160
Glassfibre	2	10	4	2	9	10	10	3	10	6	10	10	8	8	6	1	7	800	720	240
Phosphor based FRs	10	3	3	1	7	8	10	10	6	10	10	10	2	5	3	1	6	600	420	120
Colourants	3	8	8	8	10	10	4	10	10	10	10	2	6	2	9	4	2	720	288	144
-Contaminants																				
Brominated FRs	10	4	4	1	7	3	10	10	7	10	10	2	5	3	10	6		1000	300	200
Restricted substances	5	6	5	3	10	7	10	10	10	10	10	2	3	5	10	3		600	420	120
-Irreversible chemical changes																				
Oxidation	7	5	5	2	6	10	8	10	10	10	10	7	5	5	7	3	3	490	294	245
Chain scission	8	8	8	8	10	10	10	10	10	6	10	6	10	8	3	2	9	720	720	432
Crosslinking	7	6	6	3	8	10	10	4	10	10	10	10		9	5	2	7	630	504	252

Column groups: Influences on Quality | Probability of presence | Probability of non-detection | Severity on quality | Risk Priority

FIGURE 11.4 Mendeleev-like FMEA table of the quality assessment of the mixed plastic flakes from WEEE (Wagner et al., 2020). Typically, the periodic table of Mendeleev enjoys three awesome potentials. First, it makes it possible to give each category and type of element a unique place, which can be accessed by the user in a quick yet precise way. Second, the columns and rows in such a table are in a meaningful way interrelated based on some elementary rules, like the size, the molecular weight, the polarity, etc. Third, the table predicts a possible place for the elements that are not explored/discovered yet. The present table provides the user a basic classification angle, but still a long way should be paved to make it comprehensive and systematic like the Mendeleev periodic table.

monomer can be produced in large quantities, e.g., in the case of poly-methyl methacrylate. Likewise, a fuel oil can occur in polyolefins, such as low-density polyethylene (LDPE) and high-density polyethylene (HDPE) and polypropylene (PP), together with other useful by-products (Achilias, 2015; Bennett, 2007). In order to find applications as a fuel or source of chemicals, Vasil et al. (2008) tried to improve the quality of heat-degrading oil produced by WEEE during catalytic hydrogenation. In addition to halogen removal, nitrogen and sulfur removal were also performed successfully. They have introduced a conventional acid catalyst and a neutral catalyst in the hydrogenation process such that both catalysts improved the quality of pyrolysis oil as a result of removal of oxygen, nitrogen, halogen and sulfur-containing compounds. In the case of thermoset composites, some other methods of recycling are examined. The conventional recycling approaches, including mechanical and thermal recycling, with recovery of energy and material is used (Oh et al., 2019; Zhao & Abu-Omar, 2018). However, natural fiber reinforced thermoset composites cannot be recovered easily by applying thermal processes (Nguyen et al., 2021; Rad et al., 2019).

Conventional recycling methods applied for polymer composite wastes are quite limited. Presorting flame retardant polymer composites before recycling is costly and time consuming. The process of recycling requires a lot of energy and often leads to the production of low-quality polymers; therefore, current technologies cannot be used in recycling many polymer composite materials. Recent research is paving the way for less energy-intensive chemical recycling methods, increasing the compatibility of the mixed plastic wastes by the coupling agents to avoid the need for sorting, and extending the recycling technologies over the traditional non-recyclable polymers. Chemical recycling approaches like solvolysis can be interesting, but they are not able to cope with the high content of additives or polymer mixtures commonly used in WEEE (Grause et al., 2009). Although methods used for recycling polymer composites are divers, a major challenge in recycling WEEE plastics is to produce a homogeneous fraction that earns the highest value added in the market.

11.3 DEVELOPMENT OF FLAME-RETARDANT MATERIALS FROM RECYCLED PLASTICS

Polyethylene terephthalate (PET) is the most recycled technical polymer. It comes mainly from the soda and mineral water bottles. A large proportion of this PET is recycled into new bottles or fabrics. Studies have been carried out to exploit this material in other technical applications, which require good fire performance. Laoutid et al. (2003a, 2006) focused on the development of suitable flame retardant systems for recycled PET by presenting acceptable cost and effective ways at a relatively low incorporation content to keep the material economically competitive while preserving its mechanical properties. Unfortunately, however, the multiplicity of PET recycling steps induce some PET degradation and reduction of polymer macromolecular chains, which unavoidably leads to a fall in the mechanical properties of the final material (Avila & Duarte, 2003; Mancini & Zanin, 2000). Laoutid et al. selected red phosphorus because of its effectiveness at a relatively low incorporation rate (3–5 wt.%), moderate cost and dual action in both the condensed and gaseous phases (Laoutid et al., 2003a, 2006). However, red phosphorus suffers from a major disadvantage because it releases phosphine gas (PH_3) during its thermal degradation during both storage and the extrusion process. Therefore, they sought to develop formulations, based on recycled PET, containing low red phosphorus content to avoid deterioration of mechanical properties and to ensure acceptable levels of safety without altering the fire properties of the final composite. This target was achieved by combining red phosphorus with an optimized amount of metal oxides. This basic study performed on the ground of the limit oxygen

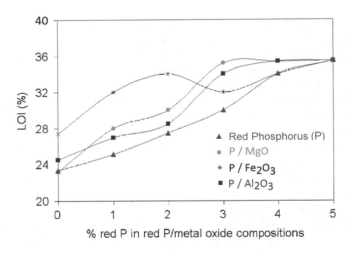

FIGURE 11.5 Evolution of the LOI value of the recycled PET-based compositions containing red phosphorus/metallic oxide combinations (Laoutid et al., 2003b). The total incorporation rate of metallic oxide/red phosphorous was set to 5 wt.%. Only aluminum oxide (Al_2O_3), magnesium oxide (MgO) and iron oxide (Fe_2O_3) were retained for the continuation of the study, while zirconium oxide (ZrO_2), manganese oxide (MnO_2) and calcium oxide (CaO) were discarded because of their weak effects. Combining red phosphorus with Al_2O_3, Fe_2O_3 or MgO enabled reaching higher LOI values than when only red phosphorous was used.

index (LOI) provided some useful insights about the selection of the most effective metallic oxides, whose combination with red phosphorus led to an improvement in the LOI value of the blend (Figure 11.5).

The strategy of combining red phosphorous with the selected metallic oxide was confirmed by the cone calorimetry test (Laoutid et al., 2003b). The results highlighted diverse and specific modes of action for each oxide. Combining red phosphorous with Al_2O_3 enabled the promotion of char formation through the development of thermally stable aluminum phosphate species, which reinforced the cohesion of the char layer formed during the combustion. The combination of red phosphorus with iron oxide boosted the flame-retardant effect of the red phosphorous through an oxidative effect of iron on the phosphorous. The analysis of the combustion residue by x-ray diffraction unveiled the presence of a mixture of Fe_3O_4 and FeO in the char layer, resulting from the reduction in the Fe_2O_3 during the combustion in the presence of red phosphorus. This observation suggested that Fe_2O_3 acts as a phosphorus oxidizing agent, thus promoting the formation of a more consistent and efficient carbon layer. This oxidizing effect of Fe_2O_3 leads to a significant improvement in the fire behavior of the composite, as reflected in an increase in the LOI value. In agreement with such outcome, cone calorimetry parameters were changed, such that the time to ignition (TTI) was significantly increased, while a substantial reduction in the peak of heat release rate

Recycled Polymer Composites 207

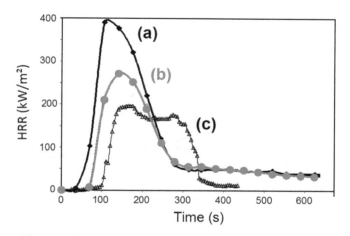

FIGURE 11.6 Heat release rate (HRR) curves of the pristine recycled PET (a), containing 5 wt.% red phosphorous (b), and a blend of 3 wt.% red phosphorous with 2 wt.% Fe_2O_3 (c) (Laoutid et al., 2006). Looking at the shape and broadness of the curves, one may come to the conclusion that using TTI or pHRR alone would end in an incorrect assessment for the flame retardancy performance. The use of FRI enables explanation of the performance of flame retardants, particularly when a hybrid system is used.

(pHRR) was additionally achieved (Figure 11.6). In a recent publication, we defined and applied a *flame retardancy index* (*FRI*) for the analysis of classification of the performance of flame retardants in polymer composites (*Poor, Good,* or *Excellent*) (Vahabi et al., 2019). This dimensionless criterion successfully made possible classification of flame retardancy performance of both thermoplastic (Fouad Laoutid et al., 2020; Vahabi et al., 2019) and thermoset (Jouyandeh et al., 2019; Movahedifar et al., 2019) polymer composites. Accordingly, it can be concluded that the combination of additives can even result in *Excellent* flame retardant properties of recycled polymer composites.

On the other hand, in the case of a red phosphorous/MgO combination, the enhancement of flame-retardant performance was not induced by any interaction between the two additives in spite of the chemical reaction between MgO and PET via ionic bonding at the chain ends. This interaction promoted the formation of carbonized structures contributing to the enhancement of both the quantity and quality of the protective carbon layer, leading to lower pHRR compared to phosphorous/Fe_2O_3 combination at identical loading (Figure 11.7).

In conclusion, the combination of metallic oxides and red phosphorus proved to be very effective, but its action was nevertheless limited by the low viscosity of the recycled PET, which hindered the formation of a stable char (Figure 11.8). Inert mineral fillers (talc or glass fibers) were added at a limited content to overcome such problem (Laoutid et al., 2003b). In this way, the importance of melt strength of the recycled PET on the flame-retardant performance was demonstrated (Figure 11.8).

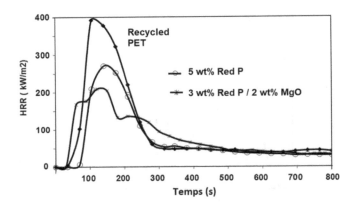

FIGURE 11.7 HRR curves of the pristine recycled PET (a), containing 5 wt.% red phosphorous (b) and a blend of 3 wt.% red phosphorous with 2 wt.% MgO (c) (Laoutid et al., 2006). Compared to the previous case (Figure 11.6), this combination prolonged burning time and a multi-modal HRR curve formed. Such a complementary effect caused by the combination of flame retardants can also be detected and visualized by the use of FRI.

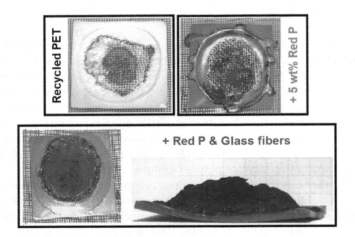

FIGURE 11.8 Photographs of the combustion residues of the various flame-retardant compositions used in recycling of PET for flame retardancy purpose (Laoutid et al., 2003b, 2006). The incorporation of this new mineral phase enabled strong reinforcement of the char mechanical properties, leading to the formation of intumescent layer as well as reduction in the pHRR from ca. 400–150 kW/m².

Improving the flame-retardant behavior of recycled PET by the incorporation of another recycled polymer into the system was studied by Swoboda et al. (2007). Polycarbonate (PC) was selected due to its high fire resistance and also its affinity with PET. The incorporation of PC into the PET resulted in a slight

Recycled Polymer Composites

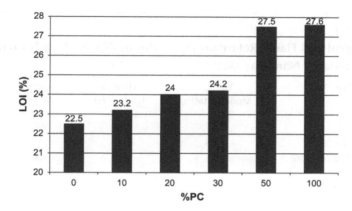

FIGURE 11.9 Limited oxygen index (LOI) levels of recycled PET/PC blends (Swoboda et al., 2007). As it can be seen, there is a need for high loading of PC to improve flame retardancy of recycled PET. Above 30 wt.% of PC, there is a jump in the value of LOI, but the author did not report compositions between 30 and 50 wt.% to specify the percolation threshold.

increase in the LOI and a slight reduction in the pHRR in the PET-rich blends (Figure 11.9).

Several strategies have been adopted for improving the flame-retardant properties of PET/PC blends: The first method attempts to enhance the compatibility between PET and PC using transesterification catalysts. This has been shown to increase the self-extinguishing properties of the blends. However, catalyzed blends present lower TTI than that of the non-catalyzed ones (Swoboda et al., 2007). The second approach is based on the incorporation of clay nanoplatelets previously modified with a phosphonium ion, i.e., phosphorylated nanoclay (MMT-P) (Swoboda et al., 2009) that further improves the flame-retardant properties of PET/PC blends. The phosphorylation step is necessary for the fact that clays, despite being known for pHRR reduction via the formation of a barrier layer during cone calorimeter test, are detrimental to LOI and UL-94 test rating. The presence of phosphonium ions during combustion helps to decrease the flammability of the material due to the condensed phase action of phosphorus. However, the flame-retardant effect of this clay is strongly dependent on the composition of the mixture. This is obviously severe in a recycled PET (RPET) matrix alone, interesting for the 80/20 RPET/PC composition, but very weak or even non-existent for a 50/50 blend of RPET/PC (Table 11.1).

The third method is based on the incorporation of triphenyl phosphite (TPP) to promote both PET chain extension and the formation of a carbon layer during the combustion (Swoboda et al., 2008). By controlling melt blending time, it is possible to obtain a blend with acceptable melt strength. It has been reported that TPP induces two antagonistic reactions: the chain-extension reaction that is predominant in the first phase (RPET) and the degradation reaction that becomes predominant in the second phase (PC) (Swoboda et al., 2009). It is therefore

TABLE 11.1
Mechanical and Flame-Retardant Properties of PET/PC Blends Containing Phosphorylated Nanoclay (MMT-P)

Composition	Young's Modulus (MPa)	Time to Ignition (s)	pHRR (kW/m^2)	LOI (%)
RPET	2017	71	700	22
RPET/PC (80/20)	2004	66	550	24
RPET/PC (80/20) – 5% MMT-P	2788	62	400	27
RPET/PC (50/50)	2102	95	420	27
RPET/PC (50/50) – 5% MMT-P	2642	82	450	31

RPET in the table means recycled PET

necessary to stop the mixture before the degradation reactions take place. Incorporating 4 wt.% TPP into an 80/20 PET/PC blend significantly reduced the pHRR in the cone calorimeter test (Figure 11.10), and also increased the LOI by 24–28%.

The fourth method combines phosphorylated clay and TPP (Swoboda et al., 2009). The hybridization of the last two strategies has improved the mechanical properties of the blends as well as their flame-retardant properties, such that a V0 classification in the UL-94 test was achieved.

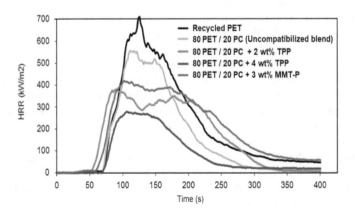

FIGURE 11.10 HRR curves during the cone calorimeter test (50 kW/m²) on RPET/PC blends. Recycled PET as the control sample and a series of 80/20 RPET/PC blends without and with variable content of chain extender and flame retardant (2 wt.% and 4 wt.% of TPP chain extender and 3 wt.% of MMT-P flame retardant additive) underwent testing (Swoboda et al., 2009).

11.4 ULTIMATE PROPERTIES OF THE RECYCLED FLAME RETARDANT POLYMER COMPOSITES

With comparison to landfill and incineration of plastic wastes, recycled plastics can replace some of the new materials in the plastics industry, which not merely reduces the material costs, but is also effective for resource recycling (Cao et al., 2019; Mwanza et al., 2018; Stichnothe & Azapagic, 2013). Flame retardant polymer wastes are useful materials for fabricating wood plastic composites (Najafi et al., 2006). However, because of a difference in chemical structure of the used plastics and their miscibility, the poor compatibility of plastics appears as a limiting factor in the recycling process, which can affect the ultimate performance of the final modified materials (Garcia & Robertson, 2017). Development of recycled flame retardant polymer composites with high fire performance as well as acceptable mechanical properties and a low environmental impact is a scientific debate. The recycling of most of flame retardant polymer composites can be performed without significant loss in their quality (Dawson & Landry, 2005). Although most of the reports show the use of a single matrix component from the waste product (Rahman et al., 2013), to compensate for the reduction of mechanical properties the waste polymer can be combined with different polymer wastes or even mix the waste polymer with a virgin polymer (Sinha et al., 2010) and also can be incorporated with fillers and additives (Suppakarn & Jarukumjorn, 2009).

The use of date palm fibers in enhancing the mechanical and flammability of recycled PP/LDPE/HDPE ternary blends containing $Mg(OH)_2$ as a flame retardant was reported by Zadeh et al. (2017). Their flame retardant polymer composite was low cost, less energy consuming and environmental friendly because it contained the recycled polymers and the fibers obtained from the date palm leaves. Though the $Mg(OH)_2$ flame retardant diminished the mechanical properties, the addition of the palm fibers were associated with to strengthen the composite, resulting in achievement of simultaneous flame retardancy and mechanical properties. Zadeh (2015) also developed a mechanically strong, thermally insulating and flame retardant: date palm leaf fiber filled composites based on recycled HDPE, LDPE and PP. It was found that maleated PP (MAPP) and maleated polyethylene (MAPE) as coupling agents in the ternary blend could improve the mechanical properties of a flame retardant system. The addition of 1 wt.% of MAPE or MAPP as nucleating agents improved the tensile strength of the system from 20 to 22.7 MPa or 21 MPa, respectively, through an increase in the crystallinity of the blends. In addition, they observed that the system containing 10 wt.% of $Mg(OH)_2$ showed a 22% improvement in the LOI.

Jarukumjorn et al. (2016) investigated the effect of alkali treatment and addition of flame retardants (ammonium polyphosphate, aluminum trihydrate and zinc borate) on the mechanical and flame retardant properties of sawdust/recycled HDPE composites. They found that alkali treatment enhanced mechanical properties and flame retardancy of the composite. However, the addition of flame retardants decreased the tensile strength, elongation at break and also the impact strength of the composites, while increased the tensile modulus and flame retarding properties. Li et al. (2020) studied the flame retardancy and mechanical properties of the recycled polyvinyl chloride (R-PVC)/ABS blends in the presence of the chlorinated polyethylene (CPE) and antimony trioxide (Sb_2O_3). For the 40/35/25 (w/w/w) ABS/R-PVC/PVC blends, the tensile strength was close to that of the neat ABS. The optimal properties were obtained by the addition of 8 wt.% CPE and 6 wt.% Sb_2O_3, so that the tensile strength of 63.5 MPa, the notched impact strength of 9.2 KJ m^{-2}, the limiting oxygen index of 31.3% and the vertical burning grade of the V0 level were achieved.

A flame-retardant panel was developed by Zhang et al. (2018) by combining different contents of recycled thermoset phenolic foam waste particles and natural wood fibers. The mechanical properties of flame retardant composites were decreased by increasing both the content and size of phenolic foam particles. The flexural strength and modulus of flame retardant panels were higher for the composite containing phenolic foam particle with sizes of 80–120 mesh (120–180 μm) compared to those of composites containing fillers with particle sizes of 60–80 mesh (180–250 μm) and 40–60 mesh (250–380 μm). As shown in Figure 11.11a–c, the larger particles affected the overlapping between fibers that could be considered a failure initiation site during the tensile or shear forces. Moreover, it was found that the pHRR of the composites containing 30 wt.% of 80–120 mesh size phenolic foam particles was reduced by 52.1% and the LOI increased by 41.3% due to the char forming mechanism, as shown in Figure 11.11d–g. They indicated that phenolic foam particles are good candidates for reuse in flame-retardant medium-density fiberboard (MDF).

11.4 CONCLUSION

This chapter aimed to assess the recyclability of flame retardant polymer composites containing flame retardants additives. First, the most commonly used approaches practiced in recycling of polymer composites were reviewed, where emphasis was placed on the mechanical, thermal and chemical recycling methods. Second, development of flame retardant composites from the recycled materials and investigations on the flame retardancy and mechanical properties of the resulting products were addressed. A part of this chaλpter was also dedicated to the WEEE plastics, which is a notable concern in the field of plastic recycling. It was observed that the composition of blend, the type of parent polymers and their thermal history, the amount of additives used and the processing conditions are all determining factors controlling the flame retardancy performance of the recycled blends. We also discussed how fast the environmental concerns are progressing about the recycled flame retardant composites and how strict the new fire safety instructions are in the

Recycled Polymer Composites

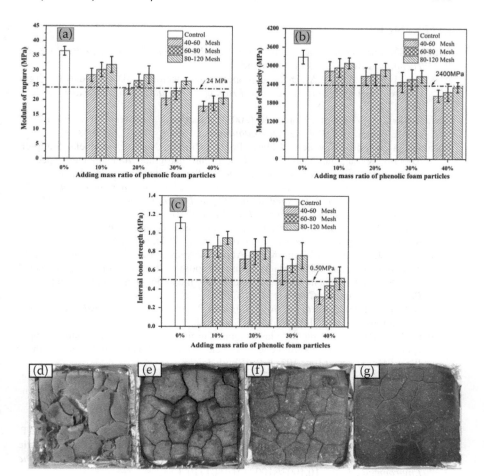

FIGURE 11.11 Effect of the content and the phenolic foam particle size on the (a) modulus of rupture, (b) modulus of elasticity, (c) internal bond strength of flame retardant panel composites and residual carbon of composites with different content of 80–120 mesh particles (d–g): (d) 0%, (e) 10%, (f) 20% and (g) 30%. From the integrity of char in image (g), it is evident that incorporation of more particles improved flame retardant properties of the recycled blend (Zhang et al., 2018).

field of polymer composites recycling. It conclusion, we should indicate that there is a growing demand for development of experimental analyses to define some novel, yet practical and feasible, approaches for recycling flame retardant polymer composites and recycled flame retardant polymer composite products. Of note, we did not find any theoretical work on the prediction of flame retardancy performance of such systems.

REFERENCES

Achilias, Dimitris (2015). *Recycling Materials Based on Environmentally Friendly Techniques*. BoD–Books on Demand.
Anderson, Isabelle (2017). Mechanical properties of specimens 3D printed with virgin and recycled polylactic acid. *3D Printing and Additive Manufacturing*, 4(2), 110–115.
Avila, Antonio F., & Duarte. Marcos V. (2003). A mechanical analysis on recycled PET/HDPE composites. *Polymer Degradation and Stability*, 80(2), 373–382.
Bacaloni, Alessandro, Cavaliere, Chiara, Foglia, Patrizia, Nazzari, Manuela, Samperi, Roberto, & Laganà, Aldo (2007). Liquid chromatography/tandem mass spectrometry determination of organophosphorus flame retardants and plasticizers in drinking and surface waters. *Rapid Communications in Mass Spectrometry: An International Journal Devoted to the Rapid Dissemination of Up-to-the-Minute Research in Mass Spectrometry*, 21(7), 1123–1130.
Balat, M. (2008). Mechanisms of thermochemical biomass conversion processes. Part 1: Reactions of pyrolysis. *Energy Sources, Part A*, 30(7), 620–635.
Bayat, Sahar, Jazani, Omid Moini, Molla-Abbasi, Payam, Jouyandeh, Maryam, & Saeb, Mohammad Reza (2019). Thin films of epoxy adhesives containing recycled polymers and graphene oxide nanoflakes for metal/polymer composite interface. *Progress in Organic Coatings*, 136, 105201.
Bennett, Gary F. (2007). *Feedstock Recycling and Pyrolysis of Waste Plastics*, J. Scheirs, W. Kaminsky (Eds.), John Wiley & Sons Ltd.
Biddle, Mike (2012). Closing the plastic loop: Turning the supply chain into a supply cycle by mining plastics from end-of-life electronics and other durable goods. In *E-Waste Management* (pp. 107–130). Routledge.
Cao, Zhiguo, Chen, Qiaoying, Li, Xiaoxiao, Zhang, Yacai, Ren, Meihui, Sun, Lifang, Wang, Mengmeng, Liu, Xiaotu, & Yu, Gang (2019). The non-negligible environmental risk of recycling halogenated flame retardants associated with plastic regeneration in China. *Science of The Total Environment*, 646, 1090–1096.
Chauliac, Diane, Pullammanappallil, Pratap C., Ingram, Lonnie O., & Shanmugam, K. T. (2020). A combined thermochemical and microbial process for recycling polylactic acid polymer to optically pure l-lactic acid for reuse. *Journal of Polymers and the Environment*, 1–10. 10.1007/s10924-020-01710-1
Dawson, Raymond B., & Landry, Susan D. (2005). Recyclability of flame retardant HIPS, PC/ABS, and PPO/HIPS used in electrical and electronic equipment. *Proceedings of the 2005 IEEE International Symposium on Electronics and the Environment*, 2005.
Delgado, Clara, Barruetabeña, Leire, & Salas, Oscar (2007). *Assessment of the Environmental Advantages and Drawbacks of Existing and Emerging Polymers Recovery Processes*. JRC Institute for Prospective Technological Studies.
Dias, Pablo, Bernardes, Andréa Moura, & Huda, Nazmul (2018). Waste electrical and electronic equipment (WEEE) management: An analysis on the australian e-waste recycling scheme. *Journal of Cleaner Production*, 197, 750–764.
Dias, Pablo, Machado, Arthur, Huda, Nazmul, & Bernardes, Andréa Moura (2018). Waste electric and electronic equipment (WEEE) management: A study on the Brazilian recycling routes. *Journal of Cleaner Production*, 174, 7–16.
Ebrahimi Jahromi, Reza, Hamid, Bakhshandeh, Golam Reza, Jahromi, Ali Ebrahimi, Saeb, Mohammad Reza, Ahmadi, Zahed, & Pakdel, Amir Saeid (2017). A comparative study to assess structure–properties relationships in (acrylonitrile butadiene rubber)-based composites: Recycled microfillers versus nanofillers. *Journal of Vinyl and Additive Technology*, 23(1), 13–20.

Garcia, Jeannette M., & Robertson, Megan L. (2017). The future of plastics recycling. *Science*, *358*(6365), 870–872.
Grause, Guido, Sugawara, Katsuya, Mizoguchi, Tadaaki, & Yoshioka, Toshiaki (2009). Pyrolytic hydrolysis of polycarbonate in the presence of earth-alkali oxides and hydroxides. *Polymer Degradation and Stability*, *94*(7), 1119–1124.
Gu, Yifan, Wu, Yufeng, Xu, Ming, Mu, Xianzhong, & Zuo, Tieyong (2016). Waste electrical and electronic equipment (WEEE) recycling for a sustainable resource supply in the electronics industry in China. *Journal of Cleaner Production*, *127*, 331–338.
Hadavand, Behzad Shirkavand, Jouyandeh, Maryam, Paran, Seyed Mohamad Reza, Khalili, Reza, Vahabi, Henri, Bafghi, Hamed Fakharizadeh, Laoutid, Fouad, Vijayan, P. Poornima, & Saeb, Mohammad Reza (2020). Silane-functionalized Al2O3-modified polyurethane powder coatings: Nonisothermal degradation kinetics and mechanistic insights. *Journal of Applied Polymer Science*, *137*(45), 49412.
Imai, Takaretu, Hamm, Stephan, & Rothenbacher, Klaus P. (2003). Comparison of the recyclability of flame-retarded plastics. *Environmental Science & Technology*, *37*(3), 652–656.
Irvine, D. J., McCluskey, J. A., & Robinson, I. M. (2000). Fire hazards and some common polymers. *Polymer Degradation and Stability*, *67*(3), 383–396.
Işıldar, Arda, van Hullebusch, Eric D., Lenz, Markus, Laing, Gijs Du, Marra, Alessandra, Cesaro Panda, Sandeep, Akcil, Ata, Kucuker, Mehmet Ali, & Kuchta, Kerstin (2019). Biotechnological strategies for the recovery of valuable and critical raw materials from waste electrical and electronic equipment (WEEE) – A review. *Journal of Hazardous Materials*, *362*, 467–481.
Jarukumjorn, Kasama, Jarapanyacheep, Rapisa, & Kenkhokkruad, Harutape (2016). Mechanical properties and flame retardancy of sawdust/recycled high density polyethylene composites: Effects of alkali treatment and flame retardant. *Walailak Journal of Science and Technology (WJST)*, *13*(12), 1017–1024.
Jouyandeh, Maryam, Hadavand, Behzad Shirkavand, Tikhani, Farimah, Khalili, Reza, Bagheri, Babak, Zarrintaj, Payam, Formela, Krzyszof, Vahabi, Henri, & Saeb, Mohammad Reza (2020). Thermal-resistant polyurethane/nanoclay powder coatings: Degradation kinetics study. *Coatings*, *10*(9), 871.
Jouyandeh, Maryam, Rahmati, Negar, Movahedifar, Elnaz, Hadavand, Behzad Shirkavand, Karami, Zohre, Ghaffari, Mehdi, Taheri, Peyman, Bakhshandeh, Ehsan, Vahabi, Henri & Ganjali, Mohammad Reza (2019). Properties of nano-Fe$_3$O$_4$ incorporated epoxy coatings from Cure Index perspective. *Progress in Organic Coatings*, *133*, 220–228.
Kobes, M., Groenewegen, K., & Duyvis, M.G. (2009). Consumer fire safety: European statistics and potential. NETHERLANDS INSTITUTE FOR SAFETY NIBRA
Laoutid, F., Ferry, L., Lopez-Cuesta, J. M., & Crespy, A. (2003a). Red phosphorus/aluminium oxide compositions as flame retardants in recycled poly (ethylene terephthalate). *Polymer Degradation and Stability*, *82*(2), 357–363.
Laoutid, F., Ferry, L., Lopez-Cuesta, J. M., & Crespy, A. (2006). Flame-retardant action of red phosphorus/magnesium oxide and red phosphorus/iron oxide compositions in recycled PET. *Fire and Materials: An International Journal*, *30*(5), 343–358.
Laoutid, F., Ferry, L., Lopez-Cuesta, J. M., & Crespy, A. (2003b). Red phosphorus/aluminium oxide compositions as flame retardants in recycled poly(ethylene terephthalate). *Polymer Degradation and Stability*, *82*(2), 357–363. 10.1016/S0141-3910(03)00213-1. http://www.sciencedirect.com/science/article/pii/S0141391003002131.
Laoutid, Fouad, Vahabi, Henri, Movahedifar, Elnaz, Laheurte, Pascal, Vagner, Christelle, Cochez, Marianne, Brison, Loïc, & Saeb, Mohammad Reza (2020). Calcium carbonate and ammonium polyphosphate flame retardant additives formulated to

protect ethylene vinyl acetate copolymer against fire: Hydrated or carbonated calcium? *Journal of Vinyl and Additive Technology*. 10.1002/vnl.21800

LeBel, Guy L., & Williams, David T. (1986). Levels of triaryl/alkyl phosphates in human adipose tissue from Eastern Ontario. *Bulletin of Environmental Contamination and Toxicology*, *37*(1), 41–46.

Li, Yingchun, Lv, Lida, Wang, Wensheng, Zhang, Jiaoxia, Lin, Jing, Zhou, Juying, Dong, Mengyao, Gan, Yuanfa, Seok, Ilwoo, & Guo, Zhanhu (2020). Effects of chlorinated polyethylene and antimony trioxide on recycled polyvinyl chloride/acryl-butadiene-styrene blends: Flame retardancy and mechanical properties. *Polymer*, *190*, 122198. 10.1016/j.polymer.2020.122198. http://www.sciencedirect.com/science/article/pii/S0032386120300422.

Lu, Shui-Yu, & Hamerton, Ian (2002). Recent developments in the chemistry of halogen-free flame retardant polymers. *Progress in Polymer Science*, *27*(8), 1661–1712.

Maher, K. D., & Bressler, D. C. (2007). Pyrolysis of triglyceride materials for the production of renewable fuels and chemicals. *Bioresource Technology*, *98*(12), 2351–2368.

Mancini, Sandro Donnini, & Zanin, Maria (2000). Consecutive steps of PET recycling by injection: Evaluation of the procedure and of the mechanical properties. *Journal of Applied Polymer Science*, *76*(2), 266–275.

Marosi, Gy, Beáta Szolnoki, K. Bocz, & Toldy, Andrea (2017). Fire-retardant recyclable and biobased polymer composites. In*Novel fire retardant polymers and composite materials* (pp. 117–146). Elsevier.

Meyer, J., & Bester, Kai (2004). Organophosphate flame retardants and plasticisers in wastewater treatment plants. *Journal of Environmental Monitoring*, *6*(7), 599–605.

Monopoli, Antonio, Nacci, Angelo, Calò, Vincenzo, Ciminale, Francesco, Cotugno, Pietro, Mangone, Annarosa, Giannossa, Lorena Carla, Azzone, Pietro, & Cioffi, Nicola (2010). Palladium/zirconium oxide nanocomposite as a highly recyclable catalyst for cc coupling reactions in water. *Molecules*, *15*(7), 4511–4525.

Movahedifar, Elnaz, Vahabi, Henri, Saeb, Mohammad Reza, & Thomas, Sabu (2019). Flame retardant epoxy composites on the road of innovation: An analysis with flame retardancy index for future development. *Molecules*, *24*(21), 3964.

Mwanza, Bupe Getrude, Mbohwa, Charles, & Telukdarie, Arnesh (2018). Strategies for the recovery and recycling of plastic solid waste (PSW): A focus on plastic manufacturing companies. *Procedia Manufacturing*, *21*, 686–693.

Najafi, Saeed Kazemi, Hamidinia, Elham, & Tajvidi, Mehdi (2006). Mechanical properties of composites from sawdust and recycled plastics. *Journal of Applied Polymer Science*, *100*(5), 3641–3645.

Nguyen, Quoc-Bao, Vahabi, Henri, Anda, Agustín Rios de, Versace, Davy-Louis, Langlois, Valérie, Perrot, Camille, Nguyen, Vu-Hieu, Naili, Salah, & Renard, Estelle (2021). Dual UV-thermal curing of biobased resorcinol epoxy resin-diatomite composites with improved acoustic performance and attractive flame retardancy behavior. *Sustainable Chemistry*, *2*(1), 24–48.

Oh, Yuree, Lee, Kyoung Min, Jung, Doyoung, Chae, Ji Ae, Kim, Hea Ji, Chang, Mincheol, Park, Jong-Jin, & Kim, Hyungwoo (2019). Sustainable, naringenin-based thermosets show reversible macroscopic shape changes and enable modular recycling. *ACS Macro Letters*, *8*(3), 239–244.

Paran, Seyed Mohammad Reza, Vahabi, Henri, Jouyandeh, Maryam, Ducos, Franck, Formela, Krzysztof, & Saeb, Mohammad Reza (2019). Thermal decomposition kinetics of dynamically vulcanized polyamide 6–acrylonitrile butadiene rubber – halloysite nanotube nanocomposites. *Journal of Applied Polymer Science*, *136*(20), 47483.

Pearce, Eli. (2012). *Flame-Retardant Polymeric Materials.* Springer Science & Business Media.
Peeters, Jef R., Vanegas, Paul, Tange, Lein, Houwelingen, Jan Van, & Duflou, Joost R. (2014). Closed loop recycling of plastics containing flame retardants. *Resources, Conservation and Recycling, 84*, 35–43.
Rad, Elaheh Rohani, Vahabi, Henri, de Anda, Agustin Rios, Saeb, Mohammad Reza, & Thomas, Sabu (2019). Bio-epoxy resins with inherent flame retardancy. *Progress in Organic Coatings, 135*, 608–612.
Ragaert, Kim, Delva, Laurens, & Van Geem, Kevin (2017). Mechanical and chemical recycling of solid plastic waste. *Waste Management, 69*, 24–58.
Rahman, Khandkar Siddikur, Islam, Md Nazrul, Rahman, Md Mushfiqur, Hannan, Md Obaidullah, Dungani, Rudi, & Abdul Khalil, H. P. S. (2013). Flat-pressed wood plastic composites from sawdust and recycled polyethylene terephthalate (PET): Physical and mechanical properties. *SpringerPlus, 2*(1), 629.
Reijnders, L. (2011). Recycling of elastomeric nanocomposites. In *Recent Advances in Elastomeric Nanocomposites* (pp. 179–198). Springer.
Reijnders, L. (2014). Safe recycling of materials containing persistent inorganic and carbon nanoparticles. In *Health and Environmental Safety of Nanomaterials* (pp. 222–250). Elsevier.
Rodil, Rosario, Quintana, José Benito, Concha-Graña, Estefanía, López-Mahía, Purificación, Muniategui-Lorenzo, Soledad, & Prada-Rodríguez, Darío (2012). Emerging pollutants in sewage, surface and drinking water in Galicia (NW Spain). *Chemosphere, 86*(10), 1040–1049.
Saeb, M. R., Vahabi, Henri, Jouyandeh, Maryam, Movahedifar, Elnaz, & Khalili, R. (2017). Epoxy-based flame retardant nanocomposite coatings: Comparison between functions of expandable graphite and halloysite nanotubes. *Progress in Color, Colorants and Coatings, 10*(4), 245–252.
Saito, I., Onuki, A., & Seto, H. (2007). Indoor organophosphate and polybrominated flame retardants in Tokyo. *Indoor Air, 17*(1), 28–36.
Seidi, Farzad, Jouyandeh, Maryam, Taghizadeh, Ali, Taghizadeh, Mohsen, Habibzadeh, Sajjad, Jin, Yongcan, Xiao, Huining, Zarrintaj, Payam, & Saeb, Mohammad Reza (2020). Polyhedral oligomeric silsesquioxane/epoxy coatings: A review. *Surface Innovations, 9*(1), 3–16.
Seidi, Farzad, Jouyandeh, Maryam, Taghizadeh, Mohsen, Taghizadeh, Ali, Vahabi, Henri, Habibzadeh, Sajjad, Formela, Krzysztof, & Saeb, Mohammad Reza (2020). Metal-organic framework (MOF)/epoxy coatings: A review. *Materials, 13*(12), 2881.
Shamsi, Ramin, Sadeghi, Gity Mir Mohamad, Vahabi, Henri, Seyfi, Javad, Sheibani, Reza, Zarrintaj, Payam, Laoutid, Fouad, & Saeb, Mohammad Reza (2019). Hopes beyond PET recycling: Environmentally clean and engineeringly applicable. *Journal of Polymers and the Environment, 27*(11), 2490–2508.
Sinha, Vijaykumar, Patel, Mayank R., & Patel, Jigar V. (2010). PET waste management by chemical recycling: A review. *Journal of Polymers and the Environment, 18*(1), 8–25.
Stapleton, Heather M., Klosterhaus, Susan, Eagle, Sarah, Fuh, Jennifer, Meeker, John D., Blum, Arlene, and Webster, Thomas F. (2009). Detection of organophosphate flame retardants in furniture foam and US house dust. *Environmental Science & Technology, 43*(19), 7490–7495.
Stichnothe, Heinz, & Azapagic, Adisa (2013). Life cycle assessment of recycling PVC window frames. *Resources, Conservation and Recycling, 71*, 40–47.
Sundkvist, Anneli Marklund, Olofsson, Ulrika, & Haglund, Peter (2010). Organophosphorus flame retardants and plasticizers in marine and fresh water biota and in human milk. *Journal of Environmental Monitoring, 12*(4), 943–951.

Suppakarn, Nitinat, & Jarukumjorn, Kasama (2009). Mechanical properties and flammability of sisal/PP composites: Effect of flame retardant type and content. *Composites Part B: Engineering, 40*(7), 613–618.

Swoboda, B., Buonomo, S., Leroy, E., & Lopez Cuesta, J. M. (2007). Reaction to fire of recycled poly (ethylene terephthalate)/polycarbonate blends. *Polymer Degradation and Stability, 92*(12), 2247–2256.

Swoboda, B., Buonomo, S., Leroy, E., & Lopez Cuesta, J. M. (2008). Fire retardant poly (ethylene terephthalate)/polycarbonate/triphenyl phosphite blends. *Polymer Degradation and Stability, 93*(5), 910–917.

Swoboda, B., Leroy, E., Laoutid, F., & Lopez-Cuesta, J.-M. (2009). *Flame-Retardant PET-PC Blends Compatibilized by Organomodified Montmorillonites.* ACS Publications.

Tikhani, Farimah, Hadavand, Behzad Shirkavand, Bafghi, Hamed Fakharizadeh, Jouyandeh, Maryam, Vahabi, Henri, Formela, Krzyszof, Hosseini, Hossein, Paran, Seyed Mohammad Reza, Esmaeili, Amin, & Mohaddespour, Ahmad (2020). Polyurethane/silane-functionalized ZrO2 nanocomposite powder coatings: Thermal degradation kinetics. *Coatings, 10*(4), 413.

United Kindgom. (2002). Department for environment, food & rural affairs. *Emissions Trading Scheme Auction.* http://www.defra.gov.uk/environment/climatechange/trading/auctionwin.htm.

Vahabi, Henri, Dumazert, Loïc, Khalili, Reza, Saeb, Mohammad Reza, & Cuesta, José-Marie Lopez (2019). Flame retardant PP/PA6 blends: A recipe for recycled wastes. *Flame Retardancy and Thermal Stability of Materials, 2*(1), 1–8.

Vahabi, Henri, Jouyandeh, Maryam, Cochez, Marianne, Khalili, Reza, Vagner, Christelle, Ferriol, Michel, Movahedifar, Elnaz, Ramezanzadeh, Bahram, Rostami, Mehran, & Ranjbar, Zahra (2018). Short-lasting fire in partially and completely cured epoxy coatings containing expandable graphite and halloysite nanotube additives. *Progress in Organic Coatings, 123*, 160–167.

Vahabi, Henri, Kandola, Baljinder K., & Saeb, Mohammad Reza (2019). Flame retardancy index for thermoplastic composites. *Polymers, 11*(3), 407.

Vahabi, Henri, Laoutid, Fouad, Mehrpouya, Mehrshad, Saeb, Mohammad Reza, & Dubois, Philippe (2021). Flame retardant polymer materials: An update and the future for 3D printing developments. *Materials Science and Engineering: R: Reports, 144*, 100604.

Vahabi, Henri, Laoutid, Fouad, Movahedifar, Elnaz, Khalili, Reza, Rahmati, Negar, Vagner, Christelle, Cochez, Marianne, Brison, Loic, Ducos, Franck, & Ganjali, Mohammad Reza (2019). Description of complementary actions of mineral and organic additives in thermoplastic polymer composites by Flame Retardancy Index. *Polymers for Advanced Technologies, 30*(8), 2056–2066.

Vahabi, Henri, Movahedifar, Elnaz, Ganjali, Mohammad Reza, & Saeb, Mohammad Reza (2021). Polymer nanocomposites from the flame retardancy viewpoint: A comprehensive classification of nanoparticle performance using the flame retardancy index. In *Handbook of Polymer Nanocomposites for Industrial Applications* (pp. 61–146). Elsevier.

Vahabi, Henri, Saeb, Mohammad Reza, Formela, Krzysztof, & Cuesta, José-Marie Lopez (2018). Flame retardant epoxy/halloysite nanotubes nanocomposite coatings: Exploring low-concentration threshold for flammability compared to expandable graphite as superior fire retardant. *Progress in Organic Coatings, 119*, 8–14.

Vahabi, Henri, Shabanian, Meisam, Aryanasab, Fezzeh, Mangin, Rémy, Laoutid, Fouad, & Saeb, Mohammad Reza (2018). Inclusion of modified lignocellulose and nanohydroxyapatite in development of new bio-based adjuvant flame retardant for poly (lactic acid). *Thermochimica Acta, 666*, 51–59.

Vahabi, Henri, Wu, Hao, Saeb, Mohammad Reza, Koo, Joseph H, & Ramakrishna, Seeram (2021). Electrospinning for developing flame retardant polymer materials: Current status and future perspectives. *Polymer*, 123466.

Vasile, C., Brebu, M. A., Totolin, M., Yanik, J. A. L. E., Karayildirim, T. A. M. E. R., & Darie, H. (2008). Feedstock recycling from the printed circuit boards of used computers. *Energy & Fuels*, *22*(3), 1658–1665.

Vecchiato, Sara, Skopek, Lukas, Jankova, Stepanka, Pellis, Alessandro, Ipsmiller, Wolfgang, Aldrian, Alexia, Mueller, Bernhard, Acero, Enrique Herrero, & Guebitz, Georg M. (2018). Enzymatic recycling of high-value phosphor flame-retardant pigment and glucose from rayon fibers. *ACS Sustainable Chemistry & Engineering*, *6*(2), 2386–2394.

Wagner, Florian, Peeters, Jef R., Ramon, Hans, De Keyzer, Jozefien, Duflou, Joost R., & Dewulf, Wim (2020). Quality assessment of mixed plastic flakes from Waste Electrical and Electronic Equipment (WEEE) by spectroscopic techniques. *Resources, Conservation and Recycling*, *158*, 104801. 10.1016/j.resconrec.2020.1 04801. http://www.sciencedirect.com/science/article/pii/S0921344920301221.

Wang, Feng, Huisman, Jaco, Meskers, Christina E. M., Schluep, Mathias, Stevels, A. B., & Hagelüken, Christian (2012). The Best-of-2-Worlds philosophy: Developing local dismantling and global infrastructure network for sustainable e-waste treatment in emerging economies. *Waste Management*, *32*(11), 2134–2146.

Wang, Zhuoran, Wang, Heng, Liu, Bin, Qiu, Wenzhe, Zhang, Jun, Ran, Sihan, Huang, Hongtao, Xu, Jing, Han, Hongwei, & Chen, Di (2011). Transferable and flexible nanorod-assembled TiO2 cloths for dye-sensitized solar cells, photodetectors, and photocatalysts. *ACS Nano*, *5*(10), 8412–8419.

Yasin, Sohail, Behary, Nemeshwaree, Curti, Massimo, & Rovero, Giorgio (2016). Global consumption of flame retardants and related environmental concerns: A study on possible mechanical recycling of flame retardant textiles. *Fibers*, *4*(2), 16.

Zadeh, Khadija M., Ponnamma, Deepalekshmi, & Al-Maadeed, Mariam Al Ali (2017). Date palm fibre filled recycled ternary polymer blend composites with enhanced flame retardancy. *Polymer Testing*, *61*, 341–348.

Zadeh, Khadija M., Ponnamma, Deepalekshmi, & Al-Maadeed, Mariam Al Ali (2017). Date palm fibre filled recycled ternary polymer blend composites with enhanced flame retardancy. *Polymer Testing*, *61*, 341–348. 10.1016/j.polymertesting.2017.05 .006. http://www.sciencedirect.com/science/article/pii/S0142941817302210.

Zadeh, Khadija Morad Shik (2015). *Tensile, Thermal and Flammability Properties of Date Palm Fiber Filled Recycled Ternary Blends and Composites*. Universiti Teknologi Malaysia.

Zhang, Longfei, Liang, Shanqing, & Chen, Zhilin (2018). Influence of particle size and addition of recycling phenolic foam on mechanical and flame retardant properties of wood-phenolic composites. *Construction and Building Materials*, *168*, 1–10. 10.1 016/j.conbuildmat.2018.01.173. http://www.sciencedirect.com/science/article/pii/ S0950061818301971.

Zhao, Shou, & Abu-Omar, Mahdi M. (2018). Recyclable and malleable epoxy thermoset bearing aromatic imine bonds. *Macromolecules*, *51*(23), 9816–9824.

12 Mechanical and Tribological Properties of Scrap Rubber/Epoxy-Based Composites

L.M.P. Ferreira[1,2], I. Miskioglu[3], E. Bayraktar[2], and D. Katundi[2]
[1]Federal University of Southern and Southeastern Pará – UNIFESSPA, Brazil
[2]ISAE-SUPMECA-Paris, School of Mechanical and Manufacturing Engineering, France
[3]Michigan Technological University, Houghton, Michigan, USA

CONTENTS

12.1 Introduction 221
12.2 Materials and Method 222
12.3 Results and Discussion 224
12.4 Conclusions 231
References 231

12.1 INTRODUCTION

In engineering applications, there are many possibilities for usage of recycling of scrap rubber and most of them are used in automotive industry and domestic area, etc. Other main areas such as the aerospace and microelectronics industries have enormous demand for high performance (ductile and high toughness) structural adhesive systems like epoxy and/or elastomers reinforced composites (Bayraktar et al., 2006, 2015; Bessri et al., 2010; Kaynak et al., 2001; Papadopoulos, 2005; Yesilata & Turgut, 2007; Zaimova et al., 2012, 2014, 2015). Today, additionally, the main component of these waste rubbers is styrene–butadiene rubber (SBR) and, in spite of the different uses for recycling it, the research for new applications is still a need because of the extremely high amount of waste rubber produced every year (Bayraktar et al., 2006, 2008; Bessri et al., 2010; Irez & Bayraktar,

2019; Irez et al., 2017, 2019, 2020; Kaynak et al., 2001; Luong et al., 2007; Masoud et al., 2020; Nacif et al., 2013; Pacheco-Torgal et al., 2012; Yesilata & Turgut, 2007; Zaimova et al., 2015, 2016).

Additionally, these materials are commonly used for long-term applications at ambient or at moderately elevated temperature conditions. Conforming to these needs, elastomers (rubbers) should be used by simple processing with various materials in different conditions by the addition of new alloying elements (Bayraktar et al., 2006, 2008, 2015; Bessri et al., 2010; Irez et al., 2019, 2020; Luong et al., 2007; Nacif et al., 2013; Pacheco-Torgal et al., 2012; Zaimova et al., 2015).

The present work reviews manufacturing facilities of scrap elastomers (SBR-rubbers) + epoxy resin composites with different proportions of particulate reinforcements. The main objective of this research was to determine the ductility and toughness of scrap elastomer (SBR rubber) matrix composites containing GF, TiO_2, B, etc. as basic reinforcements. Basically, these composites are aimed for use in automotive and aeronautics applications as bumpers and as internal furniture as ductile and tough and sound materials. Scanning electron microscopy (SEM) was used to study the microstructure and fracture surfaces of these composites.

12.2 MATERIALS AND METHOD

The composite materials used in this study were designed with a rubber/epoxy blend as the matrix, fibers and other reinforcing elements were then added to this matrix. In order to facilitate the bonding of rubber with epoxy, scrap rubber powder was chemically treated with toluene, acrylic acid and vinyltrimethoxysilane (2%). The treated rubber powder was dried in an oven to eliminate any residual trace of the chemicals. Then the rubber powder was blended with epoxy and the blend was used as the matrix for the composites. The final constitution of the matrix was 60% rubber and 40% epoxy by weight. The composites were manufactured by adding reinforcing fibers and particles to the matrix. Aluminum (Al), alumina (Al_2O_3) fiber and powder, glass fiber (GF), titanium dioxide (TiO_2), boron (B) and copper (Cu) were used as reinforcements.

The compositions of the composites and the nomenclature used to identify them are given in Table 12.1.

For all the composites considered titanium dioxide (TiO_2) was chosen as the main reinforcing element. The composites were formulated to capture the effect of increasing TiO_2 content on the mechanical and tribological properties when different fibers were used. Also, small amounts of boron (B) and copper (Cu) were added as additional reinforcing elements, along with aluminum (Al) powder.

The composites were produced as follows: First, the scrap rubber and epoxy were blended as described previously. Then the reinforcements given in Table 12.1 were added for each composition and the entire mixture was milled for 2 hours to attain a homogenous compound. Hot compacting (double uniaxial action) under a pressure of 70 MPa at 180°C was used to obtain the final form of the composite specimens. The dwell time for the compacting process was 15 minutes. After hot-compacting,

TABLE 12.1
General Compositions of the Rubber/Epoxy-Based Composites (wt.%)

Composition Name	Matrix (Rubber 60% Epoxy 40%)	Al	TiO$_2$	B	Cu	Al$_2$O$_3$ Fiber	Al$_2$O$_3$ Powder	GF
RETI-1	Balance	13	0	1	1	6	1	–
RETI-2	Balance	13	7	1	1	6	1	–
RETI-3	Balance	13	10	1	1	6	1	–
RETI-4	Balance	13	15	1	1	6	1	–
RETIG-1	Balance	7	0	1	1	–	–	7
RETIG-2	Balance	7	7	1	1	–	–	7
RETIG-3	Balance	7	10	1	1	–	–	7
RETIG-4	Balance	7	15	1	1	–	–	7

the specimens were cooled down homogenously to room temperature slowly in the device. All the specimens were then post-cured under isothermal conditions at 80°C for 48 hours. The specimens obtained by this process were 6–10 mm thick discs with a diameter of 30 or 40 mm.

For material characterization, Shore D hardness, macroindentation, dynamic compression and wear tests were performed. The Shore D hardness test was performed on the polished flat surfaces of the specimens according to ASTM D 2240 using a durometer (type HBD-100-0). The Shore D hardness test was also performed on the samples exposed to ultraviolet (UV) radiation for a total of 2 months. The Shore D hardness tests were performed every week during the 2 months to evaluate if any degradation occurred because of the UV exposure.

Compression tests by macroindentation were carried out at room temperature using a 3 mm diameter stainless steel ball. Tests were done on 10 specimens for each composition with a diameter of 40 mm. Thickness of the specimens was variable from 6 up to 10 mm. Specimens were loaded to fracture and maximum force (F_{max}) at failure was recorded for each test. Fracture surfaces were evaluated by SEM analysis. The failure of the specimens were basically due to the three-dimensional state of stress developed under the steel ball.

Dynamic compression (drop weight) tests were carried out using a universal drop weight impact tester (Dynatup Model 8200). The drop weight used was 1.9 kg and drop height was 600 mm. The impact velocity was about 3 m/s.

The scratch function of a nanoindenter was utilized to evaluate the wear resistance of the composites. The tests were performed with a conical tip that had a 90-degree apex angle under 20 mN and 50 mN normal loads. The load was applied over a linear track of 500 μm for 50 cycles. One cycle is defined as a pass and return of the tip over the track; the total distance for one test was 0.050 m. The speed of the tip during wear tests was 50 μm/s. A total of 10 wear tests were performed for each sample.

12.3 RESULTS AND DISCUSSION

The microstructure of the RETI-(alumina fibers) and RETIG-(glass fibers) compositions are shown in Figures 12.1 and 12.2. The distribution of the constituents are fairly homogeneous with very little agglomeration observed. Longer milling time during the production of material should improve homogeneity and help to eliminate the agglomerations.

Figure 12.3 shows the Shore D hardness values for all compositions before and after they were exposed to UV radiation. The measurements were performed at 7-day intervals, and the results show no significant change in the Shore D hardness values due to UV exposure. This indicates that these materials can be suitable for external parts in transportation industry [7–13, 17–22].

Macroindentation tests were carried out to fracture the specimens and the average maximum force obtained for each composite and the results are shown in Figure 12.4. For RETI composites, the addition of TiO_2 drastically reduced F_{max} when combined with Al_2O_3 fibers; on the other hand, for RETIG composites the addition of TiO_2 with glass fibers showed an increase in F_{max} with the increase in the TiO_2 content.

The SEM images of the fracture surfaces of RETI and RETIG samples from the macro indentation tests are shown in Figures 12.5 and 12.6. For the RETI

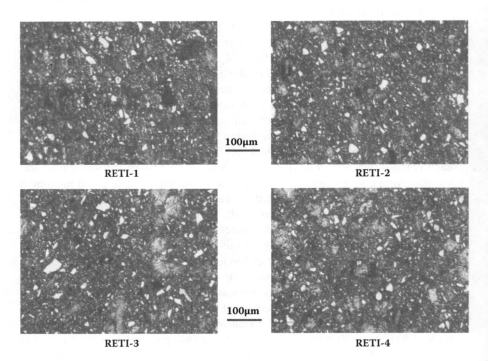

FIGURE 12.1 Microstructure of RETI-compositions in transversal section.

Epoxy-Based Composites 225

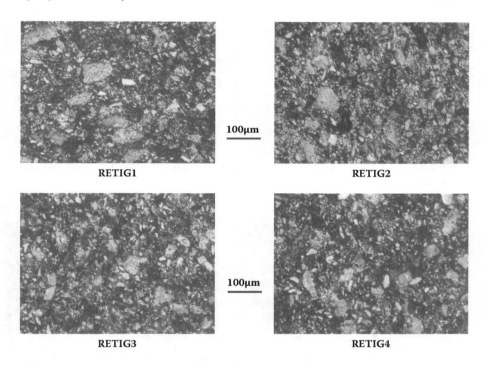

FIGURE 12.2 Microstructure of RETIG-compositions in transversal section.

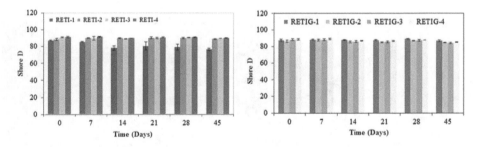

FIGURE 12.3 Comparison of Shore D results after UV exposure for 7, 14, 21, 28 and 45 days.

samples shown in Figure 12.5, a brittle structure is observed with some localized porosity. No decohesion of the particulate reinforcements were observed. Fracture surfaces of the RETIG samples shown in Figure 12.6 exhibit a ductile structure without internal defects such as porosity, decohesion of particulate reinforcements. The ductility is observed to increase with the increase in TiO_2 content.

Dynamic compression (drop weight) tests were conducted to study the effect of dynamic loading on failure of these composites. The test conditions were

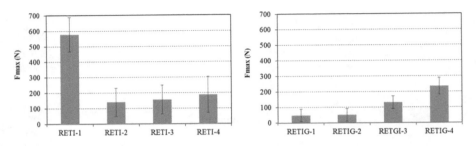

FIGURE 12.4 Force to fracture for each composition obtained by macro indentation test.

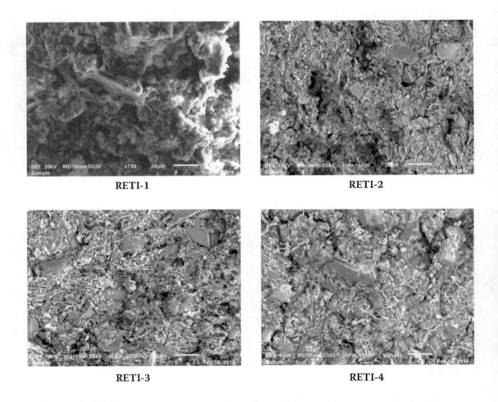

FIGURE 12.5 Fracture surfaces of the RETI samples after macro indentation tests.

described before and the absorbed energy was calculated for each composite with four specimens via Matlab. Absorbed energy for the RETI and RETIG specimens are shown in Figure 12.7. Figure 12.8 depicts the absorbed energy as a function of TiO_2 contents. The trends observed in Figures 12.7 and 12.8 follow

Epoxy-Based Composites

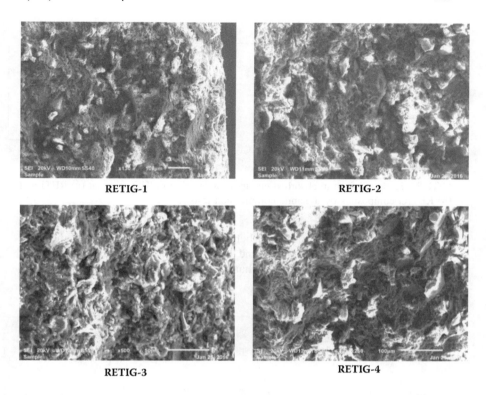

FIGURE 12.6 Fracture surfaces of the RETIG samples after macro indentation tests.

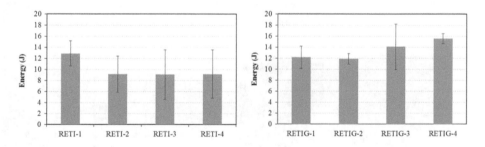

FIGURE 12.7 Absorbed energy for RETI and RETIG composites by dynamic compression tests.

the trends in Figure 12.4; F_{max} is obtained from macro indentation. When the TiO_2 content is increased in the RETI specimens with Al_2O_3 fibers, more brittle behavior is observed. Alumina fibers, alumina and boron added play a role in the brittle damage. For the RETIG specimens, the synergistic effect of glass fibers

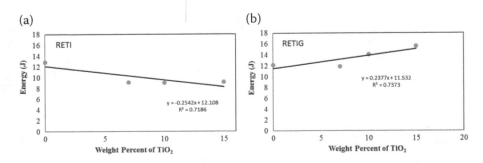

FIGURE 12.8 Change in absorbed energy as a function of TiO_2 content (a) RETI and (b) RETIG composites by dynamic compression tests.

and TiO_2 resulted in a more ductile behavior. This is also verified by the SEM observations shown in Figures 12.9 and 12.10. Also, it should be mentioned that these results are in agreement with what was reported previously [4, 14, 16–22].

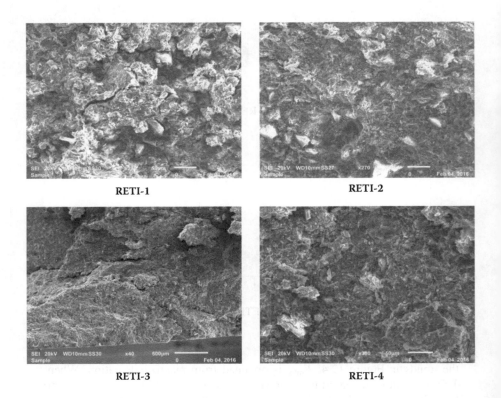

FIGURE 12.9 Fracture surface of the RETI samples from dynamic compression tests.

Epoxy-Based Composites

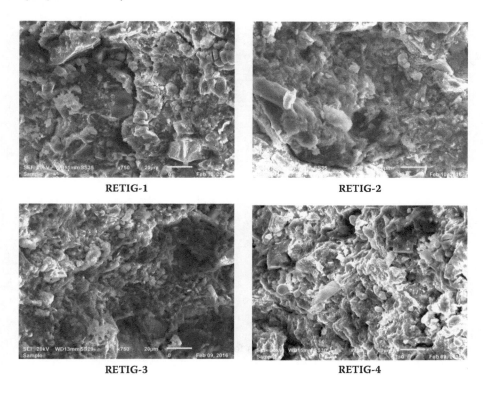

FIGURE 12.10 Fracture surface of the RETIG samples from dynamic compression tests.

Wear tests using the scratch function of a nanoindenter were performed with a conical tip that had a 90-degree apex angle. The tests were performed with 20 mN and 50 mN normal load over a 500 mm linear track for 50 cycles. The data recorded for each test was the initial profile and the final residual profile over the wear track. A typical data set is shown in Figure 12.11 with an image of the corresponding wear track. Knowing that the wear tests were performed with a conical tip that had 90-degree apex angle the volume removed can be approximated by using the initial profile and the final residual profile and the conical tip profile at any point along the wear track. The average volume removed from the 10 wear tests for each composite is shown in Figure 12.12. From this figure it can be seen that any amount of TiO2 added has improved the wear response of the RETI composites to the same extend. On the other hand, for the RETIG composites, it looks like for the 7 wt.% and 15 wt.% TiO_2 content there was no appreciable change in wear response. But for 10 wt.% of TiO_2 the wear response was worse. This looks like an anomaly and will be revisited in the future.

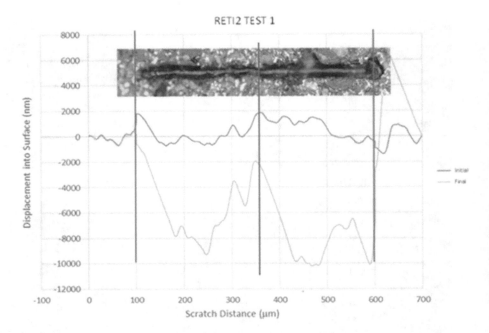

FIGURE 12.11 Initial and final profiles for the RETI-2 under 50 mN normal load.

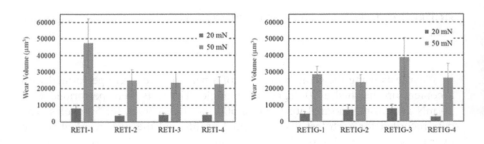

FIGURE 12.12 Wear volume results for RETI and RETIG specimens.

It should also be noted that the wear track image shows that the track is not perfectly straight; apparently the conical tip did not fracture the hard particles (alumina fiber, boron) and went around them. Also, these particles were not pulled out of the matrix, indicating that the adhesion between these particles and the matrix was very good. The same response (not perfectly straight wear track) was observed for both RETI and RETIG specimens.

12.4 CONCLUSIONS

New composites with a matrix of 60 wt.% scrap rubber and 40 wt.% epoxy were designed that can potentially be used in automotive and aeronautics industries. All composites had fixed Al, B, Cu content and the effect of TiO$_2$ was evaluated when different fibers were added. One set of composites had alumina fiber and alumina powder (RETI) added and the second set of composites had glass fibers (RETIG) added. Powder metallurgy techniques were utilized to manufacture specimens with relatively homogeneous distribution of the reinforcements. The toughness, wear resistance, hardness and resistance to UV damage were the properties considered in this study. The results indicate that all composites considered did not exhibit any degradation due to UV exposure for all practical purposes. The hardness (Shore D) values were approximately the same for all formulations. The maximum fracture under compression via macro indentation and energy absorbed during dynamic compression tests showed a decrease with increasing TiO$_2$ content when TiO$_2$ was added to the formulation with alumina fiber and alumina powder. When TiO$_2$ was added to the formulation with glass fibers, an increase in F_{max} and absorbed energy was observed. An improvement in wear response was observed for the RETI composites when TiO$_2$ was added, but there was no difference for the RETIG composites except for 10 wt.% TiO$_2$ content. In summary, RETI-1 without any fibers was able to carry the highest load before fracture (almost twice of RETIG-4), the amount of absorbed energy was less than RETIG-4. So for potential applications, RETI-1 can be recommended when higher strength and toughness is important, and for applications where toughness is important RETIG-4 can be recommended. For applications where wear could be an issue, RETI-4 can be recommended.

REFERENCES

Bayraktar, E., Antholovich, S., & Bathias, C. (2006). Multiscale observation of fatigue behviour of elastomeric matrix and metal matrix composites by x-ray tomography. *IJF, International Journal of Fatigue, 28*, 1322–1333.

Bayraktar, E., Miskioglu, I., & Zaimova, D. (2015). Low-cost production of epoxy matrix composites reinforced with scarp rubber, boron, glass bubbles and alumina. *Mechanics of Composite and Multifunctional, Materials, SEM, 7*, 163–172.

Bayraktar, E., Isac, N., Bessri, K., & Bathias, C. (2008). Damage mechanisms in natural (NR) and synthetic rubber (SBR): nucleation, growth and instability of the cavitations. *IJFSM, International Journal of Fatigue and Fracture of the Structural Materials, 31*(1), 1–13.

Bessri, K., Montembault, F., Bayraktar, E., & Bathias, C. (2010). Understanding of mechanical behaviour and damage mechanism in elastomers using x-ray computed tomography at several scales. *International Journal of Tomography and Statistics, IJTS, 14*(29–40), 2010.

Irez, A. B., Zambelis, G., & Bayraktar, E. (2019). A new design of recycled ethylene propylene diene monomer rubber modified epoxy based composites reinforced with alumina fiber: fracture behavior and damage analyses. *Materials, 12*, 2729. doi:1 0.3390/ma12172729.

Irez, A. B., Bayraktar, E., & Miskioglu, I. (2020). Fracture toughness analysis of epoxy-recycled rubber-based composite reinforced with graphene nanoplatelets for structural applications in automotive and aeronautics. *Polymers, 12*, 448. doi: 10.33 90/polym12020448.

Irez, A. B., Bayraktar, E., & Miskioglu, I. (2017). Mechanical characterization of epoxy – scrap rubber based composites reinforced with alumina fibers. *Mechanics of Composite and Multi-functional Materials, 6*, 59–70. doi: 10.1007/978-3-319-63408-1.

Irez, A. B., & Bayraktar, E. (2019). Design of a low-cost aircraft structural material based on epoxy: recycled rubber composites modified with multifunctional nano particles. *Mechanics of Composite and Multi-functional Materials, 5*, 73–80. doi: 10.1007/ 978-3-030-30028-9_11.

Irez, A. B., Bayraktar, E., & Miskioglu, I. (2019). Flexural fatigue damage analyses of recycled rubber - modified epoxy-based composites reinforced with alumina fibres. *FFEMS, Fatigue & Fracture of Engineering Materials & Structures*. doi: 10.1111/ ffe.12964.

Kaynak, C., Sipahi-Saglam, E., & Akovali, G. (2001). A fractographic study on toughening of epoxy resin using ground tyre rubber. *Polymer, 42*, 4393–4399.

Luong, R., Isac, N., & Bayraktar, E. (2007). Damage initiation mechanisms of rubber. *JAMME, Journal of Archives of Materials Science and Engineering, 28*(1), 19–26.

Masoud, F., Sapuan, S. M., Mohd Khairol Anuar Mohd Ariffin, Nukman, Y. & Emin Bayraktar. (2020). Cutting processes of natural fiber-reinforced polymer composites. *Polymers, 1332*(12). doi: 0.3390/polym12061332.

Nacif, G. L., Panzera, T. H., Strecker, K., Christoforo, A. L., & Paine, K. A. (2013). Investigations on cementitious composites based on rubber particle waste additions. *Materials Research, 16*(2), 259–268.

Pacheco-Torgal, F., Ding, Y., & Jalali, S. (2012). Properties and durability of concrete containing polymeric wastes (tyre rubber and polyethylene terephthalate bottles): An overview. *Construction and Building Materials, 30*, 714–724.

Papadopoulos, A. M. (2005). State of the art in thermal insulation materials and aims for future developments. *Energy and Buildings, 37*, 77–86.

Yesilata B., & Turgut, P. (2007). A simple dynamic measurement technique for comparing thermal insulation performances of anisotropic building materials. *Energy and Buildings, 39*, 1027–1034.

Zaimova, D., Bayraktar E., Katundi, & Dishovsky, N. (2012). Elastomeric matrix composites: effect of processing conditions on the physical, mechanical and viscoelastic properties. *JAMME, Journal of Achievements in Materials and Manufacturing Engineering, 50-52*, 81–91.

Zaimova, D., Bayraktar, E., & Miskioglu, I. (2014). Manufacturing and damage analysis of epoxy resin-reinforced scrap rubber composites for aeronautical applications. *Experimental Mechanics of Composite, Hybrid, and Multifunctional Materials, SEM, 6*, 65–76.

Zaimova, D., Bayraktar, E., Miskioglu, I., & Katundi, D. (2015). Manufacturing of new elastomeric composites: mechanical properties, chemical and physical analysis. *Composite, Hybrid, and Multifunctional Materials, SEM, 4*, 139–150.

Zaimova, D., Bayraktar, E., Miskioglu, I., & Dishovsky, N. (2014). Optimization and service life prediction of elastomeric based composites used in manufacturing engineering. *Experimental Mechanics of Composite, Hybrid, and Multifunctional Materials, SEM, 6*, 157–166.

Zaimova, D., Bayraktar, E., & Miskioglu, I. (2015). Characteristics of elastomeric composites reinforced with carbon black and epoxy. *Mechanics of Composite and Multifunctional, Materials, SEM*, 191–202.

Zaimova, D., Bayraktar, E., Miskioglu, I., & Katundi, D. (2015). Study of influence of SiC and Al$_2$O$_3$ as reinforcement elements in elastomeric matrix composites. *Composite, Hybrid, and Multifunctional Materials, SEM, 4,* 129–138.

Zaimova, D., Bayraktar, E., & Miskioglu, I. (2016). Design and manufacturing of new elastomeric composites: Mechanical properties, chemical and physical analysis. *International Journal of Composites Part B, 105,* 203–210.

13 Design for Recycling Polymer Composites

Siti Norasmah Surip[1], Hakimah Osman[2], and Engku Zaharah Engku Zawawi[3]
[1] Eco-Technology Program, Faculty of Applied Sciences, Universiti Teknologi MARA, Selangor, Malaysia
[2] Faculty of Chemical Engineering Technology, Universiti Malaysia Perlis, Perlis, Malaysia
[3] Polymer Technology Program, Faculty of Applied Sciences, Universiti Teknologi MARA, Selangor, Malaysia

CONTENTS

13.1 Introduction 235
13.2 Product Design for Recycling Composites 236
13.3 Environmental and Sustainability Issues 241
13.4 Material Selection for Recycling 243
13.5 Eco-Design in Composites Recycling 245
13.6 The Effect of the Recycling Process on Polymer
 Composite Properties 245
13.7 Eco-Design Approaches in Composites Recycling 246
13.8 Conclusion 247
References 248

13.1 INTRODUCTION

Population growth, modernization and improved living standards of people have made the product designers look for an alternative to conventional materials *viz.*, metals, ceramics and woods for daily use. In this regard, composites of mainly carbon fiber reinforced polymers (CFRP) and glass fiber reinforced polymers (GFRP) are found to be attractive and versatile options due to durability with high specific tensile strength, high modulus, lighter weight, outstanding wear resistance and easy processability (Shan-Shan Yao et al., 2018). The materials are commonly used in well-established sectors such as electrical engineering, electronics, building and civil engineering, rail, road and marine, aerospace technique and aeronautical, etc. Hence, usage of the composites in day-to-day life has increased for the last decade, with global demand for carbon fibers (CFs) expected to increase from 72,000 tons to 140,000 tons in 2020. Inevitably, these results are with more waste accumulating every year from various manufacturing

DOI: 10.1201/9781003148760-13

and applications containing valuable carbon fibers, glass fibers, etc. (Gopalraj & Timo, 2020).

In the industrial reality, several companies exist globally in recycling composites by exploiting the technologies from research and development activities and selling valuable recyclates to new markets. Recyclate is the term used in this section to describe a recycled material obtained from recycling composites activities. An annual report from agencies in the United Kingdom have identified several companies commercially active like ELG Carbon Fibre, UK CFK Valley Stade Recycling, Germany Carbon Conversions (formerly MIT-RCF), South Carolina, USA Karborek, Italy Carbon Fiber Recycle Industry Co Ltd. and Japan (may be still at pilot scale). Recyclate are commonly resins and fibers that can be used to develop products that fit easily into manufacturing processes (stampable reinforced polymer sheets, sheet molding compound, pelletized injection molding compounds) (Stella et al., 2016). When recyclates are used for the manufacture of composites, these materials do not require extensive preparation. This greatly reduces the potential cost of manufacturing (Rowell et al., 1991).

Developing product requires product design. For recycling composites subject matter, the composite designer must work closely with the material expert to understand how those materials work together, and hence allow them to choose from a database on the most suitable combination of fiber and resin system (Valentina et al., 2013). Only a material expert knows the qualities of the recyclate, particularly for the reason that different recycling techniques produce different shapes, lengths, densities and other physical properties of the recovered fibers. For example, processing of CFRC using a thermo-mechanical technique produces short and randomly distributed recyclate *viz.* CF, making them not suitable for use in high structural applications to avoid deterioration in mechanical properties. An alternative to this, recyclate *viz.* CF with short fiber fraction, could be used in sheet molding compound and bulk molding compound in replacement of virgin CF. Consequently, product design optimization viz. shape and as well as a classification according to length appears to be essential (Perry et al., 2012).

13.2 PRODUCT DESIGN FOR RECYCLING COMPOSITES

A composite product is a versatile material that can be designed in various configurations from simple bonded laminates to multidimensionally reinforcing fabric composites and advanced sandwich construction. Not many products made from recycled composite are available on the market even though recyclates are commercially available. Amongst the commercially available products using recyclate from fiber is the construction industry for reinforcement in polymer lumber, improving the strength of asphalt for bridge decking, incorporated in between two layers of virgin glass sheet molding compound (SMC) and mixing operation of bulk molding compound (BMC). All of these applications accept small amounts of contamination with no compromise about strength (Amanda et al., 2006).

Design for Recycling Polymer Composites 237

On the other hand, composites made from recyclates in aerospace and automotives are not allowed due to superior design requirements; aerospace design is determined by conservative tolerate (load cases), automotive design appears to be determined by process and manufacturing rate (design life). New aircraft have a design life of more than 35 years, whereas for automobiles it is considerably shorter. The geometric tolerances for assembly in the automotive sector are tighter than for aerospace because the shape of an automobile is more design driven, while the shape of an aircraft is functionality driven (Snudden et al., 2014). Different perspectives in product design of both industries has led to a collaboration agreement between BMW and Boeing to participate in joint research for carbon fiber recycling as well as share manufacturing knowledge and explore automation opportunities (Bret, 2012). The challenge for making part of the car from recyclates has come to the end when the BMW company moves forward toward greater efficiencies, lower costs and higher quality by using recycled carbon fiber to produce a composite roof for their i3 life module (passenger cell) (Jeff, 2014).

The purpose of this section is to describe product design key elements for producing composites from recyclates *viz.* resin and fiber. Even though the recyclates are not meant for all purposes, future planning is an urgent call to create new standards and certification requirements; a huge design database must be created for various composite materials based on recyclates and low-tech to high-tech applications. The design procedures, rules and design data that are well established for virgin material structures *viz.* polymer, metal, ceramic, cannot be directly transplanted for recycled composite product design. With proper solutions to the identified bottleneck of the composite materials, more use of strong and lightweight composites for various applications are needed. For example, it might be necessary to develop new fiber re-alignment and techniques to make discontinuous fiber yarns because certain processing of recycled composites like CFRC use a thermo-mechanical technique that tends to produce short and randomly distributed fibers. The characterization of the fiber surface and of the fiber–matrix adhesion also requires further investigation (Geraldine et al., 2015).

Design strategy, design methods and design tools are the key elements of product design. First, for the design strategy, an appropriate model with multiple options available at each stage allows frequent processing, as shown schematically in Figure 13.1. Second, design methods for composite structures range from robust and deterministic to probabilistic approaches. These methods determine fracture mechanics-based damage tolerance, which is based on strength, stiffness and stability, particularly required in critical applications such as aerospace vehicles. Third, several design tools have been developed and categorized as composite configuration tools, basic laminate analysis, FE software for more complex structures and production design tools. The composite designer must select the appropriate tool for the stage of the design (Snudden et al., 2014). All the elements are highly dependent on several factors including functional specification, material selection, manufacturing process, property estimation, reliability and cost (Gopal, 2016).

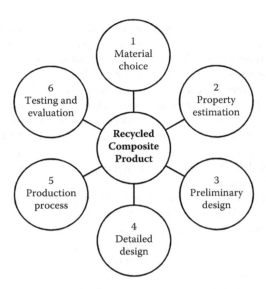

FIGURE 13.1 Design strategy of recycled composite product development.

There are certain criteria referenced for the design strategy of recycled FRP, in order to create a new composite product to be economically and technically viable (Amanda et al., 2006).

- The use of ground FRP should be beneficial to the product, i.e., the FRP should have either a structural/reinforcing role or weight saving role, not just act as an inert filler.
- The mix of materials should be synergistic.
- The product should not have to be reinforced with other material or made thicker to compensate for some deficiency caused by inclusion of ground FRP.
- It should not be merely a novel disposal method.
- The reuse method should be realistic in respect to the likely volumes of recyclate available.
- The reuse of FRP should not make the ultimate recycling of the product difficult (current types of "plastic wood" made from post-consumer HDPE and wood fiber can be easily recycled or burned without pollution).
- The product should not pose environmental problems or health and safety problems in use, e.g., abrasion, wear-related loss of glass fibers, or during cutting and drilling.
- The product should not be a substitute for something that is made from a more sustainable material in the first instance, such as plantation timber.
- The combination of ground FRP with some other waste material should not divert this waste from an existing higher-end reuse chain.

Design for Recycling Polymer Composites

A development of a new product using conceptual design/design concept or conceptualization beginning phase of the design process is widely applied in polymeric composites, such as making a bumper. A bumper system is composed of three main elements: fascia, energy absorber and bumper beam. In this section, the bumper beam is further discussed as it is a major damping structure component in passenger cars. Besides, two energy absorbers damp both the low- and high-impact energy by elastic deflection between two traverse-fixing points and crushing process, respectively. Due to safety requirements in developing the bumper beam, careful design, optimized structure, high quality and consistent manufacturing must be considered. Hence, other product requirements such as structural energy absorption, material consumption and cost can be improved (Cheon et al., 1995; Nishino et al., 2003; Hambali et al., 2009). As a beginning phase of the conceptual design, it normally starts with a customer requirement "voice of the customer" to identify functional design problems, and finally use to create a product design specification (PDS) through the evaluation of design concept selection (DCS). Many tools are developed for this purpose, such as Fuzzy ANP-based, Analytical Hierarchy Process (AHP), Multi Criteria Decision Making (MCDM), Multi-attribute Decision-Making (MADM) and Technique of Order Preference Similarity to The Ideal Solution (TOPSIS). Figure 13.2 shows the bumper beam PDS with selected parameters evaluated by using TOPSIS. Related to that, a complete conceptual design, as suggested by Davoodi et al. (2011), is illustrated in Figure 13.3.

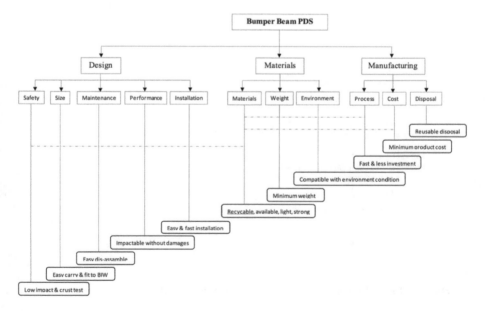

FIGURE 13.2 Selected parameters for bumper beam PDS.

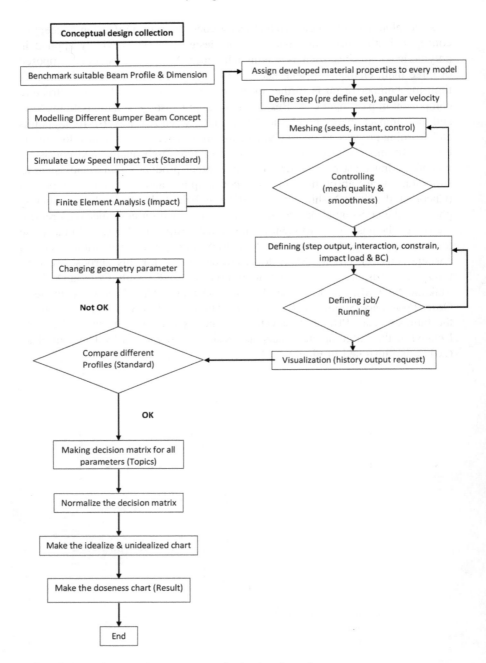

FIGURE 13.3 Bumper beam conceptual selection flow chart.

13.3 ENVIRONMENTAL AND SUSTAINABILITY ISSUES

The problem of dumping plastic garbage and polymer composites is a major issue in the world today as it affects the environment. This is unavoidable, due to the increasing population of the world. Modern lifestyles contributed to that require the use of food items that are easily accessible and stored. Therefore, people will choose to buy food items that are readily packaged and of course their hygiene is guaranteed. This contributed to a rapidly increasing production of disposable plastic products. The plastic sheet and film for food wrap, soft drink bottle (Figure 13.4) or even diapers that we use and discard every day end up in landfill sites. In the use of other products, people will choose something durable, lightweight and alternatively cheaper, like plastic. The use of plastic material is the best option to meet this need. Plastics and polymer composites have become a crucial part of our modern lifestyle and important in most aspects of our lives, with many applications that can satisfy our needs. This has caused plastic to become a type of waste that is widely found on this earth. This factor has also contributed to the inability of plastic to decompose on its own in less than 50 years. This problem is increasing day by day.

Plastics are made from polymers that consist of long chains of carbon or hydrogen atoms, which are derived from petroleum. These are not a renewable energy source, which means plastic is not a sustainable material. Plastics are very strong; in landfills, plastics degenerate in full sunlight, but never decompose completely when buried. Therefore, we need to find the best way to control the use of plastic. A good example is by educating the younger generation, starting at the primary school level, on how to properly manage the waste disposal of plastic materials and their products such as plastic packaging of their food and drinks in a smart and manageable way; even a way that can produce something

FIGURE 13.4 Plastics that we use and discard every day.

that is more beneficial than just simply disposing it. We can teach them to reuse plastic items and also do the recycling process. Another alternative way is to use plastic from biodegradable sources, which are relatively more expensive than synthetic plastic. However, this bioplastic plant-derived material is actually not really compostable at landfill sites. Biopolymers that are in the landfill also eventually have the potential to produce methane gas under certain conditions. It will be there and begin to decompose due to the hot, cold, rainy and humid weather.

There are several options for recycling that can be done, namely reusing, mechanical recycling, chemical recycling and incineration. Regrinding is the simplest and cheapest recycling process, and basically involves cutting, grinding or chipping the waste composite into a certain size depending on the requirement to be used as filler material in new molded composite products. The regrinding process results in lower mechanical properties such as stiffness and strength as the continuous fibers were broken down into small fragments. Continuous fibers in composite reinforcement, regardless of new or recycled, will give maximum mechanical performance for most of the products.

Secondly is the physical or mechanical recycling. In this process, the plastic is sorted, cleaned, crushed and then reprocessed and compounded to produce new products. This method benefits plastic industries and as well as the environment. That means reduced costs and more plastic waste materials can therefore be used without remarkably reducing the material performance. However, there are problems with mechanical recycling of plastics; for example, it is labor intensive during the sorting, cleaning and separation process. The plastic waste also contains mixed polymers with a wide range of materials. There is also a possibility of contamination of material from the previous use or during production processes. The contamination may be of a chemical or microbiological nature during waste management, e.g., contact with acidic substances, UV radiation and extreme temperatures. These inherent properties of plastics have led to their classification as non-permanent materials (although they are highly resistant to complete degradation in the environment). Other than that, the addition of some types of plastic additives, e.g., antioxidants and stabilizers, form intended reaction products, thereby losing their original function (Geueke et al., 2018). Recycled plastics can be blended with virgin plastic and used in an application different than the original use without reducing properties of the products. These recycled plastics can be used to make fences, deck and outdoor playgrounds or even for construction (Gu & Ozbakkaloglu, 2016). It provides low maintenance and can replace natural wood. Plastic from soft drink and water bottles can be spun into fiber for the production of carpet or made into new food bottles.

The third method is chemical recycling. In this method, the polymer waste is turned back into its oil or hydrocarbon component. In the case of polyolefin and monomers, polyesters and polyamides can be used as raw materials for new polymer production and petrochemical industry, or into pure polymers using suitable chemical solvents (Solis & Silveira, 2020). However, this recycle method is not environment friendly; some of them use chemicals that impact nature and cause another problem.

The fourth method is incineration. This is the most effective way to reduce the volume of organic materials. This method produces substantial energy from polymer waste. Burning plastic waste can be converted to electrical energy. This could reduce the burning of fuels. This method is ideal for recycling medical applications and packaging of hazardous goods. Ash from incineration can be used for construction and building roads. However, the disadvantage of this method is that it has an effect on health and environmental problems from airborne toxic substances such as dioxin, nitrogen oxide, particulates and acid gas. For example, halogenated polymers, such as PVC, when burned will produce hydrochloric acid. Burning polymers will also produce CO_2 gas, which contributes to greenhouse gases (Sara & Ovi, 2018). Besides, the installation and maintenance of an incineration plant is an expensive process. The recyclables could also end up in the incinerator if it is not properly sorted.

Furthermore, people can help to recycle plastics to give a better environment and lead to a better future. This is because we cannot stop from using plastics since they offer a lot of advantages (inexpensive, lightweight and durable) that make them suitable to be used in most applications. Because of that, the production of plastics has increased significantly year by year. Among the above techniques, the chemical recycling is an acceptable technique according to the principal of sustainable development. This recycling process is the best technique to treat waste polymer products compared to old-style methods, which is combustion of polymers or burying underground which leads to negative influences in the environment via the formation of dust, fumes and toxic gases.

13.4 MATERIAL SELECTION FOR RECYCLING

The use of plastic is very important. It is inevitable. As can be seen in the current state of the COVID-19 pandemic, plastics are widely used in making personal protective equipment (PPE) such as gowns for medical front-liners, face masks and face shields. A non-woven plastic material made of polypropylene (PP) is largely used in making PPE and it can be used only once. Therefore, it is also important to determine the way these materials should be disposed. So, that's why biopolymers are introduced to the world. This type of polymer not only reduces the dependent to the petrochemical supply but it also reduces the recycle process because of the biodegradable properties. However, currently biopolymers will not be able to compete/replace with commodity plastics such as polyethylene and polypropylene which have great properties. They're lightweight, strong, have great barrier properties and are cheap, as these materials are a by-product of the oil industry.

Therefore, it is important to determine the selection of the appropriate material before deciding to choose a plastic material as a product. For example, food packaging and drinking water or other single-use products can use bioplastics or blend with synthetic thermoplastic polymers. Blending can reduce the amount non-degradable part of polymers in the landfill. The most popular plastic used in the packaging industry is from the thermoplastic group, such as high density and linear low densities polyethylene (HDPE, LLDPE) and polypropylene (PP). These materials are classified as semi-crystalline, while others, such as polystyrene and ABS, are considered amorphous. Thermoplastics are easily recycled as they are naturally softened and melted when they are heated, allowing for remolding/reshaping into new products. Fatemeh et al. (2016) investigated the effect of adding recycled (reprocessed) polypropylene on the mechanical behavior of polypropylene samples at room temperature. Virgin, recycled and fiber resin pallets of polypropylene were blended and then injection molded to produce tensile test samples. The results showed that using a higher percentage of recycled material does not have a significant effect on the mechanical properties of polypropylene. However, in another work on the recyclability of PP composites, it was investigated by the effect of recycling on the properties of PP-based composites (ethylene propylene diene monomer (EPDM)/PP and talc/PP) using an extrusion process. The mechanical, rheological and structural properties of the composites were determined and compared after each extrusion cycle. The results showed that the melt viscosity of the composites decreases with processing number (Bahlouli et al., 2012).

The engineering and construction application can use recyclable thermoplastic materials and for composites using filler or reinforcement materials from natural sources such as from plant sources, for example in car bumper making. Physical and mechanical properties of recycled PP composites reinforced with rice straw lignin was investigated by Karina et al. (2017). The composites of recycled PP reinforced with rice straw lignin were prepared via melt blending in a double screw internal mixer. Their results of physical and mechanical properties of composites made from a combination of recycled polypropylene (RPP) and maleic anhydride polypropylene compatibilizer (MAPP) showed no substantial differences when compared to PP and MAPP-based composites. This indicated that the recycled polypropylene has the potential to substitute virgin PP in composites reinforced with natural plant sources with the addition of MAPP applications.

Polymer composites, which are plastics reinforced with fibers (FRPs), commonly glass or carbon, and some thermoplastic materials have superior properties such as high strength, stiffness, abrasion resistance and light weight, making them useful across a wide range of industries. Therefore, polymer composites are widely utilized in the manufacturing of high-performance engineering components because of their outstanding strength to weight ratio. Their use is increasing around the world, with the global market for products expected to reach US$95 in 2020 (Jennifer, 2016). The effect of filler on the recycled PP was studied by Hadi and Mohamed (2017). Authors have investigated the flow, thermal and mechanical properties of composites made of waste polypropylene (WPP) reinforced with silica (SiO_2) nanoparticles (NPs) using a twin screw extruder. They observed that the crystallinity level and the crystallinity temperature decreased with a SiO_2 NPs concentration increasing while impact and hardness increasing.

Thermosetting is the most widely used polymer in the manufacturing of polymer composites. This is due to the characteristic properties of thermoset, such as the existence of chemical cross-linking that increases the strength and stiffness and also reduces the susceptibility to creep as compared to thermoplastics. However, these properties make them classified amongst the foremost difficult materials to be recycled, or even considered impossible due to high thermal and mechanical stability. The increase in products made out of these materials also means there is an increased need for sustainable recycling options for these materials.

However, not all recycle methods introduced are environment friendly. Therefore, for now, biodegradable polymers are the best alternative way, especially as single-use plastics. These plastics can be degraded in the environment by four mechanisms, i.e., photodegradation, thermo-oxidative degradation, hydrolytic degradation and biodegradation by microorganisms (Siracusa, 2019). Recent trends demonstrate a substantial increase in the rate of recovery and recycling of plastic wastes. Many researchers have proposed many solutions that are environmentally friendly (Solis & Silveira, 2020).

There is still a need for more research to develop more accurate methods by understanding the structures of the many different types of polymers used today and the mechanisms of their degradation. It will make it possible to optimize the decomposition of plastics waste materials to give useful products.

13.5 ECO-DESIGN IN COMPOSITES RECYCLING

Recycling of composites and composite materials waste aims to find ways to recover and reuse these materials into useful and profitable applications, and reduce the environmental burden. According to sustainable development principles, an eco-design of a composite product must be in compliance with the health protection, environmental preservation and quality assurance, as illustrated in Figure 13.5.

Eco-design should not only focus on the performance and aesthetic requirements of products, but should also consider the needs of the future application of the recycled materials. It is estimated that approximately 80% of the environmental burden of a product is determined during the design stage, when most of the decisions are made regarding the selection of materials and processes for the new product (Subic et al., 2009).

Thus, in modern design, environmental issues are given high priority, which has resulted in the development and application of eco-design methods that are applicable to recycled polymer composites.

13.6 THE EFFECT OF THE RECYCLING PROCESS ON POLYMER COMPOSITE PROPERTIES

Many industries faced a problem with the quality of the recycled polymer, which is lower than the primary material. This is due to polymer degradation over many

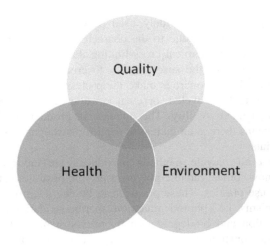

FIGURE 13.5 Relationship between quality, health and environment in an eco-design concept.

years of use and the products exposed to light, heat or water. Products made from recycled polymer composite can be less durable and in limited application since the composite formulations are very specific and contain up to 60% of reinforcement, filler, stabilizers, pigments or other additive. The formulation of the secondary polymer can hence not be adjusted freely to meet the required specification. Normally, recycled polymer composites are also heavier since a higher percentage of polymer materials is needed to encounter their mechanical properties that are affected after several processes when recycled. This is again a big disadvantage if the products will be used in a mobile industry because this will lead to higher fuel consumption.

The industry faces a challenge to dispose of the composite products in an environmentally friendly way. Composites are more expensive to recycle and reprocess than primary materials. The cost of recycling composite material is not competitive against the cost of using new material. For these reasons, the current practice for disposing of most composite products is landfill disposal. This will cause an environmental problem because of the tons of composite products that occupy the landfill, and on the other hand polymers and fibers from the composites products will take a very long period to degrade naturally. Despite the challenges with recycling, the composite industry has come up with few solutions in the development of reprocessing techniques, which are classified as regrinding, thermal and chemical processes (Baillie, 2004; Henshaw et al., 1996).

13.7 ECO-DESIGN APPROACHES IN COMPOSITES RECYCLING

Three eco-design approaches were proposed by Dewberry (1996), which are green design, eco-design and sustainable design concept. Green design has a

Design for Recycling Polymer Composites 247

single-issue focus, perhaps incorporating the use of some new material or consider energy consumption; while eco-design adopts the product life cycle management approach, exploring and tackling all of the greatest impacts across the product's life cycle; and the sustainable design concept is much more complex and moves design concerns outwards to societal conditions, regional development and ethics.

Eco-design and green design are widely used in sport products. Plenty of commercial brands also adopted the eco-design approach that mainly utilizes life cycle assessment (LCA) as a tool for product design and development. The environmental considerations should be addressed as a LCA practice. The eco-design approach and design strategy for improved environmental performance has its technological possibilities for product innovation in product concept or structure.

The sustainable design concept includes design for recycling and design for disassembly, widely used in industrial design and product development (Zebedin, 2000). The possibility of disassembling products into recyclable fractions and the material quality issue are the main concerns of optimal recycling. The difficulty in applying these methods to polymer composites is that there are many structural or reinforcement components in composite products that are difficult to extract and disassemble after use.

13.8 CONCLUSION

   i. Design strategy, design methods and design tools are the key elements of the product design. The final stage of completing product design is to incorporate design method and design tool. For the designer, the analysis and design methods for composite structures are more complex than the metallic counterparts cause delay in production. Therefore, design tools speed up the design process and allow several investigations of several concepts to define the optimal solution.
   ii. The composites industry is continually seeking a sustainable approach that yields better quality reusable material at a lower cost when recycled and rapidly biodegraded when disposed into landfills. Much more research and development is required in the recycling and disposal of products, especially in polymer and metal composites.
   iii. Recycling and reuse are approaches for end-of-life waste management of plastic products. These are increasingly economical as well as environmentally safe. Recent trends demonstrate a substantial increase in the rate of recovery and recycling of plastic wastes.
   iv. Since the world is very concerned by the eco-design strategy and the implementation of an environmental management system, scientists and researchers are therefore encouraged to move towards the optimization of the eco-design function, especially in recycled composites.

REFERENCES

Amanda Conroy, Sue Halliwell, and Tim Reynolds. (2006). Composite recycling in the construction industry. *Composites: Part A*, 1216–1222. 10.1016/j.compositesa.2005.05.031

Bahlouli, Pessey, Raveyre, Guillet, Ahzi, and Dahoun. (2012). Recycling Effects on The Rheological and Thermomechanical Properties of Polypropylene Based Composites. *Material & Design*, 33, 451–458. 10.1016/j.matdes.2011.04.049.

Baillie, C. (2004). *Green Composites: Polymer Composites and the Environment*. Woodhead Publishing: Cambridge.

Bret, Jensen. (2012). BMW Group and Boeing to collaborate on carbon fiber recycling. *BMW Group Corporate Communications*. Article published on 12 Dec 2012, Last seen on 27 Aug 2020.

Cheon, S. S., Choi, J. H., and Lee, D. G. (1995). Development of the composite bumper beam for passenger cars. *Compos Struct*, 32(1–4), 491–500. 10.1016/0263-8223(95)00078-X.

Cunliffe, A. M., Jones, N., Williams, P. T. (2003). Pyrolysis of composite plastic waste. *Environmental Technology*, 24, 653–663.

Davoodi, M. M., Sapuan, S. M., Ahmad, D., Aidy, A., Khalina, A., and Mehdi Jonoobi. (2011). Concept selection of car bumper beam with developed hybrid bio-composite material. *Materials and Design*, 32, 4857–4865. 10.1016/j.matdes.2011.06.011.

Dewberry, E. (1996). Ecodesign-Present Attitudes and Future Directions: Studies of UK Company and Design Consultancy Practice (PhD Thesis). Open University: UK.

Fatemeh, Yongsheng, Cagri, Keith, and Kajsa. (2016). ffects of Recycling on the Mechanical Behavior of Polypropylene at Room Temperature Through Statistical Analysis Method. *Polymer Engineering and Science*, 5(11), 1283–1290. 10.1002/pen.24363.

Geraldine Oliveux, Luke O. Dandy, and Gary A. Leeke. (2015). Current status of recycling of fiber reinforced polymers: Review of technologies, reuse and resulting properties. *Progress in Materials Science*, 72, 61–99. 10.1016/j.pmatsci.2015.01.004

Geueke, Groh, and Muncke. (2018). Food Packaging in The Circular Economy: Overview Of Chemical Safety Aspects for Commonly Used Materials. *Journal of Cleaner Production*, 193, 491–505. 10.1016/j.jclepro.2018.05.005

Gopal, K. V. N. (2016). Product design for advanced composite materials in aerospace engineering. In Advanced composite materials in aerospace engineering-Processing, properties and application. *Woodhead Publishing*, 2016, 413–428. 10.1016/B978-0-08-100037-3.00014-6.

Gu, L. & Ozbakkaloglu, T. (2016). Use of recycled plastics in concrete: A critical review. *Waste Management*, 51, 19–42. 10.1016/j.wasman.2016.03.005.

Hadi, N. J., & Mohamed, D. J. (2017). Study the Relation Between Flow, Thermal and Mechanical Properties of Waste Polypropylene Filled Silica Nanoparticles. *Key. Eng. Mater*, 724, 28–38. 10.4028/www.scientific.net/KEM.724.28.

Hambali, A., Sapuan, S., Ismail, N., and Nukman, Y. (2009). Application of analytical hierarchy process in the design concept selection of automotive composite bumper beam during the conceptual design stage. *Sci Res Essays*, 44, 198–211.

Henshaw, J. M., Han, W., Owens, A. D. (1996). An overview of recycling issues for composite materials. *Journal of Thermoplastic Composite Materials*, 9, 4–20. 10.1177/089270579600900102

Jeff Sloan. (2014). The making of the BMW i3. *Composites World*. Article published on 23 Apr 2014, Last seen on 27 Aug 2020.

Jennifer McDowall. (2016). How Recyclable Are Composites Plastics, Plastic (taxanomy/term/18). https://resource.co/article/how-recyclable-are-composite-plastics-11261.

Karina, M., Syampurwadi, A., Satoto, R., Irmawati, Y. & Puspitasari T. (2017). 2017. Physical and mechanical properties of recycled PP composites reinforced with rice straw lignin. *BioResources, 12*(3), 5801–5811. 10.15376/BIORES.12.3.5801-5811.

Marion Meunier, White Young Green, Simon Knibbs, and Cenit. (2012). Design tools for fiber reinforced polymer structures composites network best practice guide. *National Composites Network.*

Nishino, T., Hirao, K., Kotera, M., Nakamae, K., Inagaki, H. (2003). Kenaf reinforced biodegradable composite. *Compos Sci Technol, 639*, 1281–1287. 10.1016/j.cirp. 2012.03.081

Perry, N., Bernard, A., Laroche, F., and Pompidou, S. (2012). Improving design for recycling – Application to composites. *CIRP Annals - Manufacturing Technology, 61*, 151–154. 10.1016/j.cirp.2012.03.081.

Roger M. Rowell, John A. Youngquist, and Dobbin Mcnatt. (1991). Composites from recycled materials. *Proceedings of the 25th International Particleboard/Composite Materials Symposium. Pullman, WA.* Pullman. WA: Washington State University, 301–314.

Sankar Karuppannan Gopalraj, and Timo Karki. (2020). A review on the recycling of waste carbon fibre/glass fibre-reinforced composites: fibre recovery, properties and life-cycle analysis. . *SN Applied Sciences, 2*, 433. 10.1007/s42452-020-2195-4

Sara, and Ovi. (2018). Waste Incineration: Advantages and Disadvantages. https://greentumble.com/waste-incineration-advantages-and-disadvantages/

Shan-Shan Yao, Fan-Long Jin, Kyong Yop Rhee, David Hui, and Soo-Jin Park. (2018). Recent advances in carbon-fiber-reinforced thermoplastic composites: A review. *Composites Part B: Engineering, 142*, 241–250. 10.1016/j.compositesb.2017.12.007.

Siracusa. (2019). Microbial Degradation of Synthetic Biopolymers Waste. *Polymers, 11*, 1–18. 10.3390/polym11061066.

Snudden J. P., Ward C., and Potter K. (2014). Reusing automotive composites production waste (part 3). *Materials Today*, Article published on 8 Dec 2014, Last seen on 27 Aug 2020.

Solis, M. & Silveira, S. (2020). Technologies for chemical recycling of household plastics-A technical review. *Waste Management, 105*, 128–138. 10.1016/j.wasman.2020.01.038

Stella Job, Gary Leek, Paul Tarisai Mativenga, Geraldine Oliveux, Steve Pickering, and Norshah Aizat Shuaib. (2016). Composites recycling: Where are we know?. *Composites UK.*

Subic, A., Mouritz, A., Troynikov, O. (2009). Sustainable design and environmental impact of materials in sports products. *Sport Technology, 2*, 67–79. 10.1002/jst.117

Valentina Rognoli, Elvin Karana, and Owain Pedgeley. (2013). Natural fiber composites in product design: An investigation into material perception and acceptance. *Proceeding of International Conference on Designing Pleasurable Products and Interfaces, 36*, 1–4. 10.1145/2347504.2347543.

Zebedin H. (2000). Design for Disassembly and Recycling for Small and Medium Sized Companies for the Next Generation. In: Kopacek P., Moreno-Díaz R., Pichler F. (eds) *Computer Aided Systems Theory - EUROCAST'99. EUROCAST 1999. Lecture Notes in Computer Science*, vol 1798. Springer, Berlin, Heidelberg. 10.1007/10720123_24

14 Effect of Heat Treatment Modification on the Tensile Strength and Microstructure of X7475 Al-Alloy Fabricated from Recycled Beverage Cans (RBCs) for Bumper Beam Applications

A. Kazeem[1,2], N.A. Badarulzaman[1], W.F.F. Wan Ali[3], M.Z. Dagaci[4,5], and S.S. Jikan[5]

[1]Nanostructure and Surface Modification Focus Group (NANOSURF), Faculty of Mechanical and Manufacturing Engineering, Universiti Tun Hussein Onn Malaysia (UTHM), Johor, Malaysia

[2]Department of Science Policy and Innovation Studies, National Centre for Technology Management (NACETEM), North Central Zonal Office, Abuja, Nigeria

[3]Faculty of Mechanical Engineering, Universiti Teknologi Malaysia (UTM), Skudai, Johor

[4]Department of Chemistry, Ibrahim Badamasi University, Niger State, Lapai, Nigeria

[5]Faculty of Applied Science and Technology, Universiti Tun Hussein Onn Malaysia, Johor, Malaysia

CONTENTS

14.1 Introduction..252
14.2 Materials and Experiments..253
      14.2.1 Materials...253
      14.2.2 Methodology...253
14.3 Results and Discussion...255
      14.3.1 Ultimate Tensile Strength of Experimental Alloys after Annealing...255
      14.3.2 Microstructure Study..257
14.4 Conclusions...260
Acknowledgment..260
References..261

14.1 INTRODUCTION

In an effort to recover more aluminium from RBCs, Juniarsih et al. (2019) stated that 97.49% of the aluminium was recovered with 240 minutes holding time. The experiment showed that an increase in temperature and holding time during the recovery of aluminium form RBCs was key in reducing the wt.% of Mg and other constituents. At present, bumper beams are designed using steel, aluminium alloys and composites (Francis et al., 2016; Tanlak, 2012). The use of aluminium alloys for bumper beam applications requires some modifications in the heat treatment process. The modification resulted in the alteration of principles of precipitation. This justification was due to the supposed appearance of the vacancy-rich clusters (VRCs) at an early stage during the quenching process. Hence, the industrial application of alloys in the typical T6 condition required additional treatment. The two-stage artificial aging leads to the proposal of a primary sequence for formation of n' as thus: solid solution VRC-GP(ll)•n' against the conventional precipitation process often represented by the simplified sequence (solid solution) GP-zones-n' ($-n-MgZn_2$). This sequence followed when the alloy was quenched above 450°C. At a temperature below 450°C, GP(II) was not detected, and it was concluded that VRC was not produced (Waterloo et al., 2001). These are modifications to the conventional precipitation hardening heat treatment. In fact, multiple steps of retrogressions and re-solution were evaluated in order to achieve a precise forming and performance improvement (Choudhry & Ashraf, 2007; Li et al., 2017; Rometsch et al., 2017). The main aim of all modifications was to obtain suitable mechanical properties for engineering applications. For instance, Kremmer et al. (2020) reported the effect of interrupted cooling of alloys at temperatures ranging from 100°C to 200°C and continued with a very slow cooling to room temperature following rapid cooling from the solution heat treatment temperature. The final material's condition was achieved whether or not there was subsequent artificial aging. An improved strength-toughness trade-off was reported through heat treatment modification. In the retrogression (Li et al., 2016) treatment process deployed to obtain improved strength-corrosion trade-off, samples of 7xxx alloys were treated at 200°C for 5 minutes in a salt bath. An additional T73 temper was obtained by a two-step aging at 115°C for 8 hours and then 165°C for 16 hours. For the re-solution temper, the specimen with T73 temper went through a re-solution treatment including hot insulation at 470°C for 50 minutes and rapid water quenching (Li et al., 2017). Previous reports on the modifications of the heat treatment process were conducted using commercial 7xxx alloys and samples cast using pure aluminium. The novelty of the current study is the fabrication of an experimental X7475 alloy from RBCs and the modification of the heat treatment sequence. The modification was done by conducting annealing between solution heat treatment and the artificial aging. The intent of the annealing was to relieve the stresses due to the machining (Tabei et al., 2018) and internally induced stress after quenching (Choudhry & Ashraf, 2007) with the aim of reporting the tensile strength and microstructure transformation following the variation in the wt.% Zn from 4.0–5.0 in the novel alloy.

14.2 MATERIALS AND EXPERIMENTS

14.2.1 Materials

The raw materials used in fabricating the new X7475 alloy are Al, Zn, Mg, Mn and Cu. Almost 80% of the raw materials were obtained from secondary sources, except Mg that was supplied. About 5 kg of RBCs were collected from recycle points in Universiti Tun Hussein Onn Malaysia (UTHM). Primary aluminium ingots were cast using a graphite crucible with the aid of a gas-fired furnace. Zn and Mn were retrieved from used *Hawk* batteries of the GB/T 8897.2–2008. In their raw form, Zn was in flake form and cast to ingots using a *JT0332* portable induction electric melting furnace, while Mn was obtained in powder form.

14.2.2 Methodology

In the description of the methodology in Figure 14.1, it shows that the recycled Cu sourced by unwinding the coils of a table fan was cast into 70% Cu–30% Al to support the ease of melting Cu in the alloy. Nine samples (S1–S9) were fabricated, varying the Zn Zn from 4.00 to 5.00 wt.%. Other constituents like Mg, Mn and Cu were held at 1.5 wt.%, 1 wt.% and 0.35 wt.% respectively.

A portable mechanical mixer of the *TAC 1803-Pentec* was used when the speed was set at a 7th speed of ±550 rpm. Pouring was done at 710 ± 10°C. Casting began by introducing Al, Zn, 70% Cu–30% Al, Mn and Mg in that order. Samples used for the tensile test were machined using the ASTM E8/E8M-11. The *Mazak Nexus 100-II* universal CNC lathe machine was used for the machining. Variations in Zn contents, annealing temperature and artificial aging parameters are presented in Table 14.1.

Heat treatment was achieved using the Nabertherm B180 MB2 furnace.

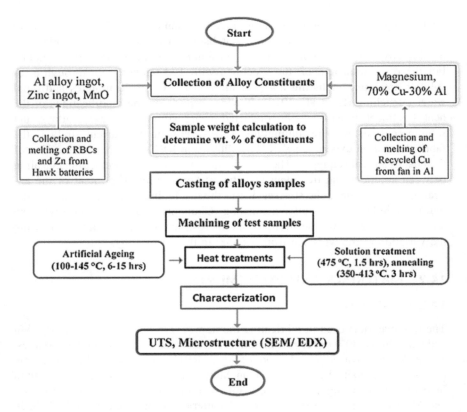

FIGURE 14.1 Methodology flow chart.

TABLE 14.1
The Variation in wt.% Zn, Annealing Temperature and Artificial Aging Parameters

Sample No.	Variation in Zn (wt.%)	Annealing Temperature (°C)	Artificial Aging Temperature (°C)	Artificial Aging Time (Hours)
S1	4	350	145	6
S2	4	350	100	6
S3	4	350	145	15
S4	4.5	380	120	10.5
S5	4.5	380	120	10.5
S6	4.5	413	145	6
S7	5	413	145	15
S8	5	413	100	15
S9	5	413	100	6

Bumper Beam Applications

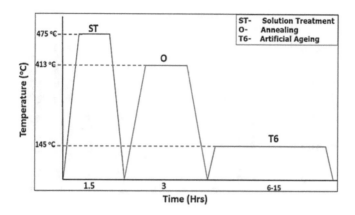

FIGURE 14.2 Schematic representation of heat treatment profile.

The solution treatment was done at 475°C, quenched in clean water. Annealing and artificial aging were performed in accordance with Figure 14.2 and specifications in Table 14.1. The temperature specifications used for artificial aging were 100, 120 and 140°C. The time corresponding to the temperatures were 6, 10 and 15 hours. ASTM B 557M – 02a was adopted in conducting the tensile test on a Gotech AI-7000 LA5 Servo controlled machine. Load resolution of 1/200,000 was maintained at a test speed of 0.5 mm/min.

14.3 RESULTS AND DISCUSSION

14.3.1 Ultimate Tensile Strength of Experimental Alloys after Annealing

There are three categories of alloys, each fabricated using 4.0 wt.% (S1–S3), 4.5 wt.% (S4–S6) and 5.0 wt.% Zn (S7–S9). The results presented in Figure 14.3 depict an alloy (S1) fabricated using 4.0 wt.% Zn, 1.5 wt.% Mg, 1.0 wt.% Mn and 0.35 wt.% Cu recorded an ultimate tensile strength (UTS) of 353 MPa following annealing at 350°C and artificial aging at 145°C for 6 hours. In S2, the annealing temperature and aging time were maintained at 350°C and 6 hours, but aging temperature was reduced to 100°C. A corresponding decrease of 132 MPa was observed in UTS of S2 when the 221 MPa was compared with the UTS of S1. The result of UTS in S3 suggested that prolonged aging time after the annealing was detrimental to the UTS; hence, 251 MPa was obtained in the alloy. The alloy becomes softer by reducing solid solution, coalescing precipitates and for relieving residual stresses (Garber et al., 2004). The study by Zaid et al. (2011) reported that after retrogressive aging and re-aging of 7xxx alloys, a reduction of 10–15% was observed. This trade-off was relevant in applications involving impact forces like collision.

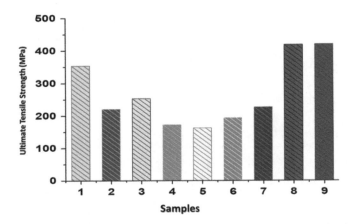

FIGURE 14.3 Ultimate tensile strength (MPa) of samples after heat treatment.

In furtherance to assessing the modification in the heat treatment process, samples S4–S6 were cast using 4.5 wt.% Zn, 1.5 wt.% Mg, 1.0 wt.% Mn and 0.35 wt.% Cu. Annealing temperature was increased from 350°C to 380°C, artificial aging temperature and time were both maintained at 120°C and 10.5 hours in the duo. An UTS of 172 MPa and 162 MPa were observed in both alloys. In alloy S6, the annealing temperature was increased to 413°C and artificial aging temperature was also increased from 120°C in S5 to 145°C, whereas the aging time was decreased from 10.5 hours to 6 hours. The UTS of 194 MPa was obtained from the sample. The results suggest that a decrease in the aging time was favorable to the increase in tensile strength of the alloy. This is in contrast with the findings of Zaid et al. (2011) where an increase in retrogression time and/or retrogression temperature was detrimental to the hardness of 7079-T651 (Al–Zn–Mg) alloys that were exposed to retrogression heat treatment for various retrogressed times (20, 40 and 60 minutes) at 190, 200 and 210°C and then re-aged at 160°C for 18 hours, respectively. The report attributed to the precipitation hardening (formation of η′- semi-coherent $MgZn_2$ phase).

In the alloys fabricated using 5 wt.% Zn, S7–S9 (Figure 14.1), the annealing temperature was held at 413°C, while artificial aging temperature was 145°C in S7, but reduced to 100°C in both S8 and S9. The 226 MPa observed in S7 against the 416 MPa obtained in S8 was an indication that annealing of the cast alloys following quenching at 475°C may relieve the stress due to machining. More recently, "interrupted aging" of 7xxx recorded simultaneous improvement in the tensile properties, ductility, reduction in residual stress (ASM International, 1984) and fracture toughness (Marlaud et al., 2010).

However, the artificial aging temperature had a significant effect on the tensile strength of the cast alloys from RBCs. Both samples were aged for 15 hours. The difference of 45°C in the temperature suggested the increase in UTS. The reason was because hardness increased with heat treatment, but declined due to the suggested reversal in hardening precipitates when the temperature or time

Bumper Beam Applications

increased. This was the case in the study reported by Barkov et al. (2018) where an increase in temperature and time resulted in a decrease in UTS.

The result equally revealed that when wt.% Zn was maintained, artificial aging parameters had an effect on the alloys more than annealing. This was because during annealing, for 7xxx series alloys, even slow cooling does not provide complete precipitation to remove solid solution effects (ASM International 1984; Marlaud et al., 2010). The maximum UTS of 418 MPa was obtained when the aging temperature was 100°C and time was 6 hours.

The result suggest that reduction in aging temperature had more effect against reduction in time in the alloy fabricated using RBCs. The above result is in agreement with Zaid et al. (2011), who attributed the improvement in impact energy with retrogression times to the growth of equilibrium phases at grain boundaries, which improved the alloys' stress corrosion cracking resistance during the retrogression treatment and maintained high strength during the re-aging.

14.3.2 Microstructure Study

The microstructure in Figure 14.4a–d was observed in alloy S1 after annealing at 350 °C and artificial aging at 145°C for 6 hours. Grain boundaries were evidenced as the α-Al formed the alloy matrix. Precipitates of secondary phases were observed as dispersals in the alloy. In Figure 14.4b, micro zones in the form of precipitate free zones (PFZs) were observed along the grain boundaries. The size of the boundaries were moderate as an indication of the effect of annealing on the microstructure of the alloy. The minimal segregations of the grain

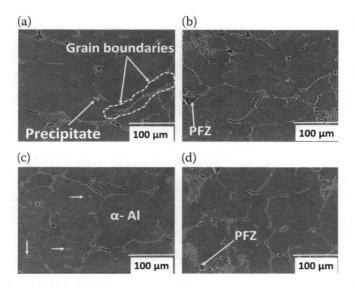

FIGURE 14.4 Microstructure of alloy S1 fabricated using Al–4.5Zn–1.5Mg–1.0Mn–0.35Cu, annealed at 350°C and artificial aging at 145°C for 6 hours.

boundaries indicated that the 6 hours of aging supported the formation of the η' dispersoids.

However, the undissolved Fe-bearing intermetallic compounds had persistently been found in Al–Zn–Mg–Cu alloys, and may be clustered within the region marked PFZ in Figure 14.4d. They are brittle, incoherent with the matrix, and severely damage the fracture toughness and ductility of the material by promoting earlier crack initiation. The lower plasticity, particularly in the peak-aged condition, was associated to the relatively lower grain boundary strength and coarse undissolved Fe bearing particles after the aging process (Ditta et al., 2020).

The T6 treatment is central to the formation of the hardening phases, which precipitates or transforms from GP I to GP II. For instance, the formations of $MgZn_2$, AlZnMgCu and $(AlCuMg)_2$ were reported to be formed between 100–270°C (Xu et al., 2017) in Al–Zn–Mg–Cu alloys. These phases were suggested to be formed in the microstructure presented shown in Figure 14.3, following the aging conducted at 120°C.

Precipitates shown with arrows (Figure 14.5a and b) were densely dispersed within the alloy matrix as the grains were observed with microvoids and abysses. The flake-like particulates were formations of second phases. An increase in the artificial aging temperature was suggested to increase the size of the PFZs. The microstructure showed refined texture as evidenced in annealed 7xxx alloys.

The microstructure of alloy S8 presented in Figure 14.5 showed the presence of sites of vacancy solute clusters (VRC) akin to the one formed at a temperature range of 450°C when the AA7108 and AA7030 were subjected to two-stage

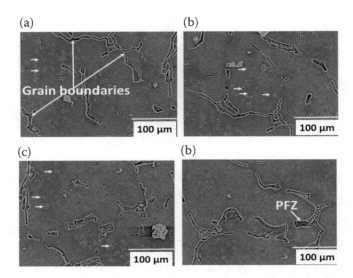

FIGURE 14.5 Microstructure of alloy S5 fabricated using Al–4.5Zn–1.5Mg–1.0Mn–0.35Cu, annealed at 380°C and artificial aging at 120°C for 10.5 hours.

Bumper Beam Applications

artificial aging, comparable to annealing in the current study, to prepare them for bumper beam applications (Waterloo et al., 2001). The PFZ in the microstructure was comparable to the one in Kremmer et al. (2020) who conducted the study on the effect of modifications in heat treatment procedures against microstructure. The report showed that the final material's condition contained the η' phase, GP (II) and metastable η-$MgZn_2$ with two-stage quenching and subsequent artificial aging. However, commercial alloys were used against the RBCs herein reported.

The microstructure in Figure 14.6 was characterized by ductility after the annealing heat treatment and prolonged aging time of 15 hours. This was suggested to be due to the increase in the volume fraction of precipitates as a result of the annealing, increase in the time and was akin to the increase in heat treatment temperature from 70°C to 135°C stepwise. The 416 MPa observed in S8 was the consequence of improved strength of the alloy fabricated using 5 wt.% Zn from RBCs as the ductility was maintained. There are precipitates located on the interior of the grains and at grain boundaries. The sizes are different; some are small and coherent with the aluminium matrix whereas the larger ones are incoherent, as shown in Figure 14.6d (Chemingui et al., 2009).

The microstructure in Figure 14.6c was suggested to be a formation of coarser precipitates, which developed at the grain boundaries as an outcome of slow quenching during annealing, which had a longer diffusion time. Precipitates were

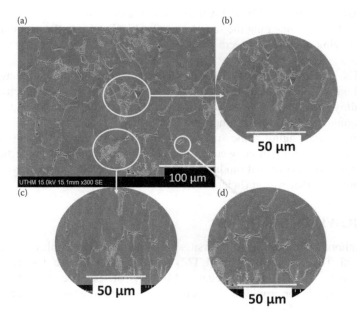

FIGURE 14.6 Microstructure of alloy S8 fabricated using Al–4.5Zn–1.5Mg–1.0Mn–0.35Cu, annealed at 413°C and artificial aging at 100°C for 15 hours.

also observed at interfaces with constituent particles and dispersoids. The presence of the precipitates was suggested to be the reason for reduction in amount of solute available, which contributes to fine-scale hardening precipitates formed in subsequent aging. The solid state remains in the 7xxx series alloys despite the slow cooling. It implies that the effect of aging is still relevant once quenching was done (ASM International 1984; Marlaud et al., 2010).

In order to improve the microstructure and properties of the alloys, dissolving unwanted coarse constituent particles through well-defined high-temperature treatments like annealing was suggested. The report observed that the development of 7xxx series alloys and processing treatments is a complex matter that evolves with the desired functionality of the alloy (Rometsch et al., 2017). Still on microstructure of modified alloys, when one-step stress relieve treatment was done at 470°C for 24 hours followed by another two-step homogenization at 300°C and 400°C after casting, the alloys were then direct water quenched. It turned out that the two-step homogenization treatment minimized the precipitate-free zones. There was a significant improvement of dispersoids distribution in an aluminium matrix (Guo et al., 2015). The previous observations were a replica of the microstructure presented herein in Figure 14.6.

14.4 CONCLUSIONS

A novel experimental 7475 aluminium alloy was fabricated from RBCs with 4.0–5.0Zn–1.5 Mg–1.0 Mn–0.35Cu as compositions. Heat treatment modification was done by annealing the samples (350–413°C) prior to the artificial aging. The following conclusions were deduced:

i. The maximum UTS of 418 MPa was obtained in an alloy fabricated using 5 wt.% Zn, annealed at 413°C, aged at 100°C for 6 hours, respectively. An alloy of 4.5 wt.% Zn had the least UTS of 162 MPa following artificial aging at 120°C for 10.5 hours.
ii. Prolonged aging time increased the precipitate-free zones (PFZs). Precipitate formations were responsible for the observed properties. This study has shown that modifications to the heat treatment process offered a property that was sufficient for bumper beam applications.

ACKNOWLEDGMENT

The authors acknowledge Universiti Tun Hussein Onn Malaysia (UTHM), Universiti Teknologi Malaysia (UTM) and National Centre for Technology Management (NACETEM), Federal Ministry of Science and Technology (FMST), Nigeria for providing the facilities used in conducting this study for and also research collaborations.

REFERENCES

ASM International (1984). Metallurgy of heat treatment and general principles of precipitation hardening. In *Aluminium Properties and Physical Metallurgy, 134*, 99. doi:10.1261/rna.065219.117.

Barkov, R. Yu, Pozdniakov, A. V., Tkachuk, E., & Zolotorevskiy, V. S. (2018). Effect of Y on microstructure and mechanical properties of Al-Mg-Mn-Zr-Sc alloy with low Sc content. *Materials Letters, 217*, 135–138. 10.1016/j.matlet.2018.01.076.

Chemingui, M., Khitouni, M., Mesmacque, G., & Kolsi, A. W. (2009). Effect of heat treatment on plasticity of Al-Zn-Mg alloy: microstructure evolution and mechanical properties. *Physics Procedia, 2*(3), 1167–1174. 10.1016/j.phpro.2009.11.079.

Choudhry, M. Arshad, & Ashraf, Muhammad (2007). Effect of heat treatment and stress relaxation in 7075 aluminum alloy. *Journal of Alloys and Compounds, 437*(1-2), 113–116. 10.1016/j.jallcom.2006.07.079.

Ditta, Allah, Wei, Lijun, Xu, Yanjin, & Wu, Sujun (2020). Microstructural characteristics and properties of spray formed Zn-Rich Al-Zn-Mg-Cu alloy under various aging conditions. *Materials Characterization, 161*(37), 110133. 10.1016/j.matchar.2020.110133.

Francis, E. D., Jayakrishna, S., & Raj, B Harish (2016). Design analysis of bumper beam subjected to offset impact loading for automotive applications. *Research Inventy: International Journal of Engineering And Science, 6*(2), 40–46.

Garber, Tal, Goldenberg, Jacob, Libai, Barak, & Muller, Eitan (2004). *Modern Physical Metallurgy and Materials Engineering Science, Process, Applications. Elsevier Science* (sixth edn, Vol. 23). Elsevier Science.

Guo, Zhanying, Zhao, Gang, & Chen, X. Grant (2015). Effects of two-step homogenization on precipitation behavior of Al3Zr dispersoids and recrystallization resistance in 7150 aluminum alloy. *Materials Characterization, 102*, 122–130. 10.1016/j.matchar.2015.02.016.

Juniarsih, A., Oediyani, S., & Zain, A. P. (2019). The effect of flux's towards Mg reduction from aluminium beverage cans. *IOP Conference Series: Materials Science and Engineering, 478*(1), 1–7. 10.1088/1757-899X/478/1/012006.

Kremmer, Thomas M., Dumitraschkewitz, Phillip, Pöschmann, Daniel, Ebner, Thomas, Uggowitzer, Peter J., Kolb, Gernot K. H., & Pogatscher, Stefan (2020). Microstructural change during the interrupted quenching of the AlZnMg(Cu) Alloy AA7050. *Materials, 13*(11), 2554. doi:10.3390/ma13112554.

Li, Bo, Wang, Xiaomin, Chen, Hui, Hu, Jie, Huang, Cui, & Gou, Guoqing (2016). Influence of heat treatment on the strength and fracture toughness of 7N01 aluminum alloy. *Journal of Alloys and Compounds, 678*, 160–166. 10.1016/j.jallcom.2016.03.228.

Li, Heng, Lei, Chao, & Yang, He (2017). Formability and performance of Al-Zn-Mg-Cu alloys with different initial tempers in creep aging process. In *Aluminium Alloys – Recent Trends in Processing, Characterization, Mechanical Behavior and Applications Developed*. 10.1016/j.colsurfa.2011.12.014.

Marlaud, T., Deschamps, A., Bley, F., Lefebvre, W., & Baroux, B. (2010). Evolution of precipitate microstructures during the retrogression and re-aging heat treatment of an Al-Zn-Mg-Cu alloy. *Acta Materialia, 58*(14), 4814–4826. 10.1016/j.actamat.2010.05.017.

Rometsch, Paul A, Zhang, Yong, & Knight, Steven (2017). Heat treatment of 7xxx series aluminium alloys – some recent developments. *Transactions of Nonferrous Metals Society of China, 24*(7), 2003–2017. 10.1016/S1003-6326(14)63306-9.

Tabei, A., Liang, S. Y., & Garmestani, H. (2018). Effects of machining on the microstructure of aluminum alloy 7075. *Applied Mechanics and Materials, 23*(4), 621–627.

Tanlak, Niyazi. (2012). Shape optimization of beams under transverse crash. In *The 15th International Conference on Machine Design and Production*, June 19–22, 2012, Pamukkale, Denizli, Turkey.

Waterloo, G., Hansen, V., Gjønnes, J., & Skjervold, S. R. (2001). Effect of pre-deformation and preaging at room temperture in Al-Zn-Mg-(Cu,Zr) alloys. *Materials Science and Engineering A, 303* (1–2), 226–233. 10.1016/S0921-5093(00)01883-9.

Xu, D., Li, Z., Wang, G., Li, X., Lv, X., Zhang, Y., Fan, Y., & Xiong, B. (2017). Phase Transformation and microstructure evolution of an ultra-high strength al-zn-mg-cu alloy during homogenization. *Materials Characterization, 131*(June), 285–297. 10.1016/j.matchar.2017.07.011.

Zaid, H. R., Hatab, A. M., & Ibrahim, A. M. A. (2011). Properties enhancement of Al-Zn-Mg alloy by retrogression and re-aging heat treatment. *Journal of Mining and Metallurgy, Section B: Metallurgy, 47*(1), 31–35. 10.2298/JMMB1101031Z.

15 Recycling of Multi-Material Plastics in the Example of Sausage Casings Wastes

Marek Szostak and Pawel Brzek
Institute of Materials Technology, Department of Mechanical Engineering, Poznan University of Technology, Poznan, Poland

CONTENTS

15.1 Introduction 263
15.2 Materials and Methodology 265
    15.2.1 Materials 265
    15.2.2 Methodology 265
15.3 Overview of Lines for Film Recycling 268
    15.3.1 Line C: GRAN – Next Generation Recycling (NGR) 268
    15.3.2 Line INTAREMA® T, TE – EREMA 268
    15.3.3 STARLINGER Recycling Lines 272
15.4 Recycling of Selected Film Waste on an Industrial Line 273
15.5 Examination of Mechanical Properties of Manufactured Regranulates 274
    15.5.1 Tensile Test 276
    15.5.2 Charpy's Notched Impact Strength 279
    15.5.3 VICAT Softening Temperature Test 279
    15.5.4 Scanning Electron Microscopy (SEM) 280
15.6 Production of Transport Pallet – Industrial Test 282
15.7 Conclusions 284
Acknowledgment 285
References 285

15.1 INTRODUCTION

Today, single-use packaging has become a big problem for our environment, especially if they are multilayered and made from various polymers. Recycling of packages made from a single polymeric material is not difficult and is now already very well established but the use of multi-layer packaging of various polymers in different layers creates a big problem for the inexperienced recyclers. We are using such packaging due to the fact that they guarantee greater product durability and less food waste. Currently, we don't have many technologies that are effective from the practical and cost point of view and that also do not harm our environment (Foltynowicz, 2020).

There are a few papers in scientific journals on recycling of mixed polyamide and polyolefin wastes (Bursa, 2004; Shanmugam et al., 2019) and only one concerns a similar topic of recycling plastic mixture wastes (PA/PP/PE) from the food packaging sector to produce composite materials with short glass fibers (SGF) (7–40 wt.%) and maleic anhydride grafted polypropylene (PPgMAH) (3–6 wt.%) as a coupling agent (Hajj et al., 2020).

This chapter presents the possibilities of using the process of recycling polyamide and polyethylene mixtures derived from sausage casings supplied by Podanfol S. A. from Podanin. The company turned to the Institute of Materials Technology of the Poznan University of Technology with a proposal to solve the problem of 300 tons of polyamide-polyethylene wastes generated annually during production of multilayer film used as sausage casings. Enterprise Podanfol S.A. was the first in Poland (since 1982) and one of the few in Europe that deals with the production of cheese and sausage casings. Over time, the company has become a leading producer of casings in the world, supplying customers in many markets of the European Union, as well as in the USA, Africa, Canada and South America. Its products have the highest quality, so they are intended for a variety of applications (Podanfol, 2020). Among the Podanfol range of products there is a group of several types of sausage casings, which include, among others:

- Pecta standard – this is the most popular casing type with medium and high barriers for the production of cooked sausages.
- Pecta smoke – the casings are obtained from polyamide and water-soluble polymers, which allow for smoke, gases and water vapor permeability. They are used for production of different types of sausages (smoked, cooked, raw and semi-dry).
- Pecta flex – the most popular casing with medium and high barriers for cooked and cured meat production.
- Pecta ring – these casings have the curved line and are also used for different types of sausages.
- Special line – the special customized casings for meat and cheese producers.

Achieving the company's goal required the selection of appropriate technologies for the production and processing of composite materials and their mixtures. To determine the degree of effectiveness and efficiency of processes, mechanical, thermal and structural properties of polymer mixtures were examined. The last stage of the work was technological tests to manufacture, from regranulate obtained from recycling, the transport and storage pallets in injection molding technology.

15.2 MATERIALS AND METHODOLOGY

15.2.1 MATERIALS

In the research, waste polymer mixtures based on PA6 (37–50%), PA6/PA6.6 (15–22%) and LDPE (11–22%) and also with some amount of adhesives (13%), dyes (9–13%) and modifiers (4%) were examined. Basic reference materials: PA6 (Ultramid B40L); PA6.6 (Akulon S240-C) and PA6/PA6.6. (Ultramid C33 01) were supplied for testing in the form of granules.

15.2.2 METHODOLOGY

Samples for mechanical testing of the supplied film wastes were made by grinding them on a high-speed mill with sieves with a diameter of 36 mm (Figure 15.1) to form flakes (Figures 15.2 and 15.3). Clusters were crushed on a milling machine and then grinded in a low-speed mill.

FIGURE 15.1 Grinder, impeller and grinder sieve used to crush the waste films.

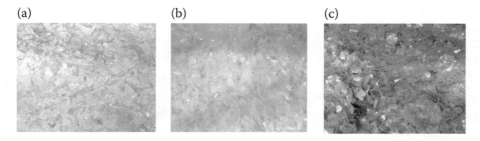

FIGURE 15.2 Shredded PA6/LDPE/PA6.6 film/tape wastes in the form of flakes: (a) white film, (b) clear film and (c) color film.

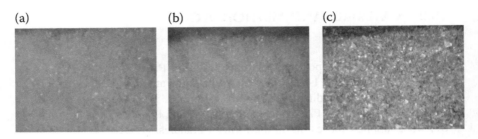

FIGURE 15.3 Shredded types of PA film in the form of flakes: (a) PA6/LDPE/PA6.6 clear film, (b) PA 6.6 clear film and (c) PA 6 color film.

Standardized test samples were made by injection molding on an ENGEL injection molding machine with a screw with a diameter of 22 mm and a ratio $L/D = 18$, type ES 80 80/20HLS. Prior to the injection process, the flakes were dried in a Binder thermal chamber at 80°C for 4 hours. Table 15.1 presents injection molding parameters of the tested materials.

The mechanical properties of the produced samples were determined in a static tensile test in accordance with PN-EN ISO 527-2: 1998. The results are shown in Table 15.2.

The final conclusion of the preliminary mechanical tests is that the studied waste materials are high-quality polymer composites with properties exceeding

TABLE 15.1
Injection Molding Parameters for Sample Production

Processing Parameters	Unit	White and Color Flakes from Films and Tapes
Temperature of injection nozzle	[°C]	270
Temperature cylinder zone III	[°C]	275
Temperature cylinder zone II	[°C]	260
Temperature cylinder zone I	[°C]	245
Injection speed	[m/s]	0.7
Injection pressure	[MPa]	80
Holding pressure	[MPa]	50
Plasticizing pressure	[MPa]	0.3
Holding time	[s]	3
Cooling time	[s]	40
Screw rotation	[rpm]	150
Screw movement	[mm]	80
Mold temperature	[°C]	30

TABLE 15.2
Mechanical Properties of Tested Materials Obtained in Tensile Tests

Material	Elongation at Break e [%]	Tensile Strength R_m [MPa]	Young's Modulus [MPa]
(No. 1) White film PA6/LDPE/PA6.6	167 ± 66	41.5 ± 2.1	1275 ± 33
(No. 2) Color tape PA6/LDPE/PA6.6	201 ± 58	40.5 ± 2.7	1153 ± 52
(No. 3) Color film PA6/LDPE/PA6.6	185 ± 72	41.9 ± 2.8	1326 ± 52
(No. 4) Clear film PA6/LDPE/PA6.6	185 ± 33	45.1 ± 5.8	1552 ± 71
(No. 5) Clusters PA6/LDPE/PA6.6	186 ± 25	40.1 ± 4.0	1269 ± 235
(No. 6) Clusters PA6.6/PA6	6 ± 2	69.3 ± 6.6	2885 ± 85
(No. 7) Color film PA6.6/PA6	14 ± 2	74.7 ± 1.4	2779 ± 53
(No. 8) White film PA6	80 ± 85	70.1 ± 1.5	2459 ± 79
PA6 Original	67 ± 68	74.4 ± 0.5	1723 ± 119
PA6.6 Original	33 ± 11	77.8 ± 1.9	1928 ± 237
PA6/6.6 Original	82 ± 52	65.1 ± 1.8	1451 ± 163

the initial expectations of the supplier, which, after finding suitable applications for them, should not cause any problems during their future use.

Samples for mechanical testing of the supplied wastes materials, as mentioned previously, were made after grinding them on a high-speed mill with sieves with a diameter of 36 mm to form flakes. Clusters were crushed on a vertical milling machine and then in a low-speed mill. It should be noted that during the grinding process, the phenomenon of static electricity occurred, which caused some problems with dosing the obtained recyclate to the injection molding system for: PA6/LDPE/PA6.6 color film (No. 3), PA6.6/PA6 film (No. 7) and PA6 film (No. 8). These problems were not observed for PA6/LDPE/PA6.6 white film (No. 1), which is probably due to the addition of titanium white as a dye, which discharges electrostatic charges quite well and does not allow the film to electrify during grinding. For the PA6/LDPE/PA6.6 tape (No. 2), the phenomenon of electrostatics also occurs but in a lesser extent due to the greater orientation of the polymer in the tape and its greater rigidity.

Also, no problems were observed with the electrostatics of materials originating from the fragmentation of PA6/LDPE/PA6.6 clusters (No. 5) and PA6.6/PA6 clusters (No. 6). Pieces of material pre-cut on a milling machine and then comminuted in a low-speed mill were fed very well to the injection molding screw system.

The above technological tests and observations of the course of processes showed the usefulness of the technique of preliminary one- or two-stage grinding and then injection technology for the production of high-quality products from the tested waste clusters, tapes and casing films. However, the complicated method of dosing the recyclates obtained in the process of grinding reduces this way of recycling waste to low- and medium-series production. In order to ensure high-volume production of the highest quality and repeatability of finished injection products, it would be necessary to first produce regranulate on a fully automatic extrusion line for recycling of polymeric materials and then subject it to the injection process. An overview of several process lines that can be used is described in Section 15.3.

15.3 OVERVIEW OF LINES FOR FILM RECYCLING

There are many competing companies on the European market that offer high-performance technological lines for the recycling of plastic films. Each of the companies offers lines built of modular construction segments that allow you to create a structure tailored to individual customer needs. Austrian companies have a large market share, offering high-quality recycling machines at affordable prices. Further on, recycling lines were presented for the most popular producers in this country: NGR, EREMA and STARLINGER.

15.3.1 LINE C: GRAN – NEXT GENERATION RECYCLING (NGR)

In Figure 15.4, Line C: GRAN recycling line from Austrian company NGR is presented.

The polymer waste is crushed, concentrated and heated in a cutting chamber (Figure 15.5), then fed to a single-screw extruder. The rotating blades shred plastic waste and the specially arranged guides ensure proper thickening of the polymer material along with the direction of the rotating blades.

The high homogeneity of the processed material affects the optimal filling of the screw in the shortest possible time, and thus ensures high process efficiency. Temperature adjustment is carried out automatically using sensors. The line C: GRAN is intended for processing among others materials such as coated films, plastic waste, LDPE/HDPE films, pre-shredded plastic waste and flakes (up to max. 10% humidity). The material is transported using a roll feeder (Figure 15.6) or a conveyor belt (Figure 15.7). The line has the granulator for extruder with a hot extrusion head.

15.3.2 LINE INTAREMA® T, TE – EREMA

The INTAREMA compact recycling system (Figure 15.8) with a short, single screw from the Austrian EREMA company is available in two series: T series – without degassing the extruder, it is ideal for film materials without printing and

Sausage Casings Wastes

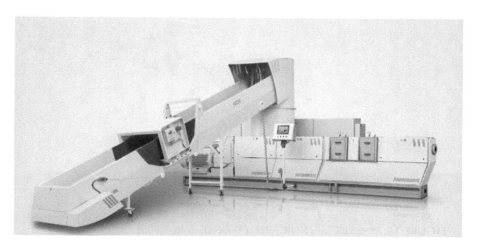

FIGURE 15.4 Recycling line C: GRAN – NGR Company (NGR – Next Generation Recycling, 2020).

FIGURE 15.5 Cutting chamber (NGR – Next Generation Recycling, 2020).

FIGURE 15.6 Roller feeder (NGR – Next Generation Recycling, 2020).

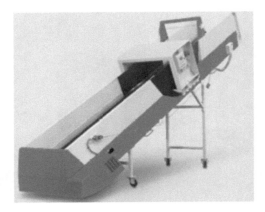

FIGURE 15.7 Conveyor (NGR – Next Generation Recycling, 2020).

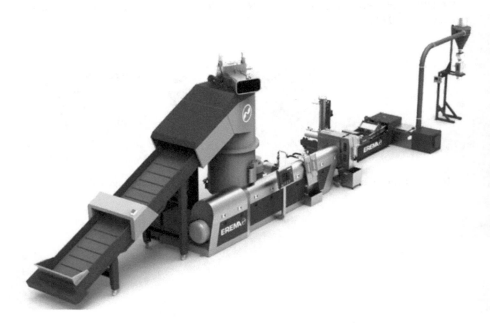

FIGURE 15.8 INTAREMA compact recycling system (EREMA, 2020).

shredded waste and TE series – offers double degassing and is intended for processing plastics with a small print or industrial waste and technical polymers.

The diagram in Figure 15.9 shows how the EREMA line works. Material transport (1) is carried out automatically according to customer requirements. The material is crushed, heated, dried, compacted and stabilized in the

Sausage Casings Wastes

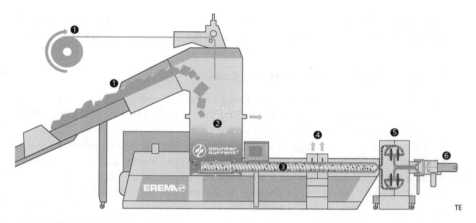

FIGURE 15.9 INTAREMA® TE: 1 – material transport, 2 – cutting and compacting chamber, 3 – extruder, 4 – degassing zone, 5 – filters, 6 – tool (EREMA, 2020).

cutting-compacting chamber (2). Then the tangentially connected extruder is filled continuously with hot, pre-compacted material. The polymeric material is plasticized and homogenized in the extruder (3) and degassed (TE) if necessary in the degassing zone (4). The molten material is then fully automatically cleaned using self-cleaning filters (5). Then it is transferred under low pressure to a special tool (6), e.g., a granulator (EREMA, 2020).

The patented DD – Double Disc technology (Figure 15.10) allows for efficient processing of materials containing up to 12% moisture, and thus affects the high quality of the material and a wide range of processing. The addition of a second disk divides the cutting and compacting chamber into: chambers 1 and 2.

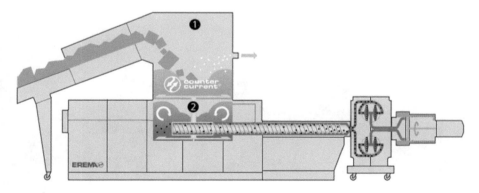

FIGURE 15.10 Double Disc technology: 1 – chamber 1 (preliminary cutting and mixing), 2 – chamber 2 (homogeneous material ready for dosing to the extruder plasticizing system) (EREMA, 2020).

Preparation of the material (cutting, mixing, drying, pre-compacting) takes place in chamber 1. The extruder is fed with material from chamber 2 (EREMA, 2020).

15.3.3 STARLINGER Recycling Lines

The STARLINGER company specializes in the technology of recycling single and multilayer films of such polymer materials as PP, PE, PA, PS, PLA, EVA, EVOH and all types of waste (post-production and post-consumer).

The company's challenge is to strive to homogenize various polymers while maintaining high-quality molten plastic. Post-production and post-consumer waste occurs in various shapes. The choice of pretreatment is defined by the type of waste and its humidity. Degassing of the extruder is required for moist and printed waste. The range of available filtration systems and agglomerators allows the processing of waste from various types of polymers. Modular design allows for flexible adaptation of input material recycling systems. The efficiency and stability of the basic production process can be achieved thanks to the high-quality regranulate. The STARLINGER company offers innovative solutions for film recycling machines (recoSTAR basic and recoSTAR universal) (Figure 15.11) for efficient processing of various types of materials. Table 15.3 compares two examples of technological solutions for film recycling.

The recoSTAR basic line is distinguished by pre-comminution and heating of the material in the agglomerator. The process of cutting and compacting the material is carried out using knives mounted on a rotating lower disk. Material temperature measurement ensures stable production conditions. An additional protective process is carried out by automatically cooling the processed material with a liquid in the lower part of the agglomerator. The recoSTAR universal line allows film to be processed without pre-grinding thanks to the use of a high-strength single shaft. The material temperature is controlled by a cooled rotor to facilitate the operation of the line. The continuous feeding of material to the

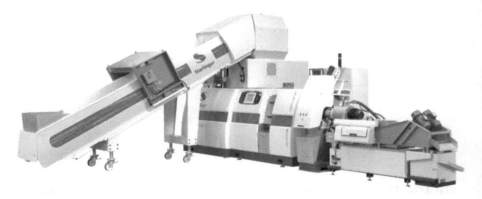

FIGURE 15.11 STARLINGER recoSTAR recycling line (STARLINGER, 2020).

TABLE 15.3
Comparison of the recoSTAR Basic and recoSTAR Universal Lines

Line	Material	Moisture Content	Shape and Form of Film	Flakes
recoSTAR basic	PE (LDPE, HDPE), PP (BOPP), PA (BOPA), PET (BOPET), PS, PLA	Up to 10%	Unbound, pre-ground, from a roll	Yes
recoSTAR universal		Up to 5%	Unbond, pre-portioned, from a roll	Yes

extruder is controlled by the feeding speed of the screw. This feature also enables precise dosing of additional substances in order to modify the produced regranulate. The hydraulic piston stabilizes the supply of material to the system.

From the lines presented above for carrying out technological tests on the film and tape wastes collected by PODANFOL, the recoSTAR Universal line from STARLINGER recoSTAR Universal has been chosen.

15.4 RECYCLING OF SELECTED FILM WASTE ON AN INDUSTRIAL LINE

We have divided the main waste streams declared by the PODANFOL company: Clusters – 20% (60 tons/year), Tapes – 20% (60 tons/year) and Films – 60% (180 tons/year) into four other groups: Group A – Clear film (No. 4); Group B – White and colored films (No. 1, 3, 7 and 8), Group C – Colored and clear tapes (No. 2) and Group D – Clusters (No. 5 and 6). Of these groups, the first three are suitable for technological industrial tests, because group D – clusters (outflows) is recycled on completely different machines. So, at the end of preliminary studies we have decided to send the waste materials from groups A and C to STARLINGER Company for industrial tests. Sent samples are shown in Figures 15.12 and 15.13.

For the industrial process of recycling of multimaterial film and tape wastes, the recoSTAR universal 65VAC Starlinger line was used. It consist of a conveyor belt with a metal detector, single shaft cutter 800, temperature-controlled shredder shaft, recoSTAR 65 VAC controlled through a dosing screw, speed-controlled extruder, VAC degassing unit, filter with backflush SPB-180-H, watering pelletizer WRP 120R CC, inline output measurement, storage silo 1 m³ and screw design Z1R-09622/Z2R-09894.

The processing parameters of the waste recycling are shown on the screen-shots from the STARLINGER machine controller. In Figures 15.14 and 15.15 the parameters for processing the transparent PA6/LDPE/PA6.6 film (No. 4) and colored and clear PA6/LDPE/PA6.6 tapes (No. 2) are shown.

FIGURE 15.12 Clear film sample (No. 4).

FIGURE 15.13 Colored and clear tape samples (No. 2).

After the test we can confirm that both materials could be handled without problems. Regranulate produced on the recoSTAR Universal recycling line is shown in Figure 15.16. The filtration with 300 μm mesh size was not a problem and the most important material after recycling was well homogenized (with standard screw design), and was shown by producing the sample film on the laboratory extruder (Figure 15.17).

15.5 EXAMINATION OF MECHANICAL PROPERTIES OF MANUFACTURED REGRANULATES

Two types of regranulate were made in the Starlinger Company on the "recoSTAR universal" line: one from clear film wastes (300 kg) and the second from colored and clear tapes wastes (900 kg). Both produced regranulate and blends of both regranulates in the ratio: 25% regranulate from films and 75% regranulate from tapes were taken for testing the mechanical properties.

Sausage Casings Wastes

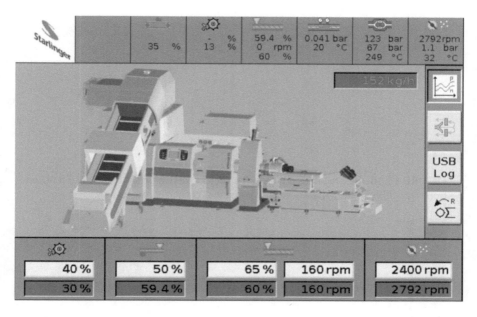

FIGURE 15.14 Processing parameters for recycling transparent film wastes PA6/LDPE/PA6.6 – screenshot from the recycling line display.

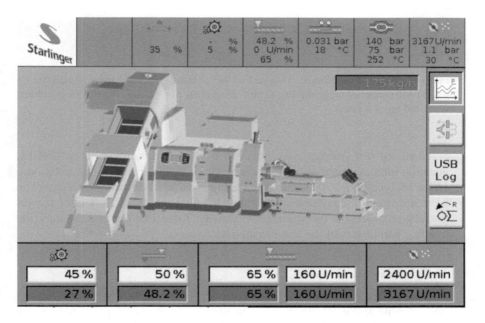

FIGURE 15.15 Processing parameters for recycling clear/colored tape wastes PA6/LDPE/PA6.6 – screenshot from the recycling line display.

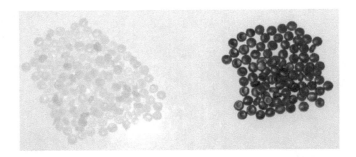

FIGURE 15.16 Regranulate produced on recoSTAR Universal recycling line (Left – from clear film wastes, Right – from colored and clear tape wastes).

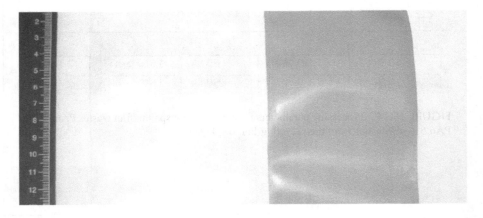

FIGURE 15.17 Film samples obtained on lab extruder produced from regranulate on recoSTAR Universal recycling line (Left – from clear film wastes, Right – from colored films and tapes wastes).

Standardized test samples were obtained by injection molding on an ENGEL injection molding machine with a screw with a diameter of 22 mm and a ratio $L/D = 18$, type ES 80/20HLS. Before the injection process, the tested regranulate were not dried but injected at the humidity that remained in it after the regranulation process at the Starlinger Company. Table 15.4 presents the injection molding parameters of the tested materials.

15.5.1 Tensile Test

The static tensile test was carried out in accordance with PN-EN ISO 527-1: 2012 (tensile speed 50 mm/min.). Ten samples from each type of regranulate were tested. The results (tensile strength, elastic modulus and elongation at break) are shown in Figures 15.18–15.20.

TABLE 15.4
Injection Conditions Used to Prepare the Standardized Samples

Processing Parameters	Unit	Regranulate from Clear Film	Regranulate from 25% Film/ 75% Tape Blend
Temperature of injection nozzle	[°C]	265	235
Temperature cylinder zone III	[°C]	260	230
Temperature cylinder zone II	[°C]	245	220
Temperature cylinder zone I	[°C]	200	195
Injection speed	[m/s]	1	1
Injection pressure	[MPa]	80	70
Holding pressure	[MPa]	70	60
Plasticizing pressure	[MPa]	0.3	0.3
Holding time	[s]	3	3
Cooling time	[s]	45	45
Screw rotation	[rpm]	180	180
Screw movement	[mm]	80	80
Mold temperature	[°C]	30	30

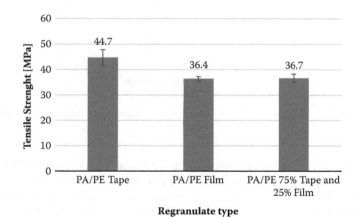

FIGURE 15.18 Tensile strength of samples produced from different types of PA/PE regranulates.

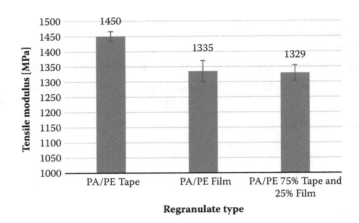

FIGURE 15.19 Tensile modulus of samples produced from different types of PA/PE regranulates.

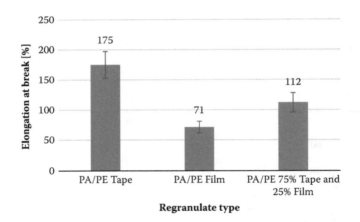

FIGURE 15.20 Elongation at break of samples produced from different types of PA/PE regranulates.

Comparing the immediate strength of the tested samples, the highest value was 44.7 MPa of the samples made of PA/PE regranulate obtained from a colored tape. Samples from regranulate derived from the waste of clear film were characterized by a maximum stress of 36.3 MPa, and for a blend (25% film/75% tape) of both regranulates it was 36.7 MPa. A similar relationship was observed for the tests of the Young's modulus, whose values were 1,450, 1,335 and 1,329 MPa for regranulate produced respectively from tapes, films and from a blend of both regranulates. As can be seen from the results, the strength values obtained for blends of regranulates are close to the values for the worst one, i.e., in the analyzed case for regranulates obtained from clear film wastes. The results of elongation for the tested regranulate and their blend are a little different. An elongation of 175% was obtained for tape regranulate, 71% for regranulate from film and 112% for a blend of both regranulates. As you can see, the elongation value for the regranulate blend is the result of both input components. The results obtained are very similar to the results obtained for the materials tested in a preliminary study that indicates that the recycling process was carried out correctly and the mechanical values did not deteriorate, despite carrying out two processing stages – extrusion on the recycling line and injection of samples for mechanical testing.

15.5.2 Charpy's Notched Impact Strength

Determination of Charpy impact strength according to PN-EN ISO 179 was performed using the PW5 INSTRON WOLPERT impact hammer.

As can be seen from Figure 15.21, the notched impact strength of samples obtained from regranulate from foil and colored tapes is much higher (56.5 kJ/m^2) than the other samples. For regranulate obtained from tapes, the notched impact strength was 47.2 kJ/m^2 and for a blend of regranulates (1 + 2) the measured impact strength was the lowest, 28.3 kJ/m^2. However, even the lowest impact strength for the tested regranulate is 2–3 times higher compared to pure polyamides.

15.5.3 VICAT Softening Temperature Test

Determination of the VICAT softening temperature according to the PN-EN ISO 2039-1:2001 standard was made on the cabinet thermostat type FWV 29/73 in Table 15.5.

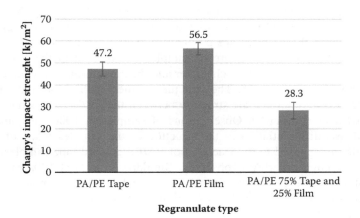

FIGURE 15.21 Notched Charpy's impact strength of samples produced from different type of PA/PE regranulates.

TABLE 15.5
Results of VICAT Softening Temperature of Tested Samples

Sample	1	2	3	Average Temperature [°C]
PA/PE film	92	94	93	93
PA/PE tape	120	120	122	121
PA/PE 25% film + 75% tape	101	99	100	100

As can be seen from the summary of the results shown in Table 15.5, the highest temperature resistance of the samples obtained from regranulates from tape wastes (121°C), then samples from a blend of both regranulates (100°C) and finally from regranulate from film wastes (93°C). Despite the significant drop in the softening point for the tested materials in relation to pure PA6 (195°C) and pure PA6.6 (245°C), they are comparable to the values obtained for high-density polyethylenes (HDPE 80–115°C) and polypropylenes (PP 95–125°C) and sufficient for the most standard industrial applications.

15.5.4 Scanning Electron Microscopy (SEM)

The Tescan Vega 5135 scanning electron microscope was used to study the surface structure of the fractures of the prepared samples. The observation of the structure has been made at two magnifications: 100× and 1000×. Figures 15.22–15.24 presented the SEM images of the fracture structures of the tested materials.

The structure of fractures of samples made of PA/PE clear film wastes and a blend of 25% film and 75% tape wastes visible at 100× magnifications is rather fine-grained and the type of fracture can be described as fragile. The white areas in the pictures are probably polyethylene and dark polyamide. However, the fracture structure of samples obtained from the PA/PE tape shows the most homogeneous character and the breakthrough is characteristic for highly elastic materials. The structure of the PA/PE 25% film and 75% tape blend is an intermediate structure between the PA/PE film structure and the PA/PE tape structure. Observations of samples at 1000× magnification showed less differentiation in the structure of the samples made of PA/PE clear film as opposed to the fractures of PA/PE colored tape samples, where greater structure ordering is observed. Similarly to tests at 100× magnification, at 1000× magnification for the tape and film blend samples an intermediate structure between the PA/PE film structure and the PA/PE tape structure is observed.

Sausage Casings Wastes 281

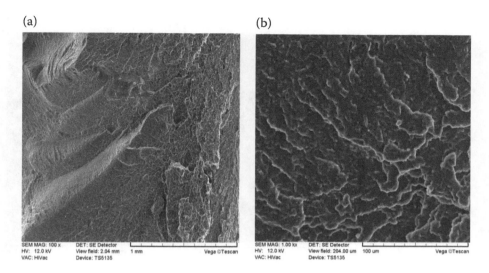

FIGURE 15.22 SEM pictures presenting the structure of the PA/PE samples obtained from clear film wastes. Magnifications: (a) 100× and (b) 1000×.

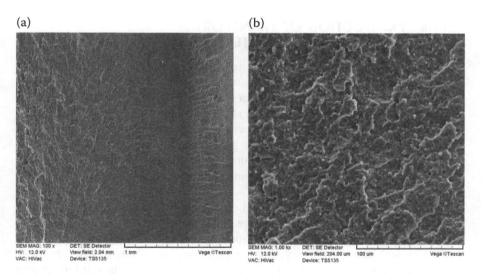

FIGURE 15.23 SEM pictures presenting the structure of the PA/PE samples obtained from clear and colored tape wastes. Magnifications: (a) 100× and (b) 1000×.

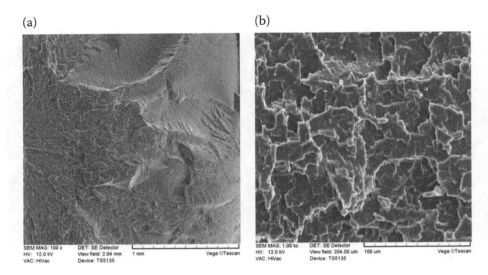

FIGURE 15.24 SEM pictures presenting the structure of the PA/PE samples obtained from (25% film/75% tape) blend. Magnifications: (a) 100× and (b) 1000×.

15.6 PRODUCTION OF TRANSPORT PALLET – INDUSTRIAL TEST

The final stage of the research was to check the possibility of using PA/PE regranulates for the production of the transport pallets. The pallet was made by the injection molding technology at the HANPLAST production plant in Bydgoszcz. This is a slot palette with slots at an angle; this design is often used by the IKEA Company. The palette was made of a blend of tested regranulate (25% film and 75% tape). Making the pallet from this blend was dictated to the greatest extent by economic and ecological considerations, as the tape wastes made of PA/PE are much more difficult to dispose of, and the profit from their reprocessing is significantly higher.

The production of pallets has been made on the 1,000 ton ENGEL injection molding machine. The processing parameters are presented in Table 15.6. During injection molding of a pallet from regranulate obtained as a result of PA/PE film and tape wastes recycling, no technological problems were observed. The material flowed very well in the mold and already with the third injection a good quality product was obtained without any surface defects or other faults. During the tests, 30 pallets were made, which were then subjected to deflection tests under the load on high storage racks in the Podanfol Company.

Figures 15.25 and 15.26 show the top and bottom view of the manufactured pallets. The dimensions of the pallet are as follows: 1,200 mm × 800 mm × 135 mm (length × width × height). The weight of the pallet was 13 kg.

TABLE 15.6
Injection Molding Parameters for Pallet Production

Processing Parameters	Unit	White and Color Flakes Obtained from Films and Tapes
Temperature of injection nozzle	[°C]	275
Temperature cylinder zone IV	[°C]	280
Temperature cylinder zone III	[°C]	275
Temperature cylinder zone II	[°C]	270
Temperature cylinder zone I	[°C]	250
Mold temperature	[°C]	60
Injection speed	[m/s]	1.0
Screw rotation	[rpm]	180
Injection pressure	[MPa]	100
Holding pressure	[MPa]	70
Plasticizing pressure	[MPa]	0.5
Holding time	[s]	15
Cooling time	[s]	60
Mold temperature	[°C]	60

FIGURE 15.25 Top view of the pallet.

The industrial test has shown that the permissible load of the pallet injected by us from a PA/PE blend placed on the high storage rack (to achieve deflection below the allowable value of 15 mm) is 2,800 kg, while for pallets of similar construction and weight (13–18 kg) made of HDPE or PP material, their maximum load may be in the range from 750 kg to 1,500 kg.

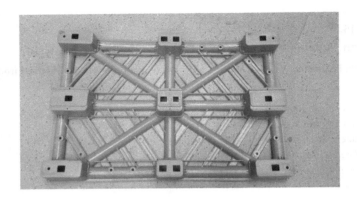

FIGURE 15.26 Bottom view of the pallet.

15.7 CONCLUSIONS

It can be concluded that:

1. All tested regranulate obtained as a result of recycling on the STARLINGER recoStar universal line have good mechanical properties.
2. The tensile strength for samples obtained from different types of regranulates was 44.7 MPa for regranulates obtained from clear and colored tapes, 36.4 MPa for regranulates from clear film and 36.7 MPa for 75%/25% blend of both of these regranulates. The tensile modulus of the samples was 1,450 MPa for tape regranulates, 1,335 MPa for film regranulates and 1,329 MPa for a blend of both regranulates, while the elongation at the break was 174%, 71% and 112%, respectively.
3. The notched impact strength for the tested regranulates was 47.2 kJ/m^2 (from clear and colored tape), 56.5 kJ/m^2 (from clear film) and 28.5 kJ/m^2 (for 75/25 blend of both these regranulates).
4. The VICAT tests of samples obtained from PA/PE wastes showed that the addition of polyethylene to polyamide significantly reduces the softening temperature (to 93–121°C) but it is still sufficient for most standard applications of plastic products.
5. The study has shown also that all tested regranulates should easily be used again in the production of responsible market products.
6. Technological injection molding tests of transport pallet production from tested regranulates carried out at the HANPLAST Company show that they have very good mechanical properties: strength, rigidity and can be successfully used as transport or storage pallets even for high storage warehouses.

ACKNOWLEDGMENT

The results of this research, executed under the project of No. 01/25/ITMat/JGN were funded by the Podanfol Company (Poland). The authors would also like to thank STARLINGER (Austria) and HANPLAST (Poland) for enabling industrial recycling and injection molding tests.

REFERENCES

Bursa, Jerzy (2004). Characteristics of PA/PE composite with polyolefine fillers. In *Scientific Papers of the Institute of Electrical Engineering Fundamentals of Wroclaw Technical University Conferences, 40*, 293–296.

EREMA. (2020). *EREMA Plastics Recycling Machines & Plastic Recycling Plant.* Accessed 18 May 2020. https://www.erema.com.

Foltynowicz, Zenon (2020). Polymer packaging materials – friend or foe of the circular economy. *Polimery, 65*(I), 3–7. doi:10.14314/polimery.2020.1.1.

Hajj, E. L., Seif, S., Saliba, K., & Zgheib, N. (2020). Recycling of plastic mixture wastes as carrier resin for short glass fiber composites. *Waste and Biomass Valorization, 11*(5), 2261–2271. doi:10.1007/s12649-018-0446-z.

NGR – Next Generation Recycling. (2020). *The Universal Solution for Industrial Plastics Wastes.* Accessed 25 May 2020. http://www.ngr-world.com/product/sgran.

Podanfol. (2020). *Podanfol Professional Packagings.* Accessed 4 May 2020. http://www.podanfol.com/en/casings.d1.

Shanmugam, K., Doosthosseini, H., Varanasi, S., Garnier, G., & Batchelor, W. (2019). Nanocellulose films as air and water vapour barriers: A recyclable and biodegradable alternative to polyolefin packaging. *Sustainable Materials and Technologies, 22.* 10.1016/j.susmat.2019.e00115.

Starlinger. (2020). *Recycling Technology.* Accessed 11 May 2020. http://www.starlinger.com/en/recycling.

UTZ. (2020). *Pallets UPAL, Stacking Frames.* Accessed 15 June 2020. https://www.utzgroup.com/en/products/pallets-upal-stacking-frames.

16 Influence of Recycled Steel Scrap in Nodular Casting Iron Properties

Marcelo Luis Siqueira[1],
Sebastião Bruno Vilas Boas[1], Fabio Gatamorta[2],
Claudney de Sales Pereira Mendonça[1], and
Mirian de Lourdes Noronha Motta Melo[1]
[1]Department of Mechanical Engineering, Federal University of Itajubá – UNIFEI, Brazil
[2]Mechanical Engineering Faculty, University of Campinas, São Paulo, Brazil

CONTENTS

16.1 Introduction 287
16.2 Experimental Conditions 288
    16.2.1 Materials and Green Compact 288
    16.2.2 Ductile Iron Production 288
    16.2.3 Characterization 288
16.3 Results and Discussion 290
16.4 Conclusions 293
References 294

16.1 INTRODUCTION

Cast iron, the first man-made composite, is at least 2,500 years old. It remains the most important casting material, with over 70% of the total world tonnage (Stefanescu, 2005). Their use is justified by the wide range of mechanical properties that can be achieved, associated with their competitive price (Theuwissen et al., 2016). Cast iron represents a family of alloys composed of graphite wrapped in a metallic matrix. It has unique properties and one of the most produced engineering materials used in the world (Olawale et al., 2016). Nodular cast iron is a class of cast iron in which graphite is in a spherical form or nodule (Andriollo et al., 2015). Since graphite is in the form of nodules, the alteration of the microstructure to meet a specific application is carried out in the matrix. The nodular cast iron, depending on the application, may present ferrite, perlite, ferrite-perlite, cementite, martensite or ausferrite in its microstructure (Čanžar et al., 2012). The different phases can be obtained in the casting process by the chemical composition of the raw material, melting temperature, casting, inoculation, nodulization and cooling rate during solidification and by heat treatments (Biswas & Monroe, 2019; Bočkus & Žaldarys, 2010; Glavaš, 2012; Herrera-Navarro et al., 2011; Kopyciński et al., 2013). Further research showed that the shape of graphite precipitates in cast iron is very sensitive to small variations in melt composition. Currently, many studies have been conducted to understand the effect of chemical elements on the microstructure and properties of nodular cast iron. The unwanted addition of small amounts of chemical elements in the process occurs through the contamination in the foundry. The recycle used in the production of foundry requires special attention in the control of elements, especially when these elements come from the use of recyclable materials. Steel scrap is one of the main raw materials used in the production. The scope of this work is to fill a gap in knowledge about the effect of small additions of chemical elements such as copper, nickel, molybdenum and chromium in the process of casting by centrifugation in metallic molds to produce nodular cast iron.

16.2 EXPERIMENTAL CONDITIONS

16.2.1 Materials and Green Compact

In this work, the material used in a foundry is scrap for automotive ductile cast iron. Its chemical composition is presented in Table 16.1. The main difference refers to the Mo, Cu, Ni and Cr alloy content.

16.2.2 Ductile Iron Production

The ductile irons were melted in a medium frequency induction furnace to a temperature of 1.450°C by the addition of steel scrap (5 wt.%), pig iron (25 wt.%), ductile iron returns (70 wt.%) and graphite, Fe–Si (75 wt.%). Spheroidizing practices were performed in a conventional sandwich method (Fe–Mg–Si 0.9 wt.%) in a 120 kg ladle capacity. Inoculation practice was performed in three parts: the first one in panela during transference of liquid metal, second one in the crucible and the last one in the in the mold. Additions used for this work are ferrochrome (Cr 56 wt.%) nickel metal, ferromolybdenum (Mo 80 wt.%) and copper (wire 98 wt.%). The metal was poured into a metallic mold in a horizontal centrifugal machine to obtain the cylindrical tubes shown in Figure 16.1. The casted dimensions have lengths of 1.243 mm, outside diameter 115 mm and inside diameter 86 mm.

16.2.3 Characterization

The graphite nodules were evaluated in terms of quantity, shape and size, according to the ASTM 247 standard. The microstructure change was determined by optical microscopy, scanning electron microscope (SEM), hardness, tensile test and Charpy test. For the microstructural analysis, the specimens were ground with

TABLE 16.1
Chemical Compositions of Ductile Cast Iron (wt. %)

MELT	C	Si	Mo	Cu	Ni	Cr	Mg
Ni					0.908		
Cr						0.742	
Mo			0.763				
Ni/Cr					0.854	0.821	
Mo/Cu			0.771	0.734			
Cu/Ni				0.712	0.924		
Mo/Ni			0.869		0.897		
Mo/Cr			0.412			0.877	
Cu/Cr				0.504		0.558	

FIGURE 16.1 DCI in cylindrical shape.

120, 320, 400, 600 and 1,200-grit SiC papers and polished with three diamond suspension, 1 μm diamond paste and colloidal silica. Then, the samples were etched with a Nital 2% solution to determine the microstructure using an optical microscope (Leica, model DMI 5000M image analyzer LAS V-4.5) and scanning electron microscope (Jeol, model JXA-840A). Tensile specimens were machined in the same direction of the long of cast (Figure 16.2). Three tensile specimens for each condition were tested in a 250 kN hydraulic EMIC DL 3000 universal testing machine using a constant cross-head travel speed of 4 mm/min and the procedure ASTM E18-12 as reference. Notched Charpy specimens, with the dimensions 55 mm × 10 mm × 10 mm were machined in the same direction of the length of the cast. The Charpy test was performed as ASTM E8M-04 in a Charpy tester machine model JB-300AI/C maximum energy of 300 Joules. The hardness was

FIGURE 16.2 Position and direction of the samples extracted from the foundry for the tensile test specimen. The samples for the tensile test are taken in the longitudinal direction of the piece.

performed using an OTTO WOLPERT-WERKE tester machine. Each hardness result was determined from an average of six measurements per sample with a load of 100 kg and a steel ball with a 10 mm diameter.

16.3 RESULTS AND DISCUSSION

The type 1 graphite nodule is the preferred form for nodular cast iron and it is observed that it is predominant for 8 of the 9 samples studied, as shown in Figure 16.3. This shape is desirable, as it is the one that best approaches a perfect

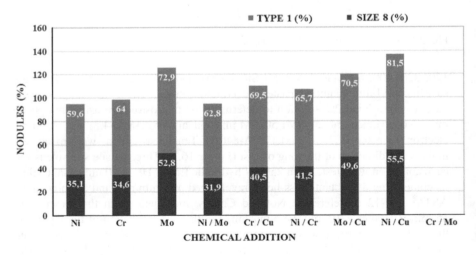

FIGURE 16.3 Nodules type and size for nodular cast iron.

Influence of Recycled Steel Scrap

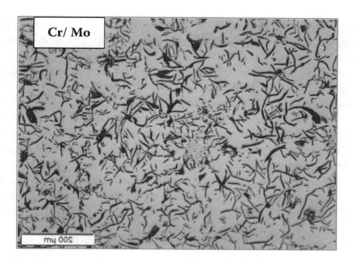

FIGURE 16.4 Graphite lamellar for nodular cast iron.

sphere and, therefore, the one that is least susceptible to concentrating tension in service.

Nodules of graphite are uniformly dispersed within the matrix, indicating a homogeneity of properties associated with this structure. The only exception refers to the Cr/Mn addition shown in Figure 16.4. In this case, graphite is present in a lamellar form, which indicates a bad nodulization performance.

Graphite ferrite was found in all samples in this study. The maximum ferrite content (40%) was observed for the sample containing Ni. The samples without the chromium addition presented a bull's-eye ferrite rim around the nodules, as shown in Figure 16.5.

Perlite was found in all samples, ranging in content from 30% to 90%. Notably, the highest percentage values of perlite were exhibited by the samples

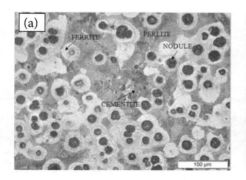 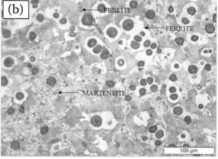

FIGURE 16.5 (a) Matrix for Ni addition and (b) matrix for Mo addition.

containing chromium, between 85% and 90%. The martensite was observed in three samples, varying in content between 35% and 50% and molybdenum is the common element for the three cases. Martensite, in general, forms carbides around the cell boundaries and, considering the cast dimension, its concentration must be limited. The hardness of a nodular cast iron is strongly affected by the microstructure obtained in the raw state of casting with the addition of chemical elements. The percentage of perlite in the matrix of nodular cast iron is directly related to the presence of chromium. The highest hardness values were observed for samples containing perlite above 85%. The hardness increased by promoting pearlite and by forming dispersed chromium carbides in the matrix.

Figure 16.6a shows a region Cu- and Mo-rich carbides. In this figure, scanning chemical microanalysis is possible to identify a clear region, rich in molybdenum between approximately 22 μm and 30 μm. In the rest of the matrix, copper and molybdenum appear homogeneously distributed. Figure 16.6a shows the sample containing Cr/Mo. The images show two distinct regions: the first being formed by molybdenum (lighter region) and the second formed by the elements chromium and molybdenum, as observed in Figure 16.6b. It means this ductile cast iron is formed by Mo/Cr carbides and Mo carbides.

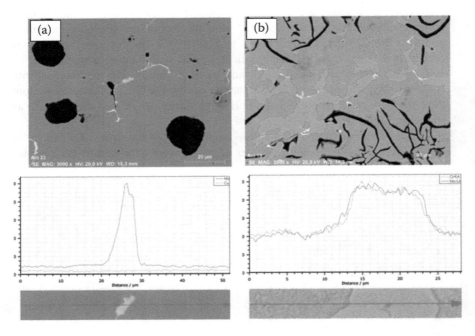

FIGURE 16.6 (a) The SEM image and chemical microanalysis for the Mo/Cu sample and (b) the SEM image and chemical microanalysis for the Mo/Cr sample.

Influence of Recycled Steel Scrap 293

FIGURE 16.7 (a) Fracture surface image for the Mo/Cu sample and (b) fracture surface image for the Mo/Cr sample.

Ferrite increases the ductility of nodular cast iron, increasing the impact resistance energy. Microstructural components of high mechanical resistance have the effect of reducing the energy of impact resistance, such as perlite and martensite. Another factor to be considered is the presence of graphite nodules. The best situation refers to the largest number, the most spherical shape (type 1) and the most homogeneous distribution of the nodules in the matrix.

In accordance with EDS profile Cr and Mo, shows peaks indicating Cr-rich carbides and Mo-rich carbides. This means a segregation of Cr and Mo during solidification.

The analysis of the fracture surface after the Charpy impact test reveals, for each condition, characteristics of brittle and ductile fractures and in some cases the presence of both patterns. Figure 16.7a show sthe surface fracture to sample Mo/Cu and Figure 16.7b shows the surface fracture to sample Mo/Cr. The fractured surface depends on the phases present; that is, samples with higher levels of ferrite tend to have ductile-type fracture characteristics with plastic deformations observed. On the other hand, martensite, perlite and carbide regions tend to present a matrix with a fracture of the type cleavage river patterns.

The individual analysis was made in the samples. The chemical microanalysis of the added elements and the behavior of each was observed. The copper was distributed evenly in the matrix of cast iron samples and did not form compounds rich in these elements; chromium and molybdenum were uniformly distributed in the cast iron matrix and concentrated in specific regions, forming structures that can be observed under microscopy.

16.4 CONCLUSIONS

- The basic chemical elements added to nodular cast iron through the use of scrap steel recycling in the cast iron alloy have no effect on the shape, size and quantity of the graphite nodules;

- The matrix undergoes changes with the addition of chemical elements and, in certain cases, there is a competition between the effects caused by these elements;
- The presence of undesirable chemical elements added to nodular cast iron through the use of scrap steel recycling modifies the alloy matrix and its mechanical properties Chromium and molybdenum form precipitates in the matrix of nodular cast iron;
- Chromium favored the formation of perlite and molybdenum the formation of martensite;
- Therefore, the quality of steel scrap for the production of nodular cast iron components must be controlled.

REFERENCES

Andriollo, Tito, Thorborg, Jesper, Hattel, Jesper. (2015). The influence of the graphite mechanical properties on the constitutive response of a ferritic ductile cast iron – A micromechanical FE analysis. In: *COMPLAS XIII: Proceedings of the XIII International Conference on Computational Plasticity: fundamentals and applications*. (pp. 632–641). CIMNE. http://hdl.handle.net/2117/81388.

Biswas, Siddhartha, & Charles Monroe. (2019). Identifying cast iron microstructure variation using acoustic resonance techniques. *International Journal of Metalcasting*, *13*(1), 26–46. doi:10.1007/s40962-018-0241-4.

Bočkus, Stasys, & Gintautas Žaldarys. (2010). Production of ductile iron castings with different matrix structure. *Materials Science (Medžiagotyra)*, *16*(4), 307–310.

Čanžar, P., Tonković, Z., & Kodvanj, J. (2012). Microstructure influence on fatigue behaviour of nodular cast iron. *Materials Science and Engineering: A*, *556*, 88–99. doi:10.1016/j.msea.2012.06.062.

Glavaš, Zoran. (2012). The Influence of metallic charge on metallurgical quality and properties of ductile iron. *Kovové materiály*, *50*(2), 75–82.

Herrera-Navarro, A., Jimenez-Hernandez, H., Peregrina-Barreto, H., Morales-Hern, L., & Manriquez-Guerrero, F. (2011). A new approach for measuring the distribution of graphite nodules based on singular value decomposition. In *2011 IEEE Electronics, Robotics and Automotive Mechanics Conference* (pp. 450–454). IEEE.

Kopyciński, D., Kawalec, M., Szczęsny, A., Gilewski, R., & Piasny, S. (2013). Analysis of the structure and abrasive wear resistance of white cast iron with precipitates of carbides. *Archives of Metallurgy and Materials*, *58*(3), 973–976. doi:10.2478/amm-2013-0113.

Olawale, J. O., Ibitoye, S. A., Oluwasegun, K. M. (2016). Processing techniques and productions of ductile iron: A review. *International Journal of Scientific & Engineering Research*, *7*(9), 397–423.

Stefanescu, D. M. (2005). Solidification and modeling of cast iron – A short history of the defining moments. *Materials Science and Engineering: A*, *413*, 322–333. doi:10.1016/j.msea.2005.08.180.

Theuwissen, K., Lacaze, J., & Laffont, L. (2016). Structure of graphite precipitates in cast iron. *Carbon*, *96*, 1120–1128. doi:10.1016/j.carbon.2015.10.066.

17 Optimization of Surface Integrity of Recycled Ti–Al Intermetallic-Based Composite on the Machining by Water Jet Cutting via Taguchi and Response Surface Methodology

M. Douiri[1], M. Boujelbene[1], O. Aslan[1], and E. Bayraktar[2]

[1]Isae-Supmeca/Paris School of Mechanical and Manufacturing Engineering, France
[2]Atilim University, Dept of Mechanical Engineering, Ankara-Turkey

CONTENTS

17.1 Introduction	296
17.2 Materials and Methods	297
17.2.1 Materials	297
17.2.2 Methodology	298
17.3 Results and Discussion	298
17.3.1 Visual Evaluation of the Geometric Structure of the Cutting Surface	299
17.3.2 The Effect of Cutting Speed on Surface Quality of Ti–Al	300
17.3.3 The Effect of Mass Flow Rate of the Material on Ti–Al Surface Quality	301

DOI: 10.1201/9781003148760-17

17.3.4 Optimization of the Quality Surface of Parts Machined by the Abrasive Water Jet Process of Ti–Al Composite Material 301
17.4 Study and Optimization of the MRR Material Removal Rate of Ti–Al Composite Material 307
    17.4.1 Effects of Cutting Conditions on Material Flow of Ti–Al Composite Material 307
    17.4.2 Effect of the MRR Ti–Al Abrasive on Material Flow Rate 307
    17.4.3 Recycled Ti–Al Composite Material Removal Rate Optimizations 308
17.5 Conclusion 311
References 313

17.1 INTRODUCTION

Abrasive water jet (AWJ) technology and its applications have been on the market for a long time. With this process, a wide range of materials are shaped for different applications with this process. According to Begic-Hajdarevic et al. (2015) and Löschnera et al. (2016), the demand for more resistant and heat-resistant materials is increasing, especially in the aerospace industries. The abrasive water jet machine is an industrial machine in which we can cut any kind of materials from softer materials and harden materials (Yuvaraj et al., 2016). The cutting principle is based on the use of abrasive grains that will be projected into the material to be cut using a water pressure of the order of a thousand bars, according to Divyansh and Puneet (2015), Andrzej (2016) and Ushasta et al. (2014).

The advantages of abrasive waterjet cutting include the ability to cut almost any material (Ramprasad et al., 2015; Krajcarz, 2014), such as Ti–6Al–4V titanium alloys that have been widely used in industries, especially in aerospace power and medical industries due to their good mechanical and chemical properties (Daymi et al., 2011). However, titanium alloys are generally difficult material to cut. Therefore, after cutting Ti–6Al–4V, the surface quality and surface roughness should be given more attention.

Cutting Ti–6Al–4V titanium alloy material with different parameters pressures, cutting feed rate and material thickness were performed by an abrasive water jet machining (AWJM) process in order to determine its machinability using a different control parameter of the AWJM Douiri process. Bayraktar et al. (2019), Cojbasic et al. (2016), Hreha et al. (2015) and Lehocka et al. (2016). The Abrasive Water Jet Machine (AWJM) removes material by the action of a focused beam of abrasive charged gas. In recent years, many research efforts have been made to understand the abrasive water jet process and improve its cutting performance such as material thickness and surface finish for various materials, according to Liu and Chen (2004), Caydas and Hascalik (2010) and Axinte et al. (2009). They used granite samples for their experimental studies and investigated the effect of process parameters on rock cutting. It has been found that the entrainment of abrasive particles increases the cutting capacity of the water jet and that increasing the pressure of the water jet results in deeper depths of cut.

Miranda (2005) experimentally studied the effect of material properties on cutting performances using calcareous stones.

It was observed that the effect of the feed rate was significant when machining ductile materials such as aluminum, affecting the roughness of the cutting surface, while travel speed affected the kerf characteristics according to Mahabalesh (2007). According to Azmir et al. (2009), the ingredients of the abrasive slurry affect the taper on the cut surface. In addition, the presence of a polymer in the slurry could improve the rate of MRR metal removal. During machining, the rotational speed of the fiber-reinforced plastic jet has been observed to affect the quality of the machined surface (Deam, 2006). Along the kerf wall produced by the AWJC method, an initial damaged region, a smooth cut region and a coarse cut region were visualized (Cosansu and Cogun, 2012). In general, garnet was used as an abrasive in machining operations, but colemanite powder could also be used to achieve better cutting characteristics to allow complete quality control of the process. AWJC and factor interactions require further study, according to Kechagias (2012).

With changing the input parameters, such as water pressure jet feed speed, abrasive mass flow rate and nozzle tilted angle, the material removal rate effect in the abrasive water jet machining process was studied by Chuanzhen et al. (2014). They found with the increasing water pressure and mass flow rate that if cutting the alumina ceramic then increase the MRR.

Jun WANG studied the effect of depth of cut in multi-pass cutting and single pass cutting in the abrasive water jet machining. with four controlled parameters, i.e., water pressure, nozzle traverse speed, nozzle stand-off distance and abrasive flow rate, Jun Wang have cut the alumina ceramic in both cutting forms. He deduced that in single pass cutting, nozzle oscillation cutting creates a scanning cutting action by the particles that not only reduces the particle interference, but also clears the target surface and in the multi-pass cutting the number of passes may be not endlessly increased to increase the total depth of cut for thick materials.

The aim of this study is to analyze the effects of parameters of cuts like the cutting speed and the mass flow rate on the quality surface and the material removal rate. The quality surface and the material removal rate were also measured and expressed as a mathematical model.

17.2 MATERIALS AND METHODS

17.2.1 MATERIALS

The workpiece materials used in all the experiments were taken from a titanium alloy Ti–Al and the chemical composition of the Ti–Al alloy is presented in Table 17.2.

17.2.2 METHODOLOGY

The experiments were carried out on an NC 3015 EB abrasive water jet cutting system capable of providing a maximum water pressure of 415 MPa.

In this experimental study, we use six constant parameters throughout the abrasive water jet (AWJ) process in Table 17.1 for cutting the recycled titanium alloy Ti–Al and its chemical compositions shown in Table 17.2.

For this study, the cutting speed (V) and the mass flow rate (F) to be machined were selected as variable process parameters in Table 17.3.

17.3 RESULTS AND DISCUSSION

The objective of this study was to examine the effects of abrasive water jet variables such as cutting speed V (mm/min) and abrasive flow rate F (g/min) to

TABLE 17.1
Chemical Composition of Recycled Ti–Al-Based Intermetallic-Based Composite

Composition	Ti	Al	Nb$_2$Al	Nb	Yttrium-Doped-Zirconia (Y-ZrO$_2$)	Mo	B	Zn–St
%	53	27	4	15	1	0.1	0.15	1

TABLE 17.2
Constant Technological Parameters

Technical Parameters	Value
Cutting length	4 mm
Water pressure	415 MPa
Type of abrasive	Mineral
Density of the abrasive	0.5 g/l
Nozzle diameter	0.76 mm

TABLE 17.3
Variable Parameters

Cutting speed V (mm/min)	100	200	300
Abrasive grain size F (g/min)	350	450	550

cut on surface roughness S_a(μm). The varied surface roughness through the depth of cut was also measured and expressed as a mathematical model.

17.3.1 VISUAL EVALUATION OF THE GEOMETRIC STRUCTURE OF THE CUTTING SURFACE

The photograph shown in Figure 17.1 represents the cut surfaces obtained with three different displacement speeds. The resulting photographs were analyzed and edited using the image manipulation software shown in Figure 17.1.

Microstructural evaluation of the cut surfaces of the samples revealed three distinct areas identified as follows:

- Zone 1, an area of initial IDR deterioration, which intersects the area at shallow angles of attack.
- Zone 2, smooth cutting zone, SCR, which cuts the zone at large angles of attack.
- Zone 3, a rough-cut zone (RCR), is the upward deflection zone of the jet.

Surface morphology in different regions of the cutting surface is generated from the instantaneous penetration of abrasive water jet. These regions are expected to change with the increasing interaction of the abrasive water jet with the material, increased overlap in any cut region.

The degree of plastic deformation increases from the top to the bottom of the cutting surfaces. The typical microstructures of these regions are shown in Figure 17.1. In Zone 1 of the cutting surface, the abrasive particles have a sufficient level of kinetic energy to destroy the material. This small, damaged region, Zone 1, is characterized by a small, rounded corner at the top edge of the part due to the plastic deformation of the material caused by the initial JEA process bombardment. As the abrasive particles penetrate the material, some of the energy is used to erode the material in Zone 2, a smooth cut zone,

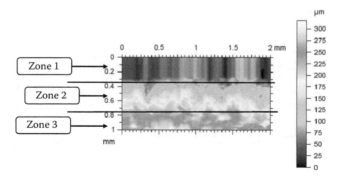

FIGURE 17.1 Topographic profile and total surface roughness, $V = 300$ mm/min, $F = 350$ g/min.

and the flow loses kinetic energy. The degree of deflection increases with transverse speed or cutting speed V. We suggest that the formation of striation results from the reduction in the energy of the jet. According to other studies reported, four main sources contribute to simultaneous striation formation, the nature of the cutting process in stages, the dynamic characteristics of the water jet, the rigidity and vibration of the system machine and material microstructure. This is because at higher cutting speeds, fewer abrasive particles get damaged on the cutting surface due to reduced interaction time between the abrasive jet and the workpiece. At lower speeds, more abrasive particles collide with the cutting surface due to the longer interaction time between the abrasive jet and the workpiece and, therefore, better observed surface finish. Increasing the cutting speed decreases the severe contamination, which was analyzed as an abrasive particle that encrusted in the cut areas of the top of the wall of the cutting surface.

17.3.2 The Effect of Cutting Speed on Surface Quality of Ti–Al

The results show that increasing the cutting speed increases the value of the arithmetic mean deviation of the surface S_a. This results in a shorter machining time but a lower surface quality. Increasing the abrasive flow increases both the depth of cut and the value of S_a, where more abrasive particles give more impact and erosion of the material. This reduces the machining time of the operation. In addition, increasing the abrasive flow rate has no significant effect on the surface finish, where the unit surface will be completely impacted by several abrasive particles, so more particles have no chance to encroach on this surface (Figure 17.2).

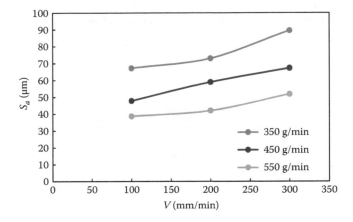

FIGURE 17.2 Effect of the cutting speed V, on the arithmetic mean deviation of the surface S_a when the mass flow rate F is fixed.

17.3.3 THE EFFECT OF MASS FLOW RATE OF THE MATERIAL ON TI–AL SURFACE QUALITY

The cut surface has a better quality in the upper region (entrance area) of the jet. From the middle of the thickness down, there is a degradation of the surface quality. As the depth of penetration of the abrasive water jet increases, the jet loses its energy due to the jet – interaction of materials and mutual impact of particles. This situation results in a characteristic surface quality in the lower region of the cutting surface. Figure 17.3 shows the dependence of average roughness S_a in the upper, middle and lower regions of the cutting surface of different values of feed rate for a material thickness of 6 mm. The results of the determination of the surface roughness in the lower region of the cutting surface relative to the material thickness, the feed rate and the abrasive flow rate are graphically shown in Figure 17.3. One may notice that the arithmetic mean area deviation increases significantly as the feed rate increases. This can be anticipated because increasing the feed rate allows less overlap and machining action and less abrasive particles to touch the surface, deteriorating surface quality. The influence of the abrasive flow is less important on the condition of cutting. The increase in the number of impacting particles contributes to the decrease in the value of S_a.

17.3.4 OPTIMIZATION OF THE QUALITY SURFACE OF PARTS MACHINED BY THE ABRASIVE WATER JET PROCESS OF TI–AL COMPOSITE MATERIAL

An analysis of variance of the surface quality S_a (μm) was carried out to analyze the influence of the cutting speed V (mm/min) and the abrasive flow F (g/min) on the results. The main effects of S/N and averages on the roughness of the S_a

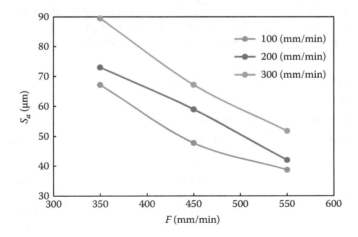

FIGURE 17.3 Effect of mass flow F, on the arithmetic mean deviation of the surface S_a when the cutting speed V is fixed.

TABLE 17.4
The Response Table for MRR's "Smaller Is Better" Signal-to-Noise Ratios

Level	V (mm/min)	F (g/min)
1	−33.96	−37.62
2	−35.41	−35.18
3	−36.62	−33.20
Delta	2.66	4.42
Rang	2	1

surface are shown in Tables 17.4 and 17.5 and Figures 17.4 and 17.5. Table 17.3 is a response table for signal-to-noise ratios (smaller is better).

Figure 17.4 shows the main effects curves for the S/N ratio for S_a against all input factors. Since it is always desirable to minimize S_a value, a better option is selected. From Figure 17.4, the lowest S_a (μm) value is reached at a cutting speed equal to 100 (mm/min) and an abrasive flow rate equal to 550 (g/min).

The S/N ratio was used to determine the optimum parameters for a good surface condition during abrasive water jet cutting of the Ti–Al composite material and according to Figures 17.4 and 17.5, the optimum level of the machining parameters is the level with the highest S/N ratio. Tables 17.4 and 17.5 show the response tables for the means and signal-to-noise ratio for S_a of the Ti–Al composite material. These response tables represent the effects of various input factors on S_a. The steeper the slope in the main effects curve, the higher the corresponding delta values in the response table. The rank directly represents the effect level of the input as a function of the delta values. Depending on the ranks, the effects of various input factors on S_a in the order of its effect are the abrasive flow then the cutting speed. Analysis of variance (ANOVA) is used to evaluate

TABLE 17.5
Response Table for Means

Level	V (mm/min)	F (g/min)
1	51.23	76.57
2	59.87	58.00
3	69.50	46.03
Delta	18.27	30.53
Rang	2	1

Taguchi and Response Surface Methodology 303

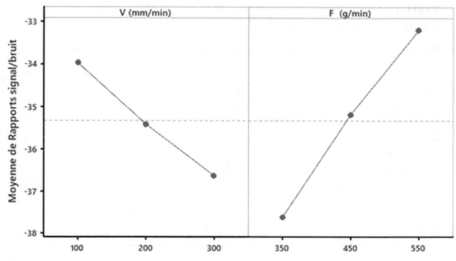

Signal/bruit : Préférer plus petit

FIGURE 17.4 The response table for MRR's "bigger is better" signal-to-noise ratios.

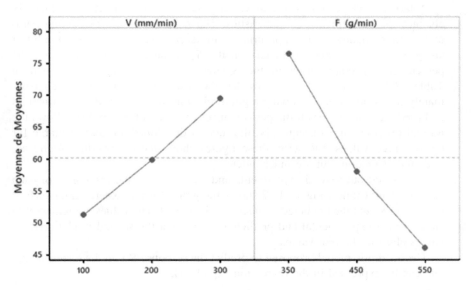

FIGURE 17.5 Principal effects of cutting speed and abrasive grain size on the surface roughness S_a (Ti–Al).

TABLE 17.6
ANOVA Result for S_a (µm)

Source	DL	SomCar Ajust	CM Ajust	F	p
Model	5	1942.37	388.47	47.13	0.005
Linear	2	1898.93	949.47	115.20	0.001
V (mm/min)	1	500.51	500.51	60.72	0.004
F (g/min)	1	1398.43	1398.43	169.67	0.001
Square	2	22.28	11.14	1.35	0.382
V(mm/min) *V(mm/min)	1	0.50	0.50	0.06	0.821
F (g/min) *F (g/min)	1	21.78	21.78	2.64	0.203
Interaction of two factors	1	21.16	21.16	2.57	0.207
V(mm/min) *F (g/min)	1	21.16	21.16	2.57	0.207
Fault	3	24.73	8.24		
Total	8	1967.10			

the developed model, as shown in Table 17.6, ANOVA analysis of variance is a result that allows us to see whether the variables give meaningful information to the model.

A larger F value indicates that varying the process parameter dramatically changes the performance characteristics. The F values of the machining parameters are compared to the appropriate confidence table. According to the F.test analysis, the significant parameters on the S_a is the abrasive mass flow. The percentages of contribution to the S_a machining parameters are shown in Table 17.3 and Figure 17.6 shows that the residuals are distributed approximately in a straight line, showing a good relationship between the experimental and predicted values for all the performances of S_a, and the variable follows the normal distribution. Consequently, the models developed are quite suitable for the observed values. Likewise, these figures show that the residues found are scattered at random but are independent.

We carried out several experiments and we measured the roughness of each sample several times. Figure 17.7 shows the graph for the experimental surface roughness S_a and the predicted value S_a for the model. The values are remarkably close for the experimental and predicted S_a and that the second model predicts values close to the real values.

The quadratic model proposed to predict the response Ra on the experimental one can be expressed in the regression Eq. (17.1).

$$Sa\ (\mu m) = 6.2 + 0.175 * V - 0.404 * F + 0.00005\ V^2 + 0.00033 * F^2 - 0.00023\ V * F \qquad (17.1)$$

Taguchi and Response Surface Methodology

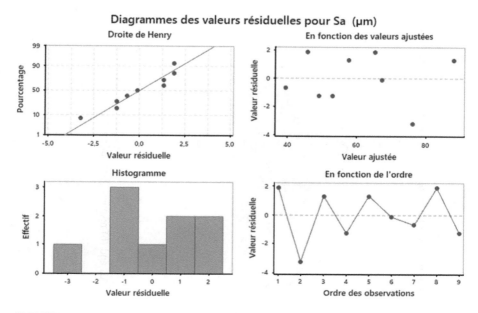

FIGURE 17.6 Residual curves for the surface roughness S_a (µm).

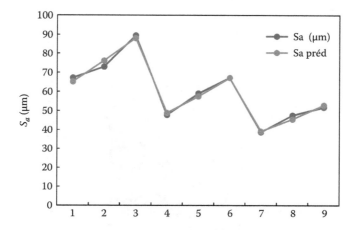

FIGURE 17.7 Comparison between measured and predicted values for S_a (µm).

Figure 17.8a and b show the contour of the surface roughness S_a as a function of the cutting speed V and the abrasive mass flow rate F. This clearly shows that the light green area is the optimum area.

306   Recycling of Plastics, Metals, and Their Composites

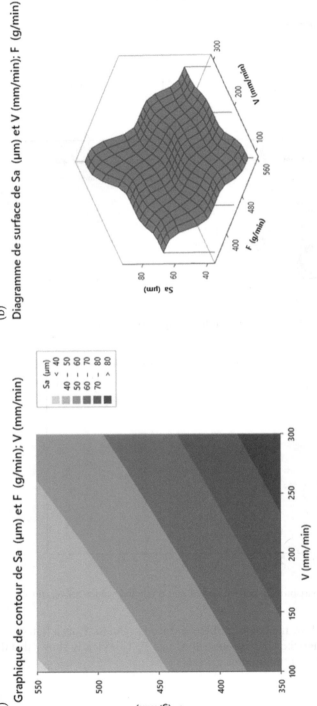

FIGURE 17.8 (a) Contour curves of the surface S_a (μm) as a function of V and F (Ti–Al) and (b) response plots of the S_a (μm).

17.4 STUDY AND OPTIMIZATION OF THE MRR MATERIAL REMOVAL RATE OF TI-AL COMPOSITE MATERIAL

17.4.1 Effects of Cutting Conditions on Material Flow of Ti–Al Composite Material

Figure 17.9 shows the evolution of the material removal rate according to JEAM parameters during the machining of Ti–Al composite material. From the curves of the figures, we notice that the MRR material removal rate increases with a large increase in the cutting speed V (mm/min). Several experiments have been conducted to find the relationship between cutting speed and MRR. During these tests, the displacement speed varied from 200 to 300 mm/min and the tests were repeated for abrasive flow rates of 350, 450 and 550 g/min.

Figure 17.10 shows the test results and their trend lines. It shows that the MRR increases with increasing cutting speed. The trend is for a polynomial function with a high R2 regression rate.

On the other hand, the MRR experiences a slight increase with the evolution of abrasive flow.

17.4.2 Effect of the MRR Ti–Al Abrasive on Material Flow Rate

Several experiments have been conducted to find the relationship between the rate of abrasive flux and MRR. During these tests, the abrasive flow varied from 350, 450 and 550 g/min and the tests were repeated for three cutting speeds. Figure 17.10 shows the test results with their trend lines. It shows that the MRR increases with increasing abrasive flow. The trend is for a polynomial function with a high regression ratio R2.

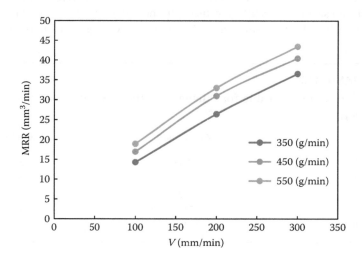

FIGURE 17.9 Effect of cutting speed on MRR (mm^3/min) material flow.

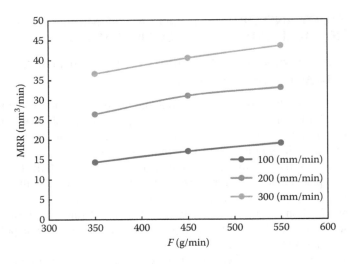

FIGURE 17.10 Effect of cutting speed on MRR material flow (mm³/min).

17.4.3 RECYCLED TI–AL COMPOSITE MATERIAL REMOVAL RATE OPTIMIZATIONS

Tables 17.7 and 17.8 show the MRR responses for the Ti–Al composite, for the averages and signal to noise ratio for the MRR of the recycled titanium composite material (a Ti–Al alloy). This response table represents the effects of various input factors on the MRR. The steeper the slope in the main effects curve, the higher the corresponding delta values in the response table. The rank directly represents the effect level of the input as a function of the delta values.

Tables 17.7 and 17.8 show the signal-to-noise ratio (SNRA) and predicted signal-to-noise ratio (PSNRA) values for the MRR of the recycled titanium aluminium based composite material. The predicted signal to noise values is

TABLE 17.7
The Response Table for "Bigger Is Better" Signal-to-Noise Ratios

Niveau	V (mm/min)	F (g/min)
1	24.43	27.60
2	29.54	28.86
3	32.06	29.57
Delta	7.63	1.97
Rang	1	2

TABLE 17.8
The Response Table for the Mean Values of MRR

Niveau	V (mm/min)	F (g/min)
1	16.77	25.77
2	30.13	29.50
3	40.20	31.83
Delta	23.43	6.07
Rang	1	2

awfully close to the calculated signal-to-noise values, hence the Taguchi analysis for signal-to-noise ratio is correct and perfect. The representation of the effects of various parameters on the MRR and the optimization of the condition is remarkably close.

Figures 17.11 and 17.12 show the influence of each cutting parameter: cutting speed and bulk abrasive grain flow rate on material removal flow rate. However, according to the effect curves, the cutting speed has the greatest influence on the MRR values and the V3 value = 300 mm/min corresponds to the greatest MRR value compared to V1 and V2 and the massive abrasive grain flow F3 = 550 g/min corresponds to the greatest MRR value compared to F2 and F1.

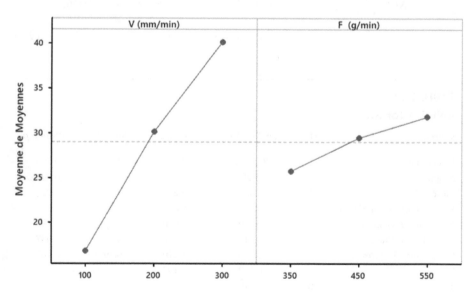

FIGURE 17.11 The effect of machining parameters on MRR.

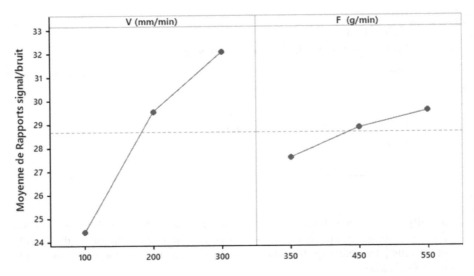

FIGURE 17.12 Effects of S/N ratios for MRR.

In Figure 17.12 the S/N ratio was used to determine the optimum parameters for a high MRR value during abrasive water jet cutting of recycled titanium-based composite material Ti–Al Table 17.9.

Eq. 17.2 show the regression models for MRR:

$$\text{MRR}\left(\frac{mm^3}{min}\right) = -22.31 + 0,1584 \times V + 0.082 \times F + 0.000165 \times V^2 + 0,0007 \times F^2 \qquad (17.2)$$

TABLE 17.9
ANOVA for MRR

Source	DL	SomCar Ajust	CM Ajust	Value F	Value de p
V (mm/min)	1	823.682	823.682	4285.05	0.000
F (g/min)	1	55.207	55.207	287.20	0.000
Carré	2	6.425	3.212	16.71	0.024
V (mm/min) *V (mm/min)	1	5.445	5.445	28.33	0.013
F (g/min) *F (g/min)	1	0.980	0.980	5.10	0.109
Interaction à 2 facteurs	1	1.210	1.210	6.29	0.087
V (mm/min) *F (g/min)	1	1.210	1.210	6.29	0.087
Erreur	3	0.577	0.192		
Total	8	887.100			

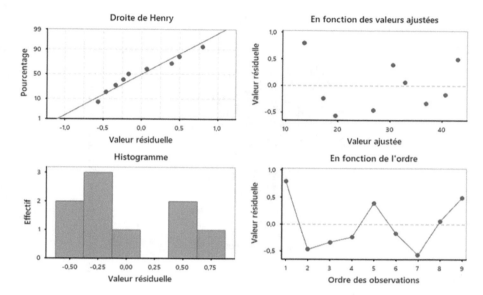

FIGURE 17.13 Residual curves for MRR.

Figure 17.13 shows that the residuals are distributed approximately in a straight line, showing a good relationship between experimental and predicted values for MRR performance, and the variable follows the normal distribution. Therefore, the model developed (Eq. 17.2) is quite suitable for the observed values. Likewise, these figures show that the residues found are scattered at random but are independent.

Figure 17.14a and b show two-dimensional and three-dimensional contours that show the effect of the tension feed rate set on the MRR removal rate. As it is clear, higher values of MRR can be obtained by selecting a high cutting speed.

The RSM model is performed with experimental tests and on each sample, we have assigned three tests of MRR value. For each combination of the input factors, the prediction value of the response Y_j, pred, is compared to the experimental value of the response Y_j, exp. We conclude that the predictive results with RSM are very close to the experimental results of MRR for the composite material Ti–Al.

17.5 CONCLUSION

In this article, we investigated the effect of machining parameters such as cutting speed and abrasive grain mass of the material on the surface quality of the Ti–Al composite material part. The experimental results are analyzed using the statistical ANOVA method, which was a reliable methodology to reduce the

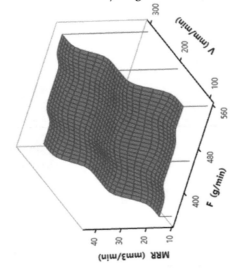
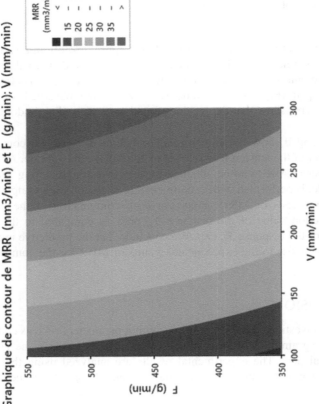

FIGURE 17.14 (a) Contour curves of the surface S_a (μm) as a function of V and F (Ti–Al) and (b) response plots of the S_a (μm).

manufacturing time and cost in the abrasive water jet machining or cutting process.

The main conclusions of this experimental work are as follows:

- Increasing the cutting speed always has the same effect to increase the surface S_a value.
- If the abrasive mass flow of material F decreases, the result is a good quality of the cutting surface obtained by the abrasive water jet cutting process.
- The developed model of surface quality can be used during abrasive water jet cutting to choose the best values of the cutting parameters in order to achieve the optimum surface quality of the cut part. This will lead to a reduction in total machining time and costs.
- Increasing the cutting speed has the same effect as increasing the abrasive mass flow to increase the surface MRR value.
- The developed model of MRR can be used during abrasive water jet cutting to choose the best values of the cutting parameters to achieve the optimum surface quality of the cutting part. This will lead to a reduction in total machining time and costs.

REFERENCES

Akkurt, A., Kulekci, M., Seker, U., Ercan, F. (2004). Effect of feed rate on surface roughness in abrasive water jet cutting applications. *Journal of Materials Processing Technology, 147*, 389–396.

Andrzej, P. (2016). Abrasive suspension water jet cutting optimization using orthogonal array design. *Procedia Engineering, 149*, 366–373.

Axinte, D. A., Srinivasu, D. S., Kong, M. (2009). Abrasive water jet cutting of polycrystalline diamond; A preliminary investigation. *International Journal of Machine Tools & Manufacture, 49*, 797–803.

Azmir, A., Ahsan, K., & Rahmah, A. (2009). Effect of abrasive water jet machining parameters on aramid fibre reinforced plastics composite. *The International Journal of Material Forming, 2*(1), 37–44.

Begic-Hajdarevic, D., Cekic, A., Mehmedovic, M., & Djelmic, A. (2015). Experimental study on surface roughness in abrasive water jet cutting. *Procedia Engineering, 100*, 394–399.

Caydas, U., & Hascalik, A. (2010). Effect of traverse speed on abrasive water jet machining of age hardened Inconel 718 nickel-based super alloy. *Material Manufacturing Process, 25*, 1160–1165.

Cojbasic, Z., Petkovic, D., & Shamshirband, S. (2016). Surface roughness prediction by extreme learning machine constructed with abrasive water jet. *Precision Engineering, 43*, 86–92.

Cogun, C. (2012). An investigation on use of colemanite powder as abrasive in abrasive water jet cutting (AWJC). *Journal Mechanical Science Technology, 26*(8), 2371–2380.

Daymi, A., Boujelbene, M., Bayraktar, E., Salem, B. A., & Katundi, D. (2011). Influence of feed rate on surface integrity of titanium alloy in high-speed milling. *Advanced Materials Research, 264*, 1228–1233.

Deam. (2006). A correlation for predicting the kerf profile from abrasive water jet cutting. *Experimental Thermal and Fluid Science, 30*(4), 337–343.

Divyansh, P., & Puneet, T. (2015). Experimental investigations of thermally enhanced abrasive water jet machining of hard-to-machine metals. *CIRP Journal of Manufacturing Science and Technology, 10*, 92–101.

Douiri, M., Boujelbene, M., Bayraktar, E., & Salem, B. (2019). Process reliability of abrasive water jet to cut shapes of the titanium alloy Ti-6Al-4V. *Mechanics of Composite, Hybrid and Multifunctional Materials, 5*, 229–236.

Douiri, M., Boujelbene, M., Bayraktar, E., & Salem, B. (2019). A study of the surface integrity of titanium alloy Ti-6Al-4V in the abrasive water jet machining process, mechanics of composite. *Hybrid and Multifunctional Materials, 5*, 221–228.

Hreha, P., Radvanska, A., Hloch, S., Perzel, V., Krolczyk, G., & Monkova, K. (2015). Determination of vibration frequency depending on abrasive mass flow rate during abrasive water jet cutting. *The International Journal of Advanced Manufacturing Technology, 77*(1-4), 763–774.

Kechagias, J. D., Petropoulos, G., & Vaxevanidis, N. M. (2012). Application of Taguchi design for quality characterization of abrasive water jet machining of TRIP sheet steels. *International Journal of Advanced Manufacturing Technology, 62*, 635–643.

Krajcarz, D. (2014). Comparison metal water jet cutting with laser and plasma cutting. *Procedia Engineering, 69*, 838–843.

Lehocka, D., Klich, J., Foldyna, J., Hloch, S., Krolczyk, J. B., Carach, & Krolczyk, G. M. (2016). Copper alloys disintegration using pulsating water jet. *Measurement, 82*, 375–383.

Liu, Y. & Chen, X. (2004). A study on the abrasive water jet cutting for granite. *Key Engineering Materials, 257–258*, 527–532.

Löschnera, P., Jarosza, K., & Niesłonya, P. (2016). Investigation of the effect of cutting speed on surface quality in abrasive water jet cutting of 316L stainless steel. *Procedia Engineering, 149*, 276–282.

Mahabalesh, P. (2007). A study of taper angles and material removal rates of drilled holes in the abrasive water jet machining process. *Journal of Materials Processing Technology, 189*(1–3), 292–295.

Miranda, Quintino. (2005). Microstructural study of material removal mechanisms observed in abrasive water jet cutting of calcareous stones. *Materials Characterization, 54*, 54370–54377.

Ramprasad, Upadhyay, Hassan. (2015). Optimization MRR of stainless steel 403 in abrasive water jet machining using anova and taguchi method. *International Journal of Engineering Research and Applications, 5*(52), 86–89.

Ushasta, A., Banerjeea, S., Bandyopadhyaya, A., & Probal, K. (2014). Abrasive Water Jet Cutting of Borosilicate Glass. Procedia Materials Science, 6, 775–785.

Yuvara, N., & Pradeep, M. (2016). Cutting of aluminium alloy with abrasive water jet and cryogenic assisted abrasive water jet: A comparative study of the surface integrity approach. *Wear, 362–363*, 18–32.

Zhongbo, Y. (2014). Optimization of machining parameters in the abrasive water jet turning of alumina ceramic based on the response surface methodology. *The International Journal of Advanced Manufacturing Technology.* 10.1007/s00170-014-5624-y

18 Wear Behavior Analysis of a AlMg1SiCu Matrix Syntactic Foam Reinforced with Boron Carbide Particles and Recycled Fly Ash Balloons

J. P. Paschoal[1], J. J. Thottathil[2], E. Daniel[3], R. C. Moraes[1], F. Gatamorta[1], E. Bayraktar[1,5], and T. V. Christy[4]

[1]University of Campinas, UNICAMP/FEM Department of Materials, Campinas, São Paulo, Brazil
[2]Amaljyothi College of Engineering, Kanjirapally, Kerala, India
[3]Karunya Institute of Technology, Coimbatore, Tamil Nadu, India
[4]PRIST University, Vallam, Tamil Nadu, India
[5]Isae-Supmeca/Paris, Mechanical and Manufacturing Engineering School, France

CONTENTS

18.1 Introduction 316
18.2 Materials and Methods 317
    18.2.1 Materials 317
    18.2.2 Methodology 317
18.3 Results and Discussion 318
18.4 Conclusion 322
References 323

DOI: 10.1201/9781003148760-18

18.1 INTRODUCTION

As a subclass of cellular metals, metallic syntactic foams are characterized by the presence of gaseous pores imbibed into a metallic matrix, guaranteed by hollow space holders (Moraes et al., 2021).

Hollow particles are introduced in the syntactic metallic foam because, in addition to decreasing the density of the final composite, they can increase the composite's capacity to absorb energy, combined with the strength of the matrix (Su et al., 2019).

Therefore, these metal matrix composites (MMCs) own particular features that turn them into interesting gammas of industrial applications, such as for energy absorbers, lightweight structures, vibration dampers, thermal and acoustic insulation, and others (Paschoal et al., 2021).

Among the various manufacturing processes available for MMC production, the stir casting process is generally accepted as a promising route. Its advantages are in its simplicity, flexibility and applicability for large-scale production. It is also attractive for allowing the use of conventional metal processing routes, which minimizes production costs. This technique is the most economical of the techniques available for the production of MMC, and allows the manufacture of large components. Therefore, to carry out this process, a high shear impeller is used to stir the molten metal inside the crucible, while the ceramic particles are slowly added to form the molten composite (Rohatgi et al., 2011).

In the stir casting process for the production of metal foam, there are factors that need considerable attention, namely, the difficulty of achieving a uniform distribution of the reinforcement materials, the wettability of the molten alloy on the surface of the solid particles and the porosity of the metallic composite while molten. Such characteristics will strongly influence the properties of the final composite.

In order to improve wear resistance, ceramic particles such as Al_2O_3 and SiC are the most used as reinforcements for aluminum matrices. As the third hardest material, behind only diamond and boron nitride, boron carbide (B_4C) can be used as a reinforcement alternative due to its high hardness (HV = 30 GPa), low density (2.52 g/cm^3), good resistance to abrasion and chemical stability.

Besides, it is noticed from the literature that fly ash balloons debris roll over the mating surface and help in reducing the coefficient of friction and wear rate, and the craters generated from balloons cracking in lubricating conditions act as reservoirs of lubricant, keeping the surface lubricated (Jha et al., 2011).

Jha et al. (2011) compared the wear behavior of a aluminum matrix syntactic foam (AMSF) containing 35 vol.% of fly ash and a Al–10 wt.% Si composite, and the results have shown that at dry conditions (2 m/s velocity), AMSF coefficient of friction behavior was compared to lubricated condition of the Al–10 wt.% Si composite. Attar et al. (2015) studied the influence of B4C addition (3–6 wt.%) on a aluminum AA7075 alloy on the wear rate, concluding that its addition decreased significantly the composite wear rate.

Wear Behavior Analysis

Therefore, at this work, boron carbide particles and recycled hollow particles known as fly ash balloons were added into a molten AlMg1SiCu alloy and processed by the stir casting technique, aiming for the production of a wear-resistant syntactic foam.

18.2 MATERIALS AND METHODS

18.2.1 Materials

In this study, an Aluminum Matrix Syntactic Foam (AMSF) composed of an AlMg1SiCu matrix reinforced with fly ash balloons (Figure 18.1), with an average diameter of 37 µm and wall thickness of 1.5 µm and boron carbide particles (Figure 18.2) was produced by the stir casting process. The alloy's nominal composition is presented in Table 18.1. The balloons are mostly composed of SiO_2, CaO and Al_2O_3.

18.2.2 Methodology

For producing the AMSF by the stir casting method, the following experiment was carried out: 1,000 g of AlMg1SiCu alloy was heated to 610°C, and stirred with a high shear impeller of four 90° blades for 10 minutes, with an average rotation of 650 rpm. During the process, 10 wt.% of fly ash balloons and 7.5 wt.% of B_4C were slowly added into the crucible. It is important to notice that the fly ash balloons had been preheated and maintained at 300°C before the process for preventing cracking of the balloons' walls by heat shock and increased wettability.

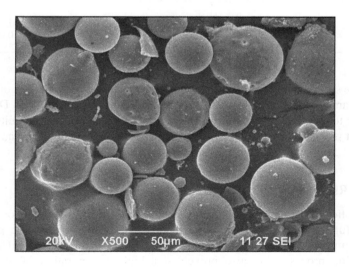

FIGURE 18.1 SEM of fly ash recycled balloons showing particles' diameter measures.

FIGURE 18.2 SEM of boron carbide particle at the AMSF with length of 5.425 μm.

TABLE 18.1
Chemical Composition of AlMg1SiCu Alloy

Elements	Mg	Si	Fe	Mn	Cu	Cr	Zn	Ni	Ti	Al
wt.%	0.95	0.54	0.22	0.13	0.17	0.09	0.08	0.02	0.01	Bal

Three specimens in the format of pins were machined with a length of 50 mm and a diameter of 10 mm for being used at the wear test named Pin on Disc, and in order to guarantee the parallelism of surfaces, the pins were submitted to a laser cut process. The average speed of the process was 2.5 m/s with an applied load of 1.5 kg and a distance covered of 4,000 m.

18.3 RESULTS AND DISCUSSION

After optical micrography analysis (Figure 18.3), it was possible to confirm the successful incorporation and homogeneous distribution of ceramic particles into the matrix. This can be attributed to the alloy's semisolid state during the production process, that also prevented a high incorporation of air during

Wear Behavior Analysis 319

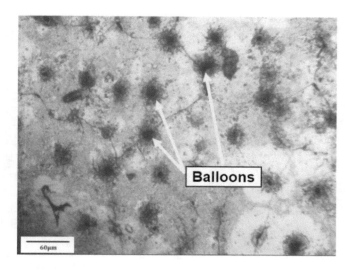

FIGURE 18.3 AMSF's optical micrograph, showing fly ash balloons and B$_4$C-rich regions in the vicinity of balloons and in grain boundaries.

the stir casting technique, avoiding the formation of macro imperfections, gaseous voids and/or cracks. These features are of high importance to the mechanical and wear resistance of the composite.

EDS was performed on the fabricated specimens (Figure 18.4), confirming the presence of elements from reinforcements into the foam. It is also understood that the reinforcement particles did not react with the aluminum matrix producing any other compound. Both B$_4$C and fly ash balloons are found to be thermodynamically stable at the casting temperature considered in the work.

In the first image from Figure 18.4 in the central region of the scanning line, a darker region is transposed, which is believed to be part of a balloon's wall with a portion of aluminum matrix inside. A peak of Si and a valley of Al between 60 μm and 70 μm point this affirmation. Besides, at proximately 45 μm, a peak of B can be observed, suggesting that darker regions surrounding fly ash balloons at Figure 18.3 comprehend higher concentrations of boron carbide particles.

Figure 18.5 represents the variation of the wear rate and the coefficient of friction, both as a function of the sliding distance. In the beginning of the experiment, the coefficient of friction increases due to the stabilizing and accommodation of the specimen surface to the countersurface. After the wear of the first contact layer of AMSF, debris from the exposed balloons' walls begin to crack into very fine particles that roll over the mating surface and favor a decrease of the coefficient of friction.

In contrast and simultaneous to the event described previously, boron carbide particles are more resistant to wear than the aluminum matrix and fly ash

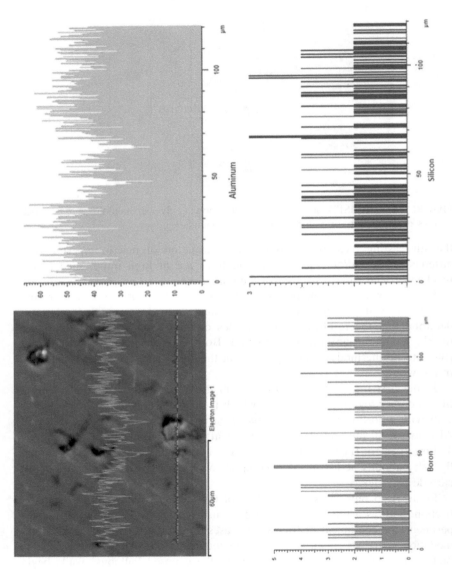

FIGURE 18.4 EDS line scanning profile of relative concentrations of Al, B and Si.

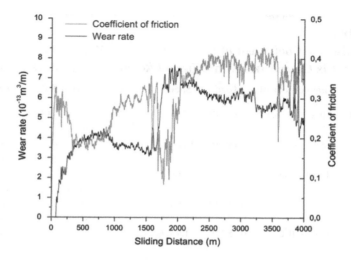

FIGURE 18.5 Wear rate/coefficient of friction versus sliding distance of a specimen from the AMSF.

balloons, so while they are eroded, the B_4C particles start protruding, scratching the counterpart and resulting in an increase in the friction coefficient and a decrease in the wear rate.

As these hard particles are pulled out of the contact surface due to the increase of the friction force, the lower presence of protuberances decreases the resistance to sliding of the pin on the disk surface, but also exposes more ductile regions, thus increasing its wear rate.

Curves from Figure 18.5 show a trend to behave in opposition to each other, in the manner that an increase of the coefficient of friction causes a decrease of the wear rate, being the opposite true. Trends of this behavior could be observed almost in a cyclic way, as for example between 500 and 1,000 meters, or between 1,500 and 2,000 meters.

From Table 18.2, it is possible to comprehend the influence of ceramic particles into an aluminum matrix on the wear rate. For example, Attar et al. (2015) comprised that by adding 6 wt.% B4C into an Al7075 matrix the wear rate decreased. Besides that, Májlinger et al. (2016) also obtained a decrease in the wear rate on its studied syntactic foam with SLC ceramic hollow spheres compared to the alloy AlSi12. Jha et al. (2011) went further and compared two LM13 matrix composites (one with 10 wt.% SiC, another with 35 vol.% fly ash), concluding that the fly ash addition guaranteed a decrease in both coefficients of friction and wear rate.

Through these findings, and the data obtained in the current work, it is assumed that the low wear rate measured in the studied syntactic foam was influenced both by the lubricating effect of the fragments of the broken walls of fly ash, and by the high hardness of the boron carbide particles.

TABLE 18.2
Comparison of Coefficient of Friction (CoF) and Wear Rate of Different Metal Matrix Composites with Different Reinforcements in Dry Condition

Alloy	Reinforcement	Load	Speed	CoF	Wear Rate	Reference
AlMg1SiCu	10 wt.% fly ash + 7.5 wt.% B$_4$C	0.19 MPa	2.5 m/s	0.08–0.45	1 – 8.5 e–13 (m^3/m)	Current work
LM13	35 vol.% fly ash	0.15 MPa	2 m/s	~0.05	~22 e–13 (m^3/m)	(Jha et al., 2011)
LM13	10 wt.% SiC	0.15 MPa	2 m/s	~0.8	~25 e–13 (m^3/m)	(Jha et al., 2011)
Al7075	–	0.15 MPa	~1.9 m/s	–	~3.0 e–6 (cm^3/mm)	(Attar et al., 2015)
Al7075	6 wt.% B$_4$C	0.15 MPa	~1.9 m/s	–	~1.5 e-6 (cm^3/mm)	(Attar et al., 2015)
AlSi12	–	0.64 MPa	0.2 m/s	0.4	~100e–13 (m^3/m)	(Májlinger et al., 2016)
AlSi12	SLG (commercial ceramic hollow sphere)	0.64 MPa	0.2 m/s	0.5	~70e–13 (m^3/m)	(Májlinger et al., 2016)

18.4 CONCLUSION

The current aluminum matrix syntactic foam, reinforced with fly ash balloons and boron carbide particles studied in this work demonstrated a superior wear resistance, and a low coefficient of friction due to the mechanisms of solid lubrication by fly ash wall's debris rollover, and also by the high hardness of B4C particles. Therefore, it is suggested that recycled and low-cost fly ash balloons have an important potential concerning the fabrications of metal matrix composites for lightweight structural and wear-resistant parts.

Although further investigation must be carried out, through this work it was possible to notice that aluminum matrix syntactic foam reinforced with boron carbide particles produced via the stir casting process, from a semisolid alloy, is promising for applications where wear-resistant materials are needed.

Finally, it is suggested that for future work, analyses of the worn surfaces, as well as of the fragmented particles of that surface, should be made, thus guaranteeing a better understanding of the mechanisms involved in the wear of the syntactic aluminum foam reinforced with ceramic particles.

REFERENCES

Aisyah, H. A., Paridah, M. T., Sapuan, S. M., Ilyas, R. A., Khalina, A., Nurazzi, N. M., Lee, S. H., & Lee, C. H. (2021). A comprehensive review on advanced sustainable woven natural fibre polymer composites. *Polymers*, *13*(3), 471. doi: 10.3390/polym13030471

Alsubari, S., Zuhri, M. Y. M., Sapuan, S. M., Ishak, M. R., Ilyas, R. A., & Asyraf, M. R. M. (2021). Potential of natural fiber reinforced polymer composites in sandwich structures: A review on its mechanical properties. *Polymers*, *13*(3) 423. doi: 10.3390/polym13030423

Asyraf, M. R. M., Rafidah, M., Ishak, M. R., Sapuan, S. M., Yidris, N., Ilyas, R. A., & Razman, M. R. (2020). Integration of TRIZ, morphological chart and ANP method for development of FRP composite portable fire extinguisher. *Polymers Composites*, *41*(7), 2917–2932. doi: 10.1002/pc.25587

Attar, S., Nagaral, M., Reddappa, H. N., Aurad, V. (2015). Effect of B4C particulates addition on wear properties of Al7025 alloy composites. *American Journal of Materials Science*, *5*(3C), 53–57. doi: 10.5923/c.materials.201502.11

Jha, N., Badkul, A., Mondal, D. P., Das, S., & Singh, M. (2011). Sliding wear behaviour of aluminum syntactic foam: A comparison with Al–10 wt.% SiC composites. *Tribology International*, *44*(3), 220–231. doi: 10.1016/j.triboint.2010.10.004

Jose, J., Christy, T. V., Eby Peter, P., John Feby, A., & Benjie, N. M. (2018). Manufacture and characterization of a novel agro-waste based low-cost metal matrix composite (MMC) by compo-casting. *Materials Research Express*. doi: 10.1088/2053-1591/aac803

Májlinger, K., Bozóki, B., Kalácska, G., Keresztes, R., & Zsidai, L. (2016), Tribological properties of hybrid aluminum matrix syntactic foams. *Tribology International*, *99*, 211–223. doi: 10.1016/J.TRIBOINT.2016.03.

Mondal, D. P., Das, S., Jha, N. (2009). Dry sliding wear behaviour of aluminum syntactic foam. *Materials & Design*, *30*, 2563–2568. doi: 10.1016/j.matdes.2008.09.034

Moraes, R. C., Paschoal, J. P., Bayraktar, E., Silva, R., Costa, R., & Gatamorta, F. (2021). Study of a semisolid processing route for producing an AlSiMg0.5Mn matrix syntactic foam via thixoinfiltration of fly ash micro balloons. In *Mechanics of Composite, Hybrid, and Multifunctional Materials*, Volume 6 *of the Proceedings of the 2020 SEM Annual Conference & Exposition on Experimental and Applied Mechanics*. doi: 10.1007/978-3-030-59868-6-16.

Nurazzi, N. M., Asyraf, M. R. M., Khalina, A., Abdullah, N., Aisyah, H. A., Rafiqah, S. A., Sabaruddin, F. A., Kamarudin, S. H., Norrrahim, M. N. F., Ilyas, R. A., & Sapuan, S. M. (2021). A review on natural fiber reinforced polymer composite for bullet proof and ballistic applications. *Polymers*, *13*(4), 646. doi: 10.3390/polym13040646

Omran, Abdoulhdi A.B., Mohammed Abdulrahman A. B. A., Sapuan, S. M., Ilyas, R. A., Asyraf, M. R. M., Rahimian Koloor, Seyed S., & Petrů, Michal. (2021). Micro- and nanocellulose in polymer composite materials: A review. *Polymers*, *13*(2), 231. doi: 10.3390/polym13020231

Paschoal, J. P., Moraes, R. C., Bayraktar, E., Sartori, J. M., Silva, R., Gatamorta, F. (2021). Compressive behavior characterization of a AlSiMg0.5Mn matrix syntactic foam produced via thixoinfiltration of fly ash micro balloons. In *Mechanics of Composite, Hybrid, and Multifunctional Materials*, Volume 6 *of the Proceedings of the 2020 SEM Annual Conference & Exposition on Experimental and Applied Mechanics*. doi: 10.1007/978-3-030-59868-6-4.

Ramachandra, M., & Radhakrishna, K. (2007). Effect of reinforcement of flyash on sliding wear, slurry erosive wear and corrosive behavior of aluminium matrixcomposite. *Wear*, *262*, 1450–1462. doi: 10.1016/j.wear.2007.01.026

Rohatgi, P. K., Gupta, N., Gupta, B. F., & Luong, D. D. (2011). The synthesis compressive properties and applications of metal matrix syntactic foams. *JOM, 63*, 36–42. doi:10.1007/s11837-011-0026-1

Sabaruddin, Fatimah A., Paridah, M. T., Sapuan, S. M., Ilyas, R. A., Lee Seng H., Abdan, Khalina, Mazlan, Norkhairunnisa, Roseley, Adlin S.M., & Abdul Khalil, H. P. S. (2021). The effects of unbleached and bleached nanocellulose on the thermal and flammability of polypropylene-reinforced kenaf core hybrid polymer bionanocomposites. *Polymers, 13*(1), 116. doi:10.3390/polym13010116

Su, M., Wang, H., Hao, H., & Fiedler, T. (2019). Compressive properties of expanded glass and alumina hollow spheres hybrid reinforced aluminum matrix syntactic foams. *Journal of Alloys and Compounds.* doi:10.1016/J.JALLCOM.2019.153233

19 Procedures for Additions of Wastes to Cementitious Composites – A Review

M.A. de B. Martins[1], F.B. Pinto[2], D. Werdine[1], L. Ramon[1], C.V. Santos[3], P.C. Gonçalves[3], M.L.M. Melo[4], and R.M. Barros[3]

[1]Physical and Chemical Institute, Federal University of Itajubá, Itajubá, Minas Gerais, Brazil
[2]Mechanical Engineering Institute, Federal University of Itajubá, Itajubá, Minas Gerais, Brazil
[3]Natural Resources Institute, Federal University of Itajubá, Itajubá, Minas Gerais, Brazil
[4]Mechanical Engineering Institute, Federal University of Itajubá, Itajubá, Minas Gerais, Brazil

CONTENTS

19.1 Introduction 326
    19.1.1 Cementitious Composite: Components and Characteristics .. 328
    19.1.2 Different Types of Additions and Waste Materials 329
19.2 Methodology 329
    19.2.1 Pre-Analyses for Waste Material Substitution or Addition in Cementitious Composites 329
    19.2.2 Influence of Aggregates and Waste Additions on Cement Composite Properties 332
    19.2.3 Applicable Trials and Tests for Cement Composites Components 334
        19.2.3.1 Standards and Techniques for Cement Composite Constituents Characterization 334
    19.2.4 Mixture Design Methods 335
        19.2.4.1 Empirical Method 338
        19.2.4.2 Compressive Strength Method 338
        19.2.4.3 Compressible Packing Method (CPM) 338
        19.2.4.4 Factorial Model Method 339

    19.2.4.5 Paste Rheology Method 339
  19.2.5 Cementitious Composite Rheology 340
    19.2.5.1 Tests for Fresh Scc Properties 340
    19.2.5.2 Common Problems in Fresh SCC......................... 343
  19.2.6 Hardened Concrete Properties 344
    19.2.6.1 Main Hardened SCC Properties............................ 344
19.3 Final Remarks.. 345
19.4 Conclusions.. 346
Acknowledgment 347
References.. 348

19.1 INTRODUCTION

Cement-based materials, or cementitious materials, such as concretes, mortars and cement pastes, can be well defined as composites (Brandt, 2009). According to the American Society for Testing and Materials, concrete can be defined as a composite material that consists of an agglomerating medium in which particles of different natures such as Portland cement, water, aggregates and, optionally, mineral additions and admixtures are agglutinated (ASTM, 2000). In order to minimize the use of natural resources, the incorporation of wastes in cementitious composites consists of the use of products from other sectors and the consequent reduction of environmental impacts. However, the substitution of natural aggregates for residues influences the properties of cementitious composites both in the fresh and in the hardened state.

  Concrete is the most used construction material in the world due to its versatility and the fact that its raw material is easily found and inexpensive (Güneyisi et al., 2015).

  Self-compacting concrete (SCC) was developed in Japan in 1988 by Okamura, stemming from the need to produce more durable concrete with greater strength and workability for applications in high-density structures needing reinforcement that are also difficult to access (Okamura & Ouchi, 2003). It is a highly fluid concrete, which can fill forms, compacting itself with its own weight without vibration, remaining homogeneous and cohesive, while also being able to pass between obstacles like reinforced bars with resistance to segregation during mixture, transportation and application (Ashish & Verma, 2019). Due to these properties, the use of SCC is advantageous for civil construction.

  The main advantages of SCC are the reduction of workforce required for concrete in the precast and construction industries, along with execution time due to the absence of vibration and consequently the final cost reduction and improvement in conditions due to noise reduction. In addition, it provides increased durability and compressive strength. The disadvantages are the rapid loss of workability and greater sensitivity to temperature and climate variation (Chen & Yang, 2015).

  The fact that cement consumes non-renewable natural materials and raises the emissions of gases in the atmosphere in its production makes it an environmental problem that requires research to optimize its production (Alsubari et al., 2016).

Cement manufacturing processes contribute to 7% of the global CO_2 emissions, which are the main cause of the greenhouse effect (SNIC, Sindicato Nacional da Industria do Cimento, 2019). In addition, concrete consumes many natural resources, such as sand and gravel, that make up about 80% of the mixture. It appears in the report of the National Association of Entities of Producers of Aggregates for Construction, ANEPAC, that in 2019, the production of aggregates in Brazil was 514 million tons (ANEPAC, 2020).

On the other hand, concrete is capable of absorbing, as a constituent material, a series of waste and by-products from industry and agriculture activities, directly collaborating to the recovery and minimization of the environmental impact of other segments of production and, indirectly, providing a reduction of pollution generated in its own productive process (Vishwakarma & Ramachandran, 2018).

Self-compacting concrete's ability to integrate industrial waste in its composition has been considered one of the great advances of civil construction in the last century (Sabir et al., 2001). The use of the wastes can be in partial replacement of the cement; the fine aggregates and/or the coarse aggregates as well as addition of fine residues are used to improve concrete viscosity. The substitutions depend on the characteristics of the waste, such as their origin and granular size. According to Gomes and Barros (2009), SCC is sustainable as much as in reducing the consumption of nonrenewable natural resources and in the possibility of using industrial by-products that harm the environment.

Environmental research on recycling and reuse of waste has gained a great level of importance in recent studies, as modern lifestyle and technological advancements have increased the amount and type of waste generated by industries and individuals, leading to a disposal crisis (Batayneh et al., 2007). These waste products can be reused by the construction industry, given that the large consumption of aggregates and reduction of natural resources has led to an increased interest in sustainable construction, the use of correct materials and more environmentally friendly concreting practices. Its reuse in concrete preserves the natural aggregates, reduces impacts on landfills, saves energy and can favor cost-effectiveness (Hebhoub et al., 2011; Kibriya & Tahir, 2017; Mondal et al., 2018).

According to Güneyisi et al. (2015), the behavior of SCC depends on its constituents. Thus, this article, based on an extensive review, aims to describe the steps for the use of mineral additives and residues and its applications in SCC. First, a preliminary analysis of the residues must be carried out to replace or add to the mixture. The study listed the most common standards and techniques for the characterization of materials that are applicable to wastes and in fresh concrete according to the American Society of Tests and Materials (ASTM) standards. A brief summary of the different mixing design methods was made.

19.1.1 CEMENTITIOUS COMPOSITE: COMPONENTS AND CHARACTERISTICS

The cementitious composite as self-compacting concrete is comprised of cement, aggregates, water and mineral additives. The use of superplasticizing additives is paramount, as these materials guarantee its fluidity.

Portland cement is a hydraulic binder in which its hardening occurs through chemical reactions when in contact with water (Neville & Brooks, 2010). The fineness of the cement is of great importance because the larger the specific area, the greater the viscosity of the mixture. Pure cement with low tricalcium aluminate (C_3A) content is the most indicated for SCC (Chen & Yang, 2015) because the high content of C_3A hinders the rheological control and retards the concrete's stiffness (Tutikian & Dal Molin, 2011). Siddique and Khan (2011) reported that the use of cementitious materials such as granulated blast furnace slag, microsilica ashes, metakaulim, limestone filler, rice husk ashes, among others, is becoming more and more commonplace. The use of such materials can be advantageous not only in reducing costs and environmental impacts, but also because it can greatly increase the final material performance of structures, such as compressive strength. In addition, some materials can help control segregation and loss of workability caused by overdosing of superplasticizers.

Fine and coarse aggregates, composed of sand and gravel, respectively, make up 60% to 80% of the mixture, in general. Fine aggregates must meet the normative requirements for concrete, ranging from 0.150 mm to 4.75 mm NBR 248 (ABNT, 2001). Okamura and Ouchi (2003) suggest that they contribute 40% to 50% of the mortar volume. It is recommended that aggregates smaller than 0.125 mm be considered part of the fine grains or fillers. However, for SCC, coarse aggregate should have a maximum dimension of 20 mm and contribute 28% to 35% of the total volume of the mixture (Tutikian & Helene, 2011).

Mineral additions are fine materials incorporated into concrete, in large quantity, to provide the characteristics of self-density, resistance to segregation, viscosity and to improve workability in the fresh state. These favor resistance to thermal cracking and sulfate attack and favor reduction of alkali-aggregate expansion in addition to reducing costs (Mehta & Monteiro, 2008). These additions can be inert, causing only physical activity giving more compactness to concrete (limestone fillers, quartz, marble and granite powders), or reactive acting as pozzolans (fly ash, rice husk ash, sugarcane bagasse ash, silica fume and metakaolin). In the environmental aspect, the parcial replacement of cement by pozzolanic materials implies the reduction of the emission of CO_2 into the atmosphere, which is considered the main cause of the greenhouse effect (Jalal et al., 2012). According to Bahoria et al. (2013), industrial waste in the form of fine aggregates brings environmental benefits and provides better performance in concrete production.

The fibers that are metallic, polymeric, glass, synthetic or natural can be used to improve the properties of the SCC. Steel fibers are normally used to improve the mechanical resistance to flexion and toughness (Altun et al., 2007). The most used are the polymeric ones that increase the cohesion and reduce the segregation of the concrete in the fresh state and reduce the shrinkage retraction (Ghernouti et al., 2015). The addition of polypropylene fibers decreases the explosive chipping that occurs in denser concretes, due to internal pressure when subjected to high temperature.

According C494 (ASTM, 2019f), admixtures are products added to concrete in small quantities in order to modify some of its properties in order to improve the conditions of concrete use. The most commonly used admixtures in SCC are superplasticizers (SP). These types of admixtures allow a reduction in concrete water volume and contribute to the improvement of slump and fluidity. Third-generation superplasticizer admixtures based on polycarboxylates help in fluidity, reduce water surface tension, cause electrostatic repulsion between particles and act as particle lubricants. It can be used in combination with set-retarding and viscosity modifier admixtures (Kumar & Roy, 2018).

The water must be a good quality, as impurities can impair the hydration of the cement, which would cause reduction of compressive strength and reinforcement corrosion (Neville & Brooks, 2010). It should be added in the amount necessary to promote a chemical reaction of the cement and fill the porosity of the aggregates and voids between them, favoring the workability (Bucher et al., 2015).

19.1.2 DIFFERENT TYPES OF ADDITIONS AND WASTE MATERIALS

Based on a systematic and critical review, Table 19.1 presents a list of studies that identify the types of waste used and the tests performed with SCC. These residues were used as additions or partial replacements of concrete components.

19.2 METHODOLOGY

19.2.1 PRE-ANALYSES FOR WASTE MATERIAL SUBSTITUTION OR ADDITION IN CEMENTITIOUS COMPOSITES

To define how the wastes will be used in cementitious composites, the material must be well characterized to obtain the maximum amount of information and to understand its influence on the concrete physical and mechanical properties; thus determining if the waste will be considered as the mixture design, as either an addition or replacement of a component (Helene & Terzian 1992). According to Tutikian (2007), most fine grains must have a larger surface area than that of the component they will replace.

A prior characterization should be made to evaluate if the material has properties such as cementitious, inert fillers, fine aggregates, coarse aggregates or fibers. The steps are shown in Figure 19.1.

TABLE 19.1
Researchers, Mineral Additions, Waste Materials and Tests Conducted

Authors	Residual and Waste Materials	Tests Conducted/Measured Properties
Alyamac et al. (2017)	Marble powder	*Slump Flow*, T 500, V Funnel
Aslani et al. (2018)	Fly ash, granulated slag, silica fume, rubber residue	*Slump flow*, T 500, J Ring, Compressive strength, splitting tensile strength
Ashish & Verma (2021)	Metacaulim, Waste Foundry Sand	*Slump Flow*, T 500, V Funnel, J Ring, Compressive, Robustness
Barluenga et al., (2015)	Reactive and non-reactive mineral additions: metakaolin, Micro silica, nanosilica, limestone filler	Ultrasonic Pulse velocity, Slump, J Ring
Barroqueiro, Silva and Brito (2020)	Fly Ash, Limestone Filler, Silica Fume, Recycled Aggregates from the Precast Industry	*Slump Flow*, T500, V Funnel, L Box, U Box, Segregated Portion, Water Absorption, capillarity, Oxygen Permeability, Chloride Migration, Electrical Resistivity, Carbonation Tests
Boukhelkhal et al. (2015)	Natural Pozzolans, Stone powder, Granulated slag	*Slump flow*, V Funnel, Compressive strength
Carro-López et al. (2015)	Civil Construction Waste	*Slump flow*, J Ring, L Box, Compressive strength, ICAR Rheometer
Chaitanya & Babu (2017)	Fly ash and steel fibers	*Slump Flow*, V Funnel, L Box, Compressive, splitting tensile and flexural strength
Chopra et al. (2015)	Rice husk ash	*Slump flow*, V Funnel, U Box, Compressive and splitting tensile strength, resistance chloride ion penetration
Craeye et al. (2010)	Limestone powder, Quartz, Fly ash	Retraction, Mercury Porosimeter, Ultrasonic Pulse velocity
Dehwah (2012)	Stone powder, Silica fume, Fly ash	Chloride Permeability, Chloride Difusion
Du et al. (2015)	Nano silica	Resistance to Water and Chloride penetration, SEM, MET, XRD, Compressive strength
Ghernouti et al. (2015)	Fibers from plastic bags	*Slump flow*, L Box, Compressive, splitting tensile and flexural strength
Hama & Nahla (2017)	Plastic waste, fly ash	*Slump flow*, T500, V Funnel, L Box
Hesami et al. (2016)	Rubber, Polypropylene fiber, Limestone powder	*Slump flow test*, L Box, Compressive,, splitting tensile and flexural strength, Modulus of elasticity, Abrasion, Ultrasonic Pulse velocity, Water Absorption

(Continued)

TABLE 19.1 (Continued)
Researchers, Mineral Additions, Waste Materials and Tests Conducted

Authors	Residual and Waste Materials	Tests Conducted/Measured Properties
Kapoor et al. (2016)	Construction waste, Silica fume, Metakaulim	*Slump Flow*, V Funnel, L Box, J Ring, Permeability, resistance to Water Penetration
Makul (2019)	Rice rusk ash, waste Foundry Sand	*Slump Flow*, T 500, V Funnel, J Ring, compressive, splitting tensile strength, Modulus of elasticity, water permeability, Drying shrinkage
Malkapur et al. (2017)	Fly ash, Polyethylene	V Funnel, L Box, SEM
Marshaline Seles et al. (2017)	Active silica, metakaulim	*Slump Flow*, T500, V Funnel, L Box, U Box, compressive, splitting tensile and flexural strength
Mastali et al. (2016)	Glass fibers, Polymers, Silica fume	*Slump flow*, T 500, Compressive and flexural strength, impact, SEM
Parashar et al. (2020)	Fly Ash, Waste Foundry Sand	*Slump Flow*, T500, V Funnel, L Box, U Box, compressive, splitting tensile strength
Pathak & Siddique (2012)	Waste foundry sand, fly ash	Compressive and splitting tensile strength, modulus of elasticity, permeability, chloride, porosity, loss of mass at high temperatures, SEM, XRD
Sadrmomtazi et al. (2016)	PET, Natural Pozzolanic, Silica fume, Fly ash	*Slump flow*, V Funnel, L Box, Compressive splitting tensile and flexural strength, Modulus of elasticity, Ultrasonic Pulse velocity
Shafigh et al. (2018)	Clinker, expanded Clay	*Slump Test*, Compressive, splitting tensile and flexural strength, modulus of elasticity
Tang et al. (2018)	Red mud (alumina)	*Slump flow*, T500, J Ring, Compressive and splitting tensile strength, modulus of elasticity, SEM, EDS, XRD
Uysal & Sumer (2011)	Fly ash, Granulated slag, Limestone powder, Basalt powder, Marble powder	*Slump flow*, T 500, V Funnel, L Box. Compressive strength, ultrasonic pulse velocity, sulphate resistance
Uysal (2012)	Limestone powder, Basalt powder, Marble powder, Polypropylene fiber	*Slump Flow*, T 500, V Funnel, ultrasonic pulse velocity, compressive strength, absorption
Younis et al. (2018)	Tire steel fibers, Silica Fume	*Slump flow*, J ring, L box, Compressive and Flexural strength

19.2.2 Influence of Aggregates and Waste Additions on Cement Composite Properties

Obtaining satisfactory results is influenced by the aggregate properties, particle size distribution, void volume, pulp consumption and viscosity (Chen & Yang, 2015). The physical, chemical and mineralogical characteristics of the cement

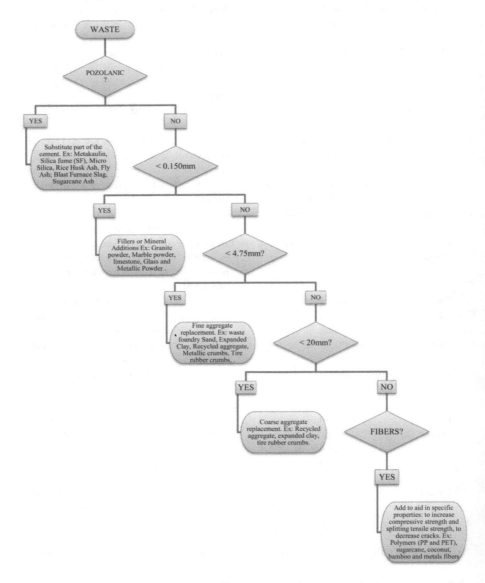

FIGURE 19.1 Sequence for initial characterization of aggregates.

composites component materials have a great influence on its mechanical properties, rheological and microstructural properties, in both fresh and hardened state (Castro et al., 2011), as seen in Table 19.2.

Okamura and Ouchi (2003) stated that any variety in the characteristics of the material can affect SCC self-compatibility. The most influential variables are water and aggregate content. The fineness modulus of the sand must be greater than 2.5 to satisfy self-compacting conditions. Very thick sand, with a fineness modulus greater than 3 can cause segregation and should be avoided. The size of the aggregate influences the segregation of the mixture. Highly graded coarse

TABLE 19.2
Characteristics of Aggregates and Waste Additions and Their Influences on Cement Composites Properties

Aggregates Characteristics	Influence on Cement Composite Properties
Grain Size Distribution Analysis	Tends to continuous. Aside from making segregation difficult, the more compact it is, the greater the reduction of permeability will be, consequently increasing durability and inhibiting the added alkali reaction.
Fineness Modulus	Very thin materials require a greater amount of water and therefore reduce mechanical resistance; on the other hand, they form a denser structure due to the reduction of pore size and improve the compressive strength and elasticity modulus.
Shape	Lamellar and cubic aggregates hamper workability, while improving roundness.
Texture	Rugged aggregates improve the adhesion of the paste while the smoother textures improve workability.
Size	Coarse aggregates form weaker transition zones, larger spaces between grains and larger pore size increase permeability and reduce durability.
Mineralogical composition	Silica aggregates provide higher strengths relative to limestone aggregates.
Chemical composition and waste classification	Inert composition contributes to compactness of the granular skeleton by promoting only physical action of pore reduction; improves the viscosity of the blend. Reactive compositions, in addition to contributing to the compacting of the mixture, act as pozzolans. That is, they contribute to the reduction of the cement and thus reduce the heat of hydration and cracking of the concrete; hazardous and non-hazardous (ABNT, 2004).

Sources: Mehta & Monteiro (2008), Neville & Brooks (2010) adapted.

aggregates and less elongated improve the workability, contributing to the reduction of waste water (Chen & Yang, 2015).

Pozzolan materials were originally associated with volcanic ash, which react with lime at room temperature and in the presence of water. At present, pozzolans are considered all the very fine silica/aluminate materials, which react with calcium hydroxide ($CaOH_2$) when in contact with water, acquire cementitious properties (Sabir et al., 2001). Natural pozzolans, metakaolin, micro and nanosilicates are commonly used in partial replacement of cement. Portland cement used with pozzolans generate more durable matrices due to the chemical reaction that consumes part of the calcium hydroxide forming compounds of the group of hydrated silicates and calcium aluminates (Silva et al., 2016). The mineral additions promote the permeability reduction of the concrete and in turn increase the durability and mechanical resistance (Dal Molin, 2005). They improve the resistance in the interfacial transition zone (ITZ), that is, at the interface between the mortar and coarse aggregates (Neville & Brooks, 2010). Additions of polymer fibers decrease explosive chipping when subjected to high temperature, increasing the cohesion of fresh concrete and reducing cracking by retraction (Castro et al., 2011).

19.2.3 APPLICABLE TRIALS AND TESTS FOR CEMENT COMPOSITES COMPONENTS

As discussed in item 3.2 material characterization is very important to evaluate its behavior in fresh cement composites and its influence on the mechanical properties in the hardened state. Thus, the more information obtained from the sample, the better the interrelationships can be understood. The tests applicable to samples of the cement composite components are shown in Table 19.3.

19.2.3.1 Standards and Techniques for Cement Composite Constituents Characterization

19.2.3.1.1 Standards for Cement Composites Components Composite Component Characterization

Material properties are obtained by tests according to ASTM standards and by means of materials characterization techniques, as described in Table 19.4.

19.2.3.1.2 Techniques for Characterization of Cement Composite Constituents

There are several techniques for characterizing materials, ranging from simple techniques at low cost to more complex and expensive assessment tools. However, analysis can be conducted according to the properties required for concrete use, such as particle size distribution of aggregates, size, fineness modulus, shape, texture, chemical composition, surface area and hazard level. All of these properties influence the fresh and hardened SCC properties, as seen in item 3.2 of this article.

TABLE 19.3
Applicable Trials for Cement Composites Components

Standards and Tests	Cementicious and Pozzolans	Additions/ Replacement Inerts	Fine Aggregates	Coarse Aggregates	Fibers
Specific gravity	X	X	X	X	
Granulometry		X	X	X	–
Bulk density		X	X	X	–
Pozzolanity	X				
Laser particle-size analyser	X	X			
XRD		X	X	X	X
SEM SE		X	X	X	X
SEM BSE		X	X	X	X
SEM EDS		X	X	X	X
X-Ray Fluorescence	X	X	X	X	X
Solid waste –Classification	X	X	X		X

SEM – Scanning Electronic Microscopy; SE – Secondary Electrons; BSE – Backscattered Electron; EDS – Energy Dispersive Spectroscopy; XRD – X- ray diffraction

The microstructural characterization of a material is closely related to its properties. A desirable microstructural characterization involves the determination of the crystalline structure, chemical composition, particle shape and size and phase distribution (Callister & Soares, 2008).

The choice of characterization techniques depends on what the researcher needs to know about the material, the availability of testing equipment and related costs. A selection of more well-known techniques is described in Table 19.5. However, to completely understand these techniques, a closer look is necessary to comprehend application, analysis and discussion of results.

19.2.4 MIXTURE DESIGN METHODS

One of the difficulties for the use of SCC is to find a good mixture design method due to the need to balance the different properties that depend heavily on the components used in blending (Su et al., 2001). For example, fresh SCC needs to have fluidity, workability and passing ability. However, to acquire all these properties, a large volume of pulp and consequently a large amount of cement will be required, which can cause shrinkage and cracking. The mix design method must meet the requirement for the hardened concrete, as compressive

TABLE 19.4
Standards for Characterizing Cement and Aggregates

Material	Properties	ASTM
Cement	Specific gravity	–
	Fineness Modulus	C 115-96a (ASTM 2018a)
	Time of Setting	C 191 (ASTM 2019b)
	Specification	C 150/C 150M-12 (ASTM 2020a)
Cementing Additions	Pozzolanity	C 311 e C 311b (ASTM 2018c)
		C 618-08 (ASTM 2019c)
Fine and Coarse	Sieve Analysis	C 136 (ASTM 2019d)
Aggregates	Materials finer than 75 μm	C 117 (ASTM 2017b)
	Bulk density and voils	C 29C/29M (ASTM 2017a)
	Sampling	–
	Specification	C 33 (ASTM 2018d)
	Lightweight content	C 123 (ASTM 2014b)
	Salt, Chloride and Sulfate Determination	C 88 (ASTM 2018f)
	Alkali-silica Reactivity aggregated	C 289 (ASTM 2007)
Fine Aggregates	Specific gravity	C 128 (ASTM 2015c)
	Absorption	C 128 (ASTM 2015c)
	Surface moisture	C 70 (ASTM 2020e)
	Swelling	–
	Organic Impurities	C 40 (ASTM 2020c)
Coarse Aggregates	Specific gravity and Absorption	C 127 (ASTM 2015b)
	Moisture content	–
Admixtures	Specification	C 494 (ASTM 2019f)

Source: Neville & Brooks 2013.

strength and durability, combined with the properties for fresh SCC such as fluidity, viscosity and stability (EFNARC, 2002).

Some authors limit the amount of aggregates by weight or volume for mixture design methods. The ranges of proportions and quantities in order to obtain self-compatibility are shown in Table 19.6. Therefore, some adjustments will be required.

According to Shi et al. (2015), mixture design methods are divided into five categories. The empirical method is used by Okamura and Ozawa (1995), Okamura and Ouchi (2003) andDomone (2007). The methodology based on the compressive strength was used by Tutikian and Helene (2011) and Repette and Melo (2005). The close aggregate packing method was used by Petersson and Blillberg (1996), Su et al. (2001), Tutikian and Dal Molin (2011), Gomes

TABLE 19.5
Characterization Techniques of Cement Composite Constituents

Techniques	Use	Authors
Laser particle-size analyser	Particle size distribution of the fine materials	Uysal & Sumer (2011); Uysal (2012); Du et al. (2015); Kapoor et al. (2016); Mastali et al. (2016)
Optical microscopy	Interface between aggregate and mass; adhesion and distribution of the aggregates in the mixture	
X-Ray Diffraction (XRD)	Distinguish the phases of chemical composition of the material	Chopra et al. (2015); Du et al. (2015); Pathak & Siddique (2012) Tang et al. (2018)
Scanning Electron Microscopy	Morphology, texture and particle size	Chopra et al. (2015); Du et al. (2015), Mastali et al. (2016); Pathak & Siddique (2012)
Energy Dispersive Spectroscopy (EDS)	Chemical elements of a point or area (qualitative)	Martins et al. (2019)
X-Ray Fluorescence	Global Chemical elements (Qualitative and quantitative)	Uysal & Sumer (2011); Uysal (2012); Carro-López et al. (2015); Dehwah (2012); Hesami et al. (2016); Mastali et al. (2016); Pathak & Siddique (2012); Sadrmomtazi et al. (2016); Kapoor et al. (2016)
Waste Classification	Toxicity and Hazard Level	Martins et al. (2019)
Thermal Analysis	Properties in Function of Temperature	Janotka et al. (2010)

TABLE 19.6
Limits of SCC Components

Ranges	Cement	Cementitious	Water	Sand	Gravel
% Volume		0.9 to 1.0		35 to 55 %Vmortar	28 to 35 % Vconcrete
Kg/m^3	200 to 450	500 to 600	160 to 185	710 to 900	750 to 920

Sources: Okamura & Ouchi (2003); Su, Hsu, & Chai (2001); Petersson & Blillberg (1996); Melo (2005); Gomes & Barros (2009); EFNARC (2002).

and Barros (2009), De Larrard (1999) and Zuo et al. (2018). The method based on statistical factorial models was adopted by Alyamac et al. (2017), Harbi et al. (2017), Kostrzanowska-Siedlarz & Gołaszewski (2016), Mastali et al. (2016),

Long et al. (2018), Sonebi et al. (2013), Khayat et al., (2000) and Khayat et al. (1999) and the method based on pulp rheology was used by Saak et al. (2001).

19.2.4.1 Empirical Method

The empirical design method is based on estimates of the proportions of the components of the mixture. The required properties are carried out through several mixes and adjustments (Shi et al., 2015).

Okamura's method is considered an empirical method based on experiments of various mixing ratios and adjustments, the method limits the aggregate to 50% of the aggregate volume; the water/fines ratio (w/f) by volume between 0.9 and 1.0 and the dosage of the superplasticizer adjusted until the fluidity required for SCC is reached (Okamura & Ouchi, 2003; Okamura & Ozawa, 1995). Okamura and Ouchi (2003) stated that it is necessary to limit the use of coarse aggregates, since they require a high paste viscosity that is achieved using SP and a low water/cement factor (w/c).

19.2.4.2 Compressive Strength Method

This mixture design method determines the amounts of the components from the desired compressive strength. This type of model is based on design methods for conventional vibrated conconcrete, CCV (Shi et al., 2015).

Tutikian's method is based on compressive strength. SCC is obtained from conventional vibrated concrete (CVC) calculated by the method of the Technological Research Institute/University of São Paulo (IPT/USP). First, one must choose and characterize all materials and measure the concrete by the desired compressive strength by setting the w/c ratio. The aggregates must be of the smallest granulometry possible. The coarse aggregates should not exceed 20 mm. Pozzolanic material must replace cement and non-Pozzolanic ones will replace the sand. The mortar content is determined for the CVC. A superplasticizer additive is added to convert to SCC and segregation correction is done with the addition of fine grains (Bernardo Fonseca Tutikian & Helene, 2011).

Repette and Melo's method is also based on compressive strength, setting the w/c factor. First, the cement paste in which the cement replacement is made by mineral additions consisting of fine grains smaller than 0.075 mm is determined. The mortar is then determined so that the ratio of the volume of fine aggregates to the mortar should be between 35% and 55%. The amount of SP admixture should result in a mortar spreading from 200 to 280 mm and V-funnel flow time from 2.5 s to 10 s. Finally, the concrete is formed through the addition of the coarse aggregate in a proportion of 27% to 33% of the concrete volume. Then, final adjustments of the SP admixture are carried out (Gomes & Barros, 2009).

19.2.4.3 Compressible Packing Method (CPM)

This method is based on the studies of Larrard (1999) on the principle of the greater amount of aggregates and the smaller space of voids between them, allowing a great granular compact, which provides a low w/c ratio, cohesive and low porosity concrete.

The first step is to determine the properties of all the materials. Then, the blends of the aggregates are made to obtain the best compacting. The paste adjustments are conducted based on fresh concrete, viscosity and yield stress (μ, $\tau 0$) and hardened concrete, compressive strength and elasticity modulus (Fcj, Ec) (Gomes & Barros, 2009). According to De Larrard (1999), the yield stress for SCC is 200 to 500 [Pa] viscosity less than or equal to 200 to 300 [Pa.s].

Petersson and Blillberg (1996) proposed a method that is based on determining the granular skeleton for better compacting of the aggregates and a minimum paste volume. The w/c factor is set for the desired compressive strength. The authors suggested that the amount of fine powders should be 500–525 kg/m^3 to obtain the mortar (Gomes & Barros, 2009).

Su et al. (2001) proposed a mixing method for SCC using a packing factor (PF) of the aggregates, which is calculated by the ratio between the masses of the compacted aggregate and the aggregate without compaction. The PF is used to control the content of fine and coarse aggregate in mixture proportion (Shi et al., 2015).

Gomes's method (Gomes & Barros, 2009) is divided into three phases: composition of the paste, definition of the granular skeleton and composition of the concrete seeking a content of paste so that the concrete has characteristics of SCC. The paste composition includes cement, pozzolans additives, water and SP. The relation w/c is defined by the desired strength, which helps set the value filler/cement (f/c). The SP dosage (SP/c) is done through trials on the Marsh Cone; the dosage of the paste is checked by the mini slump test. The granular skeleton is obtained experimentally with several proportions of mixtures choosing the one of greater density and with the smaller percentage of voids. Ideal concrete is obtained by looking for an optimum pulp content to be self-compacting and reaching the parameters indicated in the scattering tests, V-funnel and L-box.

19.2.4.4 Factorial Model Method

The design of blends based on a factorial statistical model is done mathematically using modeling techniques as a response surface method (RSM). This method is based on the effects that different parameters such as the cement contents, mineral admixtures, water/powder, aggregates volume, SP dosage, cause in the properties of fresh (workability, flowability) and hardened (compressive strength) concrete. Reasonable ranges for each parameter are determined and mixed. The proportion is calculated according to the mixture design for conventional concrete (Shi et al., 2015).

19.2.4.5 Paste Rheology Method

Saak et al. (2001) developed a mixture design method for SCC based on the rheology of paste model. The method proposes that cement paste rheology, given as a distribution of particle size and volumetric fraction of aggregates, determines the resistance to segregation and workability of fresh concrete. The theory suggests that the segregation of the aggregate is governed by yeld stress, viscosity and density of the cement paste. The applicability of the method is tested by measuring the fresh concrete flow properties (Okamura & Ozawa, 1995).

19.2.5 Cementitious Composite Rheology

Rheology is the science that studies deformation and flow of a fluid when subjected to stress. From a rheology point of view, concrete is considered an emulsion; that is, a concentration of solid particles (the aggregates) in a liquid (the cement paste). Thus, its behavior in the fresh state should be studied from the concepts of rheology in which the yield stress, plastic viscosity and thixotropy are analyzed to achieve the desired level of filling ability, passing ability and segregation resistance (Castro et al., 2011; Lu et al., 2015). The rheological behavior can be analyzed directly by the use of rheometer and viscometer equipment as in the works of Sonebi et al. (2013), Lu et al. (2018) andCarro-López et al. (2015). The surface area and the C_3A content of the cement are the main determinants of Portland cement rheology because they determine the speed of the hydration reactions and the loss of fluidity.

19.2.5.1 Tests for Fresh Scc Properties

The SCC test methods differ from those employed in the evaluation of conventional concrete only for fresh state properties. The essential characteristics of SCC can be evaluated by performing some types of tests as described below. These tests, shown in Figure 19.2, should be performed both in a laboratory and upon receipt at a worksite (Gomes & Barros, 2009; Tutikian & Dal Molin, 2008).

The most used tests methods for fresh state concrete are described in the ASTM standards available in Table 19.7.

Among the tests described in Table 19.5, the most common tests to characterize a SCC are Slump flow test, V Funnel and L Box (Khayat & Schutter 2014).

19.2.5.1.1 Slump flow test and T_{500}

This method measures SCC flowability through the spreading diameter formed by the concrete and the time the concrete takes to cover a diameter of 500 mm. Segregation can be assessed visually. The test method is to fill an unconcentrated concrete Abrams cone, placed on a metal plate, 90 by 90 cm, with the centralized

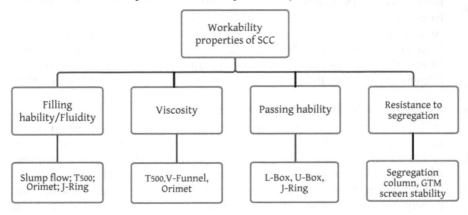

FIGURE 19.2 Requirements for fresh concrete.

TABLE 19.7
Norms for SCC Trials in Fresh State

Tests	ASTM
Specific gravity	C 138/C138 M (ASTM 6AD)
Classification, control and acceptance	C 94/C94 M (ASTM 2020f)
Scattering and flow time cone trunk Slump flow test	C 1611/C1611 M (ASTM 2018b)
Passing Ability, J Ring	C 1621 (ASTM 2017c)
Passing Ability, L Box	–
Viscosity, V Funnel	–
Segregation, column	C 1610 (ASTM 2019e)

marking of a circle of 500 mm diameter (Figure 19.3). As soon as the cone is lifted, the timer starts to measure the elapsed time to reach the diameter of 500 mm. When maximum spreading is reached, two perpendicular measurements of the final diameter, d_1 and d_2, must be performed (Tutikian & Dal Molin, 2008; Gomes & Barros, 2009). The higher the slump flow (SF) value, the greater its ability to fill formwork under its own weight. A value of at least 650 mm is required for SCC. Time T 500 is also a measure of the flow velocity and therefore of the viscosity of the SCC (EFNARC, 2006).

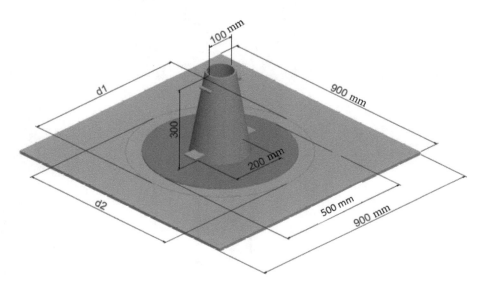

FIGURE 19.3 Abrams cone: slump flow test equipment (mm).
Source: Authors.

19.2.5.1.2 V Funnel

The V-funnel test is used to measure the viscosity and filling capacity of concrete self-compatible. The test is carried out using a V-shaped funnel, as seen in Figure 19.4. The time the material (10 liters) takes to run without segregating is measured (EFNARC, 2006).

19.2.5.1.3 L-Box Test

This test is used to measure the fluidity and ability of concrete to pass through obstacles and remain cohesive. It consists of filling concrete in the column of Box L (Figure 19.5) and after opening the gate where the steel bars are spaced. The time the concrete reaches the mark of 20 and 40 cm, T_{20} and T_{40}, respectively, is measured. It can give some indication of ease of flow. In following, the starting (H_1) and final (H_2) heights of the shape are measured. The ratio between

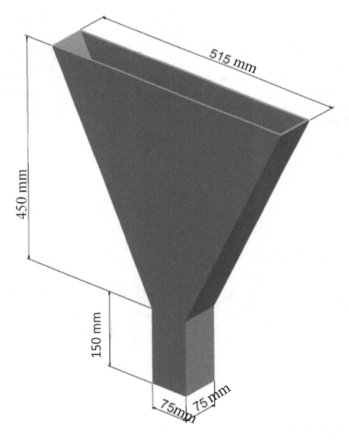

FIGURE 19.4 V funnel (mm).

Source: Authors.

Additions of Wastes to Composites

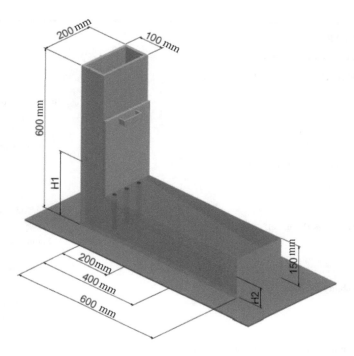

FIGURE 19.5 L box equipment (mm).

Source: Authors.

the heights indicates the fluidity and filling capacity of the concrete (Gomes & Barros, 2009). Therefore, the nearer this test value, the "blocking ratio," is to unity, the better the flow of the concrete. The authors suggested a minimum acceptable value of 0.8 (EFNARC, 2006).

The parameters for analysis the fresh SCC properties are described in Table 19.8. The results of the tests and parameters can be correlated to the behavior of the concrete and its components and have criteria for possible adjustments. That is, if the concrete is very fluid and with low viscosity, the fine grains should increase. If segregation occurs, it will be necessary to improve compacting or reduce gravel size (EFNARC, 2002).

19.2.5.2 Common Problems in Fresh SCC

During the making of the SCC, it is very common that some test results do not meet the specified requirements. This can happen due to different causes. Table 19.9 summarizes some common problems that may arise during the making of the SCC and its possible causes.

Considering the possible problems that can present the mixture of fresh SCC, some corrective actions must be taken to reach the characteristics of the concrete (Table 19.10).

TABLE 19.8
Requirements for SCC
Slump Flow Classes

Class	Slump flow in mm
SF1	550 a 650
SF2	660 a 750
SF3	760 a 850
Viscosity classes in Abrams cone	
Class	T_{500} s
VS1	<2
VS2	≥2
Viscosity classes (V funnel)	
Class	V-Funnel time in s
VF1	<9
VF2	9 a 25
Passing ability classes (L box)	
Class	Passing ability in %
PL1	≥0.80 with 2 rebars
PL2	≥0.80 with 3 rebars
Passing ability classes (J ring) with 16 rebars	
Class	Passing ability in mm
PJ1	0 a 25
PJ2	25 a 50
Resistance segregation classes	
Class	Segregation column in %
SR1	≤20
SR2	≤15

Source: NBR 15823-1 (ABNT 2017) adapted.

19.2.6 HARDENED CONCRETE PROPERTIES

The main analyzed properties of hardened concrete are compressive strength and durability. However, there are many properties such as tensile strength, elasticity, shrinkage, creep, flexural strength, electrical resistivity, thermal conductivity, permeability and other properties (Kovler, Roussel, 2011).

19.2.6.1 Main Hardened SCC Properties

Concrete properties should be evaluated according to their application.
The tests from ASTM are shown in Table 19.11.

TABLE 19.9
Problems and Causes with Fresh SCC

	Results below the Limits	Possible Cause
Slump flow	<650 mm	Viscosity; Yeld value too high
T_{500}	<2 s	Viscosity too low
V funnel	<8 s	Viscosity too low
L box (H_2/H_1)	<0,8	Viscosity too high; yield value too high; blockage
J ring	<10	Viscosity too high; yield value too high; segregation
U-Box (h 2 -h 1)	0	False result
Screen stability test	5	Viscosity too high; blockage
	Results above limits	
Slump flow	>750 mm	Viscosity too low; segregation
T_{500}	>5s	Viscosity too high; yield value too high
V funnel	> 12s	Viscosity too high; yield value too high; blockage
L box (H_2/H_1)	1	False result
J ring		Viscosity too low; segregation
U-Box (h 2 -h 1)	>30	Viscosity too high; yield value too high; blockage
Screen stability test	>15	Segregation

Source: EFNARC (2002); EFNARC (2006) adapted.

19.3 FINAL REMARKS

Among the cementitious composites, SCC is a material in which you can add a wide variety of wastes contributing to the development of a sustainable concrete, reducing the use of cement and natural aggregates, along with the amount of waste in landfills; SCC is a relatively new material when compared to CVC. The properties of the fresh concrete are fluidity, workability and viscosity depend on the characteristics and properties of its components. However, the mixture design models, in general, require intensive laboratory testing. The process of choosing the appropriate constituents and their relative quantities should aim to produce an economic SCC that meets the minimum performance characteristics of workability, strength and durability. There is not a direct and exact mixture design method that provides the ideal mixture. Therefore, knowing the influences of the components in the SCC properties is essential. Some parameters can help in the composition of the mixture.

Although there are several studies on the use of waste materials in cementitious composites, most of them analyze one or more aggregate materials as the properties of the concrete in the fresh and/or hardened state, as shown in Table 19.2. The techniques of characterization of materials are varied, so it is

TABLE 19.10
Problems and Corrective Actions in Fresh SCC

Problems	Corrective Measures
Viscosity too high	Increase w/c
	Increase paste volume
	Increase SP
Viscosity too low	Reduce w/c
	Reduce paste volume
	Reduce SP
	Increase viscosity modifying agent
	Use finer powder
	Use finer sand
Yield value too high	Increase SP
	Increase paste volume
	Increase mortar volume
Segregation	Increase pasta volume
	Increase mortar volume
	Reduce w/c
	Use finer powder
Rapid loss in workability	Use slower reacting cement type
	Use set-retarding
	Change the SP
	Exchange filler for cement
Blockage	Reduce maximum aggregate size
	Increase paste volume
	Increase mortar volume

Sources: EFNARC (2002); EFNARC (2006) adapted.

important to understand which techniques will be used. This review paper proposed the compilation of many sources of information to aid new researchers when they have some sort of waste or residual material at their disposal, in order to help them discern how to use it; that is, as an additive or substitution component to concrete. In addition they should know which analyses should be conducted to understand how this will influence fresh and hardened concrete properties.

19.4 CONCLUSIONS

Through this research study, the following points were identified:

TABLE 19.11
Technical Standards for Testing Hardened Concrete

Description	ASTM
Procedure for molding and curing test specimens	C 31 (ASTM 2021a)
Cylindrical test body compression test	C 39/C39M (ASTM 2014a)
Determination of the tensile strength in the flexion of prismatic specimens	C 78 (ASTM 2021b)
Concrete and Mortar – Determination of tensile strength by diametral compression of cylindrical test bodies	C 496 (ASTM 2017d)
Determination of water absorption and indices of voids and specific mass	C 642 (ASTM 2013)
Determination of the elastic modulus of compression elasticity	C 469 (ASTM 2014c)
Determination of electrical-volumetric resistivity	G 57 (ASTM 2020g)
Inorganic materials – Determination of abrasion wear	C 418 (ASTM 2020d)
Creep Determination – trial method	C 512 (ASTM 2015d)
Determination of ultrasonic pulse velocity	C 597 (ASTM 2016)
Determination of resistance to penetration of chloride ions	C 1202 (ASTM 2019a)
Fast determination of chloride ions in acid solution	C 1152 (ASTM 2020b)
Determination of freeze resistance	C 666 (ASTM 2015e)
Determination of penetration resistance (Windsor)	C 803 (ASTM 2018e)
Determination of the accelerated corrosion potential of steel	C 876 (ASTM 2015a)
Evaluation of hardness by the reflection sclerometer	C 803 (ASTM 2018e)

- Waste evaluation to verify if it can be used as additives and/or substitution in the cement composite constituents, contributing to the development of a sustainable concrete, reducing the use of cement and natural aggregates, along with the amount of waste in landfills;
- Distinguishing the necessary tests and trials for materials to be used in the preparation of concrete;
- Identify the influence of the materials on the concrete properties in the fresh state and apply this mix design method in the dosing process;
- Understand the implications of the residual/waste materials on the properties of hardened state concrete; carry out the necessary tests and trials in order to identify possible applications.

ACKNOWLEDGMENT

The authors would like to thank CAPES for their continued financial support for doctoral students and specifically for this research project.

REFERENCES

ABNT. (2001). "NBR NM 248: Agregados- Determinação Da Composição Granulométrica." *Associação Brasileira de Normas Técnicas*,. Rio de Janeiro, Brasil.
ABNT. (2004). "NBR 10004: Resíduos Sólidos - Classificação." *Associação Brasileira de Normas Técnicas*. Rio de Janeiro, Brasil.
ABNT. (2017). "NBR 15823-1: Concreto Autoadensável Parte 1: Classificação, Controle e Recebimento No Estado Fresco." *Associação Brasileira de Normas Técnicas*,. Rio de Janeiro, Brasil.
Alsubari, Belal, Payam Shafigh, & Mohd Zamin Jumaat. (2016). "Utilization of High-Volume Treated Palm Oil Fuel Ash to Produce Sustainable Self-Compacting Concrete." *Journal of Cleaner Production* 137: 982–996. 10.1016/j.jclepro.2016.07.133.
Altun, Fatih, Tefaruk Haktanir, & Kamura Ari. (2007). "Effects of Steel Fiber Addition on Mechanical Properties of Concrete and RC Beams." *Construction and Building Materials* 21 (3): 654–661. 10.1016/j.conbuildmat.2005.12.006.
Alyamac, Kursat Esat, Ehsan Ghafari, & Ragip Ince. (2017). "Development of Eco-Efficient Self-Compacting Concrete with Waste Marble Powder Using the Response Surface Method." *Journal of Cleaner Production* 144: 192–202. 10.1016/j.jclepro.2016.12.156.
ANEPAC. (2020). "Tendências Para o Mercado de Agregados." *Revista Areia & Brita*. São Paulo, SP. https://www.anepac.org.br/.
Ashish, Deepankar Kumar, & Verma, Surender Kumar. (2019). "Determination of Optimum Mixture Design Method for Self-Compacting Concrete: Validation of Method with Experimental Results." *Construction and Building Materials* 217: 664–678. 10.1016/j.conbuildmat.2019.05.034.
Ashish, Deepankar Kumar, & Verma, Surender Kumar. (2021). "Robustness of Self-Compacting Concrete Containing Waste Foundry Sand and Metakaolin: A Sustainable Approach." *Journal of Hazardous Materials* 401 (June 2020): 123329. 10.1016/j.jhazmat.2020.123329.
Aslani, Farhad, Guowei Ma, Dominic Law Yim Wan, & Vinh Xuan Tran Le. (2018). "Experimental Investigation into Rubber Granules and Their Effects on the Fresh and Hardened Properties of Self-Compacting Concrete." *Journal of Cleaner Production* 172: 1835–1847. 10.1016/j.jclepro.2017.12.003.
ASTM. 6AD. "C138/C138M - 17a: Standard Test Method for Density (Unit Weight), Yield, and Air Content (Gravimetric) of Concrete." West Conshohocken, PA: ASTM International.
ASTM. (2000). "C 125-00: Standard Terminology Relating to Concrete and Concrete Aggregates." West Conshohocken, PA: ASTM International. 10.1520/C0125-20.
ASTM. (2007). "C289-07: Standard Test Method for Potential Alkali-Silica Reactivity of Aggregates (Chemical Method)." West Conshohocken, PA: ASTM International.
ASTM. (2013). "C642-13, Standard Test Method for Density, Absorption, and Voids in Hardened Concrete." West Conshohocken, PA: ASTM International. DOI: 10.1520/C0642-13.
ASTM. (2014a). "C 39/C 39M – 01. Standard Test Method for Compressive Strength of Cylindrical Concrete Specimens." 10.1520/C0039.
ASTM. (2014b). "C123 / C123M - 14: Standard Test Method for Lightweight Particles in Aggregate." West Conshohocken, PA: ASTM International.

ASTM. (2014c). "C469/C469M-14e1, Standard Test Method for Static Modulus of Elasticity and Poisson's Ratio of Concrete in Compression." West Conshohocken, PA: ASTM International. DOI: 10.1520/C0469_C0469M-14E01.

ASTM. (2015a). "C 876-15: Standard Test Method for Corrosion Potentials of Uncoated Reinforcing Steel in Concrete." West Conshohocken, PA: ASTM International. 10.1520/C0876-15.

ASTM. (2015b). "C127-15 Standard Test Method for Relative Density (Specific Gravity) and Absorption of Coarse Aggregate." West Conshohocken, PA: ASTM International. www.astm.org.

ASTM. (2015c). "C128-15: Standard Test Method for Relative Density (Specific Gravity) and Absorption of Fine Aggregate." West Conshohocken, PA: ASTM International. www.astm.org.

ASTM. (2015d). "C512/C512M-15, Standard Test Method for Creep of Concrete in Compression." West Conshohocken, PA: ASTM International. DOI: 10.1520/C0512_C0512M-15.

ASTM. (2015e). "C666/C666M-15, Standard Test Method for Resistance of Concrete to Rapid Freezing and Thawin." West Conshohocken, PA: ASTM International. DOI: 10.1520/C0666_C0666M-15.

ASTM. (2016). "C597-16, Standard Test Method for Pulse Velocity Through Concrete." West Conshohocken, PA: ASTM International. DOI: 10.1520/C0597-16.

ASTM. (2017a). "C 29/C 29 M-17: Standard Test Method for Bulk Density ('Unit Weight') and Voids in Aggregate." West Conshohocken, PA: ASTM International. www.astm.org.

ASTM. (2017b). "C117-17: Standard Test Method for Materials Finer than 75-Mm (No. 200) Sieve in Mineral Aggregates by Washing." West Conshohocken, PA: ASTM International.

ASTM. (2017c). "C1621/C1621M-17: Standard Test Method for Passing Ability of Self-Consolidating Concrete by J-Ring." West Conshohocken, PA: ASTM International.

ASTM. (2017d). "C496/C496M-17, Standard Test Method for Splitting Tensile Strength of Cylindrical Concrete Specimens." West Conshohocken, PA: ASTM International. DOI: 10.1520/C0496_C0496M-17.

ASTM. (2018a). "C115/C115M -18: Standard Test Method for Fineness of Portland Cement by the Turbidimeter." West Conshohocken, PA: ASTM International. www.astm.org.

ASTM. (2018b). "C1611 / C1611M - 18: Standard Test Method for Slump Flow of Self-Consolidating Concrete." West Conshohocken, PA: ASTM International.

ASTM. (2018c). "C311/C311M-18: Standard Test Methods for Sampling and Testing Fly Ash or Natural Pozzolans for Use in Portland-Cement Concrete." West Conshohocken, PA: ASTM International. www.astm.org.

ASTM. (2018d). "C33/C33M-18: Standard Specification for Concrete Aggregates." West Conshohocken, PA: ASTM International. www.astm.org.

ASTM. (2018e). "C803 / C803M-18, Standard Test Method for Penetration Resistance of Hardened Concrete." West Conshohocken, PA: ASTM International. DOI: 10.1520/C0803_C0803M-18.

ASTM. (2018f). "C88/C88M - 18: Standard Test Method for Soundness of Aggregates by Use of Sodium Sulfate or Magnesium Sulfate." West Conshohocken, PA: ASTM International. www.astm.org.

ASTM. (2019a). "C 1202: Standard Test Method for Electrical Indication of Concrete's Ability to Resist Chloride Ion Penetration." West Conshohocken, PA: ASTM International. 10.1520/C1202-19.

ASTM. (2019b). "C 191-19: Standard Test Methods for Time of Setting of Hydraulic Cement by Vicat Needle." West Conshohocken, PA: ASTM International. www.astm.org.

ASTM. (2019c). "C 618-19: Standard Specification for Coal Fly Ash and Raw or Calcined Natural Pozzolan for Use in Concrete." West Conshohocken, PA: ASTM International. www.astm.org.

ASTM. (2019d). "C136/C136M-19: Standard Test Method for Sieve Analysis of Fine and Coarse Aggregates." West Conshohocken, PA,: ASTM International. www.astm.org.

ASTM. (2019e). "C1610/C1610M-19: Standard Test Method for Static Segregation of Self-Consolidating Concrete Using Column Technique." West Conshohocken, PA: ASTM International.

ASTM. (2019f). "C494/C494M - 19: Standard Specification for Chemical Admixtures for Concrete." West Conshohocken, PA: ASTM International.

ASTM. (2020a). "C 150/C150 M-20: Standard Specification for Portland Cement." West Conshohocken, PA: ASTM International. www.astm.org.

ASTM. (2020b). "C1152/C1152M-20, Standard Test Method for Acid-Soluble Chloride in Mortar and Concrete." West Conshohocken, PA: ASTM International. DOI: 10.1520/C1152_C1152M-20.

ASTM. (2020c). "C40/C40M - 20: Standard Test Method for Organic Impurities in Fine Aggregates for Concrete." West Conshohocken, PA: ASTM International. www.astm.org.

ASTM. (2020d). "C418-20, Standard Test Method for Abrasion Resistance of Concrete by Sandblasting." West Conshohocken, PA: ASTM International. DOI: 10.1520/C0418-20.

ASTM. (2020e). "C70-20: Standard Test Method for Surface Moisture in Fine Aggregate." West Conshohocken, PA: ASTM International. www.astm.org.

ASTM. (2020f). "C94/C94M - 20: Standard Specification for Ready-Mixed Concrete." West Conshohocken, PA: ASTM International.

ASTM. (2020g). "G57-20, Standard Test Method for Measurement of Soil Resistivity Using the Wenner Four-Electrode Method." West Conshohocken, PA: ASTM International. DOI: 10.1520/G0057-20.

ASTM. (2021a). "C31/C31M-21, Standard Practice for Making and Curing Concrete Test Specimens in the Field." West Conshohocken, PA: ASTM International. DOI: 10.1520/C0031_C0031M-21.

ASTM. (2021b). "C78/C78M-21, Standard Test Method for Flexural Strength of Concrete (Using Simple Beam with Third-Point Loading)." West Conshohocken, PA: ASTM International. www.astm.org.

Bahoria, B. V., Parbat, D. K., & Naganaik, P. B.. (2013). "Replacement of Natural Sand in Concrete by Waste Products." *Journal of Environmental Research and Development* 7 (4): 1651–1656.

Barluenga, Gonzalo, Javier Puentes, & Irene Palomar. (2015). "Early Age Monitoring of Self-Compacting Concrete with Mineral Additions." *Construction and Building Materials* 77: 66–73. 10.1016/j.conbuildmat.2014.12.033.

Batayneh, Malek, Iqbal Marie, & Ibrahim Asi. (2007). "Use of Selected Waste Materials in Concrete Mixes." *Waste Management* 27 (12): 1870–1876. 10.1016/j.wasman.2006.07.026.

Boukhelkhal, Djamila, Othmane Boukendakdji, Said Kenai, & Sarah Bachene. (2015). "Effect of Mineral Admixture Type on Stability and Rheological Properties of Self-Compacting Concrete." *HAL Archives-Ouvertes*, 8. https://hal.archives-ouvertes.fr/hal-01167737.

Brandt, A. M. (2009). *Cement-Based Composites: Materials, Mechanical Properties and Performance*. Edited by Taylor & Francis. 2nd ed. New York.
Bucher, Raphaël, Paco Diederich, Michel Mouret, Gilles Escadeillas, & Martin Cyr. (2015). "Self-Compacting Concrete Using Flash-Metakaolin: Design Method." *Materials and Structures* 48 (6): 1717–1737. 10.1617/s11527-014-0267-x.
Carro-López, Diego, Belén González-Fonteboa, Jorge De Brito, Fernando Martínez-Abella, Iris González-Taboada, & Pedro Silva. (2015). "Study of the Rheology of Self-Compacting Concrete with Fine Recycled Concrete Aggregates." *Construction and Building Materials* 96: 491–501. 10.1016/j.conbuildmat.2015.08.091.
Castro, A. L., Liborio, J. B. L., & Pandolfelli, V. C.. (2011). "Reologia de Concretos de Alto Desempenho Aplicados Na Construção Civil - Revisão." *Cerâmica* 57: 63–75.
Castro, A. L., Tiba, P. R. T., & Pandolfelli, V. C.. (2011). "Fibras de Polipropileno e Sua Influência No Comportamento de Concretos Expostos a Altas Temperaturas: Revisão." *Cerâmica* 57: 22–31. 10.1590/S0366-69132011000100003.
Chaitanya, P Venkata, & Babu, T Suresh, Prof. (2017). "A Study on Steel Fiber Reinforced Self Compacting Concrete."*International Journal of Engineering Research-Online* 5 (5): 13–19.
Chen, Zekong;, & Mao Yang. (2015). "The Research on Process and Application of Self-Compacting Concrete." *International Journal of Engineering Research and Applications* 5 (8): 12–18.
Chopra, Divya, Rafat Siddique, & Kunal. (2015). "Strength, Permeability and Microstructure of Self-Compacting Concrete Containing Rice Husk Ash." *Biosystems Engineering* 130: 72–80. 10.1016/j.biosystemseng.2014.12.005.
Craeye, B., De Schutter, G., Desmet, B., Vantomme, J., Heirman, G., Vandewalle, L., Cizer, O., Aggoun, S., & Kadri, E. H. (2010). "Effect of Mineral Filler Type on Autogenous Shrinkage of Self-Compacting Concrete." *Cement and Concrete Research* 40 (6): 908–913. 10.1016/j.cemconres.2010.01.014.
Dal Molin, D. C. C. (2005). "Adições Minerais Para Concreto Estrutural." In *Concreto: Ensino, Pesquisa e Realizações.*, edited by Ibracon, 345–379. São Paulo, SP.
Dehwah, H. A. F. (2012). "Corrosion Resistance of Self-Compacting Concrete Incorporating Quarry Dust Powder, Silica Fume and Fly Ash." *Construction and Building Materials* 37: 277–282. 10.1016/j.conbuildmat.2012.07.078.
Domone, P. L. (2007). "A Review of the Hardened Mechanical Properties of Self-Compacting Concrete." *Cement and Concrete Composites* 29 (1): 1–12. 10.1016/j.cemconcomp.2006.07.010.
Du, Hongjian, Suhuan Du, & Xuemei Liu. (2015). "Effect of Nano-Silica on the Mechanical and Transport Properties of Lightweight Concrete." *Construction and Building Materials* 82: 114–122. 10.1016/j.conbuildmat.2015.02.026.
EFNARC. (2002). "Specification and Guidelines for Self-Compacting Concrete." *European Federation of National Associations Representing Producers and Applicators of Specialist Building Products for Concrete*. Surrey,UK: European Federation of National Associations Representing producers and applicators of specialist building products for Concrete. 10.9539733.4.4.
EFNARC. (2006). "Directrices Europeas Para El Hormigón Autocompactante. Especifaciones, Producción y Uso." European Federation of National Associations Representing producers and applicators of specialist building products for Concrete.
Ghernouti, Youcef, Bahia; Rabehi, Tayeb; Bouziani, Hicham; Ghezraoui, & Abdelhadi Makhloufi. (2015). "Fresh and Hardened Properties of Self-Compacting Concrete Containing Plastic Bag Waste Fibers (WFSCC)." *Construction and Building Materials* 82: 89–100. 10.1016/j.conbuildmat.2015.02.059.

Gomes, P. C. C., & Barros, A. (2009). *Métodos de Dosagem de Concreto Autoadensável*. First. São Paulo: Editora PINI LTDA.

Güneyisi, Erhan, Mehmet Gesoglu, Asraa Al-Goody, & Süleyman İpek. (2015). "Fresh and Rheological Behavior of Nano-Silica and Fly Ash Blended Self-Compacting Concrete." *Construction and Building Materials* 95: 29–44. 10.1016/j.conbuildmat.2015.07.142.

Hama, Sheelan M., & Nahla N. Hilal. (2017). "Fresh Properties of Self-Compacting Concrete with Plastic Waste as Partial Replacement of Sand." *International Journal of Sustainable Built Environment* 6 (2): 299–308. 10.1016/j.ijsbe.2017.01.001.

Harbi, Radhia, Riad Derabla, & Zahreddine Nafa. (2017). "Improvement of the Properties of a Mortar with 5% of Kaolin Fillers in Sand Combined with Metakaolin, Brick Waste and Glass Powder in Cement." *Construction and Building Materials* 152: 632–641. 10.1016/j.conbuildmat.2017.07.062.

Hebhoub, H., Aoun, H., Belachia, M., Houari, H., & Ghorbel, E. (2011). "Use of Waste Marble Aggregates in Concrete." *Construction and Building Materials* 25 (3): 1167–1171. 10.1016/j.conbuildmat.2010.09.037.

Helene, P., & Terzian, P. (1992). *Manual de Dosagem e Controle Do Concret*. Sao Paulo: PINI.

Hesami, Saeid, Iman Salehi Hikouei, & Seyed Amir Ali Emadi. (2016). "Mechanical Behavior of Self-Compacting Concrete Pavements Incorporating Recycled Tire Rubber Crumb and Reinforced with Polypropylene Fiber." *Journal of Cleaner Production* 133: 228–234. 10.1016/j.jclepro.2016.04.079.

Jalal, Mostafa, Esmaeel Mansouri, Mohammad Sharifipour, & Ali Reza Pouladkhan. (2012). "Mechanical, Rheological, Durability and Microstructural Properties of High Performance Self-Compacting Concrete Containing SiO2 Micro and Nanoparticles." *Materials and Design* 34: 389–400. 10.1016/j.matdes.2011.08.037.

Janotka, I., Puertas, F., Palacios, M., Kuliffayová, M., & Varga, C. (2010). "Metakaolin Sand-Blended-Cement Pastes: Rheology, Hydration Process and Mechanical Properties." *Construction and Building Materials* 24 (5): 791–802. 10.1016/j.conbuildmat.2009.10.028.

Melo, K. A. (2005). "Contribuição à Dosagem de Concreto Auto Adensável Com Adição de Fíler Calcário." Universidade Federal de Santa Catarina.

Kapoor, Kanish, Singh, S. P., & Bhupinder Singh. (2016). "Durability of Self-Compacting Concrete Made with Recycled Concrete Aggregates and Mineral Admixtures." *Construction and Building Materials* 128: 67–76. 10.1016/j.conbuildmat.2016.10.026.

Khayat, K. H., Ghezal, A., & Hadriche, M. S. (2000). "Utility of Statistical Models in Proportioning Self-Consolidating Concrete." *Material and Structures* 33 (5): 338–344.

Khayat, K. H., Ghezal, A., & Hadriche, M. S. (1999). "Factorial Design Model for Proportioning Self-Consolidating Concrete." *Material and Structures* 32 (9): 679–686.

Khayat, Kamal H., & Geert De Schutter. (2014). *Mechanical Properties of Self-Compacting Concrete. RILEM State-of-the-Art Reports*. Vol. 14. Springer. 10.1007/978-3-319-03245-0.

Kibriya, Tahir, & Leena Tahir. (2017). "Sustainable Construction - Use of Masonry Construction Demolition Waste in Concrete." *World Journal of Engineering and Technology* 05 (02): 216–231. 10.4236/wjet.2017.52017.

Kostrzanowska-Siedlarz, Aleksandra, & Jacek Gołaszewski. (2016). "Rheological Properties of High Performance Self-Compacting Concrete: Effects of Composition and Time." *Construction and Building Materials* 115: 705–715. 10.1016/j.conbuildmat.2016.04.027.

Kovler, Konstantin, & Nicolas Roussel. (2011). "Properties of Fresh and Hardened Concrete." *Cement and Concrete Research* 41 (7): 775–792. 10.1016/j.cemconres.2011.03.009

Kumar, Pratyush, & Rahul Roy. (2018). "Study and Experimental Investigation of Flow and Flexural Properties of Natural Fiber Reinforced Self Compacting Concrete." *Procedia Computer Science* 125: 598–608. 10.1016/j.procs.2017.12.077.

Larrard, F. De. (1999). *Concrete Mixture Proportioning a Scientific Approach*. London: E & FN SPON.

Leite, Mônica Batista, & Denise Dal Molin. (2002). "Avaliação Da Atividade Pozolânica Do Material Cerâmico Presente No Agregado Reciclado de Resíduo de C&D." *Sitientibus* 26: 111–130. http://www2.uefs.br/sitientibus/pdf/26/avialiacao_da_atividade_pozolanica.pdf.

Long, Wu Jian, Yucun Gu, Jinxun Liao, & Feng Xing. (2018). "Sustainable Design and Ecological Evaluation of Low Binder Self-Compacting Concrete." *Journal of Cleaner Production* 167: 317–325. 10.1016/j.jclepro.2017.08.192.

Lu, Cairong, Hu Yang, & Guoxing Mei. (2015). "Relationship between Slump Flow and Rheological Properties of Self Compacting Concrete with Silica Fume and Its Permeability." *Construction and Building Materials* 75: 157–162. 10.1016/j.conbuildmat.2014.08.038.

Lu, Jinshan, Xinquan Cong, Yingde Li, Yong Hao, & Chunlei Wang. (2018). "High Strength Artificial Stoneware from Marble Waste via Surface Modification and Low Temperature Sintering." *Journal of Cleaner Production* 180: 728–734. 10.1016/j.jclepro.2018.01.181.

Makul, Natt. (2019). "Combined Use of Untreated-Waste Rice Husk Ash and Foundry Sand Waste in High-Performance Self-Consolidating Concrete." *Results in Materials* 1 (August): 100014. 10.1016/j.rinma.2019.100014.

Malkapur, Santhosh M., Divakar, L., Mattur C. Narasimhan, Narayana B. Karkera, Goverdhan, P., Sathian, V., & Prasad, N. K. (2017). "Fresh and Hardened Properties of Polymer Incorporated Self Compacting Concrete Mixes for Neutron Radiation Shielding." *Construction and Building Materials* 157: 917–929. 10.1016/j.conbuildmat.2017.09.127.

Marshaline Seles, M., Suryanarayanan, R., Vivek, S. S., & Dhinakaran, G.. (2017). "Study on Flexural Behaviour of Ternary Blended Reinforced Self Compacting Concrete Beam with Conventional RCC Beam." In *IOP Conference Series: Earth and Environmental Science*. 80. 10.1088/1755-1315/80/1/012026.

Martins, M. A. B., Barros, R. M., Silva, G., & Santos, I. F. S. (2019). "Study on Waste Foundry Exhaust Sand, WFES, as a Partial Substitute of Fine Aggregates in Conventional Concrete." *Sustainable Cities and Society* 45 (February): 187–196. 10.1016/j.scs.2018.11.017.

Mastali, M., Dalvand, A., & Sattarifard, A. R. (2016). "The Impact Resistance and Mechanical Properties of Reinforced Self-Compacting Concrete with Recycled Glass Fibre Reinforced Polymers." *Journal of Cleaner Production* 124: 312–324. 10.1016/j.jclepro.2016.02.148.

Mehta, P. K., & Monteiro, P. J. M. (2008). *CONCRETO: Estrutura, Propriedades e Materiais*. 3a. Pini Ltda. 10.1016/0009-2509(91)85060-b.

Mondal, Md. Omar Ali, Md. All Mokadim, & Abu Zakir Morshed. (2018). "Use of Supplementary Cementitious Materials in Recycled Brick." In *Proceedings of the 4 Th International Conference on Civil Engineering for Sustainable Development*, 1–12. Khulna, Bangladesh.

Neville, A. M. M., & Brooks, J. J. (2013). *Tecnologia Do Concreto*. 2a. Porto Alegre: Bookman.

Neville, A. M. M., & Brooks, J. J. J. (2010). *Concrete Technology*. Second. London: Pearson Education Limited. 10.1016/0360-1323(76)90009-3.

Okamura, H., & Ouchi, M. (2003). "Self-Compacting Concrete." *Journal of Advanced Concrete Technology* 1 (1): 5–15.

Okamura, H., & Ozawa, K. (1995). *Mix Design for Self-Compacting Concrete*. Concrete Library JSCE.

Parashar, Anuj, Paratibha Aggarwal, Babita Saini, Yogesh Aggarwal, & Shashank Bishnoi. (2020). "Study on Performance Enhancement of Self-Compacting Concrete Incorporating Waste Foundry Sand." *Construction and Building Materials* 251: 118875. 10.1016/j.conbuildmat.2020.118875.

Pathak, Neelam, & Rafat Siddique. (2012). "Properties of Self-Compacting-Concrete Containing Fly Ash Subjected to Elevated Temperatures." *Construction and Building Materials* 30: 274–280. 10.1016/j.conbuildmat.2011.11.010.

Petersson, O., Biillberg, R., & Van, B. K. (1996). "A Model for Self-Compacting Concrete." In *International Rilem Conference on ProiJuction Methods And Workabüity Of Concrete.*, edited by D. J. eds.) Bartos, P. J. M.; Marrs, D. L.; Cleand, D. J., 483–492. E & FN SPON.

Saak, A. W., Jennings, H. M., & Shah, S. P. (2001). "New Methodology for Designing Self-Compacting Concrete." *ACI Materials Journal* 98 (6): 429–439.

Sabir, B. B., Wild, S., & Bai, J. (2001). "Metakaolin and Calcined Clays as Pozzolans for Concrete: A Review." *Cement and Concrete Composites* 23 (6): 441–454. 10.1016/S0958-9465(00)00092-5.

Sadrmomtazi, Ali, Sahel Dolati-Milehsara, Omid Lotfi-Omran, & Aref Sadeghi-Nik. (2016). "The Combined Effects of Waste Polyethylene Terephthalate (PET) Particles and Pozzolanic Materials on the Properties of Selfcompacting Concrete." *Journal of Cleaner Production* 112: 2363–2373. 10.1016/j.jclepro.2015.09.107.

Shafigh, Payam, Lee Jin Chai, Hilmi Bin Mahmud, & Mohammad A. Nomeli. (2018). "A Comparison Study of the Fresh and Hardened Properties of Normal Weight and Lightweight Aggregate Concretes." *Journal of Building Engineering* 15 (December 2017): 252–260. 10.1016/j.jobe.2017.11.025.

Shi, Caijun, Zemei Wu, Kuixi Lv, & Linmei Wu. (2015). "A Review on Mixture Design Methods for Self-Compacting Concrete." *Construction and Building Materials* 84: 387–398. 10.1016/j.conbuildmat.2015.03.079.

Siddique, R., & M. I. Khan. (2011). *Supplementary Cementing Materials*. Edited by Springer Science & Business Media. 10-1007/978-3-642-17866-5.

Silva, A. R., Cabral, K. C., De, E. N., & Pinto, M. G. L.. (2016). "Substituição Parcial Do Cimento Portland Por Resídio De Cerâmica Vermelha Em Argamassas: Estudo Da Atividade Pozolânica." In *22ºCBECiMat - Congresso Brasileiro de Engenharia e Ciência Dos Materiais*, 1144–1154. Natal, RN. 10.1089/ast.2012.0877.

SNIC, Sindicato Nacional da Industria do Cimento. (2019). *Emissões de Gases de Efeito Estufa. Relatório Anual*. Sao Paulo. https://www.iba.org/datafiles/publicacoes/relatorios/iba-relatorioanual2019.pdf.

Sonebi, M., Lachemi, M., & Hossain, K. M. A.. (2013). "Optimisation of Rheological Parameters and Mechanical Properties of Superplasticised Cement Grouts Containing Metakaolin and Viscosity Modifying Admixture." *Construction and Building Materials* 38: 126–138. 10.1016/j.conbuildmat.2012.07.102.

Su, Nan, Kung-Chung Hsu, & His-Wen Chai. (2001). "A Simple Mix Design Method for Self-Compacting Concrete." *Cement and Concrete Research* 31 (12): 1799–1807. 10.1016/S0008-8846(01)00566-X.

Tang, W. C., Wang, Z., Liu, Y., & Cui, H. Z. (2018). "Influence of Red Mud on Fresh and Hardened Properties of Self-Compacting Concrete." *Construction and Building Materials* 178: 288–300. 10.1016/j.conbuildmat.2018.05.171.

Tiago Barroqueiro, Pedro R. da Silva & Jorge de Brito. (2020). "High-Performance Self-Compacting Concrete with Durability Assessment." *Buildings 2020_MDPI*, 1–23.

Tutikian, B. F., & Dal Molin, D.. (2011). "Comparativo Das Propriedades Do Concreto Autoadensável (CAA) Utilizando Areia Fina e Cinza Volante." *Revista IBRACON de Estruturas e Materiais* 4 (2): 247–276. 10.1590/s1983-41952011000200006.

Tutikian, B. F., & Dal Molin, D. C. (2008). *Concreto Autoadensável*. 1ª. Sao Paulo: PINI.

Tutikian, B. F., & Dal Molin, D. (2011). "Comparativo Das Propriedades Do Concreto Autoadensável (CAA) Utilizando Areia Fina e Cinza Volante." *Revista IBRACON de Estruturas e Materiais* 4 (2): 247–276. http://www.scielo.br/scielo.php?script=sci_arttext&pid=S1983-41952011000200006&lng=en&tlng=en.

Tutikian, Bernardo Fonseca. (2007). "Proposição De Um Método De Dosagem Experimental Para Concretos Auto-Adensáveis." Universidade Federal do Rio Grande do Sul.

Tutikian, Bernardo Fonseca, & Paulo Helene. (2011). "Dosagem Dos Concretos de Cimento Portland." In *Concreto: Ciência e Tecnologia*, edited by G. C. Isaia, 439–471. IBRACON.

Uysal, Mucteba. (2012). "Self-Compacting Concrete Incorporating Filler Additives: Performance at High Temperatures." *Construction and Building Materials* 26 (1): 701–706. 10.1016/j.conbuildmat.2011.06.077.

Uysal, Mucteba, & Mansur Sumer. (2011). "Performance of Self-Compacting Concrete Containing Different Mineral Admixtures." *Construction and Building Materials* 25 (11): 4112–4120. 10.1016/j.conbuildmat.2011.04.032.

Vishwakarma, Vinita, & Ramachandran, D. (2018). "Green Concrete Mix Using Solid Waste and Nanoparticles as Alternatives – A Review." *Construction and Building Materials* 162 (February): 96–103. 10.1016/J.CONBUILDMAT.2017.11.174.

Callister, W. D. Jr, & Soares, J. M. S. (2008). *Ciência e Engenharia de Materiais: Uma Introdução*. 7th ed. Rio de Janeiro, Brasil: LTC - Livros Técnicos e Científicos.

Younis, Khaleel H, Fatima Sh; Ahmed, & Khalid B. Najim. (2018). "Self-Compacting Concrete Reinforced with Steel Fibers from Scrap Tires: Rheological and Mechanical Properties." In *4th International Engineering Conference on Developments in Civil & Computer Engineering IEC2018 Proceedings Book*, 2018:189–203. 10.23918/iec2018.15.

Zuo, Wenqiang, Jiaping Liu, Qian Tian, Wen Xu, Wei She, Pan Feng, & Changwen Miao. (2018). "Optimum Design of Low-Binder Self-Compacting Concrete Based on Particle Packing Theories." *Construction and Building Materials* 163: 938–948.

20 Analysis of the Scientific Production of Cementitious Composites with Recycled Polymeric Materials

L.R. Roque-Silva[1], P.M. Alves[3], M.H.B. Souza[2], G.Z. Costal[1], R.M. Martins[3], P.C. Gonçalves[4], P. Capellato[1], M.G.A. Ranieri[4], R.G. Torres[3], M.L.N.M. Melo[1], and V.C. Santos[2]

[1]Institute of Mechanical Engineering, Federal University of Itajubá (UNIFEI), Itajubá, Brazil
[2]Institute of Natural Resources, Federal University of Itajubá (UNIFEI), Itajubá, Brazil
[3]Institute of Physics and Chemistry, Federal University of Itajubá (UNIFEI), Itajubá, Brazil
[4]Institute of Production and Management Engineering, Federal University of Itajubá (UNIFEI), Itajubá, Brazil

CONTENTS

20.1 Introduction 358
    20.1.1 Contextualization of Bibliometric Review and Data Analysis 358
    20.1.2 Plastic Waste 359
    20.1.3 Concrete with Plastic Waste 360
    20.1.4 Self-Compacting Concrete and Polymeric Waste ... 361
20.2 Methodology 362
20.3 Results and Discussion 363
    20.3.1 Choice of Database 363
    20.3.2 Documents per Year 365
    20.3.3 Documents by Author 368

DOI: 10.1201/9781003148760-20

20.3.4 Documents by Journal..........368
20.3.5 Documents by Institution..........370
20.3.6 Main Sponsors/Financiers..........373
20.3.7 Documents by Countries and Territories..........374
20.3.8 Documents by Area..........377
20.3.9 Documents by Type..........379
20.4 Conclusions..........380
Funding..........381
Acknowledgment..........381
References..........381

20.1 INTRODUCTION

20.1.1 Contextualization of Bibliometric Review and Data Analysis

The bibliometrics technique was developed at the beginning of the century to supply the need for studies and evaluation of communication activities and scientific production. Bibliometrics can be understood as a quantitative and statistical technique to measure production indicators and the rapid dissemination of scientific knowledge (Narin et al., 1994).

Bibliometrics use the application of statistical and mathematical techniques to describe characteristics as well as other means of communication. It is an analysis of quantitative information that was originally called "statistical bibliography" by Hulme in 1923. In 1934, Otlet created the term "bibliometrics" in his "Traité de Documentation" (Silva et al., 2020).

With the advent of technology and data storage, scientific research has undergone and is undergoing, deep transformations require the adoption of new research instruments and more careful data treatment refinements (Zhu et al., 1999). It is important to understand this new reality, its dynamics and complexity because that is the only way to produce consistent results through the indicators in a detailed and systematic way. The current technological and scientific scenario is dynamic and has countless sources of information, which in turn vary greatly in the credibility and quality of the data reported (Soares et al., 2016).

The use of a tool that optimizes searches for scientific publications in the initial study periods is extremely important so that researchers do not reproduce published results, thus saving time. Bibliometrics provide a method of analysis that provides an overview of a given topic with several relevant indicators (Su & Lee, 2010). Bibliometrics measure the contribution of scientific knowledge, based on scientific publications in certain areas of knowledge. These data can be used to map current trends, as well as to identify topics for new research (Subramanyam, 1983).

In short, bibliometrics is a network search engine that uses quantitative and statistical techniques to survey production rates in certain scientific areas, thus providing a means of quantifying data to contribute to scientific knowledge. With this, it is possible to obtain information such as who are the main authors, the years, the institutions, the countries and the areas that are related to the topic of interest (Price, 1976). The bibliometric analysis method can help in the evaluation of research data in publications and their citations, which are increasingly used in various research evaluation processes. The popularity of bibliometrics has increased in recent years due to the development of comprehensive and available databases, for example Scopus, dimensions and web of science. Scientists making use of these databases can correlate the number indicators with possible research trends (Díez-Herrero & Garrote, 2020). The high degree of bibliometric procession questions the use of expensive and time-consuming peer review processes in certain decision-making processes such as classifications of institutions and disciplines that are related to peer review assessments. A comparison of the use of bibliometrics and peer review is demonstrated by Pride and Knoth (2018). In their studies, they show that the use of bibliometrics reaches results similar or even superior to the peer review processes, which is a time-consuming and expensive process. The authors demonstrate that article citations are strongly linked to peer review assessments at the level of institutions and disciplines and processes that involve data.

20.1.2 Plastic Waste

Plastic is one of the greatest innovations of the 20th century and a ubiquitous material worldwide. Recently, in the last few decades, the consumption of plastic material has grown significantly, which has led to a considerable accumulation of plastic waste. As a result, it is possible to find plastic products in all ecosystems on the planet (Saikia & de Brito, 2012).

In the last decades, a large amount of non-degradable waste, particularly in the form of plastic waste (PW), has promoted serious environmental challenges. Furthermore, PW is considered one of the most dangerous forms of pollution, due to their chemical and physical characteristics and long life cycle (Guerrero et al., 2013;Iucolano et al., 2013;Wu, Li, & Xu, 2013; Liguori et al., 2014). In 2017, 348 million tons of plastic were produced worldwide, of which 64.4 million tons in the European continent alone. Approximately 31.1% (108 million tons) of this total was recycled and 27.2% disposed of as waste (95 million tons) (Europe, 2018).

In 2015, over 400 million tons (Mt) of plastics were produced, among which 164 Mt were destined for packaging, 36% of the total (Rhodes, 2018). The packaging of industrialized products represent approximately one-third of all used plastics, of which close to 40% is destined for the landfill. Currently, the amount of plastics actually measured in the oceans represents less than 1% of the 150 Mt estimated to have been released into the oceans over time (Rhodes, 2018). Plastic pollution has grown rapidly in all ecosystems in the world; the oceans' plastics pose a major threat to marine life (Senko et al., 2020). About 10% by mass of urban waste is made up of plastic; territorially the five most polluting countries of plastic are: China, Indonesia, Philippines, Vietnam and Sri Lanka.

TABLE 20.1
Production and Recycling of Plastic in the World in Tons

Country	Total Plastic Waste Generated	Total Incinerated	Total Recycled	Percentage Ratio of Production and Recycling (%)
United States	70.782.577	9.060.170	24.490.772	34.6
China	54.740.659	11.988.226	12.000.331	21.9
India	19.311.663	14.544	1.105.677	5.7
Brazil	11.355.220	0	145.043	1.3
Indonesia	9.885.081	0	362.070	3.7
Russia	8.948.132	0	320.088	3.6
Germany	8.286.827	4.876.027	3.143.700	37.9
United Kingdom	7.994.284	2.620.394	2.513.856	31.5
Japan	7.146.514	6.642.428	405.834	5.7
Canada	6.696.763	207.3540	1.423.139	21.3

Source: WWFBrasil, 2020.

These alone contribute 56% of the global plastic waste (Rhodes, 2018). Brazil, according to data from the World Bank, is the fourth-largest producer of plastic waste in the world, with 11.3 million tons. Still, only 145,000 tons (1.28%) are effectively recycled, according to Table 20.1, that summarizes data on the relationship between waste production and recycling in several countries (WWFBrasil, 2020).

Because the polymer has a low rate of degradation and is naturally bulky, landfills cannot be considered an ecological alternative for the disposal of PW (Faraj et al., 2020). The decomposition time for wastes, including plastics and rubber that have a polymeric nature, can reach more than 400 years. The bulky waste of the PW can obstruct groundwater flows and prevent the growth of tree roots, among other problems. The PW can contaminate the water and the soil when they are mixed with the rainwater. With this, they disperse in the environment where there are toxic elements, such as lead and cadmium (Faraj et al., 2020).

20.1.3 CONCRETE WITH PLASTIC WASTE

Concrete, after drinking water, is the most consumed material daily in the world, so research aimed at incorporating recycled waste into this cementitious composite is of great social value (Thorneycroft et al., 2018). To reduce the influence of PW on the environment in terms of energy consumption, natural resources, waste disposal, global warming and environmental pollution, recycling of PW can represent one of the best solutions (Farhad Aslani et al., 2018; Aslani & Khan, 2019). The use of PW as an aggregate in concrete is investigated in several studies that generally recommend its use for non-structural functions, in addition to describing methods of obtaining and preparing waste for use in concrete (Faraj et al., 2020; Thorneycroft et al., 2018; Farhad Aslani et al., 2018; Guendouz et al., 2016; Farhad Aslani & Kelin, 2018; Bušić et al., 2018; Sami Kohistani & Singh, 2018; Verdolotti et al., 2014).

In practical applications in civil engineering, the recycling of plastic waste can be incorporated as a base for cementitious material, such as in mixtures of mortar and concrete. Due to ecological and economic advantages, this plastic waste can replace or work together with the traditional natural aggregates of concrete (Binici & Aksogan, 2016).

Concretes with plastic waste, in general, have low specific mass and are ideal for applications of non-structural elements and concrete asphalt pavements (F. Aslani & Khan, 2019). Many studies prove the applicability of concrete with plastic waste in different situations. There are several studies on types of polymeric wastes used as concrete aggregates, for example (Binici & Aksogan, 2016) using polyvinyl chloride (PVC) waste tubes (Iucolano et al., 2013; Saikia & de Brito, 2014), polyethylene terephthalate (PET) waste (Shanmugapriya & Helen Santhi, 2017) waste from high-density polyethylene (HDPE) (Sayadi et al., 2016), expanded polystyrene foam (EPS) waste (Dalhat & Al-Abdul Wahhab, 2017), recycled thermoplastic polystyrene waste and (Yang et al., 2015) polypropylene fiber.

20.1.4 SELF-COMPACTING CONCRETE AND POLYMERIC WASTE

Among the cementitious composites, SCC has been gaining prominence for its unique qualities in the fresh state that allows its use in densely armed structures and difficult to access. Thus, its use is indicated for medium to large construction and for reinforcement and revitalization of structures (Farhad Aslani & Kelin, 2018; EFNARC, 2005).

Self-compacting concrete (SCC) fits into a special type of concrete that has the characteristic of being self-compacting; that is, due to its own weight, it can flow and fill the mold, without requiring external vibration (EFNARC, 2005; Mohamed et al., 2003; Okamura & Ozawa, 1996). However, to obtain such a result, the SCC needs to have cohesion and fluidity in equilibrium such that segregation and exudation of its constituent materials do not occur (Djelal et al., 2004). To obtain adequate workability, the SCC traces often contain the addition of superplasticizer additives, which reduce the demand for mixing water and fine material (silica fume, stone powder, fly ash and blast furnace slag) that improve stability and granulometric packing of concrete (Han & Yao, 2004). Viscosity modifying additive can also be used along with fine materials to achieve adequate cohesion and resistance to segregation (Şahmaran et al., 2006; Felekoğlu et al., 2006). SCC is frequently used to incorporate plastic waste, thus generating new characteristics for conventional self-compacting concrete. However, it is necessary to take due care to maintain the main characteristics of SCC, which in the fresh state are fluidity, stability and homogeneity, and in the hardened state are better mechanical properties and good durability (Safi et al., 2013).

Given the lack of data and studies that compile the number of studies related to the concrete theme with polymeric waste, and knowing the great importance of the research that shows the current scenario related to the theme, we tried to carry out this study through bibliometric indicators linked to the concrete with the polymeric waste theme, through a multidisciplinary and methodological approach. The use of bibliometric indicators requires users to have a comprehensive view of the area of specific knowledge and the limitations inherent to each segment.

In this sense, this work aims to make a bibliometric survey on cementitious composites with the addition of polymeric residues in order to know where these studies are more advanced, which are the most productive research groups and the number of studies that are being done. Another important point is that this type of research seeks to avoid bias or biased research that compromises the interpretation of data for the formation of standards and best practice procedures. It even helps in the definitive exclusion of certain material in some field of possible use.

20.2 METHODOLOGY

The research developed is based primarily on the choice of the engineering database that generates significant results in the search for documents related to the topic. The database chosen should initially cover much of the literature related to the topic: CC and SCC with polymeric waste. To choose the database, string refinement steps (with adjustments to keywords and Boolean operators) were carried out in a search that was inserted in five different databases.

Besides quantifying criteria concerning the coverage of articles, books and works, the selection of the database used in the bibliometric research took into account available tools to analyze the results, such as the generation of graphs and diagrams available on the database platform. Initially, several search strings were developed for five different databases: Scopus, Web of Science, Scielo, Engineering Village and Science Direct. These strings were refined and used in the same way in all selected databases. The refinement considered the number of results generated in each search.

After choosing the database and defining the search string, the database search started working in its search fields. Right from the start, the options for "documents" and search in "titles," "keywords" and "abstracts" were selected. The period considered, both for searching with CC and for SCC, was the entire period of data in the selected database, so a period of predefined years was not limited.

To assist in the analysis of data extracted from the selected database, the VOS viewer software was used. The VOS viewer software is a tool developed for the construction and visualization of a bibliometric network (van Eck & Waltman, 2010). These networks can contain data such as the names of newspapers, researchers, publications and countries and can be constructed considering criteria such as citation, academic production, co-authorship and others (van Eck & Waltman, 2010). In this study, the VOS viewer was used to analyze the keywords searched and other terms searched.

Production of Cementitious Composites

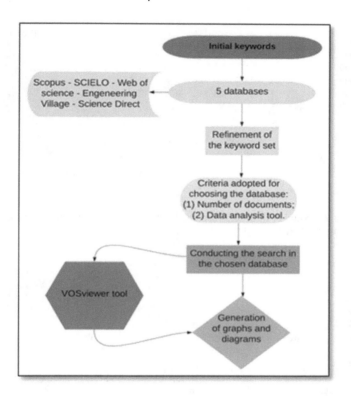

FIGURE 20.1 Flow chart of the adopted search methodology (AUTHOR, 2021).

To do so, the same search conducted in the Scopus database was used. Figure 20.1 shows a flow chart that summarizes the methodology adopted by this study, as well as the phases to get to the results.

After the generation of graphs and diagrams with bibliometric data, the results were analyzed, seeking to link the historical, political and economic events that involve the development of the use of concrete with polymeric residues and that justify exactly the behavior of the data in certain analyzed periods. Data analyses were performed in related ways; that is, the results obtained in certain analyses were used to confirm trends observed in other results.

20.3 RESULTS AND DISCUSSION

20.3.1 CHOICE OF DATABASE

Figure 20.2 shows the results of searches for databases for the CC theme with polymeric waste. Figure 20.3 shows the results of the search carried out for the SCC with polymeric waste.

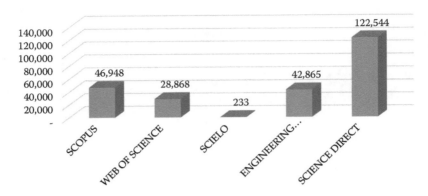

FIGURE 20.2 Number of articles searched with the string: concrete AND (plastic* OR polymer* OR RUBBER*) (AUTHOR, 2021).

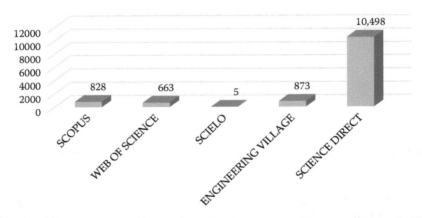

FIGURE 20.3 Number of articles searched with the string: self AND compacting AND concrete AND (plastic* OR polymer* OR rubber) (AUTHOR, 2021).

As can be seen in Figure 20.2, the bases that generated the highest results were Science Direct, Scopus and Engineering Village, respectively, for CC. For SCC, the greatest results were found in Science Direct, Engineering Village and Scopus, respectively, as can be seen in Figure 20.3.

Based on this research, we then chose to use the Scopus database, as it provides a considerable number of results for the research carried out and because it is a database that has a large number of relevant journals related to engineering, besides offering computational resources of data analysis that facilitate the organization of collected data and discussions of results. Resources similar to these are not provided by the Science Direct database that generated the most results.

After the database selection step, a further refinement was made to the search string. It was decided to introduce the keywords "waste" and "residue" to restrict

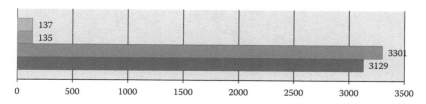

■ self AND compacting AND concrete AND (waste* OR residue*) AND (plastic* OR polymer* OR rubber*)

■ self AND compacting AND concrete AND (waste OR residue) AND (plastic* OR polymer* OR rubber*)

■ concrete AND (waste* OR residue*) AND (plastic* OR polymer* OR rubber*)

■ concrete AND (waste OR residue) AND (plastic* OR polymer* OR rubber*)

FIGURE 20.4 Document numbers in search at Scopus (AUTHOR, 2021).

the field of research to plastic waste or polymers. Figure 20.4 shows the results for both CC and SCC.

From this stage, to conduct the research, only the Scopus database was used. For this purpose, the CC research was concentrated in the period from 1954 to 2021, as this was the deadline provided by the database, knowing that this study seeks to be developed in the longest possible coverage time. For SCC, filters from 2003 to 2021 were used, which is also the period provided by the database. Authors linked to institutions of all nationalities were considered. Three thousand one hundred and twenty-nine peer-reviewed scientific articles were selected, as well as material such as books, theses and dissertations related to the CC theme with polymeric waste, and 136 SCC documents with polymeric waste.

Figures 20.5 and 20.6 show the results for the search for the most relevant keywords related to the theme of this study.

The keywords are organized to show the connection between them and in which period (2010 to 2021) the authors most cited them. In general, for CC, the keywords are recent and were mostly cited between 2016 to 2021. For SCC, there are many current cited keywords between 2019 and 2021. But others had their moments of relevance about 8 to 10 years ago, as is the case with flexural strength, waste management and crumb rubber.

20.3.2 Documents per Year

Figure 20.7a and b shows the number of documents published per year related to CC and SCC with polymeric waste (Scopus, 2020).

The number of documents per year has grown sharply since the beginning of the 21st century, around 2002. In the years 2008 to 2014, the scientific production related to the theme exceeds 100 documents per year, and from 2014 it exceeds 200 documents per year. The sharp growth continues until the moment when the number of documents approaches 500 per year.

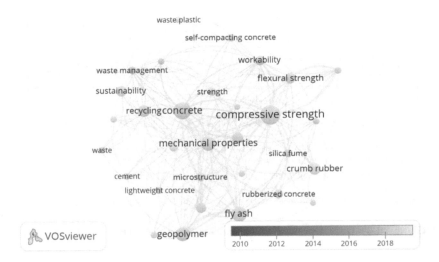

FIGURE 20.5 Relevant keywords on cement composite with recycled polymeric constituents for CC from 2010 to 2021 (AUTHOR, 2020).

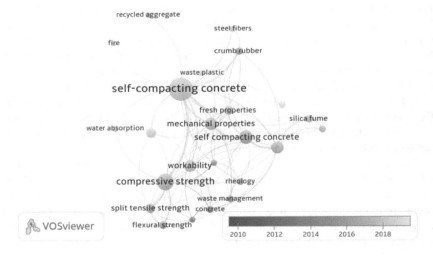

FIGURE 20.6 Relevant keywords on cement composite with recycled polymeric constituents for SCC from 2010 to 2021 (AUTHOR, 2021).

The production of SCC with waste, as shown in Figure 20.7b, has grown significantly since the last decade, reaching close to 35 documents in 2021. The number of documents related to the SCC theme represents approximately 5 to 10% of CC production in the period from 2019 to 2021 (Figure 20.7a). SCC is a special type of concrete that was developed in the 1980s in Japan by researcher Okamura and his team (Okamura & Ozawa, 1996). Nonetheless,

Production of Cementitious Composites

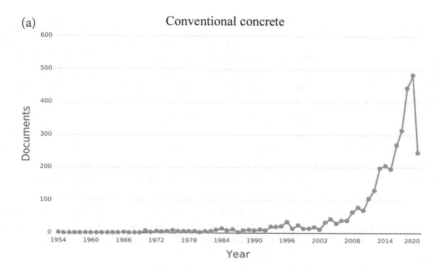

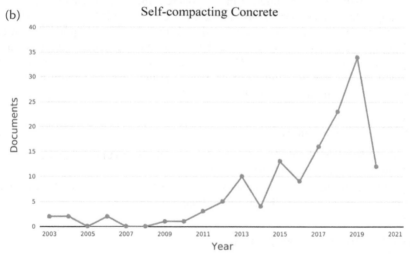

FIGURE 20.7 Documents per year on cement composite with recycled polymeric constituents for CC and SCC with polymeric waste (AUTHOR, 2021).

recycled plastic materials started to be adopted more frequently as an aggregate for both CC and SCC in the early 21st century.

The marked production of scientific studies related to the concrete theme with polymeric waste can be understood as a consequence of public policies concerning the environment that had been gradually discussed and adopted by countries in world conferences on the environment (Djelal et al., 2004). The international events that propelled the constituted powers of the countries to take mitigating measures regarding the environment follow a historical line that begins in the 1970s: The Club of Rome (1972) (Thompson, 2001), The Stockholm Conference on Environment (1972) (Grieger, 2012), A UN World Commission on Environment and Development (CMMAD) (1983/1987), The United Nations Conference on Environment and Development (UNCED) (1992), The Johannesburg World Summit on Sustainable Development (2002) (von Schirnding, 2005; Frey, 2014).

These events discuss issues related to the environmental crisis and are gaining strength and impacting the goals of the government in search of sustainable and no longer exploratory and inconsequential development (Frey, 2014). At this scenario, environmental public policies are the main stage of international economic-industrial expansion and are aimed at reducing harmful human intervention in the environment, which generates, among other problems: global warming, acid rain, contamination of water sources, desertification, deforestation, excessive CO_2 emission, species extinction, melting of polar ice caps and depletion of natural resources (*2nd International Conference on Green Buildings Technologies and Materials, GBTM 2012*, 2013).

The growth due to the interest in the concrete theme with plastic waste does not happen by chance, particularly in the early 2000s, but it follows a worldwide trend of preserving the environment and with measures that impose obligations for recycling and reuse of waste. With the development of society, there is a need for infrastructure development. New characteristics are required for concrete, and new aggregates are studied to decrease the demand for natural aggregates that are limited in nature (Gupta et al., 2016).

20.3.3 DOCUMENTS BY AUTHOR

Figure 20.8ab lists the main authors who publish documents related to CC and SCC with polymeric waste (Scopus, 2020).

Figure 20.9 shows the results for searching the VOSviewer® Software for the most relevant authors linked to the theme of this study.

20.3.4 DOCUMENTS BY JOURNAL

Figure 20.10 lists the main journals that published documents related to CC and SCC with polymeric waste (Scopus, 2020).

Through the results, it is possible to notice that the main journals are *Construction and Building Materials, Advanced Materials Research, Iop Conference Series Science and Engineering, Journal of Cleaner Production e International Journal of Civil Engineering and Technology*. The following should be highlighted: *Construction and Building Materials* and *Advanced Materials Research*, which contributed with more than 40 documents per year in the decade from 2010 to 2020. The journals *Iop Conference Series Science and Engineering* and *Journal of Cleaner Production* and *International Journal of Civil Engineering and Technology* are in the range of 20 to 40 documents produced per year.

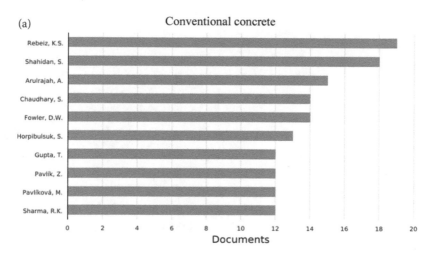

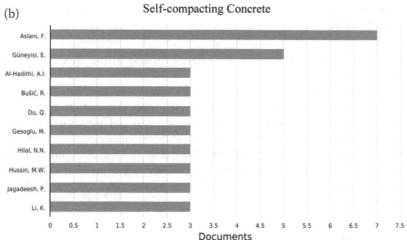

FIGURE 20.8 Documents by author for cement composite with recycled polymeric constituents on CC and SCC (AUTHOR, 2021).

Publications, in general, appear to be increasing in most newspapers. The apparent decrease shown in the 2019 to 2021 biennium should not be taken into account, as the publications for that period were probably still being updated in the database. In Figure 20.11, the results show that the journals that most publish documents related to the SCC theme with polymeric waste are *Construction and Building Materials, Internacional Journal of Civil Engineering and Technology, Journal of Cleaner Production, Journal of Building Engineering, Materials and Structural Concrete*. The *Construction and Building Materials* journal, which kept the number of publications constant from 2013 to 2021, should be highlighted. On average,

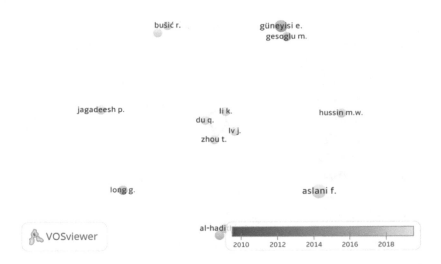

FIGURE 20.9 Relevant authors on SCC with polymeric residues from 2010 to 2019 (AUTHOR, 2021).

journals publish around 2 to 3 documents related to the SCC theme with polymeric waste per year.

Some considerations about the journal's origin can be made, considering that the main journals listed in Figure 20.11 are located in Europe. The European continent is consistent with the international trend, given that the European Union seeks to expand its collaborations with countries on the Asian continent, especially China [48]. The European Union does not appear to be seriously affected by internal and external economic problems. Investment in research comes largely from around 36% of the public sector (Zakaria & Bibi, 2019).

20.3.5 Documents by Institution

Figure 20.11 lists the main institutions that publish documents related to concrete and SCC with polymeric waste (Scopus, 2020).

The main institutions linked to the production of documents related to the concrete theme with polymeric waste are: Universiti Tun Hussein Onn Malaysia (Malasia), Universiti Teknologi Malaysia (Malasia), Universiti Sains Malaysia (Malasia), Malaviya National Institute of Technology Jaipur (India), Ceské vysoké ucení technické v Praze (Czech Republic), Universitatea Tehnica Gh. Asachi din IasI (Romania), University Of Anbar, Ministry of Education China (China), RMIT University (Australia), Southeast University, Nanjing (China). The institutions that appear as the largest producers of documents related to concrete with polymeric waste are linked to the countries with the greatest scientific production related to the topic on concrete with polymeric waste. Therefore, it is consistent that these institutions are highlighted.

Production of Cementitious Composites

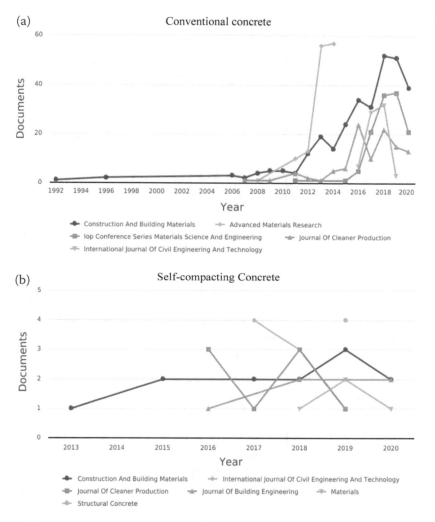

FIGURE 20.10 Documents by journals for cement composite with recycled polymeric constituents on CC and SCC (AUTHOR, 2021).

It should be noted that the Asian continent, represented in large percentages by Malaysia and China, respectively, are the countries that most produce research on the subject in question. This is certainly mostly due to infrastructure investments made in these countries over the past decade. The Chinese and Indian economies, in particular, since the late 1990s have shown annual growth above 8% (Asif & Muneer, 2007). Malaysia, which is located in Southeast Asia, is an aspiring Asian tiger country (Hong Kong, South Korea, Singapore, Taiwan) since its foreign investments come from the Asian tigers themselves that drive its development (World Bank, 2019). This group

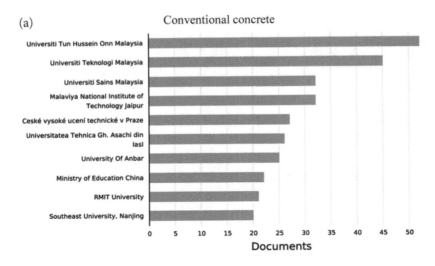

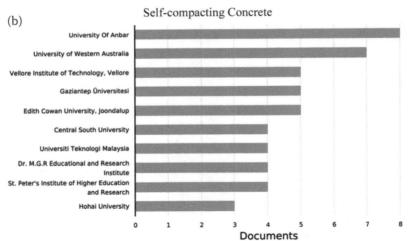

FIGURE 20.11 Documents by institutions for cement composite with recycled polymeric constituents on CC and SCC (AUTHOR, 2021).

of countries of marked economic development identified the incentive for research and development as the main factor in economic development (Zakaria & Bibi, 2019).

Regarding SCC with polymeric waste, the main institutions are: University Of Anbar (Iraq), University of Western Australia (Australia), Vellore Institute of Technology (India), Vellore Gaziantep Üniversitesi (Turkey), Edith Cowan University (Australia), Joondalup Central South University (Australia), Universiti Teknologi Malaysia (Malasia), Dr. M.G.R Educational

Production of Cementitious Composites 373

and Research Institute (India). St. Peter's Institute of Higher Education and Research Hohai University (India). The classification of institutions is certainly linked to the investment of the respective countries of origin of each institution. For example, India, China, Iraq and Australia are the countries that produce the most scientific documents related to the topic. Therefore, the institutions located in their territories are more prominent in this field of study, as with institutions linked to the research of CC with polymeric waste.

20.3.6 Main Sponsors/Financiers

Figure 20.12 lists the main sponsors/financiers that publish documents related to CC and SCC with polymeric waste (Scopus, 2020).

The main funding agencies for studies related to concrete with polymeric waste are: National Natural Science Foundation of China, Australian Research Council, European Regional Development Fund, Fundamental Research Funds for the Central Universities, European Commission, Universiti Teknologi Malaysia, Coordenação de Aperfeiçoamento de Pessoal de Nível Superior (CAPES), Grantova agentura ceske republiky, National Science Foundation (NSF), Thailand Research Fund. The main sponsors/financiers of research are largely located in Asia, with emphasis on China, which has been consistently investing in research since 1990. As a comparative measure, China's economy has grown between 9% to 10% in recent years, while investments in research have increased at the rate of 12% (Zakaria & Bibi, 2019).

For SCC with polymeric waste, the main funding agencies are: National Natural Science Foundation of China, School of Civil Environmental and Mining, Engineering University of Adelaide, Fundamental Research Funds for the Central Universities, China Postdoctoral Science Foundation, Hrvatska Zaklada za Znanost, Ministry of Education – Singapore, Coordenação de Aperfeiçoamento de Pessoal de Nível Superior (CAPES), Department of Education of Guangdong Province, Ministerio de Economia y Competitividad, Ministry of Higher Education, Malaysia. For both CC and SCC, China's funding agency, the National Natural Science Foundation of China, appears as the main sponsor. We highlight the School of Civil Environmental and Mining and the Engineering University of Adelaide, who contributed four and three studies, respectively. CAPES appears funding only one study on the topic. It is worth mentioning that the growth in research and development in Asian countries is a reflection of the rapid economic advance, population growth, and the consequent training of a generation of trained professionals (Zakaria & Bibi, 2019). Between 2003 and 2008, the number of researchers in Taiwan, China, Singapore and South Korea grew by approximately 16%. In contrast, in the USA, the number of young researchers decreased from 51% to 49%. As a comparison with the publication of articles, in Asia, in the last decade, there was an increase of 9% per year, while in the USA and the European Union only 1% (Yuan & Xiang, 2018).

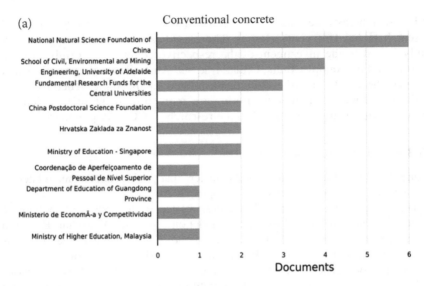
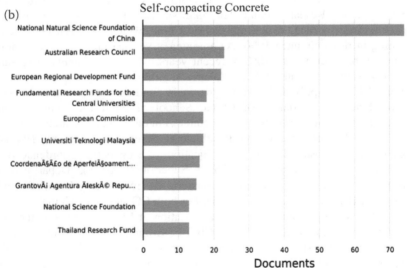

FIGURE 20.12 Documents by sponsors/financiers for cement composite with recycled polymeric constituents on CC and SCC (AUTHOR, 2021).

20.3.7 DOCUMENTS BY COUNTRIES AND TERRITORIES

Figure 20.13 lists the main countries that publish documents related to CC and SCC with polymeric waste (Scopus, 2020).

The countries India, USA, China, Malaysia and Australia produce a large number of studies related to concrete with plastic waste. It is worth highlighting

Production of Cementitious Composites

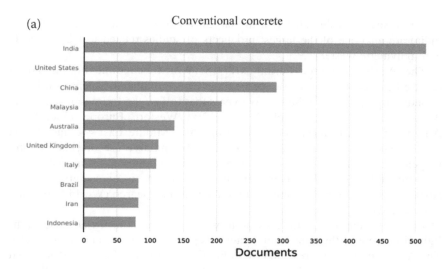

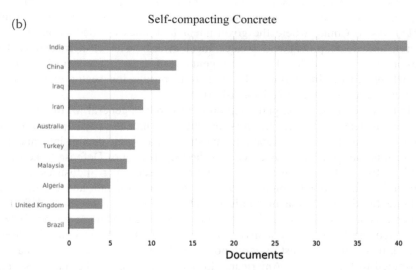

FIGURE 20.13 Documents by country or territory for cement composite with recycled polymeric constituents on CC and SCC (AUTHOR, 2021).

India, which reached the mark of 500 documents in the surveyed period, representing the sum of several countries together. Both developed countries (USA and China), as well as countries that are in the process of development (India, Malaysia and Brazil), need large volumes of concrete to develop. It is possible to draw a parallel between investment in infrastructure and the consumption of concrete (Gupta et al., 2016). It is interesting to make a correlation between the countries that produce the most polymeric wastes and the countries that produce the most studies on recycling these wastes in the form of aggregate for concrete.

It can be seen in Table 20.1 that the largest producers of waste (USA, China, India) are also one of the largest producers of studies related to CC and SCC with polymeric waste.

The World Economic Forum 2017/18 (World Economic Forum, 2018) released a classification of the percentage of accumulated expense on infrastructure investments in the last five years that follow the following sequence: Asia-Pacific 61%; USA and Canada 12%; Western Europe 11%; Latin America 5%; Central and Eastern Europe 5%; Middle East 4% and Africa 2%. These data can be used to understand the massive use of concrete and recycled material, mainly by Asian and Pacific countries, as infrastructure investment involves the investment in various aspects of society, such as sustainable development combined with scientific research (World Economic Forum, 2018).

Regarding the SCC with polymeric waste, Asian countries are also highlighted in terms of the number of documents published, as seen in Figure 20.13. The Asian countries represented by India, China and the Middle East Iraq and Iran, as well as Oceania, Australia represent regions of the world where there is a greater incentive to research and development (World Economic Forum, 2018). As in China, where the government facilitates tax deductions for research and development and creates local awards with cash incentives for authors of impact articles. The government encourages the transfer of scientific knowledge to the industrial and commercial production sector to obtain an economic return (Zakaria & Bibi, 2019).

In the case of the USA, participation in research in general for the scientific public has decreased. One reason is that the government's participation in this type of research and development investment has been decreasing, and the sector that most felt this slowdown was the Department of Defense (Zakaria & Bibi, 2019). Although government spending is decreasing, industries, private universities and NGOs continue to make their investments and try to compensate for the reduction in public incentives. For this reason, the USA, in general, is still one of the main producers of research (Yuan & Xiang, 2018). Although the USA does not invest at the pace of the Asian countries mentioned, it is still the largest research power in the world. Something that can be attributed to the fact that Americans do not appear among the main research developers in the area of SCC with plastic waste is that the USA focuses its research investments as a business product, mainly in the field of engineering. Making development and improvements in products and services that remain as a business segregate and that do not contribute significantly to the scientific community, concerning the dissemination of knowledge as belonging to the public (De Negri, 2014).

Figures 20.14 and 20.15 show the results for the search for the most relevant countries linked to the theme of this study.

There is a link or collaboration between research carried out in the main countries for both CC and SCC. The results shown in Figures 20.14 and 20.15

Production of Cementitious Composites

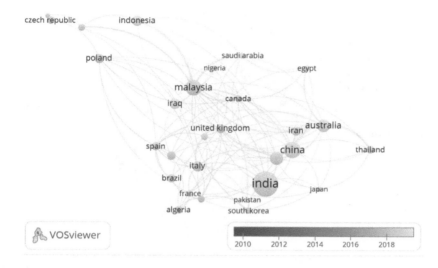

FIGURE 20.14 Relevant countries on CC with polymeric wastes from 2010 to 2021.

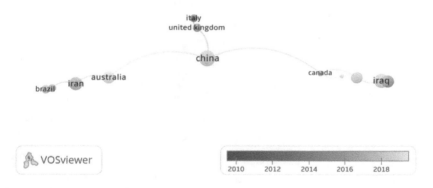

FIGURE 20.15 Relevant countries on SCC with polymeric waste period from 2010 to 2021 (AUTHOR, 2021).

collaborate with the trends shown in Figure 20.13. Furthermore, it can be seen that research with CC is current between 2016 and 2021 in practically all countries. The research with SCC is also shown to be current in most countries. However, some countries no longer publish with relevance for about 8 to 10 years, as is the case of Italy and the United Kingdom.

20.3.8 Documents by Area

Figure 20.16 shows the distribution of published documents for CC and SCC for the areas of knowledge.

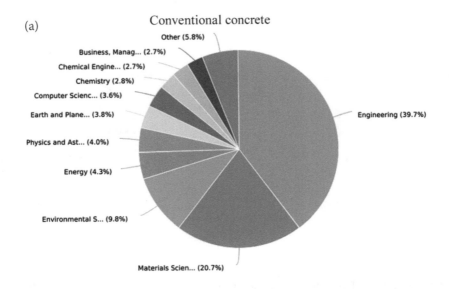

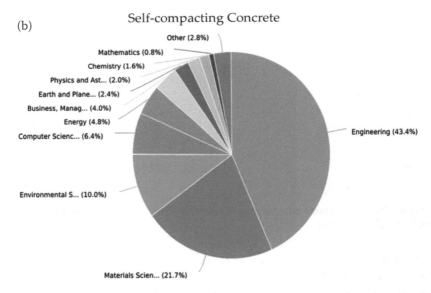

FIGURE 20.16 Documents by area for cement composite with recycled polymeric constituents on CC and SCC (AUTHOR, 2021).

It is possible to identify that the areas of engineering, materials sciences and environmental sciences were the most prominent in disseminating scientific material, both for CC and for SCC with waste. However, the areas of energy and physics appear in fourth and fifth place for CC. As for SCC, what changes is that the fourth and fifth places are areas of computer science and energy.

20.3.9 Documents by Type

Figure 20.17 shows the distribution of published documents for CC and SCC for the different types of documents.

It is possible to identify that the main means of communication for SCC are articles (56.7%), conferences or congresses (29.1%), review conferences (6.5%),

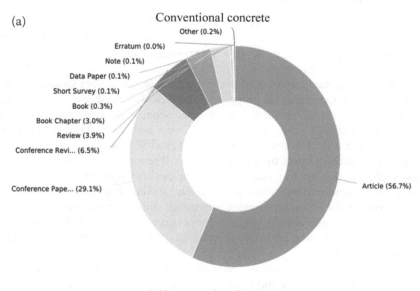

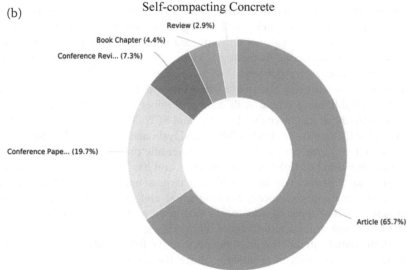

FIGURE 20.17 Documents by type for cement composite with recycled polymeric constituents on CC and SCC (AUTHOR, 2021).

review articles (3.9%) and book chapters (3.0%). For SCC, the same classification order is repeated, for articles (65.7%), conferences or congresses (19.7%) and review conferences (7.3%). However, in SCC, book chapters (4.4%) are more expressive than review articles (2.9%).

20.4 CONCLUSIONS

This study makes a bibliometric review on Portland cement composite with recycled polymeric constituents of conventional concretes and self-compacting concretes with polymeric waste instead of natural aggregates. The following conclusions can be drawn:

In general, the methods adopted to carry out the research were satisfactory and made it possible to achieve the proposed objectives. The databases that generated the most results for CC were: Science Direct, Scopus and Engineering Village, respectively. For SCC, they were Science Direct and Engineering Village and Scopus, respectively. The keywords for both CC and SCC are linked to each other, so they are cited by authors from different areas. The main keywords for CC were mostly cited in the last five years, whereas in the keywords for SCC, there is a greater dispersion over time, with some words outdated after 8 to 10 years. The main keywords for CC are concrete, compressive strength, mechanical properties, flexural strength, sustainability, fly ash, geopolymer and crumb rubber. For SCC, they are self-compacting concrete, mechanical properties, compressive strength and recycled aggregate.

The number of documents per year for both CC and SCC has grown exponentially since the 2000s and has continued to grow until the present moment. The main authors who publish in the area are segregated into individual research groups and have produced their work with greater volumes of studies in the last five years. The main journals are concentrated in European and Asian countries, and for CC, the production of the top 5 journals is over 20 documents per year, and for SCC, it is 1 to 3 per year.

The main institutions and main sponsors are from European and Asian countries. Highlighting China with a large number of institutions and a large volume of financial incentives for both CC and SCC. The countries that stand out the most for CC are India, USA, China, Malaysia and Australia. For SCC, they are India, China, Iraq, Iran and Australia. Scientific production for both CC and SCC has some kind of link between countries and has been developed mainly in the last 8 to 5 years. The areas with the largest number of publications are engineering, materials sciences, environmental and energy sciences and computer sciences. The types of documents most used as a means of publication for CC and SCC are articles, conferences and reviews.

With the data from this research, the perspective for the future of SCC and CC with polymeric residues is promising. Since the number of research in recent years is growing, there is involvement of several countries and researchers around the world. This also shows a proxy for world leaders (companies and politicians) with the environment and sustainable growth. This concern with the environment can be seen as a position in the predatory and unbalanced development of the last century.

The investigated theme is current and promising both for the development of concrete technology and for the environment with reuse of non-degradable material and replacement of natural aggregates. Since the approach of this study was carried out by the quantitative investigation of studies developed in the area of interest, thus, future studies are recommended to investigate which are the specific types of polymeric wastes most used, which are the main practical uses for concrete with polymeric wastes, and the financial and environmental impacts of the practice of reusing polymeric materials on a large scale in construction.

FUNDING

This research was funded by Coordination for the Improvement of Higher Education Personnel (CAPES) Finance Code 001. Minas Gerais Research Foundation (FAPEMIG) grant number APQ-01301-15 and National Council for Scientific and Technological Development (CNPq) grant number 308021/2018-5. The APC was funded by Federal University of Itajubá (Unifei).

ACKNOWLEDGMENT

The authors are thankful for the financial assistance granted by the Coordination for the Improvement of Higher Education Personnel (CAPES). The Federal University of Itajubá (UNIFEI) for all the support in laboratories and technicians.

REFERENCES

2nd International Conference on Green Buildings Technologies and Materials, GBTM 2012. (2013). *Advanced Materials Research*. 689.

Asif, M., & Muneer, T. (2007). "Energy Supply, Its Demand and Security Issues for Developed and Emerging Economies." *Renewable and Sustainable Energy Reviews* 11 (7): 1388–1413. 10.1016/j.rser.2005.12.004.

Aslani, F., & Khan, M. (2019). "Properties of High-Performance Self-Compacting Rubberized Concrete Exposed to High Temperatures." *Journal of Materials in Civil Engineering* 31 (5). 10.1061/(ASCE)MT.1943-5533.0002672.

Aslani, Farhad, & Jack Kelin. (2018). "Assessment and Development of High-Performance Fibre-Reinforced Lightweight Self-Compacting Concrete Including Recycled Crumb Rubber Aggregates Exposed to Elevated Temperatures." *Journal of Cleaner Production* 200: 1009–1025. 10.1016/j.jclepro.2018.07.323.

Aslani, Farhad, Guowei Ma, Dominic Law Yim Wan, & Gojko Muselin. (2018). "Development of High-Performance Self-Compacting Concrete Using Waste Recycled Concrete Aggregates and Rubber Granules." *Journal of Cleaner Production* 182: 553–566. 10.1016/j.jclepro.2018.02.074.

Binici, Hanifi, & Orhan Aksogan. (2016). "Eco-Friendly Insulation Material Production with Waste Olive Seeds, Ground PVC and Wood Chips." *Journal of Building Engineering* 5: 260–266. 10.1016/j.jobe.2016.01.008.

Bušić, Robert, Ivana Miličević, Tanja K Šipoš, & Kristina Strukar. (2018). "Recycled Rubber as an Aggregate Replacement in Self-Compacting Concrete—Literature Overview." *Materials.* 10.3390/ma11091729.

Dalhat, M. A., & Al-Abdul Wahhab, H. I. (2017). "Properties of Recycled Polystyrene and Polypropylene Bounded Concretes Compared to Conventional Concretes." *Journal of Materials in Civil Engineering* 29 (9): 4017120. 10.1061/(ASCE)MT.1943-5533.0001896.

Díez-Herrero, Andrés, & Julio Garrote. (2020). "Flood Risk Analysis and Assessment, Applications and Uncertainties: A Bibliometric Review." *Water* 12 (7). 10.3390/w12072050.

Djelal, C., Vanhove, Y., & Magnin, A. (2004). "Tribological Behaviour of Self Compacting Concrete." *Cement and Concrete Research* 34 (5): 821–828. 10.1016/j.cemconres.2003.09.013.

van Eck, Nees Jan, & Ludo Waltman. (2010). "Software Survey: VOSviewer, a Computer Program for Bibliometric Mapping." *Scientometrics* 84 (2): 523–538. 10.1007/s11192-009-0146-3.

EFNARC. (2005). "The European Guidelines for Self-Compacting Concrete." *The European Guidelines for Self Compacting Concrete,* no. May: 63. http://www.efnarc.org/pdf/SCCGuidelinesMay2005.pdf.

Europe, Plastics. (2018). "Plastics – the Facts." *Plastics – the Facts 2018,* 38.

Faraj, Rabar H., Hunar F. Hama Ali, Aryan Far H. Sherwani, Bedar R. Hassan, & Hogr Karim. (2020). "Use of Recycled Plastic in Self-Compacting Concrete: A Comprehensive Review on Fresh and Mechanical Properties." *Journal of Building Engineering* 30 (February): 101283. 10.1016/j.jobe.2020.101283.

Felekoğlu, Burak, Kamile Tosun, Bülent Baradan, Akın Altun, & Bahadır Uyulgan. (2006). "The Effect of Fly Ash and Limestone Fillers on the Viscosity and Compressive Strength of Self-Compacting Repair Mortars." *Cement and Concrete Research* 36 (9): 1719–1726. 10.1016/j.cemconres.2006.04.002.

Frey, Klaus. (2014). "Governança Urbana e Participação Pública," no. January 2007.

Grieger, A. (2012). "Only One Earth: Stockholm and the Beginning of Modern Environmental Diplomacy," Environment & Society Portal, Arcadia, no. 10. Rachel Carson Center for Environment and Society. 10.5282/rcc/3867.

Guendouz, M., Debieb, F., Boukendakdji, O., Kadri, E. H., Bentchikou, M., & Soualhi, H. (2016). "Use of Plastic Waste in Sand Concrete." *Journal of Materials and Environmental Science* 7 (2): 382–389.

Guerrero, Lilliana Abarca, Ger Maas, & William Hogland. (2013). "Solid Waste Management Challenges for Cities in Developing Countries." *Waste Management* 33 (1): 220–232. 10.1016/j.wasman.2012.09.008.

Gupta, Trilok, Sandeep Chaudhary, & Ravi K Sharma. (2016). "Mechanical and Durability Properties of Waste Rubber Fiber Concrete with and without Silica Fume." *Journal of Cleaner Production.* Elsevier Ltd.

Han, Lin-Hai, & Guo-Huang Yao. (2004). "Experimental Behaviour of Thin-Walled Hollow Structural Steel (HSS) Columns Filled with Self-Consolidating Concrete (SCC)." *Thin-Walled Structures* 42 (9): 1357–1377. 10.1016/j.tws.2004.03.016.

Iucolano, F., Liguori, B., Caputo, D., Colangelo, F., & Cioffi, R. (2013). "Recycled Plastic Aggregate in Mortars Composition: Effect on Physical and Mechanical Properties." *Materials & Design (1980-2015)* 52: 916–22. 10.1016/j.matdes.2013.06.025.

Liguori, Barbara, Fabio Iucolano, Ilaria Capasso, Marino Lavorgna, & Letizia Verdolotti. (2014). "The Effect of Recycled Plastic Aggregate on Chemico-Physical and Functional Properties of Composite Mortars." *Materials & Design* 57: 578–584. 10.1016/j.matdes.2014.01.006.

Mohamed Lachemi, Khandaker M. A. Hossain, Vasilios Lambros, & Nabil Bouzoubaa. (2003). "Development of Cost-Effective Self-Consolidating Concrete Incorporating Fly Ash, Slag Cement, or Viscosity-Modifying Admixtures." *Materials Journal* 100 (5): 419–425.

Narin, Francis, Dominic Olivastro, & Kimberly A. Stevens. (1994). "Bibliometrics/Theory, Practice and Problems." *Evaluation Review* 18 (1): 65–76. 10.1177/0193 841X9401800107.

Negri, Fernanda De. (2014). "Investimentos em P&D do Governonorte-Americano: Evolução e Principaiscaracterísticas." http://repositorio.ipea.gov.br/handle/11058/3317?mode=full. Access in Jan/2021.

Okamura, Hajime, & Kazumasa Ozawa. (1996). "Self-Compacting High Performance Concrete." *Structural Engineering International: Journal of the International Association for Bridge and Structural Engineering (IABSE)* 6 (4): 269–270. 10.274 9/101686696780496292.

Price, Derek De Solla. (1976). "A General Theory of Bibliometric and Other Cumulative Advantage Processes." *Journal of the American Society for Information Science* 27 (5): 292–306. 10.1002/asi.4630270505.

Pride, D., & Knoth P. (2018). "Peer Review and Citation Data in Predicting University Rankings, a Large-Scale Analysis." In Méndez, E., Crestani, F., Ribeiro, C., David, G., & Lopes, J. *Digital Libraries for Open Knowledge. TPDL 2018. Lecture Notes in Computer Science*, 11057, Cham: Springer. 10.1007/978-3-030-00066-0_17.

Rhodes, Christopher J. (2018). "Plastic Pollution and Potential Solutions." *Science Progress* 101 (3): 207–260. 10.3184/003685018X15294876706211.

Safi, Brahim, Mohammed Saidi, Djamila Aboutaleb, & Madani Maallem. (2013). "The Use of Plastic Waste as Fine Aggregate in the Self-Compacting Mortars: Effect on Physical and Mechanical Properties." *Construction and Building Materials* 43: 436–442. 10.1016/j.conbuildmat.2013.02.049.

Şahmaran, Mustafa, Heru Ari Christianto, & İsmail Özgür Yaman. (2006). "The Effect of Chemical Admixtures and Mineral Additives on the Properties of Self-Compacting Mortars." *Cement and Concrete Composites* 28 (5): 432–440. 10.1016/j.cemconcomp. 2005.12.003.

Saikia, Nabajyoti, & Jorge de Brito. (2012). "Use of Plastic Waste as Aggregate in Cement Mortar and Concrete Preparation: A Review." *Construction and Building Materials* 34: 385–401. 10.1016/j.conbuildmat.2012.02.066.

Saikia, Nabajyoti, & Jorge de Brito. (2014). "Mechanical Properties and Abrasion Behaviour of Concrete Containing Shredded PET Bottle Waste as a Partial Substitution of Natural Aggregate." *Construction and Building Materials* 52: 236–244. 10.1016/j.conbuildmat.2013.11.049.

Sami Kohistani, Abdul, & Khushpreet Singh. (2018). "An Experimental Investigation by Utilizing Plastic Waste and Alccofine in Self-Compacting Concrete." *Indian Journal of Science and Technology* 11 (26): 1–14. 10.17485/ijst/2018/v11i26/130569.

von Schirnding, Y. (2005). "The World Summit on Sustainable Development: Reaffirming the Centrality of Health." *Global Health* 1(8). 10.1186/1744-8603-1-8.

Sayadi, Ali A, Juan V Tapia, Neitzert, Thomas R., & Charles Clifton, G. (2016). "Effects of Expanded Polystyrene (EPS) Particles on Fire Resistance, Thermal Conductivity and Compressive Strength of Foamed Concrete." *Construction and Building Materials* 112: 716–724. 10.1016/j.conbuildmat.2016.02.218.

Scopus. (2020). "www.Elsevier.Com." 2020.

Senko, J. F., Nelms, S. E., Reavis, J. L., Witherington, B., Godley, B. J., & Wallace, B. P. (2020). "Understanding Individual and Population-Level Effects of Plastic Pollution on Marine Megafauna." *Endangered Species Research* 43: 234–252. 10.3354/esr01064.

Shanmugapriya, M., & M. Helen Santhi. (2017). "Strength and Chloride Permeable Properties of Concrete with High Density Polyethylene Wastes." *International Journal of Chemical Sciences* 15 (1): 10–17.

Silva, Ana Tereza dos Santos, Ronaldo da Silva Araújo, & Nivianne Lima dos Santos Araújo. (2020). "Bibliometric Analysis on Publications by Qualis/Capes Journals and the Web of Science: The Path of Academic Production on IPSAS and IPSASB." *Brazilian Journal of Management & Innovation* 7 (3): 100–119. 10.1822 6/23190639.v7n3.05.

Soares, Patrícia Bourguignon, Teresa Cristina Janes Carneiro, João Luiz Calmon, & Luiz Otávio da Cruz de Oliveira Castro. (2016). "Análise Bibliométrica Da Produção Científica Brasileira Sobre Tecnologia de Construção e Edificações Na Base de Dados Web of Science." *Ambiente Construído* 16 (1): 175–185. 10.1590/s1678-86212016000100067.

Su, Hsin Ning, & Pei Chun Lee. (2010). "Mapping Knowledge Structure by Keyword Co-Occurrence: A First Look at Journal Papers in Technology Foresight." *Scientometrics* 85 (1): 65–79. 10.1007/s11192-010-0259-8.

Subramanyam, K. (1983). "Bibliometric Studies of Research Collaboration: A Review." *Journal of Information Science* 6 (1): 33–38. 10.1177/016555158300600105.

Thompson, J. (2001). "Environmentalism: Philosophical Aspects." *International Encyclopedia of the Social & Behavioral Sciences* 4679–4685. 10.1016/b0-08-043 076-7/01060-3.

Thorneycroft, J., Orr, J., Savoikar, P., & Ball, R. J. (2018). "Performance of Structural Concrete with Recycled Plastic Waste as a Partial Replacement for Sand." *Construction and Building Materials* 161: 63–69. 10.1016/j.conbuildmat. 2017.11.127.

Verdolotti, Letizia, Fabio Iucolano, Ilaria Capasso, Marino Lavorgna, Salvatore Iannace, & Barbara Liguori. (2014). "Recycling and Recovery of PE-PP-PET-Based Fiber Polymeric Wastes as Aggregate Replacement in Lightweight Mortar: Evaluation of Environmental Friendly Application." *Environmental Progress & Sustainable Energy* 33 (4): 1445–1451. 10.1002/ep.11921.

World Bank. (2019). *Weathering Growing Risk. World Bank East Asia and Pacific Economic Update.* openknowledge.worldbank.org.

World Economic Forum. (2018). *The Global Competitiveness Index Report 2017.* World Economic Forum (WEF). http://ci.nii.ac.jp/naid/110008131965/.

Wu, Guiqing, Jia Li, & Zhenming Xu. (2013). "Triboelectrostatic Separation for Granular Plastic Waste Recycling: A Review." *Waste Management* 33 (3): 585–597. 10.101 6/j.wasman.2012.10.014.

WWFBrasil. (2020). "Brazil Is the 4th Country in the World That Generates the Most Plastic Waste." 2020.

Yang, Shutong, Xiaoqiang Yue, Xiaosong Liu, & Yao Tong. (2015). "Properties of Self-Compacting Lightweight Concrete Containing Recycled Plastic Particles." *Construction and Building Materials* 84: 444–453. 10.1016/j.conbuildmat.2015. 03.038.

Yuan, Baolong, & Qiulian Xiang. (2018). "Environmental Regulation, Industrial Innovation and Green Development of Chinese Manufacturing: Based on an Extended CDM Model." *Journal of Cleaner Production* 176 (March): 895–908. 10.1016/j.jclepro.2017.12.034.

Zakaria, Muhammad, & Samina Bibi. (2019). "Financial Development and Environment in South Asia: The Role of Institutional Quality." *Environmental Science and Pollution Research* 26 (8): 7926–7937. 10.1007/s11356-019-04284-1.

Zhu, Donghua, Alan Porter, Scott Cunningham, Judith Carlisie, & Anustup Nayak. (1999). "A Process for Mining Science & Technology Documents Databases, Illustrated for the Case of 'Knowledge Discovery and Data Mining.'" *Ciência Da Informação* 28 (1): 07–14. 10.1590/s0100-19651999000100002.

21 Cementitious Composites for Civil Construction Made with Marble and Granite Waste

M.G.A. Ranieri[1], P. Capellato[2],
M.A. de B. Martins[2], V.C. dos Santos[3],
P.C. Gonçalves[3], L.R.R. da Silva[4],
M.L.M. Melo[4], and A. da S. Mello[1]

[1]Institute of Production Engineering and Management – Development, Technology and Society Program (DTEcS), Itajubá Federal University, Itajubá Campus, Itajubá, Minas Gerais, Brazil
[2]Institute of Physics and Chemistry, Federal University of Itajubá, Itajubá, Minas Gerais, Brazil
[3]Natural Resources Institute, Federal University of Itajubá, Itajubá, Minas Gerais, Brazil
[4]Institute of Mechanical Engineering, Federal University of Itajubá, Itajubá, Minas Gerais, Brazil

CONTENTS

21.1 Introduction 388
21.2 Materials and Methods 389
21.3 Results 394
    21.3.1 Characterization of Marble and Granite Powder Residues 394
    21.3.2 Characterization of Specimens 396
        21.3.2.1 Water Absorption 396
        21.3.2.2 Compressive Strength 398
21.4 Conclusions 406
Acknowledgments 406
References 406

21.1 INTRODUCTION

Solid waste management is of paramount importance on a global scale and needs to be treated as a political priority (João et al., 2018). Practically all human and industrial activities generate solid waste, and the growing accumulation of this material creates grave environmental consequences which also result in serious health and economic impacts (Gomes et al., 2017). In general, solid waste material can be considered unusable or in states of disuse (Khodabakhshian et al., 2018). Environmental organizations have consistently pointed out that solid waste generation is one of the most serious environmental problems that society faces today (Gomes et al., 2017; Khodabakhshian et al., 2018; Ranieri, 2020). Recently, the European community introduced a strategy entitled "Zero Waste," where all solid waste must be converted into resources, thereby ending landfill and incineration as waste management practices. The campaign is meant to emphasize the importance of avoiding the disposal of waste materials and refocusing on the reuse, recycling and recovery of them, resorting to elimination only when all other options have been exhausted (Kore & Vyas, 2016).

Case studies in South and Central American countries on this subject are still limited. In Brazil, this tendency began to change in the first decade of the 21st century, when the National Solid Waste Policy was sanctioned in 2010 via Federal Law 12.305. The legislation set forth a series of principles, objectives and instruments for solid waste management, including hazardous waste (Gameiro et al., 2014; Gencel et al., 2012) with the overarching goal of turning solid waste management into an efficient tool for combating environmental degradation as well as a means for economic leveraging of solid waste as an alternative for substituting conventional raw materials. The construction industry is one of the greatest consumers of natural resources worldwide. In recent years, this has driven many researchers to focus their studies on this field, where different types of waste were used to manufacture raw materials to substitute cement and its aggregates.

Civil construction in Brazil is heavily reliant on concrete blocks, using them for structural purposes or simply for enclosures, depending on the block's resistance to axial compression. They favor modular coordination and rationalization in many construction phases. These blocks can be defined as a unit of masonry composed of cement, aggregates and water (ABNT NBR, 2009, 2013; ABNT, 2013; 8522 2017). The generally required conditions for standardization and quality determines the following characteristics: classification (location where they will be used), materials used, physical dimensions, manufacturing and curing methods, appearance and texture. Specific condition requirements involve resistance to compression, absorption, moisture, retraction and typology (Amit & Singh, 2013). Taking these features into account, along with accelerated urbanization, demand for construction material and the application of decorative stone slabs have seen significant growth.

Marble and granite are the most commonly used ornamental materials in civil construction due to their resistance and durability (ABIROCHAS, 2018). Processing, cutting and polishing these stones slabs for use in civil construction is done with a wet tile saw, which generates large quantities of MGR waste byproduct. These particles require correct disposal as they pose a risk to both human health and the environment. When wet, the substance contaminates soil and groundwater. Once dry, the fine particles are hazardous to human health when dispersed in the air (Khodabakhshian et al., 2018). Almost 70% of the extracted rocks are transformed into solid waste after cutting and polishing: 40% into sludge and 30% into small pieces. This high waste volume can easily reach millions of tons, and the reuse of this material can generate environmental and economic gains (ABIROCHAS, 2018; Corinaldesi et al., 2010; Fernandes et al., 2016).

Researchers have characterized and evaluated solid waste in MGR processing as ceramic raw materials alternatives for producing brick and tile. These results have shown that these materials possess physical and mineralogical characteristics similar to conventional ceramic raw material. More specifically, they have demonstrated that ceramic material is technologically similar according to the standards for ceramic bricks and tiles (Almeida et al., 2015; Hamza et al., 2011; Munir et al., 2018). Hamza et al. (2011) used marble to produce bricks, substituting up to 40% of the conventional aggregates. The samples obtained optimal results with 10% incorporation in the residue (Gencel et al., 2012). Munir et al. (2018) also researched partial substitution of clay with marble in brick production; they concluded that the addition of 15% of the weight reached satisfactory results in mechanical properties. However, the most promising results were in relation to their thermal behavior (Almeida et al., 2015; Gencel et al., 2012; Hamza et al., 2011; Junior, 2000; Munir et al., 2018). Based on the aforementioned points, the main objective of this study is to use marble and granite waste byproducts as civil construction materials as a means to preserve the environment. Powder and sludge from MGR processing were used to manufacture sustainable cement blocks for civil construction. First, the material was physically and mineralogically characterized through Differential Thermal Analysis (DTA), X-Ray diffraction (DRX), Fourier Transformed Infrared (FTIR) spectroscopy, laser particle distribution analysis, as well as fine and specific mass evaluation to assess its applicability. Once initial analysis was completed, samples were prepared with different amounts of the waste material for feasibility analysis according to industry norms and standards. Water absorption, axial compression and elastic modulus were tested. Variation analysis was used to evaluate the material's mechanical resistance and elastic modulus.

21.2 MATERIALS AND METHODS

Figure 21.1 presents the methodology used in this study.

The MGR processing residue was collected from a local marble and granite processing facility in the city of Itajubá, located in the state of Minas Gerais,

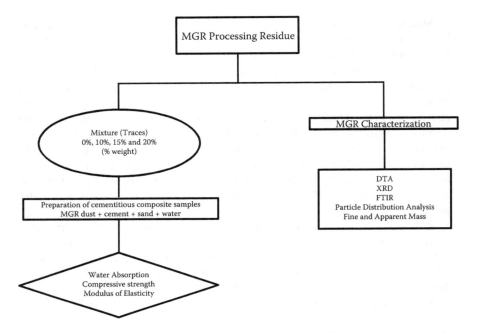

FIGURE 21.1 Current study methodology flow chart.
Source: Authors.

Brazil. As previously mentioned, this material is a waste by-product which is generated during the cutting and polishing processes. In Figure 21.2, it can be seen that the material has a sludge-like consistency due to the water used in the process.

The sludge was placed in an oven at 105°C to evaporate excess water. This residue was then sifted with a conventional construction sieve in order to remove smaller pieces of rock. The natural sand was extracted from the river and acquired from local commercial establishments, and the water was treated by the local utility company, COPASA. In order to understand the different aspects related to composition and structure of the material, a variety of characterization techniques were utilized to assess the material's feasibility in application. These techniques are presented in Table 21.1.

The parameters for calculation of the granulometric distribution were based on the Brazilian norm NBR 6502/1995 seen in Table 21.2.

Particle size distribution was evaluated in a dry sample, after having spent 24 hours in a 105°C oven, and verified using mesh sieves of 4.75, 2.36, 1.18, 0.6 and 0.3 mm. The intermediate mesh was 0.15 mm and 0.075 mm. The fineness module was determined by the sum of the percentages of the retained aggregate mass in the sieves, divided by 100.

After the characterization measurements, traces, also known as mixtures, were prepared. These were denominated T0, T10, T15 and T20, in reference to the

Cementitious Composites

FIGURE 21.2 Visual characteristics of marble and granite processing residue.
Source: Authors.

TABLE 21.1
Applied MGR Residue Characterization Techniques

Characterization Test	Analysis	ABNT Norms/ Test Data	Equipment	
Differential Thermal Analysis (DTA)	Physical and chemical reactions due to temperature variation	From 25°C to 1000°C, Rate of 10°C/min, With N_2 and O_2	Metler MT 15	
X-Ray Diffraction (XRD)		Mineral phases	Scanning from 20 to 90 degrees, 2° Theta step	Structural Characterization
Fourier Transformed Infrgared (FTIR) Spectroscopy		Atomic Structure	Scanning from 450 cm^{-1} to 40.000 cm^{-1}	Pertkin Elmer, Spectrum 100
Laser particle size distribution		Granulometry		Microtrac, S 3500
Specific Mass			NBR NM 52	Pycnometer/scale

TABLE 21.2
NBR 6502/2995 Parameters

Fraction	Diameter (mm)
Clay	d < 0,002
Silt	0.002 < d < 0.06
Fine Sand	0.06 < d < 0.2
Medium Sand	0.2 < d < 0.6
Coarse Sand	0.6 < d < 2
Fine Gravel	2 < d < 6
Medium Gravel	6 < d < 20
Heavy Gravel	20 < d < 60

Source: Authors.

TABLE 21.3
Traces Used to Prepare the Cement Cylinders

Trace	Cement	Sand	MGR Residue	Water
T0	650 g	4550 g	0	500 g
T0	1.00	7.00	0.00	0.80
T10	1.00	6.30	0.70	0.80
T15	1.00	5.95	1.05	0.80
T20	1.00	5.60	1.40	0.80

Source: Authors.

percentages (% in weight) that were substituted for MGR residue. Table 21.3 presents the exact quantities utilized for this preparation process. After each sample was duly weighed, the materials were mixed using an electric STANLEY mixer.

For each mixture, 18 samples were prepared using 50×100 (mm) metallic cylinders, as seen in Figure 21.3. The samples were removed from the molds after a 24-hour period in a humidity chamber at 90%.

Once the samples were prepared, tests were carried out to analyze the concrete quality, using the Brazilian Association of Technical Standards (ABNT) as a reference. Resistance to compression and absorption tests were carried out according to Table 21.4.

The mechanical resistance and water absorption results must be applied to the ABNT requirements to ensure they are in accordance. Table 21.5 shows the types of cylinders and their respective values according to the norms.

The norm also establishes nominal dimensions for width, height and thickness for hollow, rectangular concrete blocks. However, Frasson Junior (2000) stated

Cementitious Composites

FIGURE 21.3 Steel cylinder molds and samples prepared for this study.
Source: Authors.

TABLE 21.4
Techniques Applied to the MGR Samples: T0, T10, T15 and T20

Sample Tests	Norm (ABNT)	Samples	Cure Time
Water Absorption	NBR 13555/2012	3 cylinders Ø 5 × 10 cm	30 days
Mechanical Resistance to Compression	NBR 10836/2013	12 cylinders Ø 5 × 10 cm	7, and later 35 days
Elastic Modulus	NBR 8522/2017	12 cylinders Ø 5 × 10 cm	7, and later 35 days

Source: Authors.

TABLE 21.5
Concrete Cylinder Classifications, Compression Resistance and Water Absorption Results

Classification	Class	Resistance to Axial Compression (MPa)	Absorption % Normal Aggregate Individual	Absorption % Normal Aggregate Average	Retraction %
Structural Function	A	Fck ≥ 8.0	≤9	≤8	≤0.065
Non-Structural Function	B	4.0 ≤ Fck ≤ 8.0	≤10	≤9	
	C	Fck ≥ 3.0	≤11	≤10	

Source: ABNT 6136: 2016.

that there is a mechanical relation between hollow concrete blocks and cylindrical samples Ø 5 × 10 cm from the same trace with the same specific mass. The proportion of sample resistance Ø 5 × 10 cm corresponds to 80% of the hollow blocks (Junior, 2000).

21.3 RESULTS

21.3.1 CHARACTERIZATION OF MARBLE AND GRANITE POWDER RESIDUES

Thermal analysis techniques are widely used to determine the behavior and composition of natural and synthetic products. Differential thermal analysis (DTA) is applied to study the thermal behavior of materials such as silicates, ferrites, oxides, clays and so forth. This type of analysis provides information on fusion points, dehydration, oxidation, reduction, adsorption and solid-state reactions. Figure 21.4 shows the thermogravimetric curves for the MGR processing residue. Between 100°C and 350°C, there was a 1% loss in mass for N_2, while there was a 1.5% loss in mass with O_2, corresponding to the free water loss and water adsorption. Between 350°C and 600°C, the loss was 2% for N_2, while O_2 maintained a 1.5% loss. From 600°C to 750°C, the N_2 loss was 3.5%, while O_2 loss increased to 5%. This may be related to the decomposition of magnesium carbonate. From 750°C to 900°C, the changes are due to the decomposition of calcium carbonate, seeing that they do not decompose at the same temperature.

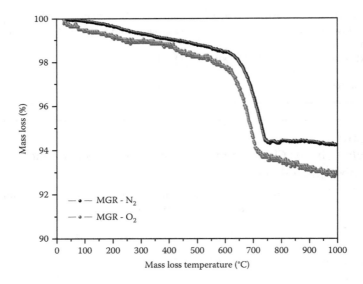

FIGURE 21.4 Thermogravimetric curve for marble and granite (MGR) processing residue.

Source: Authors.

Cementitious Composites

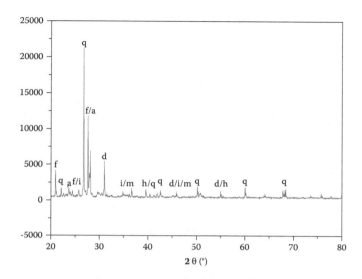

FIGURE 21.5 X-ray diffractogram, marble and granite processing residue (MGR): f = feldspar; q = quartz; a = albite; d = dolomite; i =; m =; h = hematite.

Source: Authors.

X-ray diffraction is one of the most common and efficient analyses to chemically characterize myriad materials. This technique is most suited for determining the crystalline phases in solid materials and provides information about the structural arrangement of the atoms that comprise a given solid, indicating if the sample possesses solid crystalline or amorphous characteristics based on the presence or absence of peaks in the diffractogram. Figure 21.5 presents the diffractogram for the MGR residue, and it demonstrates that dolomite $CaMG(CO_3)_2$, calcite ($CaCO_3$) and quartz (SiO_2) are present. The peaks obtained specifically indicate the presence of quartz, albite ($NaAlSi_3O_8$), calcite and dolomite. The results are similar to those found by Cosme et al. (2016), Sadek et al. (2016), Singh et al. (2017), Soltan et al. (2016) and Vardhan et al. (2015). Marble is a metamorphic rock and its residue presents calcite and dolomite as predominant minerals, typical of the carbonate rocks. Granite is an igneous rock made up principally of quartz, feldspar and mica (Cosme et al., 2016; Sadek et al., 2016; Singh et al., 2010; Soltan et al., 2016; Vardhan et al., 2015).

Fourier-transform infrared spectroscopy (FTIR) enables the identification of the types of chemical bonds and are represented by peaks and band characteristics. This information indicates how the atoms are bonded to the material molecules, which chemical functions are present and probable impurities. Figure 21.6 exhibits the spectrum obtained from the residue under study. The ranges around 990.50 cm^{-1} are attributed to the characteristic stretching vibrations from T–O and Si–O–T (T = Si or Al) of the aluminosilicates and silicates

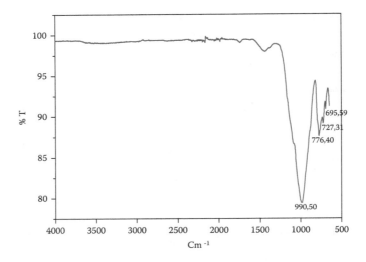

FIGURE 21.6 Marble and granite processing residue (MGR) FTIR.

Source: Authors.

present in the granite. Bonding of the silicates and aluminosilicates appears in the region around 700–656 cm^{-1}.

The results from the laser particle distribution analysis of the sand and MGR residue demonstrate that the granulometric size is similar to fine aggregates, thus making it an attractive option for substituting conventional materials. The granulometric curve shows that there are grains that vary from 0.001 mm to 0.1 mm, enabling the substitution of fine sand in the proportions of 10%, 15% and 20%. Table 21.6 presents the physical characteristics of the sand and MGR residue through granulometric, fineness modulus and specific mass analyses.

Figure 21.7 presents a granulometric curve for the MGR residue. It can be seen that the grain size varies from 0.001 mm to 0.1 mm, similar to the particle sizes of clay and fine sand.

21.3.2 Characterization of Specimens

21.3.2.1 Water Absorption

Water absorption results for the T0, T10, T15 and T20 samples are presented in Figure 21.8. The trace with 20% MGR residue increased its water absorption index. This may have occurred due to the fact that the MGR powder tends to retain water, since its molecules typically accumulate around MGR particles. This can lead to a reduction in the amount of water necessary to hydrate the cement particles, thus resulting in increased porosity. Furthermore, the MGR particles may amass around the cement grains, thus hindering their contact with water and slightly reducing hydration and increasing porosity, as mentioned

TABLE 21.6
Granulometry, Fineness Modulus, Maximum Diameter of the Sand and the Marble and Granite Residue (MGR)

Opening (mm)	% Retained and Accumulated	
	Sand	MGR Residue
4.8	8.01	0.00
2.4	21.61	0.00
1.2	40.10	1.31
0.6	63.05	3.34
0.3	87.30	5.64
0.15	97.48	15.43
0.075	99.26	41.36
Bottom pan	100.00	100.00
Fineness Modulus (FM)	3.17	0.26
Dmax (mm)	4.8	0.3

Source: Authors.

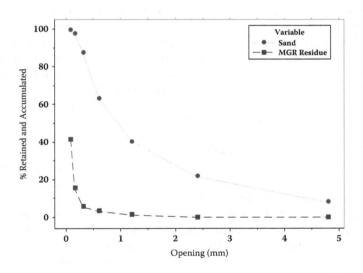

FIGURE 21.7 Granulometric curve distribution of marble and granite residue and sand (MGR).

Source: Authors.

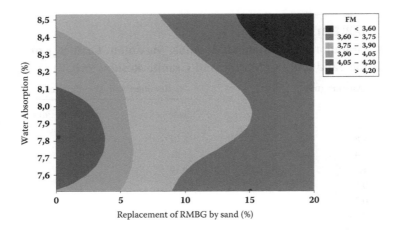

FIGURE 21.8 Water absorption relation, incorporating marble and granite residue and fineness modulus.

previously. Total fineness modulus is a proportional variation of the fineness modulus of the total fine aggregate. Figure 21.8 more clearly illustrates the degree to which the T20 sample water absorption increases. However, this does not mean that its mechanical resistance would be greater.

The results presented in Figures 21.8 confirm that there is a variation in the water absorption when MGR residue is introduced to the traces. Aside from establishing that the alteration in the fineness modulus is directly related to the absorption variable and percentage of MGR residue. The greater the degree of fineness of a mixture, the more water that will be absorbed by the mortar; this phenomenon is related to absorption through capillarity (Cosme et al., 2016), where, as the surface area of an aggregate is increased, its capacity to absorb water also increases (Sadek et al., 2016).

21.3.2.2 Compressive Strength

Different sample groupings were left to cure for 7 and 35 days in order to test their *compressive strength*. A summary of the results are seen in Table 21.7. Four specimens were used for each sample, and their results were utilized to conduct variation analysis (ANOVA) to validate the difference between the averages. In doing so, it can be determined if there is a significant difference between the averages. Furthermore, the model's linear regression was obtained by analysis of the R^2 value. In the adopted statistical method, the null hypothesis (H_0) occurs when all of the averages are equal; that is, when there is no significant difference between them. The alternative hypothesis (H_1) is validated when there is a significant difference between the averages. The significance level was set at $\alpha = 0.05$, which means that the p-values obtained below 0.05 indicate that there is a significant difference between the averages and rejects the null hypothesis.

TABLE 21.7
Results of Mechanical Strength after 7 and 35 Days of Curing for Mortar with Marble and Granite (MGR) Processing Residue

Sample	F_7	F_{35}	E_7	E_{35}
T0	5.6	9.4	554.6	906.5
T0	5.9	8.1	489.2	933.9
T0	7.1	9.0	967.5	929.8
T0	7.0	8.6	635.9	981.5
T10	8.4	11.5	721.6	885.8
T10	7.8	12.0	665.9	867.5
T10	8.5	12.2	1114.4	904.2
T10	7.4	10.8	827.2	1298.1
T15	6.7	10.6	555.8	1029.6
T15	6.1	10.8	537.8	1159.0
T15	6.2	9.9	746.4	903.9
T15	6.3	11.5	576.8	824.3
T20	5.6	11.4	702.6	915.1
T20	5.5	11.6	534.3	1026.7
T20	5.0	10.4	325.8	916.9
T20	5.9	10.3	609.9	940.2

Source: Authors.
*F_7 = mortar compressive strength in [MPa] after 7-day curing time.
*F_{35} = mortar compressive strength in [MPa] after 35-day curing time.
*E_7 = mortar elasticity modulus in [MPa] after 7-day curing time.
*E_{35} = mortar elasticity modulus in [MPa] after 35-day curing time.

Furthermore, it was assumed that the variance was equal for all the averages; that is, a single homoscedastic regime was considered.

After 7 days curing F_7, the cylinders with MGR residue demonstrate a compressive strength alteration, as seen in Figure 21.9a and Table 21.8. In relation to the reference mixture (T0), there is an increase in resistance in T10 and reduction is seen for T15 and T20. However, statistical variation should be carried out to confirm if these alterations are significant. In Table 21.9, the p-value for the ANOVA compressive strength is less than 0.05 (p-value = 0.000). This result indicates that the compressive strength of one or more of the sample groups is significantly different. The R^2 value seen in Table 21.9 expresses the degree of variance of the data, which was determined by the linear regression model as 80.55%. Thus, it can be stated that 80.55% of the resistance variation can be explained by the variation in the marble and granite residue mass. The remaining 19.45% are explained by other factors that were not considered in this study.

(a)
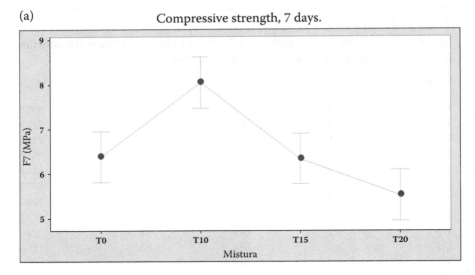

(b)
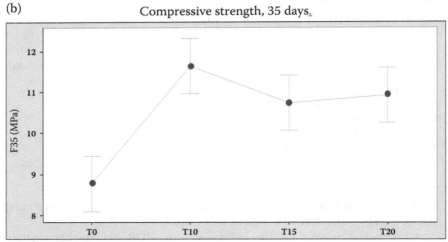

FIGURE 21.9 Relationship between resistance to compressive strength and mortar mixtures with marble and granite processing residue (95% CI for mean).

Source: Authors.

As seen in the Tukey comparison, it was possible to formally test which pairings present significant differences. Figure 21.10a includes the Tukey simultaneous confidence intervals for the difference in averages. The interval does not include zero, which indicates that there is a significant difference in the averages. The CIs for the other averages contain zero, thus indicating that the differences are not significant. Table 21.8 summarizes the Tukey comparison

TABLE 21.8
Summary of Data and Model Adopted for Statistical Analysis for Resistance to Axial Compression for Samples

	7 Days				35 Days			
Mixture	Average	SD	95% CI	Groups	Average	SD	95% CI	Groups
T0	6.38	0.78	(5.808; 6.946)	B	8.76	0.55	(8.084; 9.440)	B
T10	8.43	0.52	(7.473; 8.612)	A	11.64	0.63	(10.963; 12.318)	A
T15	6.33	0.25	(5.762; 6.900)	B	10.74	0.66	(10.057; 11.413)	A
T20	5.51	0.38	(4.939; 6.077)	B	10.93	0.65	(10.251; 11.606)	A

Source: Authors.
(CI: confidence interval and SD: standard deviation)

TABLE 21.9
Variation Analysis for Compressive Strength

7 days curing	p-value	0.000
	R^2	80.55%
35 days curing	p-value	0.000
	R^2	79.69%

Source: Authors.

data indicating that only the T10 mixture belongs to the different group (Group A) than the rest of the samples (Group B).

Upon analyzing the compressive strength results after curing for 35 days (F35), there is an alteration in the samples with MGR residue, as can be seen in Figure 21.9b and Table 21.8. In relation to the reference sample (T0), there is an increase in resistance in T10, T15 and T20. The increases observed in T10 are notedly greater than T15 and T20.

Statistical variation must be carried out to confirm if the differences are significant. In Table 21.9, the p-value for the compressive strength in MGR samples is less than 0.05 (p-value = 0.000). This result indicates that the resistance in one or more of the samples differs significantly.

The R^2 value for the 35-day aged samples expresses a data variance that 79.69% of the linear regression model can explain, Table 21.9. It can be said that 79.69% of the compressive strength variation can be explained by the addition of the MGR residue. The remaining 20.31% are explained by other variables that are not considered in this study.

Through the Tukey comparison, it was possible to formally test which groups present statistically significant differences. Figure 21.10b includes the

(a)

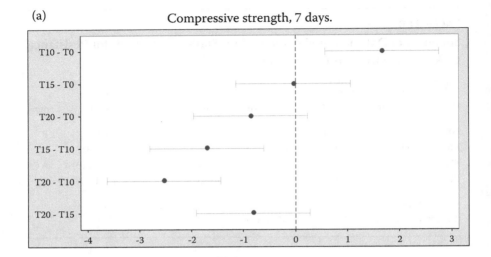

(b)

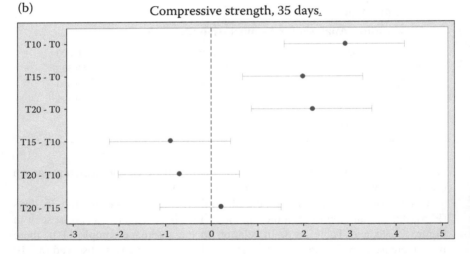

FIGURE 21.10 Differences in means with Tukey simultaneous analysis and 95% CI for resistance to compressive strength for samples with marble and granite (MGR) processing.

Source: Authors.

simultaneous confidence intervals for the difference between the averages. These intervals do not include zero, thus indicating that there are significant differences between these averages. In the case of the confidence intervals for the other pairings, the resulting averages include zero (T10 and T15, T10 and T20, T10 and T15), so it can be concluded that they do not have significant differences.

According to Figure 21.11a and Table 21.10, the elastic modulus after curing for 7 days (E7) undergoes alterations when MGR processing residue is

(a)

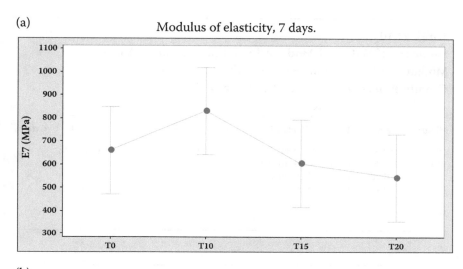

(b)

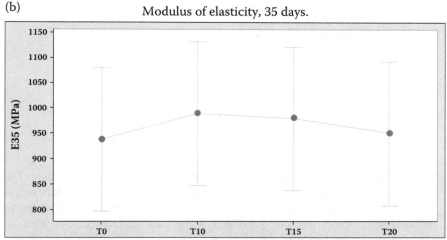

FIGURE 21.11 Relationship between modulus of elasticity for samples with marble and granite processing residue (95% CI for mean).

Source: Authors.

added to the mixture. In relation to the reference sample (T0), there is an increase in T10 and a decrease in T15 and T20. Statistical verification should be carried out to confirm if these alterations are significant. In Table 21.11, the p-value for the elastic modulus ANOVA for the sample with MGR processing residue is greater than 0.05 (p-value = 0.158). This result indicates that the elastic modulus among the varying mixture groups is not significantly different. The R^2 value seen in Table 21.11 shows that only 34.05% of the data variance is explained by the regression model. That is, only 34.05% of the elastic

TABLE 21.10
Summary of Data and Model Adopted for Statistical Analysis for Modulus of Elasticity for Samples with Marble and Granite Processing Residue after 7 Days' Curing

Mixture	7 Days Average	SD	95% CI	Groups	35 Days Average	SD	95% CI	Groups
T0	662.00	212.00	(473; 850)	A	937.90	31.40	(795.9; 1079.9)	A
T10	832.30	199.60	(643.6; 1020.9)	A	989.00	207.00	(847; 1131)	A
T15	604.20	96.10	(415.6; 792.9)	A	979.30	146.70	(837.3; 1121.3)	A
T20	543.10	160.40	(354.5; 731.8)	A	949.70	52.60	(807.7; 1091.7)	A

Source: Authors.

TABLE 21.11
Variation Analysis for Elasticity Modulus

7 days curing	p-value	0.158
	R^2	34.05%
35 days curing	p-value	0.937
	R^2	3.29%

Source: Authors.

modulus is explained by the addition of the MGR processing residue. The remaining 65.95% is explained by other factors that are not part of this study. Through the Tukey comparison for the 7-day cure samples, it was possible to formally test which pairing presented significant difference. As seen in Figure 7.14a, it can be stated that there is no significant difference between any of the pairings, seeing that all of them contain zero. Table 21.10 summarizes the data obtained using the Tukey comparison, indicating that all of the mixtures belong to the same group. Upon analyzing the results for the E35 elastic modulus after 180 days curing, it is observed that the elastic modulus sustains small alterations with the incorporation of the marble and granite processing residue. Comparing only the averages, the T10 and T15 mixtures increase slightly. However, the standard deviation of the measurements places practically all of the mixtures within the same confidence interval. Again, applying ANOVA, the p-value obtained is 0.937 (>0.05), much greater than the 5% considered to reach some level of statistical significance. This result confirms that the elastic modulus is not altered significantly for the studied samples when MGR processing residue is added. The R^2 value seen in Table 22.10 shows that only 3.29% of the elastic modulus variation can be explained by the

Cementitious Composites 405

addition of MGR processing residue. The other 96.61% are explained by other factors not considered in this study. Using the Tukey comparison, it was possible to formally test which group pairings presented significant differences. Figure 21.12b demonstrates that there is no significant difference in any of the pairings, as all of the considered pairings contain zero. Table 21.10 summarizes the date obtained through the Tukey comparison, indicating that all of the samples belong to group A.

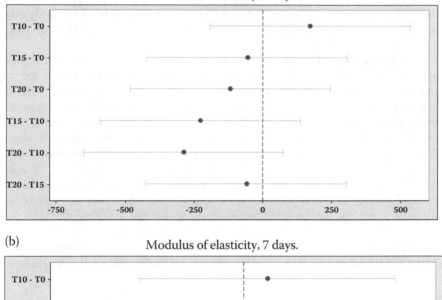

FIGURE 21.12 Differences in means with Tukey simultaneous analysis, 95% CI for modulus of elasticity with marble and granite (MGR) processing residue samples.

Source: Authors.

21.4 CONCLUSIONS

According to the DTA results, there are peaks that indicate calcium carbonate and magnesium carbonate decomposition, which is typically found in marble and granite, respectively. The X-ray diffraction demonstrates that there are minerals such as dolomite and calcite, both common in marble, and quartz, feldspar and mica, which are common granite components. Using FTIR spectroscopy, there were accentuated peaks between 600 cm^{-1} and 1,000 cm^{-1}, which are characteristic bands for materials such as marble and granite. Thus, through these mineralogic characterization methods, it can be concluded that the residues are indeed from marble and granite. Laser particle distribution analysis showed that the size of the grains is similar to fine aggregates and varies between 0.001 mm and 0.1 mm, thus being an apt substitute for fine sand in the proportions of 10%, 15% and 20% of the mass.

Upon analyzing the results for the T0, T10, T15 and T20 samples, it can be seen that the water absorption increased for the T20 sample, which led to the reduction in its mechanical resistance. Compressive strength after 7 days' curing for the T10 sample presented a 21.3% greater result than the reference (T0) sample, considering the standard deviation. The remaining samples, T15 and T20, did not present significant differences in relation to the reference. Nonetheless, the compressive strength after 35 days' curing presented a significant increase considering standard deviation. The increase for T10 was 32.8%; for T15, 22.5%; and for T20, 24.7%. For the elastic modulus after 7 days' curing time and after 35 days' curing time, there was no significant difference among the samples tested, considering the samples standard deviation. Based on these results, it is concluded that the residue from the processing of marble and granite is a viable substitute for conventional raw materials in the production of cementitious composite for civil construction, such as cement blocks, for example.

ACKNOWLEDGMENTS

The authors would like to thank the financial support of the Coordination for the Improvement of Higher Education Personnel (CAPES) and the foundation of the Ministry of Education (MEC).

REFERENCES

8522, ABNT NBR (2017). *Concreto – Determinação Dos Módulos Estáticos de Elasticidade e de Deformação à Compressão. ABNT NBR 8522.*

ABIROCHAS, Associação Brasileira da Indústria de Rochas Ornamentais (2018). *Balanço Das Exportações e Importações Brasileiras de Rochas Ornamentais Em 2017.* Vol. 55. Brasília.

ABNT (2013). *ABNT NBR 10836.*

ABNT NBR (2009). *ABNT NBR NM 52.*

ABNT NBR (2013). *ABNT NBR 13555.*

Almeida, Thiago de, Leite, Flaviane H. G., & Holanda, José N.F. de (2015). Caracterização de Resíduo de Pó de Mármore Para Aplicação Em Materiais Cerâmicos. *I Encontro de Engenharia, Ciência de Materiais e Inovação Do Estado Do Rio de Janeiro*, 1(D).

Amit, Viswakarma, & Singh, Rajput Rakesh (2013). Utilization of marble slurry to enhance soil fertility and to protect environment. *Journal of Environmental Research And Development*, 7(4), 1479–1483.

Corinaldesi, Valeria, Moriconi, Giacomo, & Naik. Tarun R. (2010). Characterization of marble powder for its use in mortar and concrete. *Construction and Building Materials*, 24(1), 113–117. 10.1016/j.conbuildmat.2009.08.013.

Cosme, Renée Jamilla, Lauret, Sudo Lutif, Emi, & Calmon, João Luiz (2016). Use of frequency sweep and MSCR tests to characterize asphalt mastics containing ornamental stone residues and LD steel slag. *122*, 556–566. 10.1016/j.conbuildmat. 2016.06.126.

Fernandes, José Luiz, Qualharini, Eduardo Linhares, Fernandes, Andrea Sousa da Cunha, & Cabral, Juliano Costa (2016). Um Estudo Sobre A Política Nacional. *Projectus*, *1*, 52–57.

Gameiro, F, Brito, J De, & Silva, D Correia (2014). Durability performance of structural concrete containing fine aggregates from waste generated by marble quarrying industry. *Engineering Structures*, 59, 654–662. 10.1016/j.engstruct. 2013.11.026.

Gencel, Osman, Ozel, Cengiz, Koksal, Fuat, & Erdogmus, Ertugrul (2012). Properties of concrete paving blocks made with waste marble. *Journal of Cleaner Production*, 21(1), 62–70. 10.1016/j.jclepro.2011.08.023.

Gomes, Carlos E, Sanchez, Rubén J., & Carvalho, Eduardo A De (2017). Microstructure and mechanical properties of artificial marble. *Construction and Building Materials*, 149, 149–155. 10.1016/j.conbuildmat.2017.05.119.

Hamza, R., El-haggar, S., & Khedr, S. (2011). "Marble and granite waste: characterization and utilization in concrete bricks." *International Journal of Bioscience, Biochemistry and Bioinformatics*, *1*(December 2015), 286–291. 10.7763/IJBBB. 2011.V1.54.

João, Tcharllis, Rodríguez, R.J.S., & Silva, F.S. (2018). "Physical and mechanical evaluation of artificial marble produced with dolomitic marble residue processed by diamond-plated bladed gang-saws." *Integrative Medicine Research*, 7(3), 308–313.

Junior, ArtêmioFrasson (2000). "Proposta de metodologia de dosagem e controle do-processo produtivo de blocos de concreto para alvenaria estrutural." Universidade Federal de Santa Catarina.

Khodabakhshian, A., Ghalehnovi, M., De Brito, J., & Elyas, Asadi (2018). "Durability performance of structural concrete containing silica fume and marble industry waste powder." *Journal of Cleaner Production*, 170, 42–60. 10.1016/j.jclepro. 2017.09.116.

Kore, S.D., & Vyas, A.K. (2016). "Impact of marble waste as coarse aggregate on properties of lean cement concrete." *Case Studies in Construction Materials*, 4(June), 85–92. 10.1016/j.cscm.2016.01.002.

Munir, M.J., Kazmi, S.M.S., Wu, Y.F., Hanif, A., & Arif Khan, M.U. (2018). "Thermally efficient fired clay bricks incorporating waste marble sludge: an industrial-scale study." *Journal of Cleaner Production*, 174, 1122–1135.10.1016/j.jclepro.2017.11.060.

Sadek, Dina M., El-Attar, Mohamed M., & Ali, Haitham A. (2016). "Reusing of marble and granite powders in self-compacting concrete for sustainable development". *Journal of Cleaner Production*, 121, 19–32. 10.1016/j.jclepro.2016.02.044.

Singh, Preetam, Park, Y.A., Sung, K.D., Hur, N., Jung, J.H., Noh, W.-S., Kim, J.-Y., Yoon, J., & Jo, Y. (2010). "Magnetic and ferroelectric properties of epitaxial Sr-doped thin films". *Solid State Communications*, *150*, 431–434. 10.1016/j.ssc.2009.12.006.

Soltan, Abdel Monem Mohamed, Kahl, Wolf-Achim, Abd EL-Raoof, Fawzia, Abdel-Hamid El-Kaliouby, Baher, Abdel-Kader Serry, Mohamed, & Abdel-Kader, Noha Ali (2016). "Lightweight aggregates from mixtures of granite wastes with clay". *Journal of Cleaner Production*, *117*, 139–149. 10.1016/j.jclepro.2016.01.017.

Vardhan, Kirti, Goyal, Shweta, Siddique, Rafat, & Singh, Malkit (2015). "Mechanical properties and microstructural analysis of cement mortar incorporating marble powder as partial replacement of cement". *Construction and Building Materials*, *96*, 615–621. 10.1016/j.conbuildmat.2015.08.071.

22 Influences of the Ceramic Inclusions on the Toughening Effects of Devulcanized Recycled Rubber-Based Composites

A.B. Irez[1] and Emin Bayraktar[2]
[1]Department of Mechanical Engineering, Faculty of Mechanical Engineering, Istanbul Technical University (ITU), Istanbul, Turkey
[2]ISAE SUPMECA – Paris, School of Mechanical and Manufacturing Engineering, France

CONTENTS

22.1 Introduction 409
22.2 Experimental Procedure 411
    22.2.1 Material Processing 411
    22.2.2 Experimental Tools 413
22.3 Results and Discussions 413
    22.3.1 Physical Characteristics and the Microstructure
           of the Composites 413
    22.3.2 Mechanical Characterization of the Composites 414
    22.3.3 Micro Scratch Tests and Tribological Assessment
           of the Composites 419
22.4 Conclusions 419
Acknowledgment 421
References 421

22.1 INTRODUCTION

Recently, due to the economic and environmental restrictions, there is an important competition between the companies in automotive, aeronautic and

DOI: 10.1201/9781003148760-22

transport industries. Regarding environmental concerns, the companies have to reduce the CO_2 emissions and fuel consumptions of the vehicles they produce. Second, vehicles' total cost must be decreased for the competitiveness in the above-mentioned industries. As a result, engineers' primary mission is to find lightweight, low-cost materials for use in the various applications of the aforementioned industries. Accordingly, material manufacturers assert that by using recycled materials, some environmental and economic advantages can be obtained. Amongst the frequently used recycled materials, rubber has an important place thanks to its extensive usage areas.

Material's properties of rubbers are improved by vulcanization process by using sulphur at a definite temperature and pressure. However, after completing their life cycle, non-used vulcanized scrap rubber causes some environmental problems, such as creating a habitat for mosquitos and soil pollution (Adhikari et al., 2000; Fiksel et al., 2011; De et al., 2005). Therefore, our first objective in this research is to devulcanize this scrap rubber to improve its reusability in the manufacture of new materials to be used in different industrial applications. Then, the devulcanized rubber is blended with a binder at elevated temperatures enabling its molding to the asked products at affordable costs. By mixing devulcanized powdered rubber with a convenient resin together with necessary reinforcing agents, a useful final product can be manufactured. In this research, the resin was chosen as epoxy due to its ease of processing, low price, environmental stability and large specific strength (May, 1988). Moreover, resin stiffness and wear resistance of the recycled rubber and epoxy blend can be ameliorated by the incorporation of some inorganic particles including titania (TiO_2) and alumina (Al_2O_3) (Arayasantiparb et al., 2001; Bittmann et al., 2012; Kim et al., 2004; Rothon, 1999; Wetzel et al., 2003; Zee et al., 1989). In the frame of this study, TiO_2 has drawn the attention because of its favorable properties in mechanical, optical, dielectric and tribological aspects (Pinto et al., 2015). TiO_2 can easily be found on the market and improves the resistance to wear (Wetzel et al., 2003). Furthermore, alumina has some decent properties such as high thermal conductivity, high adsorption, inertness to most acids and alkalis as a second reinforcement (Branch, 2011; Irez et al., 2019).

This research suggests a novel procedure to manufacture cost-efficient composites by using devulcanized scrap rubbers modified with different ceramic fillers. The purpose of this work is primarily to characterize mechanical properties and to study the effect of ceramic fillers on the fracture toughness. In the course of this study, three-point bending tests were used to determine the mechanical properties. Subsequently, fracture toughness was examined by means of a single edge notched beam (SENB) specimens. Then, to see the impact energy absorbing capacity depending on the filler contents Charpy tests were carried out. Moreover, surface wear resistance was assessed with the help of the micro scratch testing. Finally, the toughening and damage mechanisms were identified by observing the fracture surfaces by means of SEM.

22.2 EXPERIMENTAL PROCEDURE

22.2.1 MATERIAL PROCESSING

In composite manufacturing, interface quality has a paramount importance to manufacture enduring materials. Because, the stress transfer from the resin to the reinforcements can only be provided with a high interfacial adhesion between the matrix and the inclusions. In this regard, to improve the adhesion quality between the matrix and the reinforcing agents, some surface treatment techniques are utilized. In this study, our industrial partner provided us the SBR rubber waste in powder form. In the manufacturing line of shoe soles, after the cutting process, some wastes are generated from the edge of the molds. After collecting these unused clean scraps, pulverization procedure is implemented. After receiving waste rubber powder, their surface was activated by using acrylic acid solution with a small amount of benzoyl peroxide initiator. SBR rubber particles was kept in the solution until the total precipitation of all rubber particles. Then, the solution was leached and rubber particles were dried in a conditioning oven until the solution totally evaporates.

After that, devulcanization was maintained by microwave method. Because, lack of the free links on the recycled rubbers make it difficult to obtain chemical bonding between rubbers and epoxy. For this reason, manufactured composites may have some discontinuities on the interface zones. Therefore, under favor of the devulcanization process, carbon–sulphur bonds are broken and also some other links are formed. This process is maintained in an industrial microwave oven for 4 minutes by selecting the magnetron power at 900 W. Higher magnetron power and longer exposure times may degrade the main rubber chains (Hirayama & Saron, 2012; Paulo et al., 2011; Zanchet et al., 2009). Therefore, this procedure must be done in a controlled manner.

After devulcanization, surface activation of the rubbers was carried out by silanization. The rubbers were mixed with silane solution (vinyltrimethoxysilane 97% in ethyl alcohol, Sigma Aldrich), the solvents were filtered by using paper filters. Then the resting compounds were dried in a conditioning oven. After these steps, surface treatment was finalized (Shokoohi et al., 2008; Zhang et al., 2016).

After surface activation and devulcanization, powder metallurgy techniques were used in several steps to complete the manufacturing of the composites. First, recycled rubbers are milled to decrease the particle size with a toothed mill. After milling, the mixture of bisphenol-A type solid epoxy resin (Araldite GT 6099-Huntsman™) with its hardener (Desmodur N 75 (Bayer™)) and rubber powder were milled 1 hour more to obtain a homogenous compound and this mixture was used as matrix. Second, ceramic fillers (TiO_2 and Al_2O_3) were incorporated to the blend obtained in the previous step and then this mixture is homogenized by using ball milling for 2 hours. However, the temperature evolution during all milling processes should be controlled. Otherwise, binding characteristics of the epoxy and rubber can degenerate. Following the

FIGURE 22.1 Specimens after hot compaction and powder form of the composites.

preparation of a well-mixed powder compound, specimens were manufactured by hot compaction at a pressure of 70 MPa and a temperature of 180°C for 30 minutes. After hot compaction, post-cure of the specimens was done isothermally at 80°C for 24 hours. As the final step, circular disc-type specimens were cut into rectangular forms (as seen in Figure 22.1). The compositions of 4–12% TiO_2 (titania) and 10% Al_2O_3 (γ-alumina) reinforced recycled rubber-based composites (named as ALH I-II-III-IV-V hereafter) are presented in Table 22.1.

The specimens fabricated by hot compaction techniques are presented in Figure 22.1. The powder subjected to hot compaction is shown in the upper right of Figure 22.1. In addition, the beam-type specimen which is cut from circular specimens (50 mm in length) used in 3PB tests can be seen under the powder in the same figure.

TABLE 22.1
The Compositions of the Devulcanized Recycled Rubber-Based Composites

Name of the Composition	Matrix	Al_2O_3 wt.%	TiO_2 wt.%
ALH I	5% epoxy (bisphenol-A type) balance devulcanized SBR	10	4
ALH II			6
ALH III			8
ALH IV			10
ALH V			12

22.2.2 Experimental Tools

Following post-curing, the surface hardness of the polished flat specimens was measured using a Sauter Shore D hardness tester. 3PB tests are performed according to ASTM D790 by a Zwick Proline Z050TN testing system with 1 mm/min crosshead speed. Flexural strength and strain at break as well as flexural modulus, were extracted from the test results. Moreover, fractural toughness characteristics including critical strain energy release rate (G_{Ic}) and plain strain fracture toughness (K_{Ic}) were examined according to ASTM D5045 by using SENB specimens. Zwick Zwick 5102 pendulum impact tester was used to perform Charpy impact tests. In the end, wear resistance of the composites was assessed by Anton-Paar MST³ Micro Scratch tester.

22.3 RESULTS AND DISCUSSIONS

22.3.1 Physical Characteristics and the Microstructure of the Composites

To begin with, hardness results are indicated in Table 22.2. According to the table, increasing amount of TiO_2 improves the surface hardness. Surface hardness can be important for the novel composites when they are in contact with other components. In terms of order of magnitude, they are well beyond of the hardness of classical rubber-based components which is thought of as a positive effect of the ceramic inclusions.

After hardness measurement, microstructural observation of one of the compositions (ALH III) was done and given in Figure 22.2. In the micro structure, certain aggregation of the particles was observed. This may be related to the preparation conditions such as milling time and speed. In Figure 22.2, black particles show the recycled rubber and there is a general homogeneity observed in the distribution of the particles in the micro structure. Also, absence

TABLE 22.2
Shore D Surface Hardness Values of ALH Group Compositions

Hardness Measurement

Composition Name	Shore D
ALH I	71,0 ± 0,2
ALH II	72,1 ± 0,1
ALH III	72,4 ± 0,2
ALH IV	74,0 ± 0,2
ALH V	76,2 ± 0,3

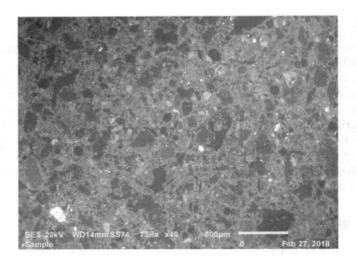

FIGURE 22.2 Microstructure of ALH III after sectioning and polishing.

of the micro cracks between hard and soft particles indicates good cohesion among the matrix and the inclusions.

The image in Figure 22.2 was supported with EDS (energy dispersive X-ray spectroscopy) mapping on the sectioned ALH III specimen. In Figure 22.3a, a reference image for the mapping zone is presented and in Figure 22.3b and c titanium and aluminum distributions are given respectively. Titanium and aluminum content dominant regions indicate the distribution of titania and alumina particles. In these maps, at the left-hand side of the images, there is a color legend that exhibits the intensity of specified element which is seen at the right bottom of the image. Upward evolution of the color signifies the increase in the intensity of the specified element. From these figures, homogeneous distribution of the reinforcement elements can be observed. However, some clusters of the inclusions are also detected.

22.3.2 MECHANICAL CHARACTERIZATION OF THE COMPOSITES

Three-point bending (3PB) tests were performed on each composition group, and the ultimate flexural stress, modulus and strain at the break are shown in Table 22.3.

Table 22.3 shows that, in general, increasing content of titania resulted in higher strength and strain in break compared to first composition (ALH I). However, elasticity modulus in bending showed certain fluctuations. This can be related to the possible agglomerations of the titania particles in the microstructure. As a result, when the manufactured composites are subjected to load, these agglomerations generate stress concentration zones that can be considered as the weak points of the material. This situation causes the failure prior to the

Influences of Ceramic Inclusions

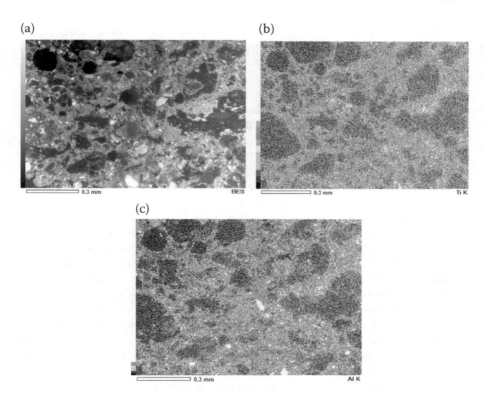

FIGURE 22.3 EDS mapping analysis on ALH III: (a) EDS reference zone, (b) titania distribution and (c) alumina distribution.

TABLE 22.3
Comparison of Mechanical Properties of ALH Compositions

Composition Name	Ultimate Flexural Strength (MPa)	Flexural Modulus (MPa)	Strain at Break (ε %)	K_{Ic} (MPa m$^{1/2}$)	G_{Ic} (kJ/m^2)
ALH I	6.46 ± 0.17	1100.52 ± 118	0.82 ± 0.07	0.570 ± 0.05	0.295 ± 0.01
ALH II	8.47 ± 0.32	1675.82 ± 221	0.75 ± 0.1	0.510 ± 0.08	0.155 ± 0.01
ALH III	6.64 ± 0.18	639.00 ± 84	1.16 ± 0.14	0.492 ± 0.06	0.379 ± 0.02
ALH IV	9.86 ± 0.44	984.62 ± 104	1.12 ± 0.09	0.378 ± 0.04	0.145 ± 0.01
ALH V	9.85 ± 0.39	1290 ± 147	1.01 ± 0.12	0.464 ± 0.06	0.167 ± 0.01

expected limits. On the other hand, if the opening micro cracks come across titania particles in the crack opening zones, the material failure is delayed, which is thought to be the root cause for the increased strain by the increasing titania content.

In terms of the fracture toughness, according to Table 22.3, by the increase in the amount of the reinforcements added in the matrix, fracture toughness values were influenced. For example, with the increase of titania content in the composition, previously mentioned local clusters may be formed and they can create possible crack nucleation sites under the stress. Similar to the fracture toughness, fracture energy (G_{Ic}) values were also degraded by the addition of more titania. The area under the stress strain curve, in other words fracture energy, becomes smaller with the ascending amount of reinforcements, except for ALH III composition.

After mechanical tests, fracture surfaces were observed using SEM to identify toughening mechanisms. First of all, it is noted that rough fracture surfaces were observed for all of the compositions. This signifies a good cohesion matrix and reinforcements. Also, regular fracture deviations are observed commonly in the fracture surfaces because of the effect of added hard particles. In Figure 22.4, some micro cracks are also indicated. These micro cracks open and propagate. However, after a while they disappear. This situation is associated with the crack pinning toughening mechanism. These cracks might be stopped when they encounter hard particles such as titania and alumina. This is an important mechanism for the hard particle reinforced composites.

In Figure 22.5, the interface between recycled rubber and epoxy is presented. Here, a decent diffusion bonding attracts attention. This situation is considered a positive effect of chemical treatment and microwave devulcanization process.

FIGURE 22.4 Micro crack observation on fracture surfaces after 3PB testing (ALH I).

FIGURE 22.5 Interface observation on fracture surfaces after 3PB testing (ALH I).

Normally, epoxy and recycled rubbers are thought incompatible due to the vulcanization process applied on rubber. Because, in the course of vulcanization, free links of the virgin rubbers are linked with sulphur atoms to improve the material properties. Then, it becomes difficult to generate chemical bonds between rubber and epoxy. However, in this study, free links on the rubbers are generated by means of pre-treatment of the rubber. Therefore, material properties of the manufactured composites are improved.

Lastly, in Figure 22.6a, cavitation of the rubber particles was observed. Certain factors such as curing pressure, resin system and environmental conditions can lead to cavitation. It is therefore challenging to obtain a composite

FIGURE 22.6 (a) Cavitation of the rubber particles and (b) debonding in the fracture surfaces.

section without a void. In the matrix, plastic deformation of the epoxy and debonding of the fillers have a significant influence on the size and the number of the voids formed. In this research, bulk moduli ($K = E/3(1 - 2\nu)$) of the titania and scrap rubber particles are quite high. For this reason, those titania and scrap SBR particles behave as rigid elastic bulks under stress and they show important resistance to any volumetric deformation. However, Poisson's ratio of epoxy, which is around 0.33, is lower than rubber. Therefore, the yielded epoxy matrix will behave relatively compliant and it can easily deform plastically. However, plastic dilatation of the matrix is hindered by rigid titania and alumina particles, unless these titania and rubber are pulled out from the matrix or if the internal cavitation of the rubber particles does not occur. By this way, some amount of the stored strain energy is dissipated by the interfacial debonding of titania and rubber, which is shown in Figure 22.6b. In Figure 22.6b, beveled edges marked with dashed lines and a red arrow indicate local debonding of the particles (Irez et al., 2018).

Subsequent to fracture surface observations, to assess the impact resistance of the manufactured composites, Charpy impact tests were conducted. The results are normalized by dividing by the impacted section and they are given in Figure 22.7.

According to Charpy impact tests, the energy-absorbing capacity of the composites were improved in general by the increasing content of titania. Adding 10% of titania and alumina (ALH IV) seems to yield the best value for the absorbed energy. As a matter of fact, absorbed energy amount for this composition is three times higher than ALH I (Irez et al., 2020; Pereira et al., 2017). Deviations in the absorbed energy values are considered to be a consequence of the local heterogeneities in the microstructure. Besides, generation of high shear stresses in the interface of epoxy and rubber can occur. Thus, a fracture occurs

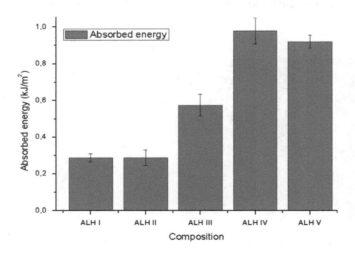

FIGURE 22.7 Absorbed energy after Charpy impact tests for ALH compositions.

only along the main crack propagation line in the absence of interfacial cracks between the epoxy and the rubber. Therefore, it is alleged that any improvement in the interface quality between epoxy matrix and rubbers can endorse higher absorbed energy in these composites.

22.3.3 Micro Scratch Tests and Tribological Assessment of the Composites

Surface wear resistance of the manufactured composites was examined by a micro scratch tester after mechanical characterization. An optical profilometer was used to detect surface damage after the scratch was created. The 3D damages are given in Figure 22.8. During this type of test, a diamond indenter produces a damage on the polished specimen surface for a definite length under two different loadings (10–15 N). After the indentation, the damage traces are evaluated to comment on the wear characteristics of the composites.

From Figure 22.8, it is seen that the damage zone topography shows a reasonable relationship between 10 N and 15 N damage traces. By the increasing forces, the damage zone expands in the vertical direction. In addition, the values of the worn volume and surface after surface scratching are indicated in Table 22.4. Wear resistance of the composites is asserted to be improved by increasing the reinforcement content. In reality, high interfacial shear stress is thought to be the primary cause of composite surface damage.

Higher force values, as expected, cause more damage on the surface, and the volume of the damaged zone is directly proportional to the force level. Besides, increasing content of titania decreases the wear resistance of the composites despite some exceptions (ALH IV-10 N). The results which are not coherent with the general trend may originate from the heterogeneities in the scratched surface (Irez et al., 2018).

22.4 CONCLUSIONS

Epoxy modified rubber-based recycled composites were designed using low-cost production methods in the framework of this research project. Five different compositions were developed for this study and characterized as low-cost composites for their possible uses. These composites suggest cost-efficient and eco-friendly solutions to manufacture different components in automotive and transport industry.

Despite the small, local agglomerations, a relatively homogeneous dispersion of the fillers is obtained. This can be improved by modifying milling parameters. 3PB tests demonstrated that a combination of alumina and titania generally improves the mechanical strength and strain at the break of the composites. On the other hand, an increasing amount of titania decreases the fracture toughness of these composites. It is stated that some clusters can be formed due to the excessive amount of the inclusions and which create crack nucleation sites upon loading.

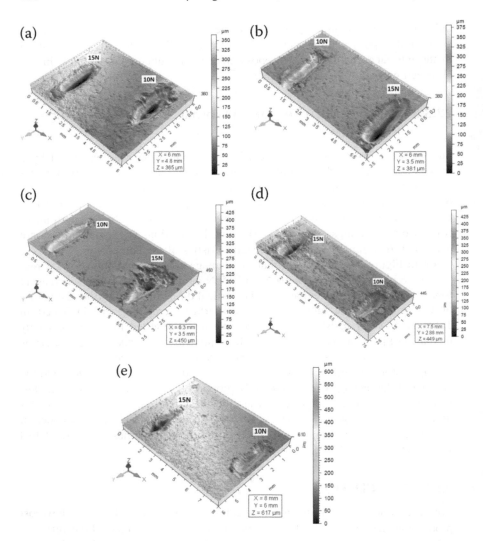

FIGURE 22.8 3D damage traces for five compositions: (a) ALH I (b) ALH II (c) ALH III, (d) ALH IV and (e) ALH V.

SEM observations on fracture surfaces revealed some toughening and damage mechanisms in these composites. Crack pinning was thought of as the main toughening mechanisms. Besides, cavitation is thought of as a mechanism providing energy dissipation leading to increased fracture toughness. According to the Charpy impact test, an increasing content of titania combined with alumina improves the energy-absorbing capability until a certain content of titania for

TABLE 22.4
Scratch Test Results for ALH Compositions

Composition Name	10 N		15 N	
	Worn Surface (mm^2)	Worn Volume (mm^3)	Worn Surface (mm^2)	Worn Volume (mm^3)
ALH I	1.03	0.083	1.11	0.096
ALH II	1.35	0.11	1.53	0.149
ALH III	1.10	0.088	1.27	0.108
ALH IV	0.66	0.051	1.29	0.137
ALH V	1.46	0.142	1.80	0.184

these composites. In addition, micro wear resistance showed fluctuations depending on the titania content.

By considering all results obtained, these composites enable the material manufacturers to produce low-cost, green composites for various components in automotive and transport industries, such as suspension pads and railway cross ties.

ACKNOWLEDGMENT

We acknowledge Institut supérieur de mécanique de Paris Research Funding that has supported this project.

REFERENCES

Adhikari, B., De, D., & Maiti, S. (2000). Reclamation and recycling of waste rubber. *Progress in Polymer Science, 25*(7), 909–948. 10.1016/S0079-6700(00)00020-4.

Arayasantiparb, D., McKnight, S., & Libera, M. (2001). Compositional variation within the epoxy/adherend interphase. *Journal of Adhesion Science and Technology, 15*(12), 1463–1484. 10.1163/156856101753213312.

Bittmann, Birgit, Frank Haupert, & Alois K. Schlarb (2012). Preparation of TiO$_2$ epoxy nanocomposites by ultrasonic dispersion and resulting properties. *Journal of Applied Polymer Science, 124*(3), 1906–1911. 10.1002/app.34493.

Branch, Maybod (2011). Preparation of nano-scale α-Al2O3 powder by the sol-gel method. *Ceramics–Silikáty, 55*(4), 378–383.

De, Sadhan K., Isayev, Avraam I., & Klementina, Khait (2005). *Rubber Recycling*. Taylor & Francis/CRC Press.

Fiksel, Joseph, Bakshi, Bhavik R., Baral, Anil, Guerra, Erika, & DeQuervain, Bernhard (2011). Comparative life cycle assessment of beneficial applications for scrap tires. *Clean Technologies and Environmental Policy, 13*(1), 19–35. 10.1007/s10098-010-0289-1.

Hirayama, Denise, & Clodoaldo, Saron. (2012). Chemical modifications in styrene–butadiene rubber after microwave devulcanization. *Industrial & Engineering Chemistry Research, 51*(10), 3975–3980. 10.1021/ie202077g.

Irez, A. B., Bayraktar, E., & Miskioglu, I. (2018). Recycled and devulcanized rubber modified epoxy-based composites reinforced with nano-magnetic iron oxide, Fe$_3$O$_4$. *Composites Part B: Engineering*, *148*, 1–13. 10.1016/j.compositesb.2018.04.047.

Irez, A. B., Bayraktar, E., & Miskioglu, I. (2019). Flexural fatigue damage analyses of recycled rubber – modified epoxy-based composites reinforced with alumina fibres. *Fatigue & Fracture of Engineering Materials & Structures*, *42*(4), 959–971 10.1111/ffe.12964.

Irez, A. B., Bayraktar, E. (2020). Design of epoxy modified recycled rubber-based composites: Effects of different contents of nano-silica, alumina and graphene nanoplatelets modification on the toughening behavior. *Gazi University Journal of Science*, *33*(1), 188–199.

Kim, Jung-il, Kang, Phil Hyun, & Nho, Young Chang (2004). Positive temperature coefficient behavior of polymer composites having a high melting temperature. *Journal of Applied Polymer Science*, *92*(1), 394–401. 10.1002/app.20064.

May, Clayton A. (1988). Introduction to epoxy resins. *Epoxy Resins Chemistry and Technology*, 1, 1–8.

Paulo, Glauco Dias, Hirayama, Denise & Saron, Clodoaldo (2011). Microwave Devulcanization of waste rubber with inorganic salts and nitric acid. *Advanced Materials Research*, *418*, 1072–1075. 10.4028/www.scientific.net/AMR.418-420.1072.

Pereira, Artur Camposo, Monteiro, Sergio Neves, Assis, Foluke Salgado de, Margem, Frederico Muylaert, Luz, Fernanda Santos da, & Braga, Fábio de Oliveira (2017). Charpy impact tenacity of epoxy matrix composites reinforced with aligned jute fibers. *Journal of Materials Research and Technology*, *6*(4), 312–316. 10.1016/J.JMRT.2017.08.004.

Pinto, Deesy, Bernardo, Luís, Amaro, Ana, & Lopes, Sérgio (2015). Mechanical properties of epoxy nanocomposites using titanium dioxide as reinforcement – A review. *Construction and Building Materials*, *95*, 506–524. 10.1016/J.CONBUILDMAT.2 015.07.124.

Rothon, R. N. (1999). Mineral fillers in thermoplastics: Filler manufacture and characterisation. In *Mineral Fillers in Thermoplastics I* (pp. 67–107). Springer. 10.1 007/3-540-69220-7_2.

Shokoohi, Shirin, Arefazar, Ahmad, & Khosrokhavar, Ramin (2008). Silane coupling agents in polymer-based reinforced composites: A review. *Journal of Reinforced Plastics and Composites*, *27*(5), 473–485. 10.1177/0731684407081391.

Wetzel, Bernd, Haupert, Frank, & Zhang, Ming Qiu (2003). Epoxy nanocomposites with high mechanical and tribological performance. *Composites Science and Technology*, *63*(14), 2055–2067. 10.1016/S0266-3538(03)00115-5.

Zanchet, Aline, Carli, Larissa N., Giovanela, Marcelo, Crespo, Janaina S., Scuracchio, Carlos H., & Nunes, Regina C. R. (2009). Characterization of microwave-devulcanized composites of ground SBR scraps. *Journal of Elastomers & Plastics*, *41*(6), 497–507. 10.1177/0095244309345411.

Zee, R. H., Huang, Y. H., Chen, J. J., & Jang, B. Z. (1989). Properties and processing characteristics of dielectric-filled epoxy resins. *Polymer Composites*, *10*(4), 205–214. 10.1002/pc.750100402.

Zhang, Guangwu, Wang, Fuzhong, Dai, Jing, & Huang, Zhixiong (2016). Effect of functionalization of graphene nanoplatelets on the mechanical and thermal properties of silicone rubber composites. *Materials*, *9*(2), 92. http://www.mdpi.com/1 996-1944/9/2/92/htm.

23 Evaluation of Mechanical and Microstructural Properties of Low-Density Concrete with Residual (Scrap) Vegetable Fiber and Blast Furnace Slag

K.M.A. Silva[1], C. Alves[1], L.M.P. Ferreira[1], and E. Bayraktar[2]
[1]Faculty of Civil Engineering, Federal University of the South and Southeast of Pará, Pará, Brazil
[2]ISAE-SUPMECA-PARIS, School of Mechanical and Manufacturing Engineering, France

CONTENTS

23.1 Introduction 423
23.2 Amazonian Vegetable Fibers 424
23.3 Methodology 426
23.4 Results and Discussion 428
23.5 Conclusions 430
References 431

23.1 INTRODUCTION

Concrete is the most popular and most adaptable building material in the world. However, it is an expensive material, because to obtain high strengths, a large amount of volume is required. It is a fact that over the years, concrete has been

DOI: 10.1201/9781003148760-23

improved and the use of vegetable fibers to reinforce the cement matrix is an example (Ferreira et al., 2017). Therefore, it is interesting for an industry to insert other materials that maintain the qualities of the concrete.

The insertion of Amazonian vegetable fibers in concrete is a way to dispose of fibrous waste and solve the problem with costs. Savastano et al. (2009) point out that the long-term durability of composites reinforced with natural fibers can be limited due to the problems generated by their insertion.

Thus, when analyzing the reinforcement of cementitious matrices, it is essential to point out the properties of the analyzed fibers and the properties of the concrete with added fibers.

The objective of the research was the essentiality of the use of new materials in the construction industry. It is known that the adhesion between the aggregates and the cement paste is an important factor of strength of the concrete (Neville, 2015). Thus, it is also intended to optimize the interface of the aggregate matrix only in the presence of blast furnace slag and plasticizers, without the need for chemical treatment.

23.2 AMAZONIAN VEGETABLE FIBERS

In this article, the fibers acai and coconut were used as study elements. Both contain different technical features due to differences in their development in the plant of origin. In this paper, the two methods of fiber insertion in the concrete were compared, to analyze which aggregate matrix ratio generates better resistance.

Coconut fiber comes from the mesocarp of the green coconut. Its extraction was carried out using the outer shell of the coconut. Therefore, the material was cut in four parts to remove the fruit part. The mesocarp was then defibrated on a mechanical stirrer to reduce the size of the fiber. Figures 23.1 and 23.2 present

FIGURE 23.1 Fibers after and before mechanical processing (Authors, 2019).

Mechanical and Microstructural Properties 425

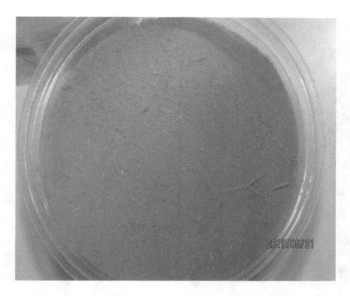

FIGURE. 23.2 Fibers after and before mechanical processing (Authors, 2019).

the fiber in the two phases of the process. It is emphasized that the fibers were separated granulometrically, since the fiber fraction used was between the sieves #4.75 mm and #1mm.

After SEM testing, it was observed that the coconut fiber has a cylindrical shape with lateral grooves that facilitate the adhesion and the mechanical fit in the aggregated matrix interface promoting the viability of the fibrous reinforcement. Figure 23.3 shows the microscopic characteristics of the fiber.

The acai fiber on the other hand is a residual fiber from the core after removal of the fruit. The extraction of açaí for consumption is done by disaggregating the fruit from the inner lump. Thus, this residual fiber was obtained in establishments selling the açaí pulp.

It is known that the fibers are attached to the lump, easily removable by hand, however, to facilitate the process in view of need for large volumes for the industry a beater was used. It should be emphasized that the mixer does not break the lumps but only scatters the fibers that are separated by the sieves. Figures 23.4 and 23.5 show the process of disaggregation. The açaí fibers were separated by the sieves #4.75 mm and # 1.0 mm, similarly to the coconut fibers.

Microscopically evaluating açaí fiber, fibrils were observed in the longitudinal direction. It is a fact that the presence of the fiber is awfully close to the fruit adds to its lignocellulose. Figure 23.6 presents the fibrils that bind the cement matrix. Concurrently, it is observed that the cylindrical shape increases the contact between the aggregate and the matrix, promoting better interaction.

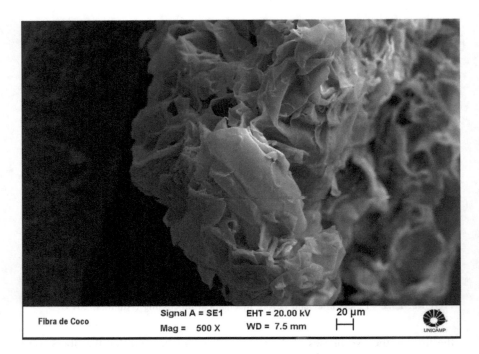

FIGURE 23.3 Scanning electron microscopy of coconut fibers (Author, 2019).

23.3 METHODOLOGY

The methodology used in this work consisted of comparing the effect on the aggregated matrix interface of the two fibers, evaluating characteristics and the mechanical properties of the concrete constituted by them.

Being as much, to evaluate the increase of the resistance over time, the axial compressive strength at 7, 14, 21 and 28 days was evaluated, defined in NBR 5739: 2007. As a result, the diametrical compression, water absorption and porosity were evaluated at 28 days in order to compare the two fibers' concrete composition. Simultaneously, the results were performed in order to validate the resistance results, ultrasound tests observing the variation over the days.

For the concrete composition of the fibers 1: 2.5: 2%: 1%: 0.45:0.001 (cement: sand: fibers: slag: water/cement: superplasticizers) was used. A concrete composition of the literature was used for the reference trait. These traits were defined based on bibliographical analyses and detailed studies, in order to optimize the quality of the material produced. It was observed that a low quantity of superplasticizers was needed to obtain good results, not generating excessive costs to the product.

In the production of the test specimens, 10 minutes of homogenization were used, the last 5 being the fiber insertion in the mortar produced, delimited by

Mechanical and Microstructural Properties

FIGURE 23.4 Fibers after and before mechanical processing (Authors, 2019).

FIGURE 23.5 Fibers after and before mechanical processing (Authors, 2019).

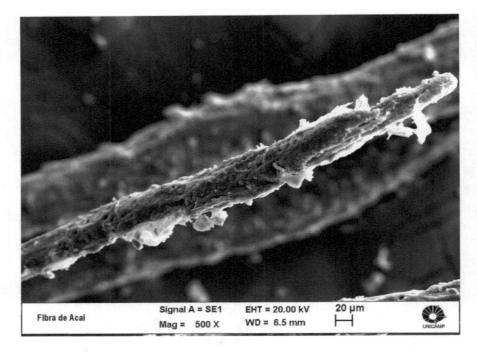

FIGURE 23.6 Fraction of the açai fiber used (Authors, 2019).

NBR 8802: 2013, which defines concrete ultrasound tests. The specimens were removed from the form with 24 hours and then submerged in alkaline solution to avoid creating cracks. Figure 23.7 shows samples of the specimens produced for the research with 24 hours.

23.4 RESULTS AND DISCUSSION

The results of axial compression were very significant, since the evolution of the resistance during the 28 days demonstrated in Graph 23.1, shows that the light concrete produced can be used with a structural function.

In the ultrasound assays, the following equations of the ABNT NBR 6118:2014 (1. e 2.) were used to convert the results into resistance information. This assay evaluates the quality of the pulse passage, so it depends on the quality of the density, internal porosity and the number of cracks.

$$E_{ci} = V^2 \cdot \rho \cdot \left(\frac{(1 + \nu) \cdot (1 - 2\nu)}{(1 - \nu)} \right) \quad (23.1)$$

$$E_{ci} = \alpha_E .5600\sqrt{f_{ck}} \quad (23.2)$$

Mechanical and Microstructural Properties 429

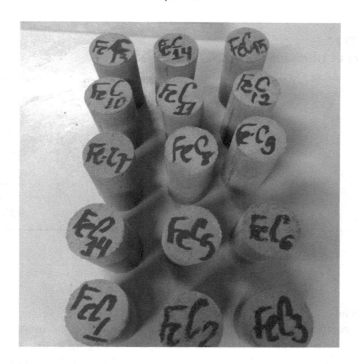

FIGURE 23.7 Sample of test specimens.

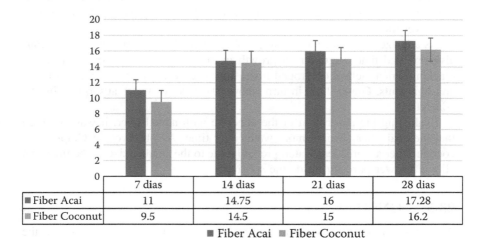

GRAPH 23.1 Results of axial compression.

TABLE 23.1
Abstract of Ultrasound Tests

Fiber	Age (Days)	Average Speed	Resistance (MPa)
Açai	7	3150	7.76
	14	3180	13.30
	21	3350	19.35
	28	3410	22.58
Coconut	7	3090	7.19
	14	3130	12.48
	21	3320	18.66
	28	3380	21.79

$$fckj = fck \cdot e^{0,25 \cdot \left(1 - \frac{28}{j}\right)} \qquad (23.3)$$

The modulus of elasticity obtained in Eq. (23.1), relates V (velocity in m/s), poisson coefficient (in this article considered 0.2) and specific mass. Equation (23.2) converts the data to resistance values considering α_E as the smallest aggregate (0.7). It is a fact that equation 3 finalizes the conversions of the ultrasound data into resistance by organizing the results according to the age of the test specimens. The factor 0.25 at Eq. (23.3) refers to the type of cement implemented in concrete. Table 23.1 summarizes the data provided by ultrasound tests over the course of days.

As analyzed by Savastano et al. (2009), the crack microstructure interactions also revealed that fatigue crack growth in the composites occurred via matrix cracking, crack deflection around fibers and crack-bridging by uncracked fibers and ligaments. Graph 23.2, in turn, shows the level of water absorption by the concrete.

Evaluating the absorption of the concrete with added fiber, it was observed that the açai fiber absorbs more water over time. As observed in Graph 23.2, coconut fiber absorbs less water, probably due to the smaller size of the fiber that allows a lower porosity inside the specimen.

23.5 CONCLUSIONS

Açai and coconut fiber are available in abundance in many places around the world, which makes it quite workable as a reinforcement material in concrete. In addition, the use of this kind of fiber in concrete can be an efficient method for the disposal of residual açai and coconut fibers; the improvement of this methodology could reduce the demand for additional waste disposal infrastructure and decrease the load on existing landfills and incinerators.

Mechanical and Microstructural Properties

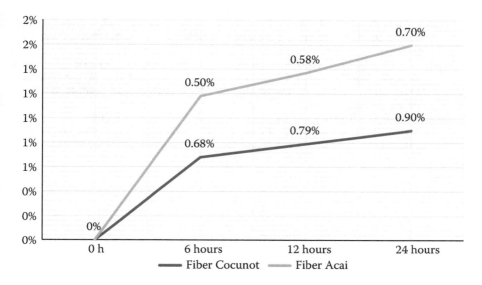

GRAPH 23.2 Results of absorption.

It is observed that coconut fibers have less porosity inside the specimen, but their resistance is lower. It is assessed that by the size of the açai fibers, they mechanically mesh better, facilitating the distribution of stresses within the composite, having better resistance.

The results obtained by compression and ultrasound tests showed an improvement of the resistance over time. This preliminary study propose that the substitution of the fibers in traces containing large aggregates to evaluate the relation of the decrease of the density with the maintenance of the resistance. Due to its properties, the concrete reinforced by coconut and açai fibers can be a good replacement for asbestos fibers, and with the advantage of being natural, it doesn't threat the environment.

REFERENCES

Associação Brasileira de Normas Técnicas (2007). *ABNT NBR 5739: Argamassas e Concretos Endurecidos–Concreto–Ensaios de Compressão De Corpos de Prova Cilíndricos.*

Associação Brasileira de Normas Técnicas. ABNT NBR 6118 (2014). *Projeto de Estruturas de Concreto. Procedimento.* Rio de Janeiro.

Associação Brasileira de Normas Técnicas. ABNT NBR 8802 (2013). *Determinação da Velocidade de Propagação de Onda Ultrassônica.* Rio de Janeiro.

Cronquist, A. *Sistema Integrado de Classificação de Plantas Florescentes.* Columbia Univ. Press.

De Oliveira, M., De Carvalho, do S. P., & Do Nascimento, José Edmar Urano (2000). Walnice Maria Oliveira. *Açaí (Euterpe oleracea Mart.).* Funep.

Ferreira, Carla Regina et al. (2017). Comparative study about mechanical properties of strutural standard concrete and concrete with addition of vegetable fibers. *Materials Research, 20,* 102–107.

Júnior, Ubirajara Marques Lima et al. (2007). *Fibras da semente do açaizeiro (Euterpe Oleracea Mart.): avaliação quanto ao uso como reforço de compósitos fibrocimentícios.* Tese de Doutorado. Pontifícia Universidade Católica do Rio Grande do Sul.

Neville, Adam M. (2015) *Propriedades do Concreto.* 5ª. Bookman Editora.

Savastano, J. R. H. et al. (2009) *Fracture and fatigue of natural fiber-reinforced cementitious composites. Cement and Concrete Composites, 31*(4), 232–243.

24 Evaluation of Reinforced Concrete (RC) with Different Scrap Coarse Aggregates

E.S. Fonseca[1], K.M.A. Silva[1], S.H.S. Santana[1], L.M. Policarpio[1], and E. Bayraktar[2]
[1]Faculdade de Engenharia Civil (FAEC), Universidade Federal do Sul e Sudeste do Pará, Pará, Brazil
[2]ISAE-SUPMECA-PARIS, School of Mechanical and Manufacturing Engineering, France

CONTENTS

24.1 Introduction 433
24.2 Materials 434
24.3 Methodology 435
24.4 Results and Discussion 436
24.5 Conclusions 440
References 440

24.1 INTRODUCTION

The useful life is known as the period in which a building maintains its performance characteristics above the minimum time required in the project. Consequently, a building material finishes its useful life when its properties, under certain conditions of use, deteriorate to the point to be considered unsafe or uneconomical.

During the design phase of the structures, the durability of the materials to be used must be evaluated with the same care as other aspects, such as mechanical properties and initial cost (Mehta & Monteiro, 2014). Durability can be understood as the resistance to degradation of products, materials, buildings and other built assets over time (Neville, 2015).

DOI: 10.1201/9781003148760-24

Therefore, when it comes to safety, the occurrence of pathological manifestations can degrade the materials that integrate the structure, resulting in a considerable decrease in its carrying capacity (Mehta & Monteiro, 2014). This work aims to evaluate the behavior of two kinds of CR, based on the type of coarse aggregate.

Analyzing concrete compounds with rubber observed that the reduction in compressive strength of concrete manufactured with rubber aggregates may limit its use in some structural applications, but rubberized concrete also has some desirable characteristics such as lower density (Siddique & Naik, 2004).

However, it is pointed out that evaluating concretes using the scrap tire particles decreases the strength performance of concrete, especially the flexura strength (Chunlin et al., 2011). Thus, the importance of verifying the influence on such properties is evaluated.

24.2 MATERIALS

The specimens were manufactured in accordance with ABNT NBR 12655, which presents the ABCP (Brazilian Portland Cement Association) dosing method. During the process, the following materials were used: hydraulic binder – Portland cement CP II E 32; coarse aggregate – pebble and gravel (19 mm), classified in accordance with ABNT NBR 7211/2007; small aggregate – medium sand classified in accordance with ABNT NBR 7211/2007 and water.

The cement used, CP II E 32, contains the addition of granulated slag from the blast furnace, responsible for conferring the property of low hydration heat. CP II E is composed of 94–56% of clinker and plaster and 6–34% of slag. This hydraulic bond is recommended for structures that require moderately slow heat release.

The data on the coarse aggregates (pebble and gravel) used were obtained through the performance of some tests, based on the standards NM 248 (ABNT, 2003), NBR 7211 (ABNT, 2009), NBR 9776/1987 and NBR 7251/1982 and NM 45/2006. Table 24.1 presents some information on the pebble and gravel used for the manufacture of RC.

TABLE 24.1
Information on the Properties of the Pebble and Gravel

Data	Pebble Aggregate	Gravel Aggregate
Maximum particle size	19 mm	19 mm
Specified mass	2,333.33 kg/m^3	2,333.33 kg/m^3
Compacted unitary mass	1,859 kg/m^3	1,584.5 kg/m^3

Evaluation of Reinforced Concrete

24.3 METHODOLOGY

In this work, two types of reinforced concrete were manufactured, as shown in Table 24.2. For the manufacture of RC, the mechanical mixture of the constituent materials was used for approximately 6 minutes, until perfect homogenization; then the concrete was cast (10 cm in diameter and 20 cm in height).

During the first 24 hours, the samples were stored in a place protected from the weather, being covered with non-reactive and non-absorbent material, avoiding loss of water from the concrete to the environment. Then, the specimens were cured until the age of 28 days in a water tank saturated with lime.

- Tensile and compression tests

In order to evaluate the strength of the concrete during the action of compressive and tensile forces, the tests were carried out in the laboratory, using the specimens manufactured for this purpose; 12 samples, 6 for each type of RC, were used.

- Capillary water absorption test

The water absorption of concrete – by capillarity – was based on NBR 9779 (Associação Brasileira de Normas Técnicas, 2012). Initially, the mass of each specimen was measured, and then they were dried in an oven with a temperature of $105 \pm 5°C$. After drying, the specimens were cooled to a temperature of $23 \pm 2°C$, and the mass of each was determined. The specimens were positioned on supports inside the test container, and the water level remained constant at 5 ± 1 mm above its lower face. Figure 24.1 shows the specimens in the test container.

The mass of the samples was determined at 3, 6, 24, 48 and 72 hours, counted from the placement of it in contact with water. The results of water absorption are expressed in g/cm², and calculated according to Eq. (24.1):

$$C = \frac{A - B}{S}, \qquad (24.1)$$

where C is the absorption of water by capillarity, in g/cm². A is the mass in grams

TABLE 24.2
Concrete Dosage

Aggregate	Compressive Strength (MPa)	A/C Ratio	Dosage	Cement Content (kg/m³)
Pebble	30	0.41	1: 081: 2.43	500
Gravel	30	0.41	1: 0.9: 2.18	500

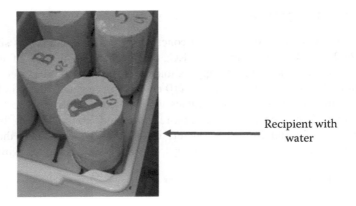

FIGURE 24.1 Capillary water test.

of the specimen that remains with one side in contact with the water for a specified period. B is the mass of the dry specimen as soon as it reaches a temperature of $(23 \pm 2)°C$, in grams. S is the cross-sectional area, in cm².

- Corrosion test

To carry out the accelerated chloride-induced corrosion test, specimens were manufactured in a cubic geometry with a 19 cm edge. Internally, four 10 mm diameter CA-50 steel bars were allocated, with 2 cm removed from the sample.

The chloride-induced corrosion acceleration process consists of carrying out analysis cycles, where each cycle presents the drying steps in an oven for 5 days and partial immersion of the cubic specimens in a solution containing chloride ions for 2 days. In the immersion stage, the samples were placed in a container with 3.5% by volume of sodium chloride. In this stage, the specimens were partially submerged up to half, to promote the entry of oxygen and electrolyte in the armature, thus favoring the electrochemical corrosion process. The chloride ions penetrate only from the side, since the upper and lower part of the concrete cube will be coated with neutrol.

During the entire experiment, measurements of the corrosion potential were carried out. To obtain the corrosion potential, a copper/copper sulfate – ESC electrode was used as reference, according to ASTM C 876 (1991). Figure 24.2a and b shows the container used to put the samples in an NaCl solution, and a specimen prepared for measurement; (c) and (d) show a multimeter used for measurement and the reference electrode, respectively.

24.4 RESULTS AND DISCUSSION

- Tensile and compressive strength results

Evaluation of Reinforced Concrete

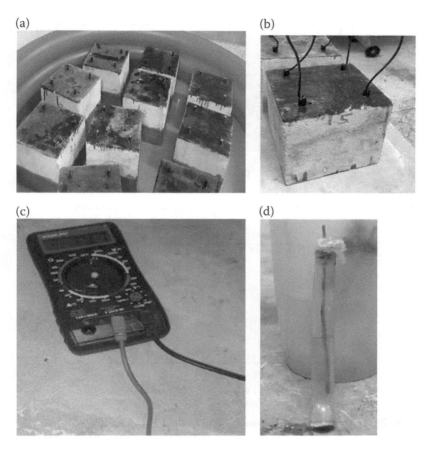

FIGURE 24.2 (a) Samples in NaCl solution, (b) sample prepared for measurement, (c) multimeter used for measurement and (d) silver chloride reference electrode.

The results obtained by the compression test for the specimens containing pebble and gravel as coarse aggregate, after age of 28 days, are presented in Figure 24.3. As it can be seen, there is some variation between the compressive strength of the specimens with gravel or pebble used as coarse aggregate. This result can be explained by the smooth surface of pebble aggregate, which hinders contact with the cement matrix, reducing the RC compressive strength.

- Tensile strength results

Although resisting tensile stresses is not the main function of concrete, it is necessary to achieve the minimum resistance to mechanical stresses like this. Therefore, the tensile test by diametrical compression was performed to define the strength of the concrete when pulled. The results obtained by the tensile test

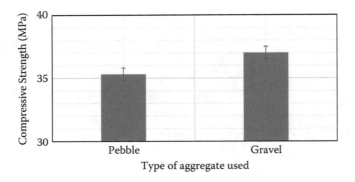

FIGURE 24.3 Compressive strength results (MPa) of the CR produced.

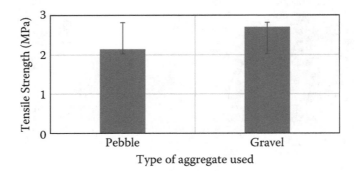

FIGURE 24.4 Tensile strength results (MPa) of the CR produced.

for the specimens containing the pebble and gravel as coarse aggregate, after age of 28 days, are presented in Figure 24.4.

Based on the work of Farias et al. (2018), where he points out that normally an acceptable result for the tensile strength in some concrete revolves around 10% of its capacity for compressive strength, it was found that both the specimens containing pebble and those containing gravel did not reach this minimum percentage. With this it is possible that cracks appear along the structure, thus compromising its physical integrity.

- Concrete exudation

The exudation process consists of the migration of water within the concrete to its surface. This process can damage the structure since it interferes in the formation of connections between the components of the concrete. Loss of water will also cause the concrete to shrink, thus creating tensile stresses within the concrete. Thus, it is necessary to analyze this phenomenon in the two studied RC.

For the RC made with pebble as coarse aggregate, the value of 0.039% of exudation was obtained, although for the gravel (broken rock), this percentage was 0.58%. Considering the work of Padilha et al. (2017), where it is defined that the limit of exudation for concrete is around 4%, it is possible to see that the studied CR does not present significant values of exudation.

- Capillarity water absorption

Water can penetrate concrete structures through the soil. It occupies the existing voids and consequently saturates the region, causing pathologies in the building. As a result of the capillarity absorption test, Tables 24.3 and 24.4 show the water absorption values of the specimens containing gravel and pebble, respectively. According to Pinheiro and Monteiro (2016), the relevant capillarity absorption values are around 1.2 and 1.4 g/cm² (12 and 14 kg/m²). In this case, it is observed that the average absorption by capillarity from CR containing pebble and in those containing gravel are well below the value presented by these authors.

Also, Cunha et al. (2015) mentions that the coefficient of water absorption by capillarity is related to the existence of smaller voids. Thus, it is possible to conclude that the studied specimens do not present significant absorption values to the point of causing possible pathologies in the building.

- Accelerated corrosion test

TABLE 24.3
Absorption in Specimens Containing Gravel

Sample	Absorption (g/cm²)	Average (g/cm²)
1	0.21	
2	0.31	0.28
3	0.31	

TABLE 24.4
Absorption in Specimens Containing Pebble

Sample	Absorption (g/cm²)	Average (g/cm²)
1	0.27	
2	0.37	0.37
3	0.47	

After 2 days partially submerged in NaCl, the samples were removed to measure the corrosion potential, with the aid of a multimeter and the reference electrode; then they were taken to the drying kiln, where it is stayed for 5 days. At the end of the cycles, the test showed a similar result to the specimens containing gravel and those containing pebble.

24.5 CONCLUSIONS

The main objective of this work was to evaluate the behavior of pebble-reinforced concrete or gravel used as graded aggregates. The reinforced concrete has been tested for its tensile resistance, compression and corrosion, as well as exudation and absorption of water by capillarity. In this regard it can be concluded that:

- The CR with two types of coarse aggregates were evaluated, but as a result, it was possible to observe only variation of compressive strength between them.
- The result of tensile strength showed a small variation between the two types of concrete produced.
- Results of capillarity water absorption, and exudation tests also are in accordance with the standards, showing no difference between the CR produced.
- The accelerated chloride-induced corrosion test showed no variation between the two types of concrete studied; however, more tests need to be conducted to consolidate this result.

REFERENCES

American Society for Testing and Materials (1991). *Standard test methord for halfcell potentials of uncoated reinforcing steel in concrete.* ASTM C 876-91.
Associação Brasileira de Normas Técnicas (2012). *NBR 9779: Argamassa e concreto endurecidos – Determinação da absorção de água por capilaridade.*
Associação Brasileira de Normas Técnicas (1998). *NBR NM 248: Agregados – Determinação da composição granulométrica.*
Associação Brasileira de Normas Técnicas (1983). *NBR 7211: Agregados para concreto – Especificação.*
Associação Brasileira de Normas Técnicas (2015). *NBR 12655: Concreto de cimento Portland – Preparo, controle e recebimento – Procedimento.*
Associação Brasileira De Normas Técnicas (1982). *NBR 7251: Agregados – Determinação da massa unitária.*
Associação Brasileira De Normas Técnicas (1987). *NBR 9776: Agregados –Determinação da massa unitária dos agregados: Método de ensaio.*
Cunha, S. R. L. et al. (2015) *Argamassas com incorporação de materiais de mudança de fase (PCM): caracterização física, mecânica e durabilidade. Revista Matéria,* 20(1), 245–261.
Chunlin, Liu; Kunpeng, Zha, & Depeng, Chen (2011). *Possibility of concrete prepared with steel slag as fine and coarse aggregates: A preliminary study. Procedia Engineering,* 24, 412–416.

Farias, L. A. et al. (2018). *Ensaios de Tração Direta em Corpos de Prova de Concreto.* p. 9.

Mehta, P. K., & Monteiro, P. J. M. (2014) *Concreto Microestrutura, Propriedade e Materiais.* 2ª Edição. Ed.: IBRACON. ISBN.:978-85-98576213. p. 751.

Neville, A. M. (2015) *Propriedades do Concreto –* 5ª Edição. [s.l.] Bookman Editora.

Padilha, F., Schimelfenig, B., & Da Silva, C. V. (2017) *Análise da utilização de endurecedores na dureza superficial de concretos para pisos. Revista de Engenharia e Pesquisa Aplicada,* 2, 3.

Pinheiro, H., & Monteiro, E. C. B. (2016) *Análise Comparativa do desempenho de concretos com adições minerais quanto à corrosão de armaduras por íons cloretos. Revista de Engenharia e Pesquisa Aplicada,* 2, 1.

Siddique, Rafat, & Tarun R., Naik (2004) *Properties of concrete containing scrap-tire rubber–an overview. Waste Management,* 24(6), 563–569.

25 Influence of Iron Content on the Microstructure and Properties of Recycled Al–Si–Cu–Mg Alloys

A.J. Vasconcelos[1], R.S.M. Silva[1], P.J. Oliveira[1], M.L.N.M. Melo[1], and O.F.L. Rocha[2]
[1]Institute of Mechanical Engineering, Federal University of Itajubá – UNIFEI, Campus Prof. José Rodrigues Seabra, Itajubá, Minas Gerais, Brazil
[2]Federal Institute of Education, Science and Technology of Pará-IFPA, Pará, Brazil

CONTENTS

25.1 Introduction 443
25.2 Materials and Methods 445
    25.2.1 Materials 445
    25.2.2 Methods 448
25.3 Results and Discussion 450
25.4 Conclusions 458
Acknowledgment 459
References 459

25.1 INTRODUCTION

The increased demand for aluminum alloys, mainly used in the automobile, aeronautic and electricity transmission industries, has increased the production of aluminum alloys, which already exceeds the production of steel and other metallic alloys (Filip Průša et al., 2017). The recycling of aluminum alloys for casting and mechanical forming processes has been reported in the scientific community for the energy to recycle is less than used to obtain the primary alloy and this has been an environmental appeal (Cameron et al., 2005; Gaustad et al. 2012). Therefore, the production from the extraction (electrolysis) of aluminum uses a lot of energy, making it highly uneconomical. Primary aluminum production (electrolysis) consumes 45 kWh/kg; however, secondary aluminum production, which is scrap recycling, consumes ~2.8 kWh/kg (Das et al., 2007; Yang et al., 2015). Thus, recycling only accounts for 6–8% of total energy compared to primary aluminum production. That is why the recycling of aluminum scrap and its alloys has become fundamental for the future processing of aluminum alloys.

Recycling industries are challenged as iron is easily incorporated into aluminum scrap as an impurity. Due to the low solubility of iron in aluminum, the formation of Fe intermetallic phases is inevitable (Khalifa et al., 2003; Kim et al., 2012; Rakhmonov et al., 2017; Wang et al., 2016). This leads to the formation of various intermetallic Fe compounds (IMCs) during solidification, such as α-Al$_8$Fe$_2$Si or α-Al$_{15}$ (Fe,Mn)$_3$Si$_2$, β-Al$_5$FeSi, Δ-Al$_4$FeSi$_2$, ω-Al$_7$Cu$_2$Fe and π-Al$_8$Mg$_3$FeSi$_6$, depending on the rate of cooling (TR) and alloy composition (Barros et al., 2019; Magno et al., 2019; Tunçay & Bayoğlu, 2017; Tupaj et al., 2016; Wang et al., 2010). It is known that the effect of these phases on the mechanical properties of aluminum alloys depend on their type, size and quantity in the molten microstructure.

Needle-shaped iron-rich phases, that is, with large longitudinal dimensions, have been characterized by high fragility (Das et al., 2007) and negatively affect the mechanical properties of alloys (Yang et al., 2015). Crystalline intermetallics or α-Fe type crystals occur during eutectic solidification with α-aluminum and usually appear with polyhedral morphology or Chinese writing. The polyhedral structure forms as a primary phase (Filip Průša et al., 2017; Gaustad et al., 2012), while the written Chinese structure forms, at a relatively low temperature, together with the eutectic structure (Filip Průša et al., 2017). As for β-Fe intermetallic, a naturally hard and brittle precipitate in the form of three-dimensional platelets, it usually acts as a stress concentrator and interferes with the flow of liquid metal in the interdendritic channels during solidification (Filip Průša et al., 2017). Its morphology is directly related to the Fe content alloy (Filip Průša et al., 2017). The dimensions of Fe-rich intermetallic phases increase with increasing iron concentration and decreasing cooling rate after solidification (Cameron et al., 2005).

Tunçay and Bayoğlu (Tunçay & Bayoğlu, 2017) concluded that increasing Fe content reduces mechanical properties, ductility, corrosion and fatigue resistance, and a considerable increase in width and length of acicular Fe intermetallics when Fe content varies from 0.2 wt.% to 1.2 wt.%. It was verified that for this same range of Fe contents, an increase in the projected area of porosity from 4.59% to 7.56% and respective reduction in the SDAS value from 17.98 μm to 14.66 μm. Finally, the authors observed a reduction in the tensile strength limit from 189 MPa to 153 MPa and elongation from 9.14% to 7.36%. Kim et al. (2012) added 0.13% by weight Mn and 0.13% by weight Cr to an alloy A356 (containing 0.2% by weight Fe) and reported that acicular β-Al5FeSi intermetallics were transformed into granular or skeletal structures α-Al (Mn, Fe) Si.

Despite the negative influence of iron on the microstructure and mechanical properties of aluminum alloys, there is still the intentional introduction of iron. For example, regardless of its harmful effects on the properties of Al–Si alloys, Fe is usually introduced in amounts above 2% to reduce the bonding effect of the castings on the metal mold walls (Tupaj et al., 2016). The addition of 0.1–0.7% by weight of Fe in aluminum alloys favors the formation of an iron-rich phase of the α-Fe type and can increase strength at high temperatures (Wang et al., 2010).

Although the literature presents studies on the mechanical characteristics of Fe-based multicomponent alloys containing Al, directional solidification studies with AlSiMgFe alloys are still scarce in the literature, such as thermal and microstructural analysis, including the effects of growth and cooling rates in the dendritic microstructure and in the mechanical properties.

Thus, this work aims to analyze the influence of iron content on the microstructure of alloy A356 under as-cast conditions, with the addition of 1% and 3% iron, and to present a study on the roles of thermal parameters, dendritic microstructure and β-Al5FeSi phase characteristic in the microhardness of a horizontally solidified Al7Si0.4Mg1.2Fe alloy.

25.2 MATERIALS AND METHODS

25.2.1 Materials

The aluminum alloy used was the commercial A356 alloy supplied by ALCOA Alumínio S.A., of chemical composition (Table 25.1) using a Spectro spectrometer, Spectro Maxx, from IMBEL - Itajubá War Material Industry.

The metallic Fe powder, obtained by atomization by the company Höganäs Brasil Ltda, has a purity of 99.5% and a particle size between 10 µm and 50 µm. The A356 alloy was remelted in a silicon carbide crucible and it was inserted in a Brasimet brand resistive electric furnace, chamber type, model K-250 and was kept at a temperature of 780°C for 30 minutes. In order to simulate the recycling process, the metallic Fe powder was then added to the still liquid metal and then homogenized. In this procedure, alloy A356 without Fe and A356 were obtained with the addition of 1% and 3% of Fe by weight.

TABLE 25.1
Nominal Chemical Composition of the A356 Alloy by wt.%

A356 Alloy	Si	Mg	Cu	Mn	Fe	Zn	Ti	Al
*Nominal (wt.%)	6.5–7.5	0.25–0.45	0–0.2	0–0.1	0–0.2	0–0.1	0–0.2	Bal.
**Sample	6.88	0.332	0.074	0.0	0.158	0.0	0.0	Bal.

Source: *Adapted from *Metals Handbook* (2004); **IMBEL (2019).

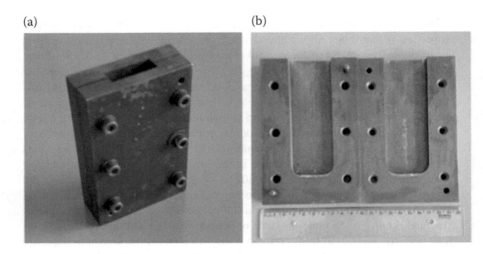

FIGURE 25.1 Metallic mold: (a) outside and (b) inside.

After remelting, the A356 alloys without Fe and with 1% and 3% Fe were cast in 1020 steel metal molds with a cavity 120 mm long, 40 mm wide and 14 mm thick, as shown in Figure 25.1a and b.

To record the temperature profiles during cooling, type K thermocouples were used. These thermocouples are coupled to a thermal data acquisition board, whose brand was National Instruments, model cDAQ-9171. To capture the signals emitted by the board, the LABTRIX software was used, which is capable of digitizing, in real time, the measurements of the thermocouples.

The ingot as cast Al–7 wt.% Si–0.4 wt.% Mg–X wt.% Fe (X = 0 e 1.2) alloys was obtained in the water-cooled horizontal solidification device. Figure 25.2 shows a complete schematic of the furnace used in the experiments, as well as the details of the water-cooled horizontal solidification device and the positioning of the thermocouples (5, 10, 15, 20, 30, 50, 70 and 90 mm), type K, from the heat transfer surface (cooled mold plate) of the horizontal ingot mold. Data generated by thermocouple readings were processed by OriginPro 2020 software and used to determine transient thermal parameters (V_L and T_R) during horizontal solidification. More details about the applied inclined mounting, as well as the working principle, have been described by other authors (Barros et al., 2019; Bouchard & Kirkaldy, 1997; Chen et al., 2014; Lima et al., 2018; Magno et al., 2019; Souza et al., 2018; Vasconcelos et al., 2014; Barbosa et al., 2019).

The horizontal device consists of a rectangular ingot structure in AISI 304 stainless steel, 150 mm long, 60 mm wide, 60 mm high and 3 mm thick. The internal faces of the ingot molds were covered with layers of alumina and an upper part placed with refractory material to avoid loss of calories to the environment.

Recycled Al–Si–Cu–Mg Alloys

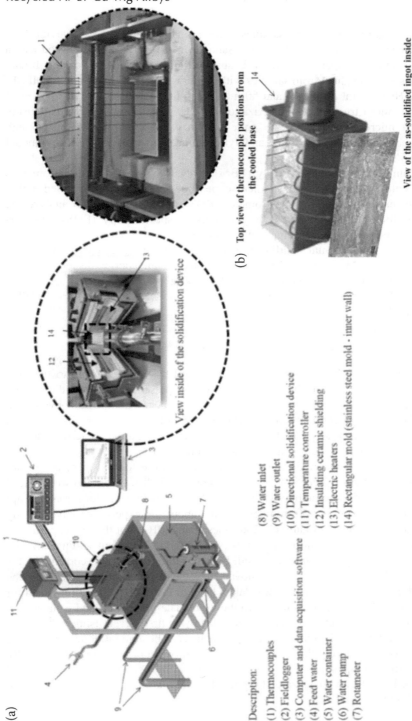

FIGURE 25.2 Schematic of the horizontal unidirectional solidification device (Lima et al., 2018; Magno et al., 2019; Vasconcelos et al., 2021).

After solidification, the samples were prepared for conventional metallographic analysis: cut, sanded with grit sandpaper up to #1200 and polished with a 3 μm diamond suspension and OP-U suspension, both from the Struders brand. Images were obtained from samples with Keller reagent etching (10 ml of HF, 15 ml of HCl, 25 ml of HNO3 and 50 ml of distilled water).

The microstructural characterization of the samples was performed using optical microscopy (OM). The optical microscope used was a Zeiss brand, Jenavert model, associated with an Olympus image acquisition accessory, model TVO.5XC-3o.

Scanning electron microscopy (SEM) analysis was performed. The images were collected by a Carl Zeiss microscope, model EVO MA15, coupled with a Bruker energy dispersive spectrometer (EDS), model XFlash. To analyze the composition of the phases, energy dispersive spectroscopy (EDS) was performed. For the microstructural characterization of the alloys obtained in the water-cooled horizontal solidification device, a scanning electron microscope (SEM TESCAM, VEGA LMU) was coupled to an energy dispersive x-ray spectrometer (EDS, AZTec Energy X-Act, Oxford).

For the microhardness measurements, a Time® Microdurometer – TH712 was used.

25.2.2 Methods

The proposal of this research will be presented briefly, according to the flow chart in Figure 25.3. In the development of the present research, a theoretical/experimental analysis methodology will be approached about the conventional and horizontal directional solidification of the A356 alloy with different iron contents, adopting sequential activities for the preparation, determination of thermal parameters and characterization of this alloy.

For casting of the A356 alloy, the specimens were formed in pieces of approximately 2 cm × 2 cm × 2 cm of the commercial alloy A356 and were placed in a silicon carbide crucible, which was later placed in a muffle furnace at approximately 780°C for 30 minutes. Details about the experiment have already been described by other authors (Rodrigues et al., 2010). For the first experiment, the alloy in its raw solidification form was remelted, without adding any elements, and then poured into the mold. For the addition of iron in contents of 1% and 3%, the remelted alloy was removed from the oven and after mixing with the aid of a pure aluminum bar, it was taken back to the oven where it remained for 15 minutes before being poured into the mold. The mold was sand isolated in a container and kept at room temperature. Iron was first sieved using a 200 mesh sieve and then weighed using a precision balance. The amount of added iron was calculated considering the iron present in the alloy, in order to total 1% and 3% of the total composition. The remelted ingots were then cut longitudinally and obtained samples with a size of 1.5 cm × 1.5 cm, totaling 16 smaller samples for each ingot.

Recycled Al–Si–Cu–Mg Alloys

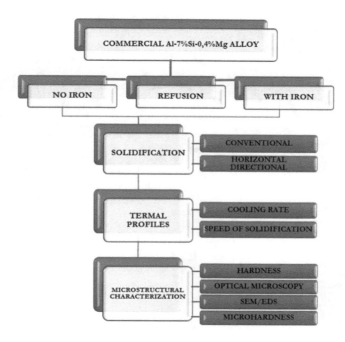

FIGURE 25.3 Flow chart of the theoretical-experimental procedure proposed for this work.

In turn, the as-cast ingot of the investigated alloys, obtained from the horizontal solidification experiment, was sectioned longitudinally, sanded to #600 and then immersed in real water (HNO_3 + HCl) to reveal the macrostructure shown in Figure 25.4a and subjected to microstructural analysis, as seen in Figure 25.4b. The microstructure was revealed with Keller's reagent. It is observed that the phase length λ_2 measurements and β-Al_5FeSi (βFe) were obtained in as-cast samples from the cooled base (5, 10, 15, 20, 30, 40, 50, 60, 70, 80, 90 mm) and, in turn, were correlated with V_L and T_R. The MOTIC image processing system and Image J software were used to measure independent λ_2 readings for each selected position. Furthermore, some samples of the studied alloy were submitted for microstructural characterization by scanning electron microscopy (SEM) and x-ray energy dispersive spectrometry (EDS). Micrographs were taken at different magnifications in order to better represent the microstructure of the samples.

The samples were submitted to Vickers microhardness (HV) tests, aiming to evaluate the remelting process of the commercial alloy and the influence of the addition of iron on the mechanical properties of the alloy. For this purpose, a load of 100 gf was applied for a period of 10 s and 20 measurements were taken in each sample. The experimental results obtained for the studied alloys provided the survey of microhardness (HV) profiles for all alloys.

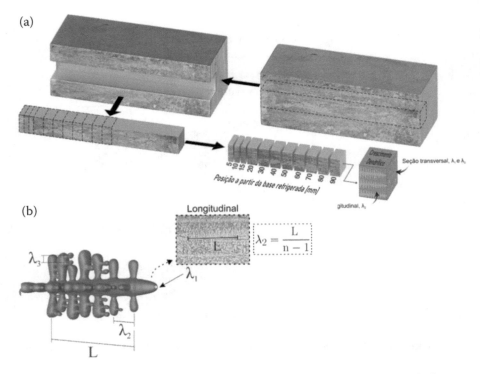

FIGURE 25.4 (a) Schematic representation of the cast ingot of the investigated alloy, showing the longitudinal section for microstructural analysis and (b) techniques applied to measure the secondary dendrite arm spacing (λ_2) (Adapted from Vasconcelos et al., 2021).

25.3 RESULTS AND DISCUSSION

Figure 25.5 shows the micrograph of the crude A356 alloy from solidification. It is possible to observe that the microstructure is formed by Al-α dendrites (Figure 25.5a) and by the binary eutectic Al–Si (Figure 25.5b).

Figure 25.6a obtained by SEM shows the typical microstructure of the alloy and matrix formed by the Al-α phase and eutectic regions. It is also possible to notice lighter regions that show the presence of intermetallic phases, which are mostly found in grain boundaries. In Figure 25.6b, the analysis of the composition of the intermetallic compounds was performed by point EDS analysis. It can be seen that the AlFeSi phase has an acicular morphology, while the AlSiMgFe phase has a complex morphology, similar to the α-AlFeSi phase (Chinese writing).

The remelted A356 alloy without the addition of iron (Figure 25.7) presented the same characteristics of the alloy as received; a matrix was formed by the Al-α phase and Al–Si eutectic regions. However, it is possible to notice that the microstructure is more refined, becoming evident when comparing

Recycled Al–Si–Cu–Mg Alloys

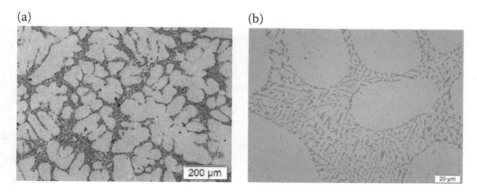

FIGURE 25.5 (a) Microstructure of the crude A356 solidification alloy and (b) magnification of the eutectic region.

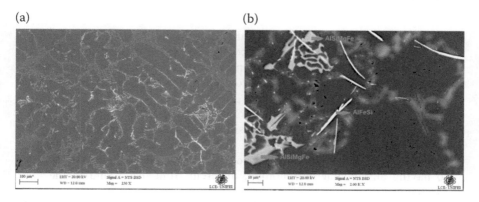

FIGURE 25.6 (a) SEM micrograph of the A356 alloy as received and (b) magnification of the intermetallic phases.

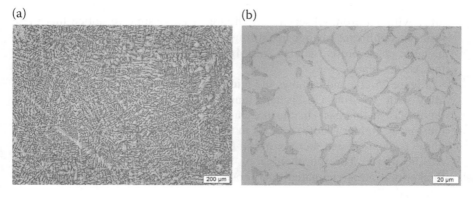

FIGURE 25.7 Alloy A356 (without addition of Fe) after remelting and solidification in a metal mold.

(a) (b)

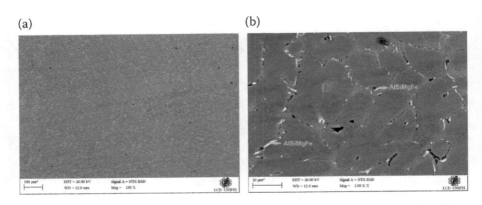

FIGURE 25.8 Alloy A356 (without addition of Fe) after remelting and solidification in a metal mold.

Figures 25.5 and 25.6, which have the same magnification. The refined microstructure may be related to rapid cooling due to the molten volume and the fact that the analyzed sample was removed from the upper surface of the remelted billet, presenting a faster cooling that resulted in a more refined microstructure.

Figure 25.8 shows the SEM micrograph of the remelted A356 alloy. By observing Figure 25.7b, it is possible to notice the presence of pores and intermetallic phase. According to EDS analysis, the present phase is composed of AlSiMgFe and is homogeneously distributed, in addition to having an acicular morphology, different from that found in the alloy as received.

The samples of alloy A356 with the addition of iron were analyzed in the SEM due to its specificity in also revealing small alterations in the microstructure, which is one of the objectives of the work.

Figure 25.9a and b shows the micrograph obtained by SEM of the A356 alloy with 1% iron. Observing the figure, it is possible to notice that there was no significant change regarding the remelted A356 alloy. The intermetallic phase present is also AlSiMgFe, with the presence of needles from the AlSiFe phase.

Figure 25.10a and b shows the micrograph obtained by SEM of the A356 alloy with 3% iron. Observing Figure 25.10a, it is possible to notice that there was an increase in the presence of intermetallic phases, due to the increase of lighter (white) regions in the image. The AlSiMgFe intermetallic phase is most present, but according to EDS there is also the presence of the AlSiFe phase. The morphology of these phases is acicular and they are thicker and more elongated than those found in the 1% Fe alloy and in the remelted alloy.

Figure 25.11 shows the typical solidification structures of the Al–7 wt.% Si–0.4 wt.% Mg–1.2 wt.% Fe alloy, at macrostructural and microstructural

Recycled Al–Si–Cu–Mg Alloys

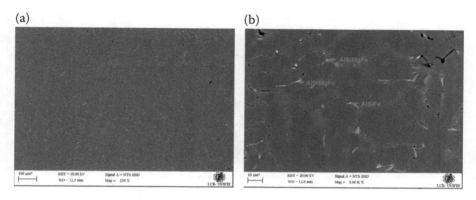

FIGURE 25.9 Alloy A356 with the addition of 1% Fe after remelting and solidification in a metal mold.

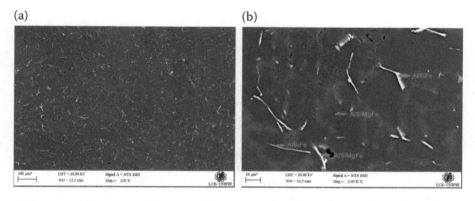

FIGURE 25.10 Alloy A356 with the addition of 3% Fe after remelting and solidification in a metal mold.

scales. A microstructural evolution along the ingot shows that the observed dendritic microstructure is constituted by a dendritic network. Therefore, finer dendritic microstructures are observed for samples as melt analyzed from the heat transfer interface.

The effect of the cooling rate had a more significant result than the iron concentration. This fact is evident in the comparison of hardness analyses in parallel with metallographic analyses, where the remelted samples presented a more refined microstructure than the sample as received. This may be related to rapid cooling due to the remelted volume, and the fact that the analyzed samples were taken from the upper surface of the billets obtained after remelting.

Figures 25.12 and 25.13 show two scanning electron micrographs with element mapping by EDS, for two as-cast samples, of the Al7Si0.4Mg1.2Fe alloy (% by weight). The final microstructure of the alloy shows the presence

FIGURE 25.11 Typical solidification structures for the Al7Si0.4Mg1.2Fe alloy (% by weight), resulting from the unsteady-state horizontal solidification (Vasconcelos et al., 2021).

of a primary phase (Al$_\alpha$) and a mixture of interdendritic eutectic phases composed of Al$_{\alpha\text{eutectic}}$ + eutectic Si + (Mg$_2$Si + β-Al$_5$FeSi) IMCs. Fine eutectic particles of β-Al$_5$FeSi are observed within the interdendritic regions surrounded by eutectic Si particles for high VL and TR and less than λ_2, as shown in Figure 25.12. Furthermore for such conditions, fibrous and type eutectic Si spheroidal was observed. This is in agreement with the results in literature (Barbosa et al., 2019; Chen et al., 2017).

On the other hand, for low V_L and T_R and higher than λ_2, the eutectic Si undergoes a morphological change from fibers or spheroidal to acicular, and the needle-like β-Al$_5$FeSi phase precipitates out of the interdendritic regions and over the primary dendritic phase, as secondary and primary eutectic reactions, respectively, as seen in Figure 25.13. Malavazi et al. (2014) described that both the increase in Fe content and the decrease in the cooling rate stimulated the platelet-like β-Al$_5$FeSi phase to crystallize independently of silicon. They also reported that the particles formed by the primary eutectic reaction are larger than those formed by the secondary eutectic reaction.

Figure 25.14 shows the results of the Vickers Microhardness (HV) measurements of the A356 alloy without Fe and with the addition of 1% and 3% of Fe. Intermetallic phases result in increased mechanical strength.

Recycled Al–Si–Cu–Mg Alloys

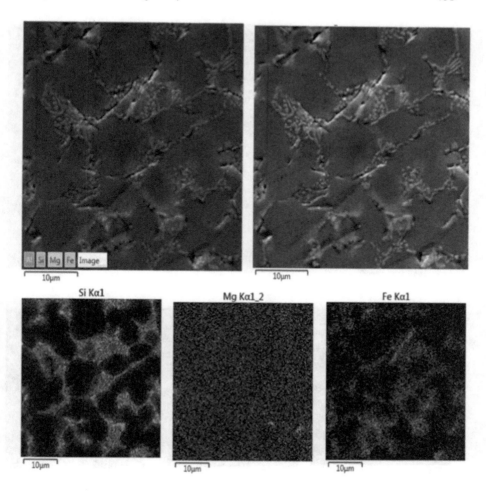

FIGURE 25.12 Scanning electron microscopy with EDS element mapping for an as-cast alloy sample investigated in the following positions at the 5 mm form the cooled base (Vasconcelos et al., 2021).

Figure 25.15 shows the Vickers Microhardness (HV) values for the alloys solidified in the horizontal solidification experiment, the Al–7% Si–0.4% Mg and Al–7% Si–0.4% Mg–1.2% Fe alloys. The Al–7% Si–0.4% Mg–1.2% Fe alloy has a large amount of Fe particles in its microstructure, so it was expected that this alloy in this condition would present a higher microhardness compared to the Al–7% Si–0.4% Mg alloy. Nevertheless, no significant variation was observed between microhardness values with the addition of 1.2% Fe in horizontal solidification. However, the microhardness values observed in the water-cooled horizontal solidification experiment were higher than conventionally solidified alloys, even for the alloy with 3% Fe by weight.

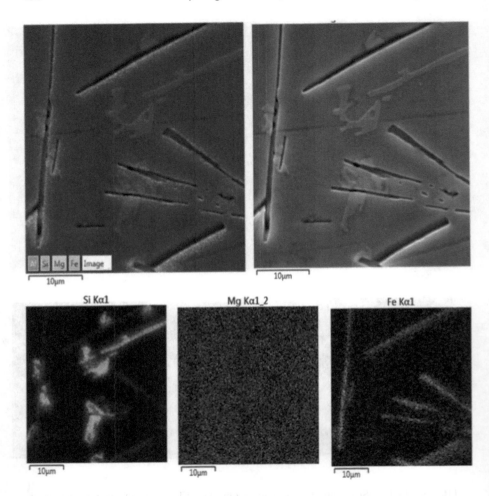

FIGURE 25.13 Scanning electron microscopy with EDS element mapping for an as-cast alloy sample investigated in the following positions at the 40 mm form the cooled base (Vasconcelos et al., 2021).

Figure 25.16 shows the results of the interconnection of secondary dendritic spacings (λ_2) with the thermal parameters of transient solidification. As expected, V_L and T_R have a great influence on the values of λ_2, since a smaller λ_2 can be observed for the positions (P) in the cast ingot where V_L and T_R become larger. Mathematical power expressions that correlate λ_2 as a function of V_L and T_R were generated, represented by general equations given by $\lambda_2 = $ constant $(V_L)^{-2/3}$ and $\lambda_2 = $ constant $(T_R)^{-1/3}$, respectively, which characterize the dependence of λ_2 as a function of the thermal parameters of solidification, as shown in Figure 25.16. It is observed that the exponents

Recycled Al–Si–Cu–Mg Alloys

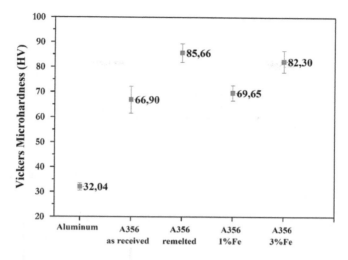

FIGURE 25.14 Vickers Microhardness (HV) measurements of the A356 alloy without Fe and with the addition of 1 wt.% and 3 wt.% Fe.

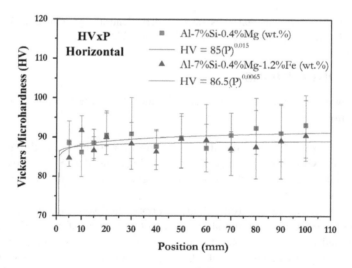

FIGURE 25.15 Vickers Microhardness (HV) measurements of Al–7% Si–0.4% Mg and Al–7% Si–0.4% Mg–1.2% Fe alloys obtained in the horizontal solidification experiment cooled to water.

1/3 and −2/3 are absolutely in agreement with the experimental predictions in literature (Lima et al., 2018; Chen et al., 2017), as well as with the theoretical prediction of Bouchard-Kirkaldy (Bouchard & Kirkaldy, 1997) who proposed a mathematical approach of $\lambda_2 = f(V_L)$ for binary alloys.

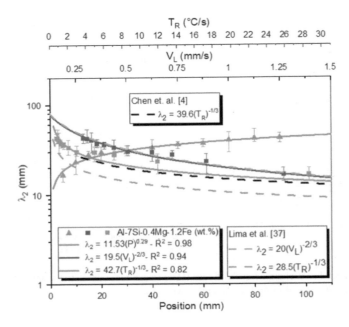

FIGURE 25.16 Experimental laws of secondary dendritic growth as a function of thermal solidification parameters (Vasconcelos et al., 2021).

25.4 CONCLUSIONS

The following main conclusions were obtained from the present experimental investigation:

1. The microstructure of the alloy containing 1% iron presented intermetallic phases with AlSiMgFe acicular morphology and AlFeSi composition and the alloy containing 3% presented a higher concentration of these phases and acicular morphology.
2. The alloy with 1.2% iron obtained in horizontal solidification, presented a microstructural evolution along the ingot consisting of a dendritic network formed by a primary phase rich in Al, and by a mixture of interdendritic eutectic phases composed of Al$_{\alpha eutectic}$ + eutectic Si + (Mg$_2$Si + β-Al$_5$FeSi) IMCs. Fine eutectic β-Al$_5$FeSi particles were observed within the interdendritic regions surrounded by fibrous and spheroidal-like eutectic Si particles for high T_R and lower λ_2.
3. The improvement in the hardness of the remelted alloys, in relation to the receiving state, is mainly due to the microstructural refinement obtained due to rapid cooling.
4. The Fe modifier element, at the tested levels, did not significantly influence the hardness values of the alloys cast and solidified in a metal

mold. A better understanding of the behavior of the mechanical properties of the studied alloy, associated with its microstructure, will be obtained by performing heat treatments and mechanical tests, such as traction, in addition to casting samples with increasing iron content and variable cooling rates.

ACKNOWLEDGMENT

The authors acknowledge the financial support provided by UNIFEI – Federal University of Itajubá, IFPA – Federal Institute of Education, Science and Technology of Pará, and CNPq – National Council for Scientific and Technological Development (Grant 302846/2017-4 and 308021/2018-5), FAPEMIG – Minas Gerais Research Foundation (Grant APQ-01301-15), and CAPES-Coordenação de Aperfeiçoamento de Pessoal de Nível Superior-Brasil-Finance Code 001.

REFERENCES

Barbosa, Carolina Rizziolli, Lima, José Otávio Monteiro de, Machado, Gabriel Mendes Hirayama, Azevedo, Hugo André Magalhães de, Rocha, Fernando Sousa, Barros, André Santos, & Rocha, Otávio Fernandes Lima da (2019). Relationship between aluminum-rich/intermetallic phases and microhardness of a horizontally solidified AlSiMgFe alloy. *Materials Research*, 22, 1–12. 10.1590/1980-5373-mr-2018-0365.

Barros, André, Cruz, Clarissa, Silva, Adrina Paixão, Cheung, Noé, Garcia, Amauri, Rocha, Otávio, & Moreira, Antonio (2019). Horizontally solidified Al-3wt%Cu-(0.5wt%Mg) alloys: tailoring thermal parameters, microstructure, microhardness, and corrosion behavior. *Acta Metallurgica Sinica*, 32, 695–709. 10.1007/s40195-018-0852-z.

Bouchard, Dominique, & Kirkaldy, John S. (1997). Prediction of dendrite arm spacings in unsteady and steady-state heat flow of unidirectionally solidified binary alloys. *Metallurgical and Materials Transactions B*, 28, 651–663. 10.1007/s11663-997-0039-x.

Cameron M. Dinnis, Taylor, John A., & Dahle, Arne K. (2005). As-cast morphology of iron-intermetallics in Al-Si foundry alloys. *Scripta Materialia*, 53, 955–958. 10.1016/j.scriptamat.2005.06.028.

Chen, Rui, Shi, Yu-Feng, Xu, Qing-Yan, & Liu, Bai-Cheng (2014). Effect of cooling rate on solidification parameters and microstructure of Al-7Si-0.3Mg-0.15Fe alloy. *Transactions of Nonferrous Metals Society of China*, 24, 1645–1652. 10.1016/S1003-6326(14)63236-2.

Chen, Rui, Xu, Qingyan, Guo, Huiting, Xia, Zhiyuan, Wu, Qinfang, & Liu, Baicheng (2017). Correlation of solidification microstructure refining scale, Mg composition and heat treatment conditions with mechanical properties in Al-7Si-Mg cast aluminum alloys. *Materials Science and Engineering A*, 685, 391–402. 10.1016/j.msea.2016.12.051.

Das, Subodh K., & Yin, Weimin (2007). The worldwide aluminum economy: The current state of the industry. *Journals of Minerals, Metals & Materials Society*, 59, 57–63. 10.1007/s11837-007-0142-0.

Das, Subodh K., Green, J. A. S., & Kaufman, J. Gilbert (2007). The development of recycle-friendly automotive aluminum alloys. *JOM: The Journal of the Minerals, Metals & Materials Society, 59*, 11, 47–51. 10.1007/s11837-007-0140-2.

Filip Průša, Vojtěch, Kučera, V., & Vojtěch, Dalibor (2017). An Al-17Fe alloy with high ductility and excellent thermal stability. *Materials & Design, 132*, 459–466. 10.1016/j.matdes.2017.07.027.

Gaustad, Gabrielle, Olivetti, Elsa, & Kirchain, Randolph (2012). Improving aluminum recycling: a survey of sorting and impurity removal technologies. *Resources, Conservation & Recycling, 58*, 79–87. 10.1016/j.resconrec.2011.10.010.

Khalifa, W., Samuel, F. H., & Gruzleski, J. E. (2003). Iron intermetallic phases in the Al corner of the Al-Si-Fe system. *Metallurgical and Materials Transactions A, 34*, 807–825. 10.1007/s11661-003-0116-y.

Kim, Bonghwan, Lee, Sanghwan, Lee, Sangmok, & Yasuda, Hideyuki (2012). Real-time radiographic observation of solidification behavior of al-si-cu casting alloys with the variation of iron content. *Materials Transactions, 53*, 374–379. 10.2320/matertrans.F-M2011834.

Lima, J. O., Barbosa, C. R., Magno, I. A. B., Nascimento, J. M., Barros, A. S., Oliveira, M. C., Souza, F. A., & Rocha, O. L. (2018). Microstructural evolution during unsteady-state horizontal solidification of Al-Si-Mg (356) alloy. *Transactions of Nonferrous Metals Society of China, 28*, 1073–1083. 10.1016/S1003-6326(18)64751-X.

Magno, Igor A., Souza, Fabrício A., Costa, Marlo O., Nascimento, Jacson M., Silva, Adrina P., Costa, Thiago S., Rocha, & Otávio L. (2019). Interconnection between the solidification and precipitation hardening processes of An AlSiCu alloy. 2019. *Materials Science and Technology, 35*, 1743–2847. 10.1080/02670836.2019.1591028.

Malavazi, J., Baldan, R., & Couto, A. A. (2014). Microstructure and mechanical behaviour of Al9Si alloy with different Fe contents. *Materials Science and Technology*, 1743–2847. 10.1179/1743284714Y.0000000659.

Rakhmonov, Jovid, Timelli, Giulio, Bonollo, Franco, & Arnberg, Lars (2017). Influence of grain refiner addition on the precipitation of Fe-rich phases in secondary AlSi7Cu3Mg alloys. *International Journal of Metalcasting, 11*: 294–304. 10.1007/s40962-016-0076-9.

Rodrigues, Jean Robert Pereira, Mirian de Lourdes Noronha Motta, & Melo, Rezende Gomes, Santos (2010). Effect of magnesium content on thermal and structural parameters of Al–Mg alloys directionally solidified. *Journal of Materials Science, 45*(9), 2285–2295, 10.1007/s10853-009-4190-4

Souza, Fabrício, Lima, José, Rizziolli, Carol, Magno, Igor, Barros, André, Moreira, Antonio, & Rocha, Otávio (2018). Microstructure and microhardness in horizontally solidified Al-7Si-0.15Fe-(3Cu; 0.3Mg) alloys. *Materials Science and Technology*, 1743–2847. 10.1080/02670836.2018.1444923.

Tunçay, Tansel, & Bayoğlu, Samet (2017). The effect of iron content on microstructure and mechanical properties of A356 cast alloy. *Metallurgical and Materials Transactions B, 48*, 2, 794–804. 10.1007/s11663-016-0909-1.

Tupaj, Mirosław, Orłowicz, Antoni Władysław, Mróz, Marek, Trytek, Andrzej, & Markowska, O. (2016). The effect of cooling rate on properties of intermetallic phase in a complex Al-Si alloy. *Archives of Foundry Engineering, 16*, 3: 125–128. 10.1515/afe-2016-0063

Vasconcelos, Angela de Jesus, Silva, Cibele Vieira Arão da, Moreira, Antonio Luciano Seabra, Silva, Maria Adrina Paixão de Sousa da, & Rocha, Otávio Fernandes Lima da (2014). Influence of thermal parameters on the dendritic arm spacing and the microhardness of Al-5.5wt.%Sn alloy directionally solidified. *REM. Revista Escola de Minas (Impresso), 67*, 173–179. 10.1590/S0370-44672014000200007.

Wang, E. R., Hui, X. D., Wang, S. S., Zhao, Y. F., & Chen, G. L. (2010). Improved mechanical properties in cast Al-Si alloys by combined alloying of Fe and Cu. *Materials Science and Engineering: A, 527*, 29, 7878–7884. 10.1016/j.msea.2010.08.058.

Wang, Qinglei, Geng, Haoran, Wang, Fulan, Lin, Xiangze, & Wang, Chongyang (2016). Effect of parameters of thermal-rate treatment of melt on iron-containing phases in alloy Al-15% Si-2.7% Fe. *Metal Science and Heat Treatment, 58*, 7–8, 405–410. 10.1007/s11041-016-0025-5.

Yang, H., Shouxung, J., & Zhongyun, F. (2015). Effect of heat treatment and Fe content on the microstructure and mechanical properties of die-cast Al-Si-Cu alloys. *Materials & Design, 85*, 823–832. 10.1016/j.matdes.2015.07.074.

26 Polymer Recycling in Malaysia
The Supply Chain and Market Analysis

K. Norfaryanti and Z.M.A. Ainun
Institute of Tropical Forestry and Forest Products
(INTROP), Universiti Putra Malaysia, Selangor, Malaysia

CONTENTS

26.1 Introduction	463
26.2 Polymers	464
26.2.1 Classification of Polymers	464
26.3 Polymer Recycling	465
26.3.1 Classification of Polymer Recycling Processes	466
26.3.2 Commonly Recycled Polymer	467
26.3.3 Recycling Process for Major Polymers	467
26.3.3.1 HDPE	467
26.3.3.2 ABS	468
26.3.3.3 PET	469
26.4 Supply Chain and Market Demand	469
26.5 Global Market	470
26.6 Status of Recycling Business in Malaysia	473
26.6.1 Plastic Recycling Plants	474
26.7 Market Potential	477
26.8 Issues and Challenges	477
26.9 Conclusion	481
Acknowledgment	482
Notes	482
References	483

26.1 INTRODUCTION

The worldwide consumption of polymers, especially plastics, is progressively increasing annually. In 2017, global production for plastic amounted to 354 million tons (Wang et al., 2019). The strength, long life span, light weight, acceptable price and ease of process have attracted commercial applications.

Despite all the advantages, plastics invite more critical issues at the global stage due to their recycling obstacles, energy recovery problems and degradable period that affect soil and water (Valentini et al., 2020). Hence, management of plastic wastes, sorting technologies and chemical compatibility are often stressed when considering recycling (Dorigato, 2021). Besides that, standardized methodologies such as life cycle assessment that proved the potential of recycled plastics to replace the virgin ones should be developed because it helps to overcome both economic and environmental aspects (Gu et al., 2017).

The aim of this chapter is to provide an understanding of the polymer recycling status in Malaysia and its market potential. A brief introduction on polymers and their classification, followed by polymer recycling and its classification, supply chain and market demand both local and global status, as well as its challenges are discussed.

26.2 POLYMERS

A polymer is a macromolecule in a long structure that is built by the repetitive joining of subunits called monomers. A monomer is a small building block of organic molecules, which in etymology derived from the Greek prefix of *meros* meaning "parts" while *poly* means "many." Due to the chemical structures of the polymer, it has unique properties that can be aligned to various uses. The process of creating polymers held together by covalent bonds is called polymerization. It may occur in two ways, namely condensation and additional polymerization (Billmeyer, 1984). The presence of many monomer subunits leads to a higher molecular mass. Hence, a polymer has high melting and boiling points. This can be seen in fiber-reinforced polymer composite products that usually involve polyester, vinyl acetate or epoxy.

26.2.1 Classification of Polymers

Polymers exist in two conditions, either naturally or man-made. A natural polymer, also known as a biopolymer, comprises cellulose, starch, lignin, DNA, protein, silk, chitin and latex. Cellulose is the most popular natural polymer on earth that can be obtained from the cell wall of plants and transformed into products like pulp, paper, textiles and cellophane. Another natural polymer is latex, which has been long applied for thousands of years due to its outstanding elasticity property. Biopolymers function as structural and functional proteins, nucleic acids, structural polysaccharides and energy storage molecules. The classes of biopolymers are grouped based on the subunit monomers and the types of polymers like lipids, proteins, nucleic acids and carbohydrates.

Polylactic acid (PLA), polybutylene adipate terephthalate (PBAT), polybutylene succinate (PBS) and polyhydroxyalkanoates (PHA) are biodegradable polymers made from renewable feedstock. One of the niche products with strong growth use is PLA. The polymer is derived from the fermentation of glucose or sucrose and refined to a higher purity before application in food packaging material (Yusoff et al., 2021), engineering plastic (Guduru & Srinivasu, 2020) and textiles (Shin et al., 2020) as well.

A man-made polymer, also called a synthetic polymer, consists of polyethylene (PE), polystyrene (PS), polytetrafluoroethylene (PTFE), polypropylene (PP), polyvinyl chloride (PVC), polyvinyl acetate (PVA) and polyethylene terephthalate (PET), which is made by humans in factories and is derived from non-renewable resource crude oil. PE is the most communal plastic, which is formed and famously used as shopping bags or storage containers while the PS is used as food packaging materials or disposable drinking cups. PTFE is generally called Teflon, which was first discovered in the 1930s and soon after became the focus of non-stick surfaces on cooking applications or as a water-resistant material on fabrics and wires. PP is a polyolefin group, which is the second most well known after PE in the production of commodity polymer or plastic. The third worldwide plastic product is PVC, which is incredibly useful in the construction trade and has been applied in making transparent and strong bottles. PVA is an important synthetic resin in paints and adhesive industries. PET is a polyester group that is suitable in making plastic containers either for food or beverage and clothing.

Synthetic polymers have three types of behaviors: elastomers, thermoplastics and thermosets. Elastomers have a rubber-like behavior in which the polymer is held by weak intermolecular forces that allow deformation and stretching movement before finally returning to its original shape. Natural rubber, polyurethanes and silicone are the most familiar samples used as tires, prosthetic items, rubber bands and more. Thermoplastics represent flexible or bondable behavior and hence, it is also called a thermosoftening plastic. It can be more or less pliable when the temperature elevates and cools, respectively. The plastics tend to be resistant to high temperature; hence, the high melting point. One disadvantage of this plastic is the effect after overheating that will spoil its behavior because of the decomposition. Thermosets, on the other hand, are rigid and most likely to retain their shape after curing by heat or radiation due to spatial crosslinking. They are beneficial in electrical installations due to their resistance to heat at high temperatures. Table 26.1 summarizes the applications commonly associated with polymers.

26.3 POLYMER RECYCLING

About 10 million tons of polymeric materials are discharged annually, consequently inflicting environmental and social difficulties. By recycling, problems such as landfill areas for decomposition purposes and pollution to the surrounding area can be decreased. Streams of waste that comprise end of waste (EOW), end of life (EOL) and post-consumer (PC) streams are produced in the production process, middle and at the end of the polymer lifetime. Based on these streams, sorting and recycling the latter would ideally appear as simpler and efficient but in actual life, this understanding is not in a good place yet, but many actions are already pointed out. The recycling methodology of polymers is usually the use of mechanical, chemical and combustion.

TABLE 26.1
Common Polymers, Specifically Plastics and Their Applications

Polymers	Applications
PE	Shopping bags, storage containers
Polyethylene low density (LDPE)	Grocery and shopping bags
Polyethylene high density (HDPE)	Detergent bottles, toys
PS	Toys, foam
PTFE	Non-stick pans, electrical insulation
PP	Carpet, upholstery
PVC	Piping, decking
PVA	Paints, adhesives
PET	Plastic container for food or beverage, clothing

26.3.1 Classification of Polymer Recycling Processes

Two standards are referred to in the classification of plastic recycling: notably ASTM D5033 and ISO 15270. The classification by ASTM used primary (mechanical), secondary (mechanical), tertiary (chemical) and quaternary (energy recovering) definitions that are also equivalent to the ISO's listing as mechanical, chemical and energy recovering (Hopewell et al., 2009).

Primary mechanical recycle process involves the utilization of discarded polymers that remains free from any contamination in the factory by having new products with the same property. In the factory, it is also termed post-industrial waste (Al-Salem et al., 2010) and needs grinding like shredding, crushing or milling for more homogenous processing. PE, PET, PP and PVC are thermoplastic polymers that can be mechanically recycled.

Secondary mechanical recycling that comprises any EOL and PC polymers with their content and purity grade are not exactly known. Thermoplastic polymers like primary mechanical recycling materials only can be remanufactured that affect their molecular weight due to contamination on the main polymer matrix. This can be solved by applying separation prior to integration using morphological observation through a microscope, chemical compositional analysis by FTIR analysis, laser sorting or electrostatic detection. Compatibilization is a secondary mechanical recycling that is improved by excluding separation (Ignatyev et al., 2014). A compatibilizer is added as the third component besides the other polymers intentionally to improve the composites' mechanical properties (Bristow, 1959) that involve physicochemical interactions between waste polymers and hybrid material.

Tertiary recycling, also called chemical or feedstock recycling, refers to the conversion of polymer chains into smaller molecules like monomers or oligomers that can be achieved via hydrolysis, pyrolysis, hydrocracking, gasification

(Brems et al., 2012), glycolysis, ammonolysis and hydrogenation. Different plastic mixtures can be used for this purpose and each conversion is unique to each other, considering their complexity of composition (Ma et al., 2016). Another polymer recycling is biological degradation, where polymers are degraded into smaller molecules by using bacteria, fungi and other microorganisms. The pressure, microorganisms and pH are among parameters that are taken into consideration during the process.

Incineration is among the polymer recycling methods to recover energy that is still being applied in most countries as a substitute to landfilling disposal (Al-Salem et al., 2010). It focuses on mixed and heavily contaminated polymers that are difficult to recycle using all-polymer recycling processes. The burnt polymers are beneficial to create heat, electricity and other forms of energy as well, but it generates consequential drawbacks on the release of hazardous gases that contribute to greenhouse gas emissions.

The final polymer recycling is by applying cross-linking that uses special agents to build bondings between polymer chains during reprocessing. The cross-linking tends to transform thermoplastic into thermoset polymers that reduce the elasticity and the ability to shape easily but can be improved by chemical and thermal binding.

26.3.2 COMMONLY RECYCLED POLYMER

Municipal solid waste (MSW), electronic equipment (WEEE) and electric waste and automobile shredder residue (ASR) are the three types of mixtures found in plastic wastes as reported by Dorigato et al. (2021). PE, PS, PP, PVC, PET, polymethylmethacrylate (PMMA), acrylonitrile butadiene styrene copolymer (ABS), and polycarbonate (PC) are the predominant polymer matrices present in such plastic waste mixture as described in Table 26.2.

26.3.3 RECYCLING PROCESS FOR MAJOR POLYMERS

There are five steps in the recycling process: collection, sorting, cleaning, reprocessing by melting and producing new products from recycled material. The first three steps are the same as recycling for most other commodities, but reprocessing and reproducing are critical. In the reprocessing phase, collected PP products are fed into an extruder where it is melted at 2400°C and turned into granules. These pellets are ready for use in the production of new products.[1]

26.3.3.1 HDPE

The HDPE recycling process starts at the collection stage. One challenge for storing HDPE is that it is often polluted and, therefore, the clean and dirty HDPE have to be separated as they are recycled differently and have different market values. Contaminated HDPE must go over a rigorous cleaning process before being recycled, to ensure the material quality is processed properly and ready for the next process. Common recycling machinery can process HDPE bottles, but HDPE films go through a different process, as it is liable to get caught up in the machines.

TABLE 26.2
Comparison of Effectiveness in the Current Recycling Process by Types of Polymer

Polymer	Effectiveness in the Current Recycling Process
LDPE	Poor rate of recovery. Especially as mixed polyolefins that can have sufficient properties for certain applications. Most post-consumer flexible packaging not recovered.
HDPE	High art of recovery for natural HDPE bottles, but more complex for opaque bottles and trays because of the various grades, color and mixtures with LDPE and PP.
PS	Poor rate of recovery. Very difficult and not cost-effective to separate from the comingled collection, but the separate collection of industrial packaging and EPS foam can be effective.
PP	Not widely recycled from post-consumer but has a potential. Needs action on sorting and separation, and further develop the outlets for recycled PP.
PVC	Poor rate of recovery because of cross-contamination with PET. PVC packages and labels are major issues with PET bottle and mixed plastics recycling.
PET	High rate of recovery with clear PET from bottles, and also colored PET is mostly used for fiber, additional issues with CPET trays, PET-G.
Recycled plastics	Considerable variability in energy, water and emissions from recycling processes as it is a developing industry and affected by the efficiency of collection, process type and product mix, etc.

Source: J. Hopewell et al., 2009.

HDPE film recycling machinery chews the plastic and breaks them into small pieces and is formed into pellets. The pellets are subsequently transformed into various things and combined with a virgin HDPE to improve the strength and its durability. The pellets are blown using various machinery and techniques to achieve a different outcome. Recycled HDPE is often used to produce piping, plastic lumber, recycling bins and rope. Recycled HDPE is often available in a dark color, such as brown or black.[2]

26.3.3.2 ABS

Unfortunately, no process is currently known for effectively and efficiently recycling ABS for subsequent use. The present process for recycling ABS is permitting used ABS to obtain a second life. The present process employs shredding, separating and blending to recycle used ABS for subsequent use in other products; for example, substrates of decorative laminates.

The compulsory separation of the assortment of plastic components prior to shredding and granulating makes recycling of commingled plastic products highly time consuming and prohibitively expensive. Existing recycling techniques are, therefore, not appropriate where the recycled goods include many components that are composed of various plastics.

For example, while it is desirable to recycle the plastic components of a computer housing (which is primarily composed of ABS, PCABS and other plastics) to take advantage of the recycled ABS, the computer housing must first be separated into various components that represent the many plastics used in the construction of the computer housing. This is highly time consuming, and it makes recycling computer housing cost prohibitive. Such limitations are commonly associated with ABS products that one might consider appropriate for recycling.

The lack of an effective recycling process for those products already manufactured from ABS allows previously used ABS products to go to waste once the useful life of the manufactured product is reached. Many materials have found a second life through recycling. However, a process for refining the used material must be developed before the material may find a second life in another product. Often, the second life of a product requires that the material be used in an environment where the aesthetic appearance of the material is not as important as when the material was used for its original purpose[3] (Gonzales et al., 2000).

26.3.3.3 PET

PET bottles are made of one of the few polymers that can be recycled into the same form – a new beverage bottle – again and again. This closes the recycling loop neatly and enables "cradle to cradle" packaging solutions.

As with virgin PET, recycled PET (rPET) can be used to make many new products, including polyester staple fiber or filament used for apparel (clothing), home textiles (duvets, pillows, carpeting), automotive parts (carpets, sound insulation, boot linings, seat covers) and industrial end-use items (geotextiles and roof insulation), as well as new PET packaging and bottles for both food and non-food products. It is generally blended in a ratio of the virgin to the recycled, depending on the application required.[4]

26.4 SUPPLY CHAIN AND MARKET DEMAND

Recycling and upcycling plastics from various products and packaging are lucrative and offer huge business opportunities due to the high consumption rate of plastics globally in various applications. To optimally seize this opportunity, it is crucial to understand the markets, material, equipment, data and environmental metrics of recycled plastics. It also requires effective collaboration among various stakeholders in the plastics value chain (Srinivasan et al., 2016).

Figure 26.1 shows the recycler value chain. The inbound scrap materials come from industrial scrap, commerce waste and residential recyclables from various countries. In spite of the opportunity, it is difficult for plastic recyclers to remain profitable due to the competition from the virgin plastics manufacturer (Gemma, 2019). Globally, only 14% of plastic packaging is collected for recycling. According to the report (World Economic Forum, 2016), after losses from

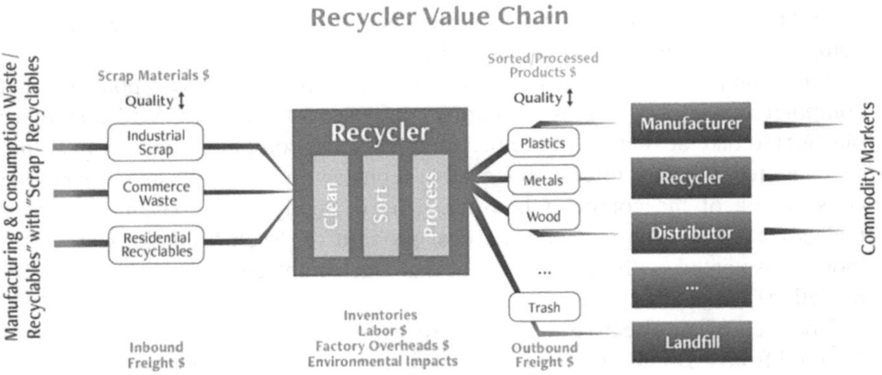

Recycler Value Chain

FIGURE 26.1 Recycler Value Chain (Adapted from Srinivasan et al., 2016[5]).

sorting and reprocessing, "only five percent of material value is retained for a subsequent use." Even though the opportunities to recycle plastics are huge, so are its challenges. The high-cost production which involved the scrap materials, its freight charges, the inventories and the factories overhead, are among the challenges in recycling the plastics compared to the low cost in producing the virgin plastics.

Srinivasan et al. (2016) highlighted the considerations that the recycler has to make to proceed with the recycling process, which includes quality of the material, level of contamination, cost of materials, feasibility of processing the materials and pricing strategies for the market. After deliberate considerations, processed products produced were transported to various industries and business activities and, finally, to the commodity market.

26.5 GLOBAL MARKET

Plastics are among the universal and multipurpose materials that have become an economy by itself. It evolves rapidly and plays a crucial part in the global economy. The applications of the materials range from industrial purposes to household products and serve every group in society. The magnitude of success for plastics is evident by the amount of global production in the past 50 years (Figure 26.2) and is projected to double in 20 years (World Economic Forum, 2016).

The huge success was contributed by utilization in various sectors. Figure 26.3 illustrates the main production of plastics from 1950 to 2015. Packaging is the most significant primary production sector, and building and construction come second. Based on this trend, it shows that LDPE, PET, HDPE and PP are among the main materials that are supposed to be highly recycled. Plastics play a significant role in packaging so as to prevent food waste, prolong food freshness, enable a wider variety of food availability and protect goods during transport and distribution (Gemma, 2019).

Polymer Recycling in Malaysia

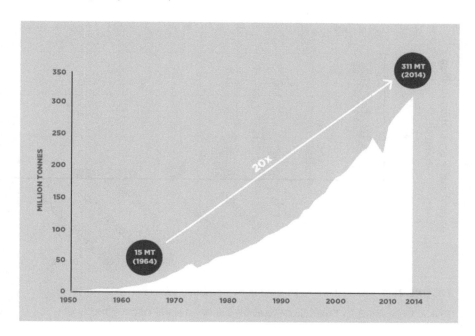

FIGURE 26.2 Growth in global plastics production 1950–2014.

Source: PlasticsEurope, Plastics – the Facts 2013 (2013); PlasticsEurope, Plastics – the Facts 2015 (2015).

Note: Production from virgin fossil-based feedstock only.

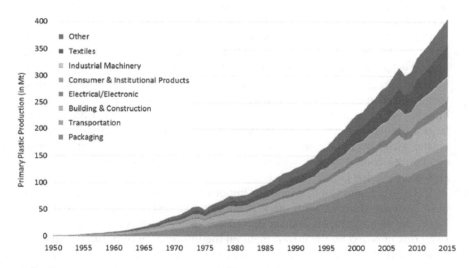

FIGURE 26.3 Global plastics production from 1950 to 2015.

Source: Organisation for Economic Co-operation and Development OECD, 2018.

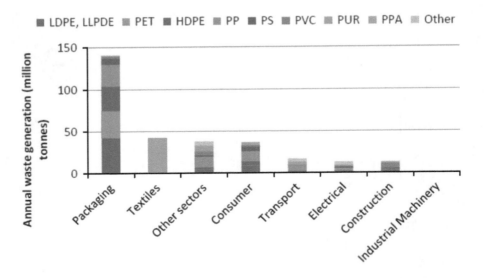

FIGURE 26.4 Annual waste generation by economic sectors.
Source: Organisation for Economic Co-operation and Development OECD, 2018.

Figure 26.4 indicates the major annual waste generation according to the economic activity sectors. Packaging generates about 140 million tons annually, and it consists of six types of polymers – LDPE/ LLDPE, PET, HDPE, PP, PS and PVC. The other economic sectors that generate less than 50 million tons of polymer waste are textiles, consumers and other sectors.

Figure 26.5 illustrates Malaysia's trading trends for products with HS code 3915 – waste, parings and scrap, of plastics for over 30 years. The export and import values increased steadily from 1989 to 2013. The decreasing pattern of trading from 2013 to 2018 was, however, unidentified. In 2018, China banned the importation of 24 types of plastic waste in support of their "National Sword Policy," which aims to reduce pollution levels.[6] Prior to 2018, China was the largest global importer of plastic waste. Some importers are able to transport plastic waste into Malaysia without proper documentation by using a different commodity code, HS3920, which does not require a permit instead of the designated code for plastic waste, HS3915.[7]

Based on Figure 26.6, Malaysia imported the waste, parings and scrap of plastics from the USA with the highest value at USD$38 million in 2018. In 2019, the import trading values decreased for all destinations due to the restrictions imposed by the government on the importation of plastic wastes. The United Kingdom and Japan are the second and third importer to Malaysia.

Based on Figure 26.7, in 2016 and 2017, Malaysia exported waste, parings and scrap of plastics to China at the highest value of USD$50–60 million. Since

Polymer Recycling in Malaysia

FIGURE 26.5 Malaysia trading trends of waste, parings and scrap of plastics 1989–2019.

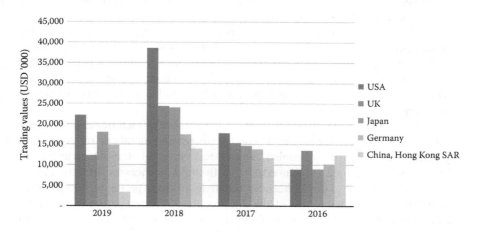

FIGURE 26.6 Major Malaysian import destinations for waste, parings and scrap of plastics in 2016–2019.

China imposed the ban in 2018, the export trade values have decreased sharply to less than USD$10 million in 2018 and 2019.

26.6 STATUS OF RECYCLING BUSINESS IN MALAYSIA

The global recycling is estimated at a rate of 20% due to the unsystematic nature of collection and recycling systems, particularly in the urban regions of lower-income countries. According to the World Bank, salvaging recyclables from

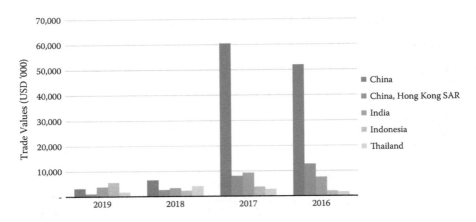

FIGURE 26.7 Major Malaysian export destination for waste, parings and scrap of plastics in 2016–2019.

waste become the means to survive among 1% of the urban population (at least 15 million people) (Gemma, 2019).

There are 38 plastic recycling plants in Malaysia. Table 26.3 presents the list of recycling plants[8] and the recycled products that they are producing. About 95% of the plants produced the recycled products into granules or pellets, where 32% were produced in the form of flakes. Both granules/pellets and flakes are produced in tiny sizes and shapes, less than 0.5 mm but present homogenously and irregular size, respectively.

Malaysia faces a significant increase in the number of factories and plants for waste recovery and recycling. There are increasing numbers of companies granted licenses for scheduled waste recovery by the Department of Environment Malaysia (Mohamed, 2019) from 52 licences in 2003 to 276 licences in 2018. Out of the 52 licences, 42 are involved in partial recovery and 10 in full recovery, whereas out of the 276 licenses, 78 have partial recovery facilities and 198 have full recovery facilities. The main difference between partial and full recovery is the processing technology and the types of machinery and facilities that are available and affordable by the industries.

A total of 111 projects were approved for recycling activities by the Malaysian Industrial Development Authority (MIDA) between 1980–2017. From 111, 80 projects are owned by Malaysians. Out of this, plastic products (38) and rubber products (32) are the highest number of recycling projects approved by MIDA.

26.6.1 Plastic Recycling Plants

Based on Figure 26.8, PP is the most acceptable materials among the 36 plastic recycling plants in Malaysia.[9] The products that are made from these materials

TABLE 26.3
List of Recycling Plants in Malaysia and Their Recycled Products

No.	Company Name	Recycled Products
1.	Aknz Resources (M) Sdn. Bhd.	Granules and/or pellets
2.	Alba Polyesters	
3.	CY Intertrade Sdn. Bhd.	
4.	Danex Plast Sdn. Bhd.	
5.	EPD Plastic Industries Sdn. Bhd.	
6.	Erapoly Marketing Sdn. Bhd.	
7.	Gold Mine Polymer Enterprise	
8.	Green Concept Technology Sdn. Bhd.	
9.	Heng Hiap Industries Sdn. Bhd.	
10.	Lida Plastics Recycling Sdn. Bhd.	
11.	Lim Seng Plastic Sdn. Bhd.	
12.	Ming Engineering Plastic Sdn. Bhd.	
13.	MJ Material Technology Sdn. Bhd.	
14.	Moldex Plastic Recycling Sdn. Bhd.	
15.	NS Plastic & Metal Trading Sdn. Bhd.	
16.	Plasticycle Industries Sdn. Bhd.	
17.	RS Poly Industry Sdn. Bhd.	
18.	SG Green Resources Sdn. Bhd.	
19.	Sheng Foong Plastic Industries Sdh. Bhd.	
20.	Soon Ye Plastic Resin Manufacturing Sdn. Bhd.	
21.	Spot Trend Sdn. Bhd.	
22.	Sri Aman Recycle Sdn. Bhd.	
23.	Sunnyjaya Plastic Industries Sdn. Bhd.	
24.	Tritex Polymer Sdn. Bhd.	
25.	TSP 3G Sdn. Bhd.	
26.	Wespack	
27.	AL Sobel Group	Granules/pellets, flakes
28.	Aun Lian Plastic Resources Sdn. Bhd.	
29.	Diyou Fibre (M) Sdn. Bhd.	
30.	Fizlestari Plastic Sdn. Bhd.	
31.	Ipoh S.Y. Recycle Plastic Sdn. Bhd.	
32.	JPT Plastics	
33.	JTE Group.	
34.	Karich Sdn. Bhd.	
35.	Plascycle Resources Sdn. Bhd.	
36.	Tai Hong Plastic Industries Sdn. Bhd.	
37.	Glowmore Express Sdn Bhd & Jupiter Privilege Sdn. Bhd.	Flakes
38.	Megatrax Plastic Industries Sdn. Bhd.	

476 Recycling of Plastics, Metals, and Their Composites

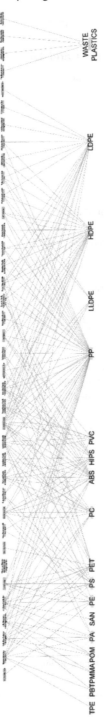

FIGURE 26.8 A mapping of the materials accepted by the recycled plastic plants in Malaysia.

include packaging trays, household products, battery cases, medical devices, etc. This could be a good sign that shows the improvement in terms of technology in recycling PP, as mentioned by Hopewell et al. (2009) that it was not widely recycled or recovered at the post-consumer level.

HDPE is the second material that is highly accepted by 23 plastic recycling plants in Malaysia. It aligns with its high effectiveness status in the recycling process for clear HDPE material, as mentioned by Hopewell et al. (2009).

LDPE is used by 19 plants and it contradicts the statement "poor recovery rates and most post-consumer flexible packaging not recovered" by Hopewell et al. (2009). It signifies that the medical sector is engaging actively with recycling programs (Table 26.4).

26.7 MARKET POTENTIAL

The Malaysian government acknowledged that the industry has contributed to economic growth, and it has estimated a value of RM 30 billion in 2018 (USD $7.1 billion).[25]

Figure 26.9 illustrates the export market countries in 2019 at a total value of USD$683 million for plastics (HS code 392690). The highest export value is Singapore for USD$222 million, USA for USD$159 million, Southeast Asia (excluding Singapore) for USD$112 million and other Asia countries has export market worth USD$111 million. Based on these values, Asia is the biggest export market for Malaysia in plastic products under the specified code.

According to the US Waterborne Import Trade (WIT) Report, there are numerous products categorized under HS 392690 and some of the products are listed in Table 26.5. The potential market for these products is moving towards an increasing trend, as illustrated in Figure 26.10.

The export market is increasing steadily, with a total export value of USD$726 million globally. About 61% of the export market is among the Asian countries, such as Singapore, Thailand, Japan, China and Indonesia (Figure 26.9).

26.8 ISSUES AND CHALLENGES

Given the versatility of plastics' properties and high dependency in various applications in our daily life, it is challenging to substitute them with other materials. Numerous efforts have been gathered by many institutions to help in managing plastics or polymer recycling. However, it has remained as one of the major global issues that requires an effective intervention by various categories of society.

According to the UNEP report (Gemma, 2019), there are four types of barriers to producing recycled plastics as secondary commodities, economic, technical, environmental and regulatory (Table 26.6).

TABLE 26.4
Materials Accepted by the Recycling Plants

Materials Accepted (No. of Recycled Plastic Plants)	Example of Products Related to the Materials
PP (36)	Packaging trays, household products, battery cases, medical devices[10]
HDPE (23)	Crates, trays, bottles for milk and fruit juices, caps for food packaging, jerry cans, drums, industrial bulk containers
LDPE (19)	Manufacturing containers, dispensing bottles, wash bottles, tubing, plastic bags for computer components, and various molded laboratory equipment[11]
ABS (18)	Automotive applications (i.e., instrument panels, pillar trim, dashboard components, door liners, handles), household appliances (shavers, vacuum cleaners, food processors), and pipes and fittings[12]
PET (13)	Mesh fabrics for screen-printing, filter for oil and sand filtration, bracing wires for agricultural applications (greenhouses, etc.), woven/knitting belt, filter cloth and other such industrial applications[13]
PC (12)	Optical media market (i.e., usage in computer and audio compact discs), automotive lightings, head lamp lenses, interior cladding, power tools, baby bottles, water dispensers, garden equipment, furniture[14]
PS (11)	Packaging materials, test tubes or Petri dishes, cups, plates, bowls, cutlery, hinged takeout containers (clamshells), meat and poultry trays and rigid food containers (e.g., yogurt)[15,16]
HIPS (11)	Food packaging (dairy packaging, meat trays, egg cartons, fruit and vegetable trays), consumer packaging (cassettes and CD shell and covers, jewelry box, etc.), industrial packaging (small, foamed shapes), TV housings, computer housings and widely used freezer and refrigerator, Petri dishes and waste canisters
PE (8)	Packaging bottles and films, pipes, hoses and fittings, tubes, wiring cables, laminates[17]
PVC (7)	Cladding, windows, roofing, fencing, decking, wallcoverings, and flooring, packaging, IV bags and medical tubing, raincoats, boots and shower curtains[18]
LLDPE (6)	Film applications such as food and non-food packaging, shrink/stretch film and non-packaging uses[19]
PA (6)	Fabric, fiber ropes, food packaging, toothbrushes, plastic fasteners and cookware
POM (4)	Automotive (modern fuel systems, interior appearances), medical and healthcare (insulin syringe, inhaler), industrial (showerhead, pipe coupling, water valve)[20]

(Continued)

TABLE 26.4 (Continued)
Materials Accepted by the Recycling Plants

Materials Accepted (No. of Recycled Plastic Plants)	Example of Products Related to the Materials
WASTE PLASTICS (4)	Shoes, clothes, gardening kit, toothbrush, bags
PBT (2)	Automotive (windshield wiper covers, mirror housings, handles and fans, fuel system components), consumer goods (oven door handles, iron handles and bases, appliance lids, ski tops, ski boots, golf cart cowls[21]
PMMA (2)	Lenses, acrylic nails, paint, security barriers, medical devices, LCD screens and furniture[22]
TPE (1)	Toys, sports equipment, packaging and hygiene articles such as toothbrushes and razors, drip chambers, seals and medical tubing[23]
SAN (1)	Mixing bowls and basins and fittings for refrigerators, outer casings of thermally insulated jugs, for tableware, cutlery, coffee filters, jars and beakers as well as storage containers, outer covers, e.g., printers, calculators, instruments and lamps[24]

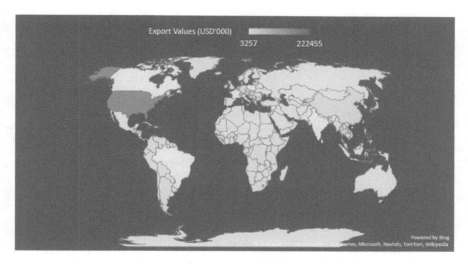

FIGURE 26.9 Top 20 export markets in 2019 for plastics (HS392690).[26]

Other than the barriers in Table 26.6, there are other challenges such as lack of information about the availability of recycled plastics and concerns over quality and suitability for specific applications that can also act as a disincentive to use recycled material. These issues have also contributed to the high cost of

TABLE 26.5
Some of the Products under the Trading Code of 392690

10 Digit HS Code	Products
3926901000	Buckets and pails, of plastics
3926901600	Pacifiers
3926902100	Ice bags, douche bags and fittings
3926902500	Handles and knobs, of plastics
3926903000	Parts for yachts or pleasure boats
3926903300	Handbags of beads, bugles and spangles
3926903500	Beads, bugles, spangles, not strung etc. and art thereof
3926904000	Imitation of gemstones
3926904510	O-rings
3926904590	Other gaskets and washers and other seals
3926909910	Laboratory ware
3926907500	Pneumatic mattresses and other inflatable art
3926906090	Belting and belts for machineries

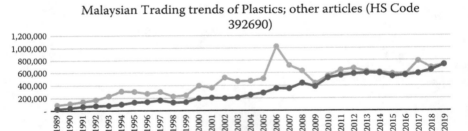

FIGURE 26.10 Malaysia trading trends of plastics over 30 years.[27]

collecting, sorting and processing plastic materials. Most of the packaging are not labeled accordingly and are limiting the early process of recycling.

There is a need for stricter enforcement by the government on policy to provide supporting tax programs for industries that drive the recycling plastics agenda and encouraging international collaboration and cooperation to promote the use of environmental design principles to accelerate the impact of plastics' recycling performance.

TABLE 26.6
Barriers to Secondary Commodities Market for Plastics

Economic	High costs incurred for collecting, sorting and processing waste plastics
	Demand for recycled plastics compared to virgin plastics
	Poor data on the structure and performance of the sector
Technical	Waste plastics are typically contaminated and mixed with other materials
	Inability to differentiate between food and non-food packaging
	Problematic additives and pigmentation
	Biodegradable plastics mixing with other plastics
	Limited collection schemes and treatment technologies for thermosets
Environmental	Hazardous additives in non-food plastics such as waste electric and electronic plastics
	Potential competition between recycling and incineration
	Concerns over environmental standards for recycling in emerging markets
Regulatory	Regulatory burden of materials classified as waste
	Illegal trafficking of waste plastics

26.9 CONCLUSION

This chapter aims to discuss the status of polymer recycling in Malaysia and its market potential. Natural and man-made polymers are the major classification of polymers, where biopolymers from renewable feedstocks (i.e., latex, cellulose and others) are among the most highly utilized and are biodegradable. While man-made or synthetic polymers possess the characteristics of elastomers, thermoplastics and thermosets, they become the limiting factors in terms of biodegradability and recyclability.

Polymers have end-of-life options: to become wastes or be recovered. In promoting circular economics for plastics, the end-of-life options must be balanced and sustainable. Presently, the recovery rate is relatively low and that is worryisome as the production of plastics is rocketing. Concerted effort is needed to ensure the recovery rate is at the optimum level as well as the waste conversion to energy.

There are three types of recycling process involved: mechanical, chemical and combustion. At the primary level, PE, PET, PP and PVC are recycled mechanically, whereas unknown or unspecified polymers are recycled by a compatibilization method at the secondary level. Chemical recycling consists of converting polymer chains to smaller molecules and different mixtures of polymers can be recycled through this process as well as the biological degradation. Incineration or combustion is employed for heavily contaminated polymers that are hard to be recycled through mechanical and chemical processes. The current advancement in recycling involves the cross-linking method for these three processes.

Based on the literature, PET and HDPE bottles are among the highly utilized materials for effective recycling rate, whereas LDPE and PP are lower. However, based on the recycling plants in Malaysia, both PP and LDPE are among the major accepted materials at the plants. It signifies the necessary improvement or the advancement in the recycling process for the two materials. In Malaysia, there is an increasing number of companies (80%) managing scheduled waste recovery since 2003 and 34% of the increments in recycling activity licenses were issued by the authority in 2017.

Trading trends for plastic waste are decreasing as the Malaysian government imposed the import ban and stricter regulations on importation. China remains the major export country even though they imposed the ban in 2017. The Asian market is identified as the market potential for plastic products (code HS 392690), with 65% of the total value of USD$683 million in 2019.

ACKNOWLEDGMENT

The authors would like to acknowledge the help of all the people involved in this chapter and, more specifically, to reviewers that took part in the review process. Without their support, this book would not have come to a fruition.

NOTES

1. https://www.thebalancesmb.com/an-overview-of-polypropylene-recycling-2877863, Rick Leblanc, 2019. An Overview of Polypropylene Recycling. Sustainable Businesses: Paper and Plastics.
2. https://www.plasticexpert.co.uk/plastic-recycling/hdpe-plastic-recycling/, Plastic Expert (na) HDPE Plastic Recycling.
3. http://polymerwastemanagement.blogspot.com/2007/11/abs-recycling.html
4. https://petco.co.za/how-is-pet-recycled/
5. https://www.recyclingproductnews.com/article/24123/profitable-and-practical-plastics-reuse-and-recycling-for-the-circular-economy
6. http://www.krinstitute.org/Views-@-Plastic-;_An_Undegradable_Problem.aspx
7. Commodity code HS 3920 (Other Plates, Sheets, Film, Foil and Strip, of Plastic) was used by some plastic waste importers instead of HS3915 (Waste, Parings and Scrap, of Plastic). Source: KAJIDATA Research (2019), Zhou (2019).
8. https://www.enfrecycling.com/directory/plastic-plant/Malaysia
9. https://www.enfrecycling.com/directory/plastic-plant/Malaysia
10. https://www.plasticpackagingfacts.org/plastic-packaging/resins-types-of-packaging/
11. https://omnexus.specialchem.com/selection-guide/polyethylene-plastic#LDPE
12. https://omnexus.specialchem.com/selection-guide/acrylonitrile-butadiene-styrene-abs-plastic
13. https://omnexus.specialchem.com/selection-guide/polyethylene-terephthalate-pet-plastic
14. https://www.bpf.co.uk/plastipedia/polymers/polycarbonate.aspx
15. https://www.creativemechanisms.com/blog/polystyrene-ps-plastic
16. https://omnexus.specialchem.com/selection-guide/high-impact-polystyrene/copy-of-applications
17. https://omnexus.specialchem.com/selection-guide/polyethylene-plastic/polyethylene-applications

18 https://www.chemicalsafetyfacts.org/polyvinyl-chloride/
19 https://www.icis.com/explore/resources/news/2007/11/06/9076161/polyethylene-linear-low-density-lldpe-uses-and-market-data/
20 https://omnexus.specialchem.com/selection-guide/polyacetal-polyoxymethylene-pom-plastic
21 https://omnexus.specialchem.com/selection-guide/polybutylene-terephthalate-pbt-plastic
22 https://www.creativemechanisms.com/blog/injection-mold-3d-print-cnc-acrylic-plastic-pmma
23 https://www.kraiburg-tpe.com/en/thermoplastic-elastomers
24 https://www.bpf.co.uk/plastipedia/polymers/SAN.aspx
25 Zuraida Kamaruddin, Minister of the Ministry of Housing and Local Government, 15 November 2018.
26 Calculation using data available https://wits.worldbank.org/trade/comtrade/en/country/ALL/year/2019/tradeflow/Imports/partner/WLD/product/392690#
27 Computed using data in UN COMTRAD.

REFERENCES

ABS Recycling. (2021). Polymerwastemanagement.blogspot.com. http://polymerwastemanagement.blogspot.com/2007/11/abs-recycling.html

Acrylonitrile Butadiene Styrene (ABS Plastic): Uses, Properties & Structure. (2021). Omnexus.Specialchem.Com. https://omnexus.specialchem.com/selection-guide/acrylonitrile-butadiene-styrene-abs-plastic

Al-Salem, S. M., Lettieri, P., & Baeyens, J. (2010). The valorization of plastic solid waste (PSW) by primary to quaternary routes: From re-use to energy and chemicals. *Progress in Energy and Combustion Science*, *36*(1), 103–129.

Billmeyer, F.W. (1984). *Textbook of Polymer Science*. John Wiley & Sons.

BPF (2021). *British Plastics Federation. Styrene Acrylonitrile (SAN) & Acrylonitrile Styrene Acrylate (ASA)*. Accessed March09. https://www.bpf.co.uk/plastipedia/polymers/SAN.aspx.

Brems, A., Baeyens, J., & Dewil, R. (2012). Recycling and recovery of post-consumer plastic solid waste in a European context. *Thermal Science*, *16*(3), pp.669–685.

Bristow, G. M. (1959). Phase separation in rubber-poly (methyl methacrylate)-solvent systems. *Journal of Applied Polymer Science*, *2*(4), 120–122.

Complete Guide on Polybutylene Terephthalate (PBT). (2021). *Polybutylene terephthalate (PBT) material guide & properties info*. Accessed March09. https://omnexus.specialchem.com/selection-guide/polybutylene-terephthalate-pbt-plastic.

Dorigato, A. (2021). Recycling of polymer blends. *Advanced Industrial and Engineering Polymer Research*. *4*, 49–132.

Gemma James. (2019). *An Investor Initiative in Partnership With UNEP Finance Initiative and UN Global Compact the Plastics Landscape: Risks and Opportunities Along the Value Chain*. UNEP.

Gonzales, James E., Krejchi, Mark T., Robson, M., Odstrcil, Kenneth W., Hoelscher, Finian E, Kendall, Eric W., Lee, Yein Ming, Cloud, Frank B , & Oriseh, Anthony S. (2000). ABS Recycling Process (United States/ EP1036641A1). European Patent Office. https://patentimages.storage.googleapis.com/21/f3/10/a5ee738d979111/EP1036641A1.pdf.

Gu, F., Guo, J., Zhang, W., Summers, P. A. & Hall, P. (2017). From waste plastics to industrial raw materials: A life cycle assessment of mechanical plastic recycling practice based on a real-world case study. *Science of the Total Environment, 601*, 1192–1207.

Guduru, K. K., & Srinivasu, G. (2020). Effect of post treatment on tensile properties of carbon reinforced PLA composite by 3D printing. *Materials Today: Proceedings, 33*, 5403–5407.

HDPE Plastic Recycling – How is HDPE Recycled? (2021). *Plastic Expert*. https://www.plasticexpert.co.uk/plastic-recycling/hdpe-plastic-recycling/

High Impact Polystyrene (HIPS): Techno Brief (2021). *High Impact Polystyrene (HIPS): Key Applications*. Accessed March09. https://omnexus.specialchem.com/selection-guide/high-impact-polystyrene/copy-of-applications.

Hopewell J., Dvorak R., & Kosior E. (2009). Plastics recycling: Challenges and opportunities. *Philosophical Transactions of The Royal Society B, 364*, 2115–2126. doi:10.1098/rstb.2008.0311

Hopewell, J., Dvorak, R., & Kosior, E. (2009). Plastics recycling: Challenges and opportunities. *Philosophical Transactions of the Royal Society B: Biological Sciences, 364*(1526), 2115–2126.

How is PET recycled? (2021). PETCO. https://petco.co.za/how-is-pet-recycled/

Ignatyev, I. A., Thielemans, W., & Vander Beke, B. (2014). Recycling of polymers: A review. *ChemSusChem, 7*(6), pp.1579–1593.

Ma, C., Yu, J., Wang, B., Song, Z., Xiang, J., Hu, S., Su, S., & Sun, L. (2016). Chemical recycling of brominated flame retarded plastics from e-waste for clean fuels production: A review. *Renewable and Sustainable Energy Reviews, 61*, 433–450.

Malaysia Recycled PE Pellet|Recycle Plastic Manufacturer Company (2021). *Epdplastic.com*. https://www.epdplastic.com/

Mohamed, A. F. (2019). *Report on Chinese Import Restriction Impact on Waste Recycling Industry in Malaysia*.

Organisation for Economic Co-operation and Development (OECD) (2018). *Improving Plastics Management: Trends, Policy Responses, and the Role of International Co-Operation and Trade. Policy Perspectives*. OECD Environment Policy Paper No.12.

Plastic Recycling Plants In Malaysia – ENF Recycling Directory (2021). *Plastic Plant Enfrecycling.com*. https://www.enfrecycling.com/directory/plastic-plant/Malaysia

Plastic Resins & Types of Packaging (2021). *Plastic Packaging Facts*. https://www.plasticpackagingfacts.org/plastic-packaging/resins-types-of-packaging/

Plastic: An Undegradable Problem (2021). KRI Institute.org. http://www.krinstitute.org/Views-@-Plastic-;_An_Undegradable_Problem.aspx

Polycarbonate (PC) (2021). British Plastic Federation. https://www.bpf.co.uk/plastipedia/polymers/polycarbonate.aspx

Polyethylene – Linear Low Density (LLDPE) Uses and Market Data (2021). *ICIS Explore*. Accessed March 09. https://www.icis.com/explore/resources/news/2007/11/06/9076161/polyethylene-linear-low-density-lldpe-uses-and-market-data/.

Polyethylene (PE) Plastic: Properties, Uses & Application (2021). Omnexus.Specialchem.Com. https://omnexus.specialchem.com/selection-guide/polyethylene-plastic#LDPE.

Polyethylene Applications (2021). *Polyethylene Applications*. Accessed March09. https://omnexus.specialchem.com/selection-guide/polyethylene-plastic/polyethylene-applications.

Polyethylene Terephthalate (PET): A Comprehensive Review (2021). Omnexus.Specialchem.Com. https://omnexus.specialchem.com/selection-guide/polyethylene-terephthalate-pet-plastic

Polypropylene Recycling – An Introduction (2021). *The Balance Small Business*. https://www.thebalancesmb.com/an-overview-of-polypropylene-recycling-2877863

Polyvinyl Chloride (PVC): Uses, Benefits, and Safety Facts (2021) *ChemicalSafetyFacts.org.* May 08, 2020. Accessed March 09. https://www.chemicalsafetyfacts.org/polyvinyl-chloride/.

Profitable and Practical Plastics Reuse and Recycling for the Circular Economy. (2021). *Recycling Product News.* https://www.recyclingproductnews.com/article/24123/profitable-and-practical-plastics-reuse-and-recycling-for-the-circular-economy

Rogers, Tony. (2021). *Everything You Need to Know About Polystyrene (PS).* Accessed March 09. https://www.creativemechanisms.com/blog/polystyrene-ps-plastic.

Shin, S., Cha, S., & Cho, G. (2020). Fabrication of electroconductive textiles based PLA nanofiber web coated with PEDOT: 4PSS. *Fashion & Textile Research Journal,* 22(2), 233–239.

Srinivasan, R., Gordon, P. J., & Fergurson, A. (2016). Profitable and practical plastics reuse and recycling for the circular economy: Pilots in healthcare and the industrial sector demonstrating how it's done. *Recycling Product News.*

Staff, Creative Mechanisms (2021). *Everything You Need To Know About Acrylic (PMMA).* Accessed March 09, https://www.creativemechanisms.com/blog/injection-mold-3d-print-cnc-acrylic-plastic-pmma.

Study Shows Low Rate of Recycling Plastic Bottles in Malaysia (2021). *Teng, C., The Star Online.* https://www.thestar.com.my/news/nation/2019/11/14/study-shows-low-rate-of-recycling-plastic-bottles-in-msia

Valentini, F., Dorigato, A., & Pegoretti, A. (2020). Evaluation of the role of devulcanized rubber on the thermo-mechanical properties of polystyrene. *Journal of Polymers and the Environment,* 28(6), pp.1737–1748.

Wang, H., Zhang, Y., & Wang, C. (2019). Surface modification and selective flotation of waste plastics for effective recycling a review. *Separation and Purification Technology,* 226, 75–94.

What We Do (2021). KRAIBURG TPE. Accessed March09. https://www.kraiburg-tpe.com/en/thermoplastic-elastomers.

World Economic Forum (2016). Ellen MacArthur Foundation and McKinsey & Company, The New Plastics Economy. *Rethinking the Future of Plastics.* http://www.ellenmacarthurfoundation.org/publications

Yusoff, N.H., Pal, K., Narayanan, T., & de Souza, F.G. (2021). Recent trends on bioplastics synthesis and characterizations: Polylactic acid (PLA) incorporated with tapioca starch for packaging applications. *Journal of Molecular Structure,* 129954.

27 Life Cycle Assessment (LCA) of Recycled Polymer Composites

H.N. Salwa[1], S.M. Sapuan[1,2], M.T. Mastura[3], M.Y.M. Zuhri[2], and R.A. Ilyas[4,5]

[1]Institute of Tropical Forestry and Forest Products (INTROP), Universiti Putra Malaysia, Selangor, Malaysia
[2]Department of Mechanical and Manufacturing Engineering, Advanced Engineering Materials and Composites Research Centre (AEMC), Universiti Putra Malaysia, Selangor, Malaysia
[3]Faculty of Mechanical and Manufacturing Engineering Technology, Universiti Teknikal Malaysia Melaka (UTeM), Melaka, Malaysia
[4]School of Chemical and Energy Engineering, Faculty of Engineering, Universiti Teknologi Malaysia (UTM), Skudai, Malaysia
[5]Centre for Advanced Composite Materials (CACM), Universiti Teknologi Malaysia, Johor, Malaysia

CONTENTS

27.1 Introduction	488
27.2 Recycling of Polymer Composites	488
27.3 Life Cycle Assessment (LCA)	490
27.3.1 LCA of Recycled Polymer Composite	492
27.3.1.1 Goal and Scope	493
27.3.1.2 Life Cycle Inventory (LCI)	496
27.3.1.3 Life Cycle Impact Assessment (LCIA)	497
27.4 Conclusion	499
Acknowledgments	500
References	500

27.1 INTRODUCTION

Plastic waste recycling started in the early 1990s when the recycled plastics were used for structural and load-bearing applications and railway sleepers (Rajendran et al., 2013). Traditionally, plastic recycling is involved in the production of second-grade pellets of a single-type polymer. The markets for most of the recycled products produced are low-pressure pipes, traffic barriers, outdoor furniture, dustbins, etc. and are becoming competitive and new applications with higher "added-value" recycled plastics are becoming an interesting research topic (Scelsi et al., 2011; Sommerhuber et al., 2016; Tshifularo & Patnaik 2020; Zander et al., 2019). Plastic recycling operations generally consist of collection, separation and cleaning, followed by melt processing steps. However, complexity of the material system is significant, where materials may be composed of one or more recycled polymers, the reinforcements/fillers and other additives such as compatibilizers, stabilizers and impact modifiers (Scelsi et al., 2011). On the other hand, the challenges in attaining an eco-friendly status of recycled plastic-based products are because of the inconsistency of plastic wastes and the low performance of secondary materials. Utilization of Life Cycle Assessment (LCA) in the product development processes helps to identify consumption of resources and environmental impacts associated directly or indirectly throughout the life cycle of the product. The LCA is defined as "a technique to compile and analyse the environmental impacts involved in all stages of the product's life cycle from raw material extraction stage to the disposal stage." This integration of LCA in the product development process will help to establish the relationship between the products (recycled) and their performance with the environmental impacts. LCA is also widely used to decide on a sustainable alternative in plastic waste management. Previous LCA studies showed that recycling resulted in lower emissions and provided benefits to the environment with the assumption that the performance of the recycled plastic materials are equivalent to those of the virgin materials (Rajendran et al., 2013).

27.2 RECYCLING OF POLYMER COMPOSITES

Recycling of plastic-based waste starts with the collection process. Commonly, used PET bottles waste are brought to a separation center where bottles are sorted out, baled and compacted. In Western Europe, PET bottle waste is collected through a mandatory deposit system, whereas in Taiwan, it is collected either together with other household waste before they are sorted out manually, or via the deposit–refund system (Shen et al., 2010). The process of recycling of plastic-based waste can be divided into two classifications: chemical recycling and mechanical recycling. Figure 27.1 illustrates in brief the process of recycling of plastic-based waste.

Mechanical recycling is the physical conversion of flakes into fiber or other products by melt-extrusion. Currently, there are two ways to produce recycled fiber from mechanical recycling: (1) directly extrude flakes into fiber; or the

LCA of Recycled Polymer Composites

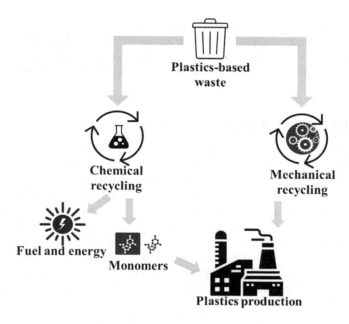

FIGURE 27.1 Recycling of plastic-based wastes.

more common method, (2) first transform flakes into pellets or chips (pelletizing) and then melt-extrude pellets or chips into fiber (Tshifularo & Patnaik, 2020). Mechanical grinding is considered a mature technology for recovery of raw materials (Cousins et al., 2019). Polymer composites can be combusted on an industrial scale to supply energy for cement kilns, and the recovered fibers can be used for other applications such as mixed with cement for building construction. On the other hand, thermal degradation-pyrolysis allows the recovery of fiber from either thermoset or thermoplastic polymer composites.

In chemical recycling, PET polymer is broken down into monomers or oligomers via several depolymerization technologies. The key benefit of chemical recycling is that the quality of virgin PET can be attained. Nevertheless, chemical recycling is more expensive than mechanical recycling and requires a large scale to become economically feasible. Glycolysis, methanolysis and alkaline hydrolysis are among the commercially available chemical recycling technologies. The glycolysis of PET yields the oligomer bis-hydroxyl ethylene terephthalate (BHET). The process is usually conducted in a temperature range between 180°C and 250°C, with excess ethylene glycol (EG) and in the presence of catalysts. After the glycolysis process, the oligomer passes through a fine filtration step before it is repolymerized into PET. The recycled polymer is then spun into fiber. Methanolysis is recycling PET back to a monomer where PET is depolymerized with methanol to dimethyl terephthalate (DMT) and EG in the presence of catalysts under a pressure of 2–4 MPa and a temperature of 180–280°C.

The reaction mix is cooled and DMT is recovered from the mix via precipitation, centrifugation and crystallization (Shen et al., 2010). The repolymerization phase is technically the same with the polymerization process leading to virgin PET.

Recent studies on recycling polymer composites are summarized in Table 27.1.

27.3 LIFE CYCLE ASSESSMENT (LCA)

Life Cycle Assesment (LCA) is defined as the "compilation and evaluation of the inputs, outputs and of the potential environmental impacts due to a product-system throughout its life-cycle" by ISO 14040:2006 (Ingrao et al., 2015). The LCA method can estimate the impact of a product across its full life cycle from resource extraction to end of life (Brundage et al., 2018). LCA too is being utilized as a supporting tool to learn and assess the technical solutions to be employed in the production process to minimize the impacts originated not only from the production itself but also from the phases of use and end of life. LCA was previously used as a reflective tool but has been extended as an action-oriented decision-making tool, to support designers and manufacturers in reducing environmental impacts. LCA could as well be the comparative assessment tool. LCA is employed for selection of product/process, design and optimization, and also sometimes attach it with simulation techniques and design tools to give insights on the environmental consequences for their decision and actions (Brundage et al., 2018). Figure 27.2 depicts the LCA framework according to ISO 14040.

An LCA involves collecting information on the inputs and outputs, such as emissions, waste and resources of a process (life cycle inventory) and translating these to environmental consequences (using impact assessment methodologies) such as contribution to climate change, smog creation, eutrophication, acidification and human and ecosystem toxicity (Mansor et al., 2015). LCA principally includes six important phases in a product life cycle: (1) materials extraction, (2) manufacturing and waste production, (3) packaging, (4) transportation, (5) product use and (6) product disposal. The Engineer Manufacture of Product activity describes that the process of making the product encompasses the acquisition of stock materials, equipment and tooling (Brundage et al., 2018). Detailed fundamental information is needed regarding the manufacturing processes, the materials and energy use to calculate the amount of emissions and waste created during the life cycle of a product (La Rosa & Cicala, 2015). In general, there are four main steps in performing the LCA technique: (1) goal and purpose, (2) life cycle inventory (LCI), (3) life cycle impact assessment (LCIA) and (4) interpretation of results. Furthermore, there are two approaches to LCA that have been developed in the past few years, namely the attributional and consequential. The aim of these approaches developed is to provide answers to questions relating to different system modeling (Brundage et al., 2018). The Attributional-LCA (A-LCA) provides information about the impacts of the processes employed to produce, consume and dispose of a product. On the other

LCA of Recycled Polymer Composites

TABLE 27.1
Recent Studies on Recycling Polymer Composites

No	Polymer Composite	Scope of Study	Discoveries	References
1	Glass fiber thermoplastic composites from disposed wind turbine blades	To study advantages and disadvantages of different recycling techniques (thermal degradation, grinding and dissolution of the polymer matrix) to recover the constituent materials	• Pyrolysis experiments required relatively low energy to decompose the polymer matrix, but lose polymer matrix and fiber mechanical property degradation • Grinding is simple and mature technology but causes reduction of fiber length • Dissolution is expensive and requires volatile solvents, but can maintain fiber length and mechanical prooerty as well as polymer matrix recovery	(Cousins et al., 2019)
2	Low-density polyethylene waste, and pine wood waste	Develop and evaluation the mechanical, morphological and thermal properties of Low-density polyethylene waste (LDPEW)/pine wood waste (PWW) composites by extrusion	• The mechanical properties of the LDPEW/PWW composites depend on the PWW content • Thermal stability of the composites is compatible with the required conditions for thermomechanical processing • Preparation of LDPEW/PWW composites showed viability for production of low-cost materials from recycled waste	(Moreno & Saron 2017)
3	Recycled polypropylene/waste paper, cardboard and wood flour composites	To study properties of the recycled polypropylene composite reinforced with cellulose waste materials for extrusion-based polymer additive manufacturing	• Recycled polymer composites had increased levels of filler incorporated in the printed parts compared to the virgin polymer composites • Tensile strength was not significantly increased with the addition of 10 wt.% cellulose	(Zander et al., 2019)
4	Biochar/recycled Polyethylene terephthalate (PET) composite	To develop a novel low cost and sustainable biochar/recycled polyethylene terephthalate (PET) composite with	• The incorporation of the biochar improved the composite mechanical, thermal and dynamic properties	(Idrees et al., 2018)

(Continued)

TABLE 27.1 (Continued)
Recent Studies on Recycling Polymer Composites

No	Polymer Composite	Scope of Study	Discoveries	References
		improved mechanical and thermal performance	• 0.5 wt.% of biochar infusion in PET resulted in 32% increase in tensile strength • 5 wt.% loading has shown 60% increase in tensile modulus over the neat PET	
5	Recycled polystyrene (rPS), coconut shell (CS), maleated polystyrene (MAPS)	Mechanical, thermal and morphological properties characterization of wood plastic composites (WPC) with expanded polystyrene (EPS)	• WPC formulated with 100 phr of rPS, 30 phr of CS, 3 phr of MAPS and 1 phr of Ultra-Plast WP516 possesses a higher modulus and tensile strength compared to the neat EPS, measured at 2.5 GPa and 27.5 MPa, respectively • WPC experienced initiation of thermal degradation at a temperature lower than neat rPS, but the thermal stability of rPS/CS composites containing varying composition of MAPS and Ultra-Plast WP516 was better at a high temperature • The glass transition temperature of the rPS/CS composite with the addition of MAPS and Ultra-Plast WP516 was found lower than the neat rPS. POLYM	(Ling et al., 2020)

hand, Consequential-LCA (CLCA) provides information about the consequences of changes in the level of output, consumption and disposal of a product, including effects both inside and outside the life cycle of the product (Ingrao et al., 2017).

27.3.1 LCA of Recycled Polymer Composite

The purpose of LCA is to understand the environmental impacts of recycled polymer composites and the results can be compared to the results of virgin

LCA of Recycled Polymer Composites

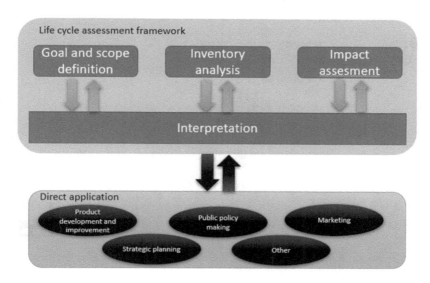

FIGURE 27.2 LCA procedure according to ISO 14040.

materials utilized. Khoo (2019) investigates eight scenarios of plastic waste management options by adopting LCA. The results exhibited different combinations of plastic valorization technologies, and associated capacities, affected the potential environmental benefits and drawbacks of plastic waste treatment systems. Gu et al. (2017), in their report, highlight LCA as a fundamental tool to assess the environmental benefits and burdens associated with waste management on mechanical recycling of waste plastics with a focus on comparing different disposal alternatives, such as incineration. Haylock and Rosentrater (2018) carried out LCA studies to compare the estimated energy intensity environmental impacts during production of composites and quantifed the environmental and economic impact. Recycled polymer and reinforced composites have been used for many years, yet information on their environmental impacts is scarce.

27.3.1.1 Goal and Scope

Shen et al. (2010) worked on a LCA of PET bottle-to-fiber recycling and they defined the goal of the LCA study as "to assess the environmental impacts of recycled PET fiber compared with virgin PET fiber." They defined a functional unit as "one metric tonne of fiber." Their study applied the common practice of the "cut-off" principle, widely applied for recycled or recovered products and is considered simple and straightforward because no data of the first life is needed. It distinguishes the first life (virgin product) and the second life (recycled product) as separate systems. Used PET bottles in this study are waste from the first life and four PET recycling cases were investigated in this study, namely mechanical recycling, semi-mechanical recycling, back-to-oligomer recycling and

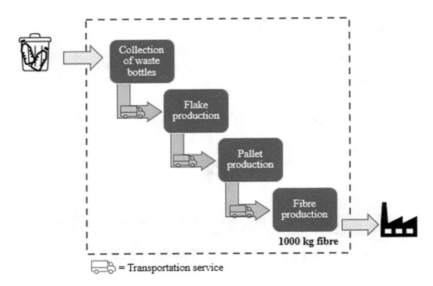

FIGURE 27.3 System boundary of cradle-to-factory gate, the second life of recycling PET fibers from waste PET bottles based on cut-off approach in Shen et al. (2010).

back-to-monomer recycling. The scope of this LCA is the cradle-to-factory gate, which in the case of a virgin product, this includes all steps from the extraction and transportation of raw materials and fuels, followed by all conversion steps until the product is delivered at the factory gate. The "cut-off" method focuses only on the recycled product and no data is required outside of the investigated product system. However, it simplifies the open-loop allocation issues, especially for the "cradle" and the "grave" stages. The system boundary determined for the cradle-to-factory gate based on the cut-off approach in this study is shown in Figure 27.3.

Al-Ma'adeed et al. (2011) studied the LCA of particulate recycled low-density polyethylene (LDPE) and recycled polypropylene (PP) reinforced with talc and fiber glass. Their goal of the study was "to analyse the environmental impact of newly developed composite materials obtained from the combination of recycled PP and LDPE with talc and glass fiber." The environmental impact of these composite materials is compared to the impact of virgin PP and PE. The functional unit for calculations and comparative purposes was defined as 1 kg material. Plastic waste (LDPE and PP) in industrial facilities is ground, washed, dried and granulated. This study considered landfilling as a reference scenario and compared it with filled recycled plastics.

Another study by Gu et al. (2017) investigated the life cycle environmental impacts of mechanical plastic recycling practice of a plastic recycling company in China. In this study, they studied waste plastics from various sources, i.e., agricultural wastes, plastic product manufacturers, collected solid plastic wastes and parts dismantled from waste electric and electronic equipment, processed in

LCA of Recycled Polymer Composites

three paths with products ending up in different markets. There are three goals proposed in this study: (1) To evaluate the environmental performance of each different waste plastic recycling route within the company, and to quantify the contribution of each process to the total environmental impacts; (2) to evaluate the environmental performance of the recycled plastics and composites produced as secondary raw materials for substituting their virgin counterparts, and to compare them with the respective virgin materials from an environmental point of view; (3) to evaluate the environmental performance of the selected plastic recycling system, and a sensitivity analysis to assess the impacts of potential changes in the operational mode of the company. The system boundary determined for the three paths of plastic waste is shown in Figure 27.4.

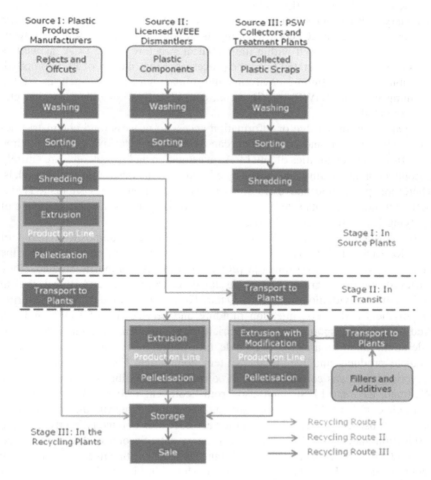

FIGURE 27.4 System boundaries of the studied mechanical plastic recycling system in Gu et al. (2017).

27.3.1.2 Life Cycle Inventory (LCI)

Life cycle inventory (LCI) is the phase to quantify the use of resources and materials and the consumption of fuel and energy, as well as the involved transport associated with a product in its life cycle of the analyzed functional unit (FU) (Ingrao, et al., 2015; Khoo et al., 2018). LCI is crucial in LCA, whereby in this phase, data is collected to calculate material use, energy input and pollutant emissions during the entire life cycle of a product or process. These data can be obtained from companies engaged in product fabrication and processing activities and also from published databases (Parameswaranpillai & Vijayan, 2014). Several databases have been developed that allow users to insert new information or data specific to their products and processes. The most comprehensive international LCI database is Ecoinvent v3 (http://www.ecoinvent.org/database/) (La Rosa & Cicala, 2015). Other processes or activities outside the LCA boundary will be ignored and thus environmental damages caused by them are not counted. A set of LCI consists of inputs and outputs that represent the flow of material and energy used within the technical structure of the LCA, as well as emissions that are generated from the processes. From the list of LCI compiled, contributions to different environmental impact categories, such as global warming potential (GWP), acidification, eutrophication or human toxicity, etc. are generated.

Apart from knowledge or information access to process (and chemistry) details, data selection and compilation can demand an extensive amount of time and effort. The reliabilities of any LCA investigation and its outcome are entirely dependent on the quality and adequacy of data used in constructing LCA models. Therefore, precaution and extra steps in data selection is advisable. It is important to note that data selectivity, quality and adequacy is the first challenge of applying LCA (Khoo et al., 2018).

In Al-Ma'adeed et al. (2011)'s LCA study of new composite material of recycled LDPE and PP with talc and glass fiber as fillers, the data needed for the inventory were based on material proportion of each composite studied. They retrieved this from various supporting databases (Ecoinvent and Buwal 250) in GaBi software with modification for Qatar. These data include the manufacturing of polymers and fibers, injection molding and service life of LDPE and PP. Additionally, transportation and electricity systems were also taken into consideration. Shen et al. (2010) include transportation of raw materials, intermediate products and fuel in the system boundaries. For PET bottle-to-fiber recycling, data was collected from three recycled PET fiber producers, namely Wellman, LJG and FENC; and also, from Ecoinvent v2.0 based on literature.

In Gu et al.'s (2017) work, primary and secondary data from different sources have been used. Operational data of the year 2016 was obtained directly from the production manager of the company as primary data. Secondary data is referred to in the LCI data sets, which cover a wide range of most existing materials and energy supply processes. From the recycling system shown in Figure 27.4, they employed a standard waste plastic washing process employed in the LCA calculation.

LCA of Recycled Polymer Composites

The sorting processes do not impart any environmental impacts because the sorting process was done by manual labor. The shredding process increased the bulk density of waste plastics, which facilitates the transport process by increasing the payload per trip. All the shredded pieces and granulates are transported to the recycling plants using lorries for further processing or sale and no immediate reuse took place. The distances between the sources and the recycling plants are within a radius of 200 km, ranging from 10 km to 187 km, but 80 km was used in this LCA calculation and the modeling of the transport used the data set transport, Lorry 22 tons including fuel.

27.3.1.3 Life Cycle Impact Assessment (LCIA)

The third phase of the LCA framework is the life cycle impact assessment (LCIA), which focuses on evaluating and understanding the environmental impacts established by the LCI analysis. ISO 14040 outlines the impact assessment phase with the following components: (1) classification: assigning inventory results to impact categories; (2) characterization: modeling inventory data within impact categories; the characterization process involves defining characterization factors to convert each pollutant emission into equivalent potentials represented by a reference substance (e.g., CO_2 equivalent); (3) weighting: aggregating inventory data in very specific cases to combine the impact categories into a single score (Parameswaranpillai & Vijayan, 2014). Impact categories represent the negative effects to the environment due to substances emitted and resources used that caused the occurring damage. The impact categories are grouped and termed "Damage Categories" and they represent the environmental sectors suffering from the damage.

The LCIA represents all emissions released by the product system to the environment and all raw material requirements converted into environmental impact categories; the results are referred to as LCA midpoint results. In Shen et al.'s (2010) study, the environmental indicators are NREU (non-renewable energy use), GWP (global warming potential) (IPCC, 2007) and the indicators from the CML 2 baseline 20013 impact assessment method, namely abiotic depletion, acidification, eutrophication, human toxicity, fresh water aquatic ecotoxicity, terrestrial ecotoxicity and photochemical oxidant formation.

There are several impact assessment methodologies (EDIP, Eco-Indicator, Environmental Priority Strategies and CED23) that allow aggregating the results into a single score. Rajendran et al. (2013) carried out on the comparison of LCA results between individual and aggregated impacts and integration of performance of recycled plastics in the LCA. The study evaluates mechanical recycling and energy recovery with the CML 2001 baseline method, EDIP 2003 and EI99. The basic difference between these methods is that CML 2001 and EDIP 2003 follow a problem-oriented approach, while Eco-Indicator 99 follows a damage-oriented approach. These methodologies are described in Table 27.2.

TABLE 27.2
Summary of LCIA Methodologies

LCIA Method	Model	Impact Categories	Sub-Divided Categories	Results
Eco-Indicator 99	Endpoint	Ecosystem quality, resource use and human health.	Acidification, eutrophication, ecotoxicity, ozone layer depletion, radiation effect, respiratory effect, climate change, land use and land conversion	Aggregated to get a single score
Centrum voor Milieukunde Leiden (CML)	Midpoint	Abiotic depletion, acidification, eutrophication, global warming, ozone layer depletion, human toxicity, fresh water ecotoxicity, marine aquatic ecotoxicity, terrestrial eco-toxicity and photochemical oxidation	–	Individual impact scores
Environmental Development of Industrial Products (EDIP)	Midpoint	Global warming, ozone depletion, ozone formation, acidification, terrestrial eutrophication, aquatic eutrophication, human toxicity, ecotoxicity, hazardous waste, radioactive waste, slags/ashes, bulk waste and resources	–	Aggregation of impact categories to a single score using distance
ReCiPe 2008	Midpoint and Endpoint	Two sets of impact categories with associated sets of characterisation factors. Eighteen impact categories are addressed at the midpoint level (climate change, ozone depletion, terrestrial acidification, freshwater eutrophication,	–	Midpoint indicators focus on single environmental problems. Endpoint indicators show the environmental impact on three

(Continued)

TABLE 27.2 (Continued)
Summary of LCIA Methodologies

LCIA Method	Model	Impact Categories	Sub-Divided Categories	Results
		marine eutrophication, human toxicity, photochemical oxidant formation, particulate matter formation, terrestrial ecotoxicity, freshwater ecotoxicity, marine ecotoxicity, ionising radiation, agricultural land occupation, urban land occupation, natural land transformation, water depletion, mineral resource depletion, fossil fuel depletion). At the *endpoint* level, most of these midpoint impact categories are further converted and aggregated into three endpoint categories i.e., damage to human health, damage to ecosystem diversity, damage to resource availability.		higher aggregation levels 1. effect on human health, 2. biodiversity, 3. resource scarcity.

Source: From Rajendran et al. (2013) and Goedkoop et al. (2009).

27.4 CONCLUSION

Polymer-based products are discarded after use and large volumes of disposal can be avoided by converting this waste to valuable resources. Due to the complex nature of plastic waste mixtures, often accompanied by the presence of impurities, the recycling technologies to be employed are being studied for the most feasible and efficient. LCA methodology is a useful tool to drive the choice of materials and processes toward a more sustainable production system. Researchers suggested that LCA studies should focus on specific recycled composites and include the applications of the recycled composite materials that would be appropriate. In concept, polymer-based products can be recycled numerous times before being finally converted into fiber. The environmental impact of such recycling systems and the effect of the number of cycles can also be further investigated.

ACKNOWLEDGMENTS

The authors wish to express their highest appreciation to the Public Service Department (JPA), Malaysia for the study sponsorship to the main author and financial support from the Ministry of Education Malaysia through Universiti Putra Malaysia Grant Scheme HiCOE (6369107).

REFERENCES

Al-Ma'adeed, Mariam, Ozerkan, Gozde, Kahraman, Ramazan, Rajendran, Saravanan, & Hodzic, Alma (2011). Life cycle assessment of particulate recycled low density polyethylene and recycled polypropylene reinforced with talc and fiberglass. *Key Engineering Materials, 471–472*, 999–1004. 10.4028/www.scientific.net/KEM.471-472.999.

Brundage, Michael P., Bernstein, William Z., Hoffenson, Steven, Chang, Qing, Nishi, Hidetaka, Kliks, Timothy, & Morris, K. C. (2018). Analyzing environmental sustainability methods for use earlier in the product lifecycle. *Journal of Cleaner Production, 187*, 877–892. 10.1016/j.jclepro.2018.03.187.

Cousins, Dylan S., Suzuki, Yasuhito, Murray, Robynne E., Samaniuk, Joseph R., & Stebner, Aaron P. (2019). Recycling glass fiber thermoplastic composites from wind turbine blades. *Journal of Cleaner Production*. 10.1016/j.jclepro.2018.10.286.

Goedkoop, Mark, Heijungs, Reinout, Huijbregts, Mark, Schryver, An De, Struijs, Jaap, & Zelm, Rosalie Van (2009). ReCiPe 2008. *Potentials*.

Gu, Fu, Guo, Jianfeng, Zhang, Wujie, Summers, Peter A., & Hall, Philip (2017). From waste plastics to industrial raw materials: a life cycle assessment of mechanical plastic recycling practice based on a real-world case study. *Science of the Total Environment*. 10.1016/j.scitotenv.2017.05.278.

Haylock, Randall, & Rosentrater, Kurt A. (2018). Cradle-to-grave life cycle assessment and techno-economic analysis of polylactic acid composites with traditional and bio-based fillers. *Journal of Polymers and the Environment*. 10.1007/s10924-017-1041-2.

Idrees, Mohanad, Jeelani, Shaik, & Rangari, Vijaya (2018). Three-dimensional-printed sustainable biochar-recycled PET composites. *ACS Sustainable Chemistry and Engineering*. 10.1021/acssuschemeng.8b02283.

Ingrao, Carlo, Gigli, Matteo, & Siracusa, Valentina (2017). An attributional life cycle assessment application experience to highlight environmental hotspots in the production of foamy polylactic acid trays for fresh-food packaging usage. *Journal of Cleaner Production, 150*, 93–103. 10.1016/j.jclepro.2017.03.007.

Ingrao, Carlo, Giudice, Agata Lo, Bacenetti, Jacopo, Khaneghah, Amin Mousavi, Sant'Ana, Anderson de Souza, Rana, Roberto, & Siracusa, Valentina (2015). Foamy polystyrene trays for fresh-meat packaging: life-cycle inventory data collection and environmental impact assessment. *Food Research International, 76*, 418–426. 10.1016/j.foodres.2015.07.028.

Ingrao, Carlo, Tricase, Caterina, Cholewa-Wójcik, Agnieszka, Kawecka, Agnieszka, Rana, Roberto, & Siracusa, Valentina (2015). Polylactic acid trays for fresh-food packaging: a carbon footprint assessment. *Science of the Total Environment, 537*, 385–398. 10.1016/j.scitotenv.2015.08.023.

Khoo, Hsien H. (2019). LCA of plastic waste recovery into recycled materials, energy and fuels in singapore. *Resources, Conservation and Recycling*. 10.1016/j.resconrec.2019.02.010.

Khoo, Hsien H., Isoni, Valerio, & Sharratt, Paul N. (2018). LCI data selection criteria for a multidisciplinary research team: LCA applied to solvents and chemicals. *Sustainable Production and Consumption, 16*, 68–87. 10.1016/j.spc.2018.06.002.

Ling, Sing Li, Koay, Seong Chun, Chan, Ming Yeng, Tshai, Kim Yeow, Chantara, Thevy Ratnam, & Pang, Ming Meng (2020). Wood plastic composites produced from postconsumer recycled polystyrene and coconut shell: effect of coupling agent and processing aid on tensile, thermal, and morphological properties. *Polymer Engineering and Science*. 10.1002/pen.25273.

Mansor, Muhd Ridzuan, Salit, Mohd Sapuan, Zainudin, Edi Syam, Aziz, Nuraini Abdul, & Ariff, Hambali (2015). Life cycle assessment of natural fiber polymer composites. In Khalid Rehman Hakeem, Mohammad Jawaid, & Othman Y. Alothman (Eds.), *Agricultural Biomass Based Potential Materials*. Springer. 10.1007/978-3-319-1384 7-3_6.

Moreno, Diego David Pinzón, & Saron, Clodoaldo (2017). Low-density polyethylene waste/recycled wood composites. *Composite Structures*. 10.1016/j.compstruct.201 7.05.076.

Parameswaranpillai, J., & Vijayan, D. (2014). *Life Cycle Assessment (LCA) of Epoxy-Based Materials. Micro and Nanostructured Epoxy/Rubber Blends*. Wiley-VCH Verlag GmbH & Co. 10.1002/9783527666874.ch21.

Rajendran, S., Hodzic, A., Scelsi, L., Hayes, S., Soutis, C., AlMa'adeed, M., & Kahraman, R. (2013). Plastics recycling: insights into life cycle impact assessment methods. *Plastics, Rubber & Composites, 42*(1), 1–10. 10.1179/1743289812 Y.0000000002.

Rosa, A.D. La, & Cicala, G. (2015). LCA of fiber-reinforced composites. In Subramanian Senthilkannan Muthu (Ed.), *Handbook of life cycle assessment (LCA) of textiles and clothing*. (pp. 301–323). Woodhead Publishing Series in Textiles. 10.1016/B978-0-08-100169-1.00014-9.

Scelsi, L., Hodzic, A., Soutis, C., Hayes, S. A., Rajendran, S., AlMa'adeed, M. A., & Kahraman, R. (2011). A review on composite materials based on recycled thermoplastics and glass fibers. *Plastics, Rubber and Composites*. 10.1179/174328911 X12940139029121.

Shen, Li, Worrell, Ernst, & Patel, Martin K. (2010). Open-loop recycling: a LCA case study of PET bottle-to-fiber recycling. *Resources, Conservation and Recycling, 55*(1), 34–52. 10.1016/j.resconrec.2010.06.014.

Sommerhuber, Philipp F., Wang, Tianyi, & Krause, Andreas (2016). Wood-plastic composites as potential applications of recycled plastics of electronic waste and recycled particleboard. *Journal of Cleaner Production*. 10.1016/j.jclepro.2016.02.036.

Tshifularo, Cyrus A., & Patnaik, Asis (2020). Recycling of plastics into textile raw materials and products. In *Sustainable Technologies for Fashion and Textiles*. 10.1 016/b978-0-08-102867-4.00013-x.

Zander, Nicole E., Gillan, Margaret, Burckhard, Zachary, & Gardea, Frank (2019). Recycled polypropylene blends as novel 3D printing materials. *Additive Manufacturing*. 10.1016/j.addma.2018.11.009.

Index

Note: Page numbers in *Italic* refer to figures; and in **Bold** refer to tables

A

A356 alloy, 445
  chemical composition, **445**
  conventional and horizontal directional solidification of, 448–449, *450*, *453*
  horizontal solidification device, 446, *447*
  iron content influence, on microstructure of, 445
    under as-cast conditions, 445, 449
    β-Al$_5$FeSi phase, 454
    conclusions, 458–459
    dendritic microstructures, 453
    effect of cooling rate, 453–454
    element mapping by EDS, with SEM, 453–454, *455*, *456*
    intermetallic phases, 452
    materials used, 445–448
    methods used, 445–*450*
    micrographs, 450, *451*–*453*, 452–453
    microhardness values, 455
    microstructure of crude, 450, *451*
    remelted, without addition of iron, 450, *451*, *452*
    secondary dendritic growth, 456, *458*
    theoretical/experimental procedure, 448, *449*
    thermal parameters of transient solidification, 456–457
  microstructural characterization of, 448
  Scanning electron microscopy (SEM) analysis, 448
  Vickers microhardness (HV) tests, 449, 455, *457*
Abbas, A., 46
Abdullahi, I., 100
Abidoye, J. K., 100
Ab Kadir, M. I., 42, 101
abrasive water jet (AWJ) technology; *see also* Ti–Al-based intermetallic-based composite, recycled
  advantages of, 296
  cutting performances, 296–297
    cutting speed, effect of, 300, *300*
    degree of plastic deformation, 299
    formation of striation, 300
    mass flow rate, effect of, 301, *301*
    methodology, 298, **298**
    microstructural evaluations, 299, *299*
    surface roughness, 301
  of Ti–Al alloy material, 296–297
  effects of parameters of cuts, 297
  machining process, 297
  materials used, Ti–Al alloy, 297, **298**
  multi-pass/single pass cutting, 297
  optimization of quality surface of parts, 301–305
  surface roughness, and depth of cut analysis, 298–306
Abu-Sharkh, B. F., 130
açai fiber, 425, *428*, 429, 430–431
Acrylonitrile-butadienestyrene (ABS), 12
  recycling process, 468–469
admixtures, 329
AISI 304 stainless steel, 446
Al7Si0.4Mg1.2Fe alloy, 453–*454*
Al-Alimi, S., 17
Alberto, F., 11
alloys
  defined, 110
  recycled/virgin characteristics, 159
Al-Ma'adeed, M., 494, 496
AlMg1SiCu alloy, 317, **318**
Al–Si alloys, 445
AlSiMgFe alloys, 445
aluminium metal matrix composites (AMMCs), 22
aluminum, and alloys of
  AA7075 alloy, 316
  for bumper beams applications, 252–260
  composites, demand/recycling of, 5–6, 443
  extraction (electrolysis) of, 443–444
  negative influence of iron on microstructure and mechanical properties of, 444–445
  production, 444
  recovery form RBCs (*see* X7475 alloy, from RBCs)
aluminum-based hybrid composites, 60, 69
  Abaqus®/Explicit use, 62–63
  composition preparation, 60–61, **61**, **62**
  experimental conditions

503

composition preparation, 60–61, **62**
experimental setup, 61–62
finite element modeling, 62–63
high-energy milling process, 61
Johnson-Cook fracture model use, 61–63
low velocity impact tests, 67–69
  TiB$_2$-CNTs-GNPs effects, *67*
  TiC-CNTs-GNPs effects, *68*
microstructures analyses, 63–*64*
quasi-static compression tests, 65–67
  TiB$_2$-CNTs-GNPs effects, *66*, 67
"SEM" and "EDS" chemical analyses, 61–62, **62**, 63, *65*
sinter and forging methods use, 60, 61, 69
aluminum matrix composites, publications, **81–82**
aluminum matrix syntactic foam (AMSF), 316, 322
  AlMg1SiCu alloy matrix, for composing, 317, **318**
  boron carbide particle at, 317, *318*
  coefficient of friction, and wear rate study, 319, *321*, **322**
  EDS analysis, 319, *320*
  fly ash balloons addition, 316, *317*
  optical micrograph, 318, *319*
  stir casting method, 316–317
  wear behavior, comparison, 316
Alyamac, K. E., 337
Amazonian vegetable fibers, 424–425
  açai fiber, 425, *428*, 429–431
  coconut fiber, 424–425, *424–426*, 428, 430–431
  in concrete, composition/properties analysis, 426–428
    absorption capacity, graph, 430, 431
    axial compression results, graph, 428, 429
    mechanical processing, before and after, *424*, *425*, *427*
    modulus of elasticity, 430
    processing, *427*
    quality of pulse passage, 428
    ratio, 426
    ultrasound assays, 428, **430**
American Society for Testing and Materials (ASTM), 326
Andrzej, P., 296
Anton-Paar MST^3 Micro Scratch tester, 413
aqueous solution, extraction method, 149–*151*
Aslan, A., 42, 100, 158
ASTM 247 standard, 288
Attar, S., 316, 321
Atzori C., 38

Awotundea, M. A., 41
Axinte, D. A., 296
Azmir, A., 297

B

Badarulzaman, N. A., 101
Bahoria, B. V., 328
Bakshi, S. R., 21
bamboo fibers, 188
banana fiber, 187
Barkov, R., 257
Barros, A., 327, 337
Bayoğlu, S., 444
Bazarnik, P., 48
Begic-Hajdarevic, D., 296
Bhadra, J., 10
Bhouri, M., 6, 96, 101
bibliometric indicators, 362
bibliometrics, 358–359
  analysis method, 359
  defined, 358
  network search engine, 358
  peer review and, 359
  popularity of, 359
  survey on cementitious composites
    choice of databases, 363–365, *364–365*
    documents by area, 377–379, *378*
    documents by author, 368, *369–370*
    documents by countries and territories, 374–377, *375*, *377*
    documents by institutions, 370–373, *372*
    documents by journal, 368–370, *371*
    documents by type, 379–380, *379*
    documents per year, 365–368, *367*
    funding agencies for studies, 373, *374*
    methodology, 362–*363*
    relevant keywords related to theme, *366*
  use of, 358
Biillberg, R., 336, 339
biodegradable polymers, 245, 464
biodegradation, by microorganisms, 245
bioplastic plant-derived material, 242
bioplastics, 243
biopolymers, 242–243, 464
bond exchange response (BER) process, 13
boron carbide (B$_4$C), 316
Brazil, civil construction in, 388–389;
  see also marble and granite residue (MGR) processing
Brazilian Association of Technical Standards (ABNT), 392
Bulei, C., 19
bulk molding compound (BMC), 236

Index

bumper system
  bumper beam; (*see also* X7475 alloy, from RBCs)
    aluminium alloys use for, 252–253
    conceptual selection flow chart, *240*
    PDS, selected parameters for, *239*
  elements of, 239
Byard, D. J., 14

C

Canakci, A., 42
carbides and oxides, 76
carbon fiber composites, 18
carbon fiber reinforced polymers (CFRP), 4, 18, 235
  recycling, *5*
  waste salvaging process, 20
carbon fibers (CFs), 236
Carro-López, D., 340
Casati, R., 23
casting by centrifugation
  casted dimensions, 288
  characterization, 288–290
    Charpy test, 289
    hardness test, 289–290
    microstructural analysis, 288–289
    tensile test, 289, *290*
  Cu- and Mo-rich carbides, microanalysis, *291–292*, *293*
  ductile iron production, 288, *289*
    chemical composition, **289**
  graphite ferrite, and perlite analysis, *291–292*
  inoculation practice, 288
  materials, 288
  nodules type and size for nodular cast iron, 290–291, *290–291*
  spheroidizing practices, 288
cast iron, 287–288
Caydas, U., 296
cellulose, 464
cementitious composite, and materials, 326, 328
  bibliometric survey on, 362
  rheology, 340–344
cement manufacturing processes, 327
centrifugal casting process, 146, *147*
ceramic fillers, 410
ceramic matrix composites (CMCs), 78
Charpy impact strength test, 279, *279*, 410, 413
chemical recycling, 7, 10–11, 15, 85, 242–243
Chemical Recycling Monomer (CRM), 11–12
Chen, X., 296
Chen, Z., 12
Cherrington, R., 17

Chmura, W., 158
Chuanzhen, 2014 [not found in ref.], 297
closed die forging, 49
closed-loop process design, 13
CO_2 emissions, 86, 198, 200, 327, 410
coated fillers, 186–187
Coates, G. W., 11
coconut fiber, 424–425, *424–426*, 428, 430–431
Cojbasic, Z., 296
cold forging, 49
compatibilizer, 466
compocasting method, 80
composites
  aluminum matrix, publications about, **81–82**
  cementitious materials as, 326
  CFRP and GFRP, uses in, 235
  chain of operations to recycle, 87
    availability of the composite scrap, *87*
  classification, *3*, 77, *78*, 79
    CMCs/ MMCs, 78
    PMCs, 78–79
  defined, 77
  environmental aspect of, 76
  environmental burden of, 3
  formation of, 110
  form recyclates, product design key elements
    bumper beam, 239, *239–240*
    bumper system, elements of, 239
    conceptual design/design concept, 239
    design concept selection (DCS), 239
    design methods and tools, 237, *238*
    design strategy, 237, *238*
    product design specification (PDS), 239
  hardwood, 187
  higher mechanical properties, 77
  history of, 77
  made from recyclates, used in, 237
  made from waste polypropylene (WPP), 244
  materials
    formation, 110
    with short glass fibers (SGF), 264
  metal matrix composite (MMC), 76
    recycling processes of, 84–86
  product design, 236
  production methods of, 76, 80
    liquid state processes, 80, 82–83
    solid-state processes, 83–84
  products disposing, landfill disposal, 246
  recycled polymer, 185–186, 193;
    (*see also* recycled polymer composites)
  recycling, 236
    challenges with, 246
    companies globally active, 236

costlier venture, 12, 29, 246
eco-design, approaches in, 245–247, *246*
environmental and sustainability issues, 240–243
importance, 2, 86, 88
material expert role, 236
methods comparison, of different, 85, *86*
product design for, 236–240
reinforced with natural fibers, 424
reinforcement materials, 76
as replacement of conventional materials, 76–77, 87
thermal properties of, 189–190
usage in day-to-day life, 235–236
composite scraps, dumping of, 109
compressible packing method (CPM), 338–339
concrete
defined, 326
hardened properties of, 344, **347**
pebble/gravel-reinforced; (*see* reinforced concrete (RC) with aggregates, evaluation of)
with plastic waste, 360–361
popular, building material, 423
requirements for fresh, 340
role in pollution reduction, 327
technical standards for testing, 344, **347**
vegetable fibers, insertion in, 424–425; (*see also* Amazonian vegetable fibers)
concrete theme with polymeric waste, 362, 367–368, 370
construction industry, 188, 236, 327, 388, 424
conventional material, 76
conventional vibrated concrete (CVC), 338
COPASA, 390
Cosme, R. J., 395
COVID-19 pandemic, 243
Cruz Sanchez, F. A., 9
crystalline intermetallics, 444
Cu- and Mo-rich carbides, 292–293, *292–293*
Cunha, S. R. L., 439

D

Dal Molin, D., 336
Datta, J., 10
Davoodi, M. M., 239
De Larrard, F., 337–339, 338
Dertinger, S. C., 11
devulcanized recycled rubber-based composites, study of
3D damage traces of, 419–*420*
ceramic fillers (TiO_2 and Al_2O_3) inclusion, 411–414

Charpy impact strength test, 418, *418*
hot compaction techniques, 412
material processing, 411–*412*
composition, **412**
powder metallurgy techniques, 411
tests, and tools, 413
mechanical characteristics and microstructure, 414–419, *414*, **415**, 416
cavitation study, 417, *417*
EDS mapping analysis, *415*
energy-absorbing capacity, 418, *418*
fracture surfaces and toughness, observation, **415**, 416, *416*
three-point bending (3PB) testing, 414, *416*, *417*
micro scratch tests, 419, *420*, **421**
physical characteristics and microstructure, 413–414, **413**, *414*
tribological assessment of, 419
Dewberry, E., 246
Dhanesh, S., 22
dielectric analysis (DEA), 165
differential scanning calorimetry (DSC), 165, 188
differential thermal analysis (DTA), 188, 389, 394
diffusion bonding technique, 83
Dikici, T., 42
dilatometry (DIL), 165
direct-chill casting (DC) method, 144–*145*
displacement reaction method, 118
Divyansh, P., 296
DLP 3D printing, 12
Dolatkhah, A., 45
Domone, P. L., 336
Dorigato, A., 467
Douiri, M., 296
Double Disc technology, 271, 271–272
durability, defined, 433–434
dynamic mechanical analysis (DMA), 13
dynamic mechanical calorimetry (DMA), 165
dynamic thermal analysis (DMA), 188

E

eco-design concept, 245–247, *246*
Eco-Indicator, 99, 497
economic effects, of composite recycling, 8
EDIP 2003, 497
elastomers, 222, 465
electrolysis process, 150
ELG Carbon Fibre, 236
Emadi Shaibani, M., 42
end-of-life (EOL), 3, 7, 20

Index

energy dispersive analysis (EDS), 118
ENGEL injection molding machine, 266, 276, 282
Engineering Village, 362, 364, 380
Enginsoy, H. M., 99
epoxy adhesives, 192
epoxy and/or elastomers, reinforced composites, 221–222
epoxy-based composites, 130; *see also* scrap rubber/epoxy based composites
equal channel angular pressing (ECAP), 45–47
   advantages of, 45
   MMC fabricated by, 45–46
   schematic of, 45, 46
EREMA company, 268
expanded polystyrene, 171
extruded polystyrene, 171

F

factorial statistical model, 339
Faculty of Science and Technology (FST), 103
failure mode and effect analysis (FMEA), 203, *204*
Farias, L. A., 438
Fatemeh, Y., 244
ferrite, 293
ferrous scrap composite, 122
fiberglass, 186
fiber-reinforced composite, 4, 79
filament deposition modeling (FDM), 14
flame retardancy index (FRI), 207
flame retardant polymer composites, 197–198
   chemicals present in, 198
   development, from recycled plastics, 205–211; (*see also* Polyethylene terephthalate (PET))
   difficulty in separating, from polymers, 198–199
   environmental safety issues, 198
   flame-retardant panel, 212
   large quantities of end of- life, wastes, 198
   phenolic foam particles use, 212, *213*
   recycled, 198, 211–212
   recycling of, 199–200
      difficult due to dissimilar materials in, 198–199
      flame retardant textiles, 200, *201*
      waste management hierarchy, *199*
   use and wastes disposal issue, 198
flame retardant textiles, 200, *201*
fluidized beds pyrolysis, 10
fly ash, 189, 190
Fogagnolo, J. B., 158

Fourier transform infrared spectroscopy (FTIR), 13, 389, 406
FPF/FGF 3D printer, 14
friction stir processing (FSR), *44*, 44–45
functionally graded material (FGM), 146
fused granular fabrication (FGF), 14–15
fused particle fabrication (FPF), 14–15
Fuzzy ANP-based, Analytical Hierarchy Process (AHP), 239

G

Gatamorta, F., 99
gel permeation chromatography, 13
Germany Carbon Conversions, 236
Ghambari, M., 42
Ghanbari, D., 45
Gigabot X, 14, 15
glass and CFRP recycling method, 8
glass fiber, 186
   -reinforced composites, 18
   reinforced polymers (GFRP), 235
   reinforcement, 187
global habitat loss, 142; *see also* recycling; wastes
glycolysis, 7, 489
Gołaszewski, J., 337
Gomes, P. C. C., 327, 336, 339
Goñi, J., 142
Goussous, S., 46
granite, 389, 395
GRAN-NGR recycling line, 268, *269–270*
graphene oxide (GO), 192
graphite, 287
graphite ferrite, 291
green design, 246–247
green fab labs, 14
Groenewoud, W. M., 189
Gronostajski, J., 158
Gu, F., 8, 493, 494, 496
Güneyisi, E., 327
Gupta, M., 86
GWP (global warming potential), 497

H

Hadi, N. J., 244
Hamza, R., 389
HANPLAST production plant, 282, 284
Harbi, R., 337
HarmoniX-Atomic Force Microscopy nanomechanical test, 15
Hart, K. R., 13
Hascalik, A., 296

Hatti-Kaul, R., 13
Hayat, M. D., 22
Haylock, R., 493
HDPE recycling process, 467–468
heat flow meters (HFM), 165
heat treatment process, 5
heavy metals, and environmental disasters, 142
Helene, P., 336
high-energy ball milling mixing and sintering, 94
high pressure torsion (HPT), 47–48
  limitations of, 47
  MMC fabricated by, 47–48
  schematic diagram of, *47*
Hopewell, J., 477
Hosseinzadeh, A., 45
hot extrusion method, 42–44
  extruded product properties, and parameters, 43
  schematic diagram of, 43
hot iso-static pressing (HIP), 39, 76
hot press forging
  deformation process, 48
  open or closed die forging, 49
  schematic diagram, *48*
  warm, and cold, 49
hot pressing (HP), 39
Hreha, P., 296
Hsu, C. J., 45
Hu, M., 43
Hunt, E. J., 10
Huntley, S., 9
hybrid composite materials, 60–61
  intermetallic, 60
    low velocity impact tests on, 67–69, *67*, *68*
    quasi-static compression tests on, 65–67, *66*
hybrid fibers, 191
hybrid metal matrix composites, 22
hydrolysis (water), 7, 85
  alkaline, 489
hydrolytic degradation, 245

I

Image J software, 449
impression die forging, 49
Inácio, André L. N., 132
incineration method, 14–16, 164, 200, 211, 243, 388, 467, 481
infiltration casting process, 118
infiltration process, 82
in-situ processing, 82

Institute of Microengineering and Nanoelectronics (IMEN), 103
INTAREMA compact recycling system, 268, *270–271*, 271
interfacial shear strength (IFSS), 132
inter-laminar shear strength (ILSS), 132
intermetallic Fe compounds (IMCs), 444
international events, for environment, 367–368
International Islamic University Malaysia (IIUM), 103
Italy Carbon Fiber Recycle Industry Co Ltd. 236

J

Jacobs, M., 16
Jahedi, M., 48
Jarukumjorn, K., 211
Jha, N., 316, 321
Jin, K., 15
Joharudin, N. F. M., 98, 100
Johnson-Cook constitutive model, 62, *63*, 69
Johnson-Cook damage model, 61
Joseph, P. V., 134
Juniarsih, A., 252
Junior, A. F., 394
Jute fabrics, 189

K

Kaewunruen, S., 20
Kaiser, K., 9
Kamble, A., 100
Karadağ, H. B., 94
Karayannis, V. G., 19
Karina, M., 244
Kärki, T., 18
Karuppannan Gopalraj, S., 18
Kazemi, M. E., 23
Kechagias, J. D., 297
Khajouei-Nezhad, M., 56
Khan, M. I., 328
Khayat, K. H., 338
Khoo, H. H., 493
Khushbu Dash et al. (2012) [not found in ref.], 41–42
Kim, B., 444
Kim, K., 102
Knoth, P., 359
Kopczyńska, P., 10
Korznikova, G., 48
Kostrzanowska-Siedlarz, A., 337
Kremmer, T. M., 252, 259
Krishnan, P. K., 100
K thermocouples, 446

Index

Kulkarni, S. G., 100
Kumar, H., 44

L

Lajis, M. A., 99
lamina, defined, 79
laminate, defined, 79
laminates-reinforced composite, 79
landfill disposal, 129, 164, 200, 211, 241–242, 246, 345, 359–360, 467
Laoutid, F., 205
laser flash technique (LFA), 165
laser particle distribution analysis, 389
latex, 464
Lazzaro, G., 38
leaf fiber, 189
Lehocka, D., 296
Li, F. X., 48
Li, Y., 212
Liao, P., 20
Life Cycle Assessment (LCA), 3, 247, 488, 490, 492
    attributional-LCA (A-LCA), 490
    consequential-LCA (CLCA), 492
    defined as, 488, 490
    Engineer Manufacture of Product activity, 490
    framework according to ISO 14040, 490, *493*
    phases in, product life cycle, 490
    of recycled polymer composite, 492–495
        life cycle impact assessment (LCIA), 497, **498–499**
        life cycle inventory (LCI), 496–497
    steps in performing, technique, 490
life cycle impact assessment (LCIA), 497, **498–499**
life cycle inventory (LCI), 496–497
Lin, C. B., 158
lines for film recycling *see* plastic film recycling
Ling, P. S., 86
liquid state processes, 80
    compocasting method, 80
    infiltration process, 82
    in-situ processing, 82
    spray deposition, 82
    squeeze casting method, 80
    stir casting method, 80
    ultrasonic assisted casting, 83
lithium-ion battery anodes, 18–19
Liu, T., 5
Liu, Y., 296
Long, W. J., 338
Lopez-Urionabarrenechea, A., 20

Löschnera, P., 296
low-value EOL options, 8
Lu, J., 340
Lundquist, N. A., 14

M

Ma, Z. Y., 45
Mahabalesh, P., 297
Májlinger, K., 321
Malavazi, J., 454
Malaysia
    polymer recycling business in, 473–477, *476*, 481
    acceptable materials, by recycling plants, 474, *476*, 477, **478–479**
    export destinations, 472, *474*
    import destinations, 472, *473*
    market potential, 477, *479*
    recycling plants, 474–477, **475**
    trading trends of plastic, 472, *473*, *480*, 481
    export market, 477
    issues and challenges, 477, 479–480, **481**
    products under trading code of 392690, 477, **480**
Maleic acid polyethylene (MAPE), 187
maleic anhydride grafted polypropylene (PPgMAH), 264
maleic anhydride grafted PP (MAPP), 131, 191
man-made polymer, 464–465
marble, 389, 395
marble and granite residue (MGR)
    processing, 389
    materials and methodology, 389–394, *390–391*, *393*, **393**
        absorption tests, 392, **393**
        characterization techniques, 390, **391**
        NBR 6502/2995 parameters, 390, **392**
        resistance to compression test, 392, **393**
        traces/mixtures preparation, 392, **392**
    residues characterization, 394–396
        FTIR, 395–396, *396*
        granulometric curve, 396, **397**
        granulometric curve distribution, *397*, **397**
        laser particle distribution analysis, 396
        thermogravimetric curves, 394, *394*
    solid waste in, 389
    specimens characterization, 396–406
        compressive strength, 398–405, **399**, **400**, **401**, **402**
        elastic modulus, 402–403, *403–405*
        Tukey comparison, 400–402, 404–405, **404**, *405*
        water absorption, 396–*398*

thermal analysis techniques, 394
visual characteristics of, *391*
X-ray diffractogram, 395, *395*
Marlaud, T., 257
martensite, 292, 293
Mastali, M., 337
Mativenga, P. T., 86
matrix, defined, 77
meal ready to eat (MRE) pouches, 13
mechanical recycling process, 7, 15, 84–85, 242
Melo, K. A., 336
Melt Flow Index (MFI), 11
Memon, H., 20
Mendeleev periodic table, potentials, *204*
Mendonça, C., 16
metal composites recycling
  aluminium metal matrix composites (AMMCs), 22
  aluminum alloy or silicon carbide MMCs, 19
  in aluminum composites, 17
  carbon fiber composite, 18, 20–21
  carbon nanotube reinforced MMCs, 21
  CFRP-type poly-benzoxazine waste, pyrolysis of, 20
  chipped stainless steels, 16–17
  discarded CFRP composites, 21
  glass fiber-reinforced composites, 18
  hybrid MMCs, 22–23
  hybrid titanium composite laminates, 23
  lightweight composites, 19
  Mg-xGNPs, 25, 29
  MMCs use, and SPD approaches, 17
  nanoparticle reinforced MMCs, 23, 25
  nickel based composites, 19
  pyrolysis method, 16
  railway turnout system, materials in, 20
  report work on, **26–28**
  silicon or silicon oxide composites, 18–19
  studies and investigation on, 16–23
  titanium MMCs, 22
  wind turbine blades composites, 17–18
metal matrix composites (MMCs), 6, 78
  Al MMCs, pie chart, *143*
  challenges, in recycling, 109–110, 142–143
  distribution of reinforcements, ways of, 110–111
  eco-friendly/low-cost recycling processes, 143–144
  formation, with types of reinforcements, 109
  metallic syntactic foams, 316
  microstructure overview, 110–112, *111*
  with particle reinforcements, 76
  powder metallurgy for recycling, 96
  processing, and reclamation of, 144
  production from recycled materials

  centrifugal casting method, 146, *147*
  direct-chill casting (DC), 144–*145*
  scrap aluminum alloy wheels (SAAWs) using, 146
  stir-squeeze casting method, 146, *148*
  techniques used in, and applications, 155, **156–157**
  recycled, 109–110, 123
    with aluminium matrix, 112–120
    fabrication of, 95–96
    with magnesium matrix, 121–122
    mechanical properties of, 99–101, **102**, 155, **156–157**, 158
    with other matrix materials, 122–123
    physical properties of, 96, **98**, 99, **156–157**
    thermal properties of, 101–**103**
  re-melting-casting, 76, 85
  Scopus search, 142, *143*
  separation techniques, 142, 147, *148*, 149
    aqueous solution, 150–*151*
    classification of, *148*
    electrolysis process, 150, *150*
    nozzle filtration method, 152–*153*
    salt fluxing separation approach, 153–155, *154*
    simple remelting method, *149*, 149–150
    supergravity centrifugal separation, *151–152*
  stirring casting for, 76, 95, *96*, 316–317
  techniques, of recycling, 76, 84, 109
    chemical process, 76, 85
    cold press and hot extrusion, 158
    hot isostatic pressing process, 158
    mechanical process, 84–85
    in situ reaction, 142
    thermal, 85–86
  utilization, and production of, 109
  waste metals/alloys, potential matrix materials, 143, *144*
metals, natural resource, 141–142
metal strength, enhancement, 110
Methacanon, P., 132
methanolysis, 489
microwave pyrolysis, 10
microwave sintering, 94
milling process, 16–17
Mindivan, H., 43
Miranda, Q., 297
Mizumoto, M., 152
Mohamed, D. J., 244
Mohd Joharudin, N. F., 6
Monteiro, E. C. B., 439
Moreno, D. D. P., 132
MOTIC image processing system, 449

Index

Multi-attribute Decision-Making (MADM), 239
Multi Criteria Decision Making (MCDM), 239
Mumbach, G. D., 13
Munir, K., 33
Munir, M. J., 389
Mzali, F., 6, 96

N

nanocomposites, consolidation techniques for, 41
nanofillers, 185
National Association of Entities of Producers of Aggregates for Construction (ANEPAC), 327
National Solid Waste Policy, Brazil, 388
"National Sword Policy," China's, 472
natural fiber reinforced composites *see* recycled thermoplastic composites
Natural fiber reinforced polymer composites (NFRPCs), 130, 131
natural fibers, 186, 190
natural polymer, 464
newspaper fibers, 186
Next Generation Recycling (NGR), 268
nickel-based composite, 123
nodular cast iron, 287–288
  casting process, to produce (*see* casting by centrifugation)
  chemical elements effect, on microstructure, 288
  experimental conditions, 288–290
  graphite lamellar for, *291*
  nodules type and size for, *290*
nonconventional materials, 76
non-used vulcanized scrap rubber, 410
nozzle filtration method, 152–*153*
NREU (non-renewable energy use), 497

O

Okamura, H., 326, 328, 333, 336, 338, 366
one-use plastics, 128
OpenSCAD scripts, 10
open-source additive manufacturing (AM), 9
organoclay Cloisite 15A, 189
original equipment manufacturer (OEM) companies, 86, 88
OTTO WOLPERT-WERKE tester machine, 290
Ouchi, M., 328, 333, 336, 338
Ozawa, K., 336

P

Padilha, F., 439

Paraskevas, D., 101
particle-reinforced composite, 79–80
passenger cars, and pollution, 77, 86, 88
paste rheology method, 339
Peeters, J. R., 201
perlite, 291–292, *293*
personal protective equipment (PPE), 243
Petersson, O., 336, 339
PET polymer, 489
PET recycling process, 469
petroleum-based polymers, 198
photodegradation, 245
physical vapor deposition technique, 83–84
Pietroluongo, M., 21
pineapple leaf fiber, 187
Pinheiro, H., 439
Pinho, A. C., 12
plastic film recycling; *see also* regranulates
  Double Disc technology, 271, 271–272,
  GRAN-NGR recycling line, 268, *269–270*
  INTAREMA compact recycling system, 268, *270–271*, 271
  lines for, overview, 268
  Starlinger recycling lines, 272–273, *272*, **273**
  processing parameters, 273–274, *274–275*
plastics; *see also* recycling; WEEE plastic recycling
  annual waste generation, *472*
  barriers to secondary commodities market for, 477, 479, **481**
  -based wastes, recycling, 488–489
  from biodegradable sources, 242
  chemical recycling, 242–243
  China's ban on importation of waste, 472–473, 482
  contamination possibility, 242
  control on usage, 241–242
  critical issues, at global stage, 463–464
  environmental effects, of recycled, 8
  export markets for, 477, *479*
  global market for, 470–473, *471*;
    (*see also* Malaysia)
  incineration method, 243
  material selection, 243–244
  non-permanent materials, 242
  in packaging, 470
  personal protective equipment (PPE), 243
  pollution, in ecosystems, 359–360
  recycling, 128–129, 185, 242–243, 488
    classification standards, ASTM D5033 and ISO 15270, 466–467
    physical or mechanical, 242
    production and, **360**
    supply chain, market demand, 469–470

regrinding process, 242
single-use, 243, 245, 263–264
thermoplastic group, in packaging industry, 244
usage, and discarding of, *241*, 243
plastic waste (PW), 185, 198, 359–360; *see also* plastics; wastes
concrete with, 360–361
environmental challenges, 359–360
estimate, 359, **360**
recycling, 360–361, 488, *489*
plate-like fillings, 186
PME/FPF 3D printers, 15
PM titanium alloy, 4
Podanfol S. A., 264, 273
pollution, and passenger cars, 77, 86, 88
polyamide-polyethylene wastes recycling, 264; *see also* plastic film recycling
materials, 265
injection molding parameters of, **266**
mechanical properties of, 267, **267**
methodology, 265–268, *265–266*
polybutylene adipate terephthalate (PBAT), 464
polybutylene succinate (PBS), 464
Polyethylene (PE), 130
chemical structure of, 167, *169*
recycled polymer composites based on, **170–171**
thermal properties of, 167, *169*
polyethylene (PE), 465
polyethylene terephthalate (PET), 203, 465
chemical structure of, 166, *167*
flame retardant systems, for recycled, 205
most recycled, 205
recycled PET (RPET), 205
flame-retardant behavior improvement of, 208–209
heat release rate (HRR) curves of, *207*
incorporation of PC into, 208–209
LOI value evolution, 205–206, *206*
PET/PC blends, strategies for improving flame-retardant properties, 209–**210**, *210*
red phosphorus with metal oxides, 205–207, *208*
time to ignition (TTI) increased, 206, *207*
use and studies on, 166, 167
recycled polymer composites based on, **168–169**
recycling steps, multiplicity of, 205
thermal properties of, 166, *167*
polyhydroxyalkanoates (PHA), 464
polylactic acid (PLA), 9, 203, 464
polymer and metal composites recycling

impact, on environment, and economy, 8, 164–165
methods of, 4, 7–8, 10, 29
polymer, and metal waste risks, 29, 164
technologies, and issues, 2–3, *4*
polymer composite, defined, 164–165
polymer composites recycling, 10–11, 129, 488–490, **491–492**
chemical recycling, to monomer, 10, 489
codes for distributed manufacturing, 10
costlier venture, 12, 29, 246
eco/green design for, difficulty in applying, 247
extended polystyrene (EPS) wastes, 13
FPF/FGF method, 14–15
and green fab labs, 14
high-density polyethylene (HDPE), 11
inferior functional properties, 130
investigation, and studies on, 9–16
life cycle assessment (LCA) of
goal and scope, 493–495, *494–495*
life cycle impact assessment (LCIA), methodologies, 497, **498–499**
life cycle inventory (LCI), 496–497
plastic waste management options using, 492–493
longitudinal data use, 11–12, 15
meal ready to eat (MRE) pouches, 13
mechanical, 129, 488
mechanical grinding, 489
methods, 16, 129
multilayered packaging, 9
natural fiber reinforced (NFRPCs), 130
in open-source additive manufacturing (AM), 9
in open-source context, 11
PET bottles waste, 488
plastics reinforced with fibers (FRPs), 244
polymer-based multilayer packaging, 9
process effect on properties, 245–246
process of, *4*
product design for, 236
recent studies on, **491–492**
report work on, **24–25**
solvent-based, 129
studies, on effectiveness evaluation, 10–11, 15–16
sulfur polymers and, in AM, 14
thermal process, 129
thermosetting, 12–13, 245
treatment with BMI, 15–16
using water-jet tape deposition method, 9
utilized in, engineering components, 244
from waste and current trends, 11
windshield blade case study, 10

Index

polymeric materials, thermal analyses of, 165–166
polymerization process, 464
polymer matrix composites (PMCs), 78–79
polymers, 463–464, **466**; *see also* polymer composites recycling
  applications, 465, **466**
  blend, filler or additives, for enhancing properties of, 189
  classification of, 464–465
  combination of fillers and, 164
  commonly recycled, 467
  defined, 464
  essential physical properties of, classified, 188–189
  organic and inorganic fillers, 164
  production, and recycling technology, 2
  recycler value chain and market demand, 469–470, *470*
  recycling, 465
    ABS, 468–469
    biological degradation, 467
    classification, 466–467
    cross-linking method, 467
    effectiveness comparison, **468**
    HDPE, 467–468
    incineration method, 467
    PET, 469
    primary mechanical, 466
    process, 467–469, **468**, 481
    secondary mechanical, 466
    steps in process, 467–469
    tertiary, 466–467
  thermoplastic, 164
  thermosetting, 164
  use in daily life, and waste, 185
  variables to identify, polymeric system, 189
  waste processing, 10
polypropylene (PP), 465
  applications, and fully recyclable, 171
  chemical structure of, 171, *172*
  fibers, 329
  recycled polymer composites based on, **173–174**
  -rice husk composites, 192
  thermal properties of, 171, *172*
polystyrene (PS), 465
  applications, 171
  chemical structure of, 171, *174*
  recycled polymer composites based on, **175**
  solid or rigid foamed version, 171
  styrofoam, major waste, 171, 174
  thermal properties of, 171, *174*
polytetrafluoroethylene (PTFE), 465
polyvinyl acetate (PVA), 465

polyvinyl chloride (PVC), 465
  chemical structure of, 171, *172*
  fully recyclable, 169
  recycled polymer composites based on, **172**
  rigid or flexible form, 169
  thermal changes, sensitive to, 169
  thermal properties of, 171, *172*
  thermoplastic polymer, 169
Portland cement, 326, 328, 334
powder blending and consolidation technique, 83
powder metallurgy (PM), 94, 96
  aluminium/graphite composites fabrication, 96, *97*
  basic steps in, 38–42, *39*
  expensive method, 96
  fabricated MMC by, 41–42
  HIP process in, *41*
  HP used in, 39–40, *40*
  uni-axial die compaction in, 39–40, *40*
powder metallurgy/solid-state processing, 94
pozzolans, 334
Pride, D., 359
process chain recycling, 9
Puneet, T., 296
Pusat Citra Universiti, 103
pyrolysis, 8, 10, 16
  -based composites reprocessing method, 20
  recycling composites by, 16
pyrolyzed rice husk, 189

R

Rady, M. H., 5
Rajendran, S., 8, 497
Rayon fibers, 200
reactive compression molding, 14
recoSTAR basic/universal line, *272*, **273**, 274, *276*
recyclate, defined, 236
RecycleBot open-source filament system, 11
recycled aluminium composite
  casting technique use, 112
  cold forging method, followed by sintering, 119
  defect of oxide film formation, 115
  friction stir extrusion, 120
  hot pressing use, 120
  increased holding time, influence, 115
  in-situ formation of TiB_2 particles, 112, *114*
  made with, chips of aluminium/tin, 119–120
  microstructure, and fluidity analysis, 114–120
    illustration of, *113*, *114*, *116*, *117*
  reinforced with in-situ ZrB_2 particles, 113

with scrap aluminium beverage cans,
  115–116
 scrap aluminium composites, melted,
  113–115
 through infiltration technique, 119
 using scrap aluminium alloy, 118
recycled FRP, design strategy, 237, *238*
recycled high-density polyethylene (RHDPE)
  composites, 192
recycled magnesium composite, 121–122
 repeated plastic working (RPW)
  method, 121
 solid-state processing, 121
recycled metal composites
 with aluminium as matrix material, 112–120
 with magnesium as matrix material,
  121–122
 morphological analysis, 109–110
  critical observations in, 123–124
  microstructure overview, 110–112
  outcomes of, material adding, 110, *111*
  reinforcements particles distribution,
   ways of, 110–111, *112*
 with other matrix material, 122–123
 properties, 5–6
recycled poly(vinyl chloride) (rPVC), 169, 171,
  172, **172**, 190
recycled polymer composites
 by-products, used as fillers, 187–188
 case studies of
  Polyethylene (PE), 167, *169*, **170–171**
  Polyethylene terephthalate (PET),
   166–*167*, **168–169**
  Polypropylene (PP), 171, *172*, **173–174**
  Polystyrene (PS), 171, *174*, **175**
  Polyvinyl chloride (PVC), 169, 171, *172*,
   172, 190
 chemical treatment, importance, 190
 decrement in crystallinity, 130
 DMA analysis, 165–166
 filler types, for enhancing thermal
  properties, **191**
 general steps to reusing, 199–200
 overview of, 185–188
 quality issue, 245–246
 SEM analysis, 191
 TGA and DSC methods, of characterization,
  165–166
 thermal analysis techniques, fundamentals
  of, 188–189
 thermal properties of, 189–190, 193
  analyses, 165–166, 175, 190
  composition affecting, 192
  effects of fibers and binding agents, 192
  factors influencing, 190–193

fillers role in, 190, **191**, 193
 weathering characteristic of, 130
recycled polymer matrixes, with suitable
  fillers, **191**
recycled polypropylene (RPP), 171, *172*,
  173–174, 191
recycled thermoplastic composites
 natural fiber reinforced, aging/weathering
  effect, 128, 130, 132, 135–136
 fiber-matrix de-bonding, 135
 fiber-matrix interfacial bonding, 132
 lowdensity polyethylene (LDPE),
  132, *133*
 on mechanical properties, 131–135
 PP/ethylene vinyl acetate, 132, *134*
 recycled PP composites, 134, *135*
 recycled PP/wood, 132, *134*
 tensile strength impact, 132
 on thermal properties, 130–131
recycling; *see also* wastes
 agricultural waste, 188
 of aluminum alloys, 443–444
 of aluminum matrix composites, 86
 biodegradable polymers and, 245
 carbon fiber reinforced polymers (CFRP), 5
 chemical or feedstock, 466
 chipped stainless steels, 16
 of composites, chain of operations, 87
 ecological benefits of, 142
 economic aspects of, 86–88, 158
 environmental aspects of, 86–88, 94, 142,
  163–164, 185, 241–243, 410
 extended polystyrene (EPS) wastes, 13
 factors for achieving success, 203–205
 of feedstock, 10
 flame retardant polymer composites,
  199–200
 difficult, due to dissimilar materials in,
  198–199
 flame retardant textiles, 200, *201*
 waste electrical and electronic equipment
  (WEEE), 200–201, *202*, 203–205
 of glass fiber reinforced plastic, 86
 material selection for, 243–245
 meal ready to eat (MRE) pouches, 13
 of metal and alloy chips, 94
 remelting technique, problems, 38–39
 metal process, **95**
 methods, 4, 10, 12, 16, 29
  chemical recycling, 7, 10, 16, 29
  of different composites, summary, 85, *86*
  mechanical recycling, 7, 16, 29
  mechanical reprocessing
   (downgrading), 16
  thermal recycling, 8, 10, 29

Index

micro- and nanocomposite, 10
of MMCs, properties, 94, 142;
   (see also metal matrix composites (MMCs))
multilayered packaging products, methods, 9
plastic materials, 128–129, 488–489;
   (see also plastic)
   slower degradation, 163, 185, 241, 359–360
   of plastic mixture wastes, 264
polymeric products, waste management of, 164, 174–175
of polymers (see polymers)
process, problems of, 85
railway turnout, process, 20
role in energy conservation, 109
scrap materials, as matrix material, 143, 159
scrap rubber, 221
of selected film waste, on industrial line, 273–274
of single-use packaging, 243, 245, 263–264
stagnant progress, reasons of, 2–3
for sustainable world, 76–77;
   (see also composites)
system creation, consideration in, 84
types of, 10
of waste, disposal challenges, 142
regranulates
   mechanical properties of, 274–276, *276*
      Charpy notched impact strength, 279, *279*, 284
      elongation at break of samples, *278*
      injection molding parameters, **277**
      tensile strength, 276, *277–278*, 278–279, 284
      VICAT softening temperature, 279–**280**, 284
   processing parameters, *276*
   production of transport pallets, 282–**283**, *284*
   SEM images, PA/PE film structure, 280, *281–282*
Reich, M. J., 15
reinforced concrete (RC) with aggregates, evaluation of
   ABCP dosing method, 434
   accelerated corrosion test, 439–440
   capillarity water absorption, 439, **439**
   concrete exudation, 438–439
   materials used, 434, *437*
      pebble and gravel, properties of, **434**
   methodology, **435**–436
      capillary water absorption test, 435–436, *436*
      corrosion potential, measuring, 436, *437*
      corrosion test, 436, *437*
      tensile and compression tests, 435
      tensile and compressive strength, 436–438, *438*
reinforcement
   defined, 77, 110
   distribution ways of, 110–111, *112*
   material
      ceramic particles, 80
      particles, 76
re-melting process, 85
   disadvantages of, 38–39
Repette (2005), 336
   [not found in ref.]
resin identifying codes, 10
response surface method (RSM), 339
Rheology, defined, 340
rice husk filling fibers, 187
rice straw fibers, 187
Rojas-Díaz, L. M., 96
Rosentrater, K. A., 493
rotary-die equal channel angular pressing (RD-ECAP), 46
rotary salt furnace technology, 148

S

Saak, A. W., 338, 339
Sadek, D. M., 395
salt fluxing separation approach, 153–155, *154*
salt or other fluxing methods, 148
Samal, C. P., 41
Saravanan, C., 83
sausage casings (Podanfol S.A.), 264
Sauter Shore D hardness tester, 413
Savastano, J. R. H., 424, 430
Schuster, D. M., 85, 158
Scielo, 362
Science Direct, 362, 364, 380
scientific research, 358
Scopus, 362–364, *365*, 380
   search, recycling of Al MMCs, 142–143
scrap aluminum alloy wheels (SAAWs), 146
scrap rubber/epoxy based composites
   absorbed energy, for RETI and RETIG composites, 226, *227*, *228*
   compositions of, **223**
   ductility and toughness determination, 222, 224–228
   dynamic compression (drop weight) tests, 223, 225, *225*, *227*–229
   hot compacting, 222–223
   macroindentation tests, and SEM images, 223–224, *226*, *227*
   materials used in, 222–223
   reinforcements used, 222

RETI-(alumina fibers), microstructure of, 224, *224*
RETIG-(glass fibers), microstructure of, 224, *225*
Shore D hardness test, 223
Shore D results after UV exposure, comparison of, 224, *225*
wear tests, RETI and RETIG specimens, 229–230, *230*
self-compacting concrete (SCC), 326, 345, 361
  additions, and waste materials used in, 329, **330–331**
  advantages, 326
  behavior depending on constituents, 327
  cementitious composite as, 328
    components and characteristics, 328–329
    constituents characterization, 334–335, **336**
    fibers to improve properties of, 328–329
    fine and coarse aggregates, 328
    mineral additions, 328
    superplasticizers (SP) in, 329
    waste addition in (see waste addition in, cementitious composites)
  developed in Japan, 326
  disadvantages, 326
  integration of industrial waste, 327
  limits of components, **337**
  mixture design method, categories, 335–339
  norms for trials, in fresh state, **341**
  packing factor (PF) of, 339
  partial replacements, 329
  polymeric waste and, 361–362
  problems, and corrective actions in fresh, 343, 345, **346**
  requirements for, **344**
  sustainable, 327
  tests for properties, *340*
    L-box test, 342, *343*
    norms for trials, **341**
    slump flow test, 340
    T 500 test, 341
    V-funnel test, 342, *342*
  used to incorporate plastic waste, 361
  waste used, and tests performed with, 329, **330–331**
separation techniques
  reinforcements separation, from matrix materials, 142, 147, *148*, 149
    aqueous solution, 150–*151*
    electrolysis process, 150
    nozzle filtration method, 152–*153*
    salt fluxing separation approach, 153, *154–155*
    simple remelting method, *149*, 149–150

supergravity centrifugal separation, *151–152*
sepiolite, 189
sever plastic deformation (SPD), 17, 38
Shamsudin, S., 43
Shanmugam, V., 14
Sharma, A. K., 22
sheet molding compound (SMC), 236
Shehab, E., 18
Shen, L., 493, 496, 497
Sherafat, Z., 43
Shi, C., 336
Shial, S. R., 42
Shuaib, N. A., 86
Siddique, R., 328
silane reinforcement, 186
silicon carbide (SiC), 76
Silva, C. B. da, 131, 132
simple remelting method, *149*, 149–150
Singh, P., 395
single edge notched beam (SENB), 410
single-use packaging, 263
sinter and forging integrated method, 60
solid-phase process, 96
solid polymer waste (SPW), 12
solid-state
  processes, 94
    diffusion bonding, 83
    physical vapor deposition, 83–84
    powder blending and consolidation, 83
  recycling techniques, 38, 60
    equal channel angular processing (ECAP), 45–46
    friction stir processing (FSR), 44–45
    high pressure torsion (HPT), 47–48
    hot extrusion (HE) method, 42–44
    hot press forging, 48–49
    powder metallurgy (PM), 38–42
    summary of, features on processing of, 49–50
solid waste management, 388, 389
Soltan, A. M. M., 395
solvolytic processes, 7, 85
Sonebi, M., 338, 340
Soufiani, A. M., 42
South Carolina, USA Karborek, 236
Spark plasma sintering (SPS) techniques, 39, 41–42, 94
spent aluminum catalysts (SACs), 146
spherical fillers, 186
sport products, 247
spray deposition, 82
squeeze casting method, 80, 85
Sri-Lanka project, 187
Srinivasan, R., 470

Index

S–S metathesis reaction, 14
starch gum (SG), 191
statistical bibliography, 358
steel scrap, 288
stir casting method, 19, 76, 80, 83, 94, *95*, 316
styrene–butadiene rubber (SBR), 221–222
styrofoam, 171
Su, N., 336, 339
Sugiyama, S., 43
Sun, N., 17, 142
supercritical fluids (SCFs), 7
supercritical water (SCW), 7
supergravity centrifugal separation, *151–152*
supergravity technology, 17
superplasticizers (SP), 329, 361
surface coating of particles, 76
sustainable design concept, 246–247
sustainable development, concept of, 198
sustainable world, and issues, 76, 241–243
Swoboda, B., 208
synthetic polymer, behaviors types, 465

T

Tarverdi, K., 88
Technique of Order Preference Similarity to The Ideal Solution (TOPSIS), 239
Teflon, 465
tensile strength test, 276, *277–278*, 278–279
 casting by centrifugation, process of, 289, *290*
 regranulates, 276, *277–278*, 278–279, 284
 reinforced concrete (RC) with aggregates, evaluation of, 435, 436–*438*
Tescan Vega 5135, SEM, 280
thermal analysis; *see also* recycled polymer composites
 role in material characterization, 188
 to study thermal behavior, 165–166, 175, 188
 techniques, 165, **166**
  commonly used, *189*
  fundamentals of, 188–189
thermal recycling, 8, 10
thermogravimetric analysis (TGA), 165, 188
thermogravimetry (TG), 188
thermomechanical analysis (TMA), 165, 188
thermo-mechanical technique, 236
thermo-oxidative degradation, 245
thermoplastics, 244, 465
 -based composites, 130
 polymers, 164, 466
 polymer composites, 130–131
thermosets, 465
thermosetting, 245

polymers, 164
 recycling, 12–13
thermosoftening plastic, 465
3D printing, 12
 expanded use of polymers in, 15
 process, 9
Ti–6Al–4V titanium alloys, 296
Ti–Al-based intermetallic-based composite, recycled
 ANOVA analysis of variance, 302, 304, **304**, 311
 chemical composition of, **298**
 contour curves of surfaces, *306*
 cutting conditions, effects of, 307, *307*
 degree of plastic deformation, 299
 formation of striation, 300
 material removal rate optimizations, **308**–311
  ANOVA for, **310**
  contours, showing effect of set tension feed rate, 311, *312*
  cutting parameter, influence, 309, *309–310*
  machining parameters effect, on MRR, *309*
  residual curves, *311–312*
  response table for mean values, **309**
  SNRA and PSNRA values, **308**
  S/N ratio effects, for MRR, *310*
 microstructural evaluation, 299, *299*
 relationship, rate of abrasive flux and MRR, 307, *308*
 surface quality, and conditions, 301–302, **304**, 305, *305*
  effect of cutting speed on, 300, *300*
  effect of mass flow rate on, 301, *301*
  effects of S/N, 301–302, **302**, *303*
Tian, S., 131
titanium alloys, use of, 4–5
titanium matrix composite, 122
transport pallets production, 282–**283**, *284*
 processing parameters, **283**
 top/bottom view, *283–284*
Tunçay, T., 444
Tutikian, B. F., 329, 336
two 2D nozzle angle analysis, 9

U

UK CFK Valley Stade Recycling, 236
ultrasonic assisted casting, 83
ultrasonic vibration, 76
Ushasta, A., 296
Ustundag, M., 4
US Waterborne Import Trade (WIT) Report, 477

V

Valerio, O., 12
Vardhan, K., 395
Varol, R., 4
Varol, T., 42
Vasile, C., 204
Vecchiato, S., 200
Vedani, M., 23
vegetable fibers, 186, 423–425, *424–426*
Verma, P., 42
VICAT softening temperature test, 279–**280**
Viswanathan, V., 41
Vo Dong, P. A., 8
VOS viewer software, 362
vulcanization process, 410
VWR-French Branch/Chemical Company, 61

W

Wagner, F., 203
Wan, B., 45
Wang, J., 297
warm forging, 49
waste addition in, cementitious composites
　aggregates influence, on properties,
　　332–334, **333**
　ASTM standards, 334, **336**
　design methods, 335–338
　　compressible packing method (CPM),
　　　338–339
　　compressive strength, 338
　　factorial statistical model, 339
　　Gomes's method, phases, 339
　　Okamura's method (empirical
　　　method), 338
　　paste rheology method, 339
　　Repette and Melo's method, 338
　　response surface method (RSM), 339
　　Tutikian's method, 338
　initial characterization, sequence, 329, *332*
　techniques/standards for characterization,
　　334–335, **337**
　trials, 334, **335**
waste electrical and electronic equipment
　(WEEE), 200
waste management hierarchy, *199*
waste paper-based secondary fiber, 186
waste-printed circuit board powders
　(WPCBP), 131
wastes; *see also* plastics; recycling
　burning plastic, 243
　carbon fibers (CFs), huge demand and, 236
　concrete, capacity of absorbing, 327
　disposable plastic products, 241
　disposal of, 142, 198
　flame retardant products, 198
　metallic chips, 94, 103, 144
　mixed polyamide and polyolefin, 264
　plastic, 128–129, 185, 359–360, 488–489
　　concrete with, 360–361
　　mixture (PA/PP/PE), 264
　polymeric, and recycling, 164
　products reused, by construction
　　industry, 327
　Rayon fibers, and WEEE, 200
　rubber, SBR, 221–222, 410
　solid waste, in MGR processing, 388–389
water-jet tape deposition method, 9
weathering effect, on performance
　recycled and virgin forms, of wood-polymer,
　　130–131
　thermal analysis, 130
Web of Science, 362
WEEE plastic recycling
　chemical approaches, 205
　conventional methods of, 205
　FMEA, for risk assessment, 203, *204*
　heat-degrading oil produced by, 204
　mechanical, 203
　Mendeleev-like FMEA table use, *204*
　phosphorus-based flame retardants in,
　　200–201
　size reduction based, 201, *202*
　success in, depends on factors, 203
wind turbine blades composites, 17–18
wood dust, 189
Wu, H., 18
Wzorek, Łukasz, 43

X

X7475 alloy, from RBCs
　artificial aging temperature, effect on tensile
　　strength, 256–257
　heat treatment
　　profile, schematic diagram, 255
　　sequence, modification, 252–253,
　　　255, 260
　materials used, in fabricating, 253
　methodology, 253–255, *254*, *255*
　microstructure study, 257–260, *257–259*
　ultimate tensile strength (UTS), after
　　annealing, 255, *256*, 257
　variations in Zn, temperature/aging
　　parameters, 253, **254**
X-Ray diffraction (DRX), 389, 395, 406
Xu, Q., 46

Xu, W., 45

Y

Yamagiwa, K., 146
Yang, Y., 85, 87
Yapici, G. G., 45
Yasin, S., 200
Yoshikawa, N., 157

Z

Zadeh, K. M., 211
Zaid, H. R., 255–257
Zander, N. E., 15
Zero Waste, 388
Zhang, L., 212
Zhou, M. Y., 23
$ZnCl_2$/ethanol catalyst method, 6
Zuo, W., 337